DYNAMICS OF MULTIBODY SYSTEMS

Second Edition

Ahmed A. Shabana
University of Illinois at Chicago

CAMBRIDGE
UNIVERSITY PRESS

PUBLISHED BY THE PRESS SYNDICATE OF THE UNIVERSITY OF CAMBRIDGE
The Pitt Building, Trumpington Street, Cambridge, United Kingdom

CAMBRIDGE UNIVERSITY PRESS
The Edinburgh Building, Cambridge CB2 2RU, UK
40 West 20th Street, New York NY 10011–4211, USA
477 Williamstown Road, Port Melbourne, VIC 3207, Australia
Ruiz de Alarcón 13, 28014 Madrid, Spain
Dock House, The Waterfront, Cape Town 8001, South Africa

http://www.cambridge.org

First published 1989 by Wiley & Sons
Published by Cambridge University Press 1998
First paperback edition 2003

Typeset in Times Roman

A catalogue record for this book is available from the British Library

Library of Congress Cataloguing-in-Publication Data

Shabana, Ahmed A., 1951–
Dynamics of multibody systems / Ahmed A. Shabana. – 2nd ed.
p. cm.
Includes bibliographical references (p. 359–366) and index.
ISBN 0 521 59446 4 hardback
1.Dynamics. 2. Kinematics. I. Title.
QA845.S45 1998
531′. 11 – DC21
97-31088
CIP

ISBN 0 521 59446 4 hardback
ISBN 0 521 54411 4 paperback

To my father

and to

the memory of my mother

Contents

Preface

The methods for the nonlinear analysis of physical and mechanical systems developed for use on modern digital computers provide means for accurate analysis of large-scale systems under dynamic loading conditions. These methods are based on the concept of replacing the actual system by an equivalent model made up from discrete bodies having known elastic and inertia properties. The actual systems, in fact, form multibody systems consisting of interconnected rigid and deformable bodies, each of which may undergo large translational and rotational displacements. Examples of physical and mechanical systems that can be modeled as multibody systems are machines, mechanisms, vehicles, robotic manipulators, and space structures. Clearly, these systems consist of a set of interconnected bodies which may be rigid or deformable. Furthermore, the bodies may undergo large relative translational and rotational displacements. The dynamic equations that govern the motion of these systems are highly nonlinear which in most cases cannot be solved analytically in a closed form. One must resort to the numerical solution of the resulting dynamic equations.

The aim of this text, which is based on lectures that I have given during the past twelve years, is to provide an introduction to the subject of multibody mechanics in a form suitable for senior undergraduate and graduate students. The initial notes for the text were developed for two first-year graduate courses introduced and offered at the University of Illinois at Chicago. These courses were developed to emphasize both the general methodology of the nonlinear dynamic analysis of multibody systems and its actual implementation on the high-speed digital computer. This was prompted by the necessity to deal with complex problems arising in modern engineering and science. In this text, an attempt has been made to provide the rational development of the methods from their foundations and develop the techniques in clearly understandable stages. By understanding the basis of each step, readers can apply the method to their own problems.

The material covered in this text comprises an introductory chapter on the subjects of kinematics and dynamics of rigid and deformable bodies. In this chapter some background materials and a few fundamental ideas are presented. In Chapter 2, the kinematics of the body reference is discussed and the transformation matrices that define the orientation of this body reference are developed. Alternate forms of the transformation matrix are presented. The material presented in this chapter is essential for understanding the dynamic motion of both rigid and deformable bodies. Analytical techniques for deriving the system differential and algebraic equations of motion of a multibody system consisting of rigid bodies are discussed in Chapter 3. In Chapter 4, an introduction to the theory of elasticity is presented. The material covered in this chapter is essential for understanding the dynamics of deformable bodies that undergo large translational and rotational displacements. In Chapter 5, the equations of motion of deformable multibody systems in which the reference motion and elastic deformation are coupled are derived using classical approximation methods. In Chapters 6 and 7, two finite element formulations are presented. Both formulations lead to exact modeling of the rigid body inertia and lead to zero strains under an arbitrary rigid body motion. The first formulation discussed in Chapter 6, which is based on the concept of the intermediate element coordinate system, uses the definition of the coordinates used in the conventional finite element method. A conceptually different finite element formulation that can be used in the large deformation analysis of multibody systems is presented in Chapter 7. In this chapter, the absolute nodal coordinate formulation in which no infinitesimal or finite rotations are used as element coordinates is introduced.

I am grateful to many teachers, colleagues, and students who have contributed to my education in this field. I owe a particular debt of gratitude to Dr. R.A. Wehage and Dr. M.M. Nigm for their advice, encouragement, and assistance at various stages of my educational career. Their work in the areas of computational mechanics and vibration theory stimulated my early interest in the subject of nonlinear dynamics. Several chapters of this book have been read, corrected, and improved by many of my graduate students. I would like to acknowledge the collaboration with my students Drs. Om Agrawal, E. Mokhtar Bakr, Ipek Basdogan, Michael Brown, Bilin Chang, Che-Wei Chang, Koroosh Changizi, Da-Chih Chen, Jui-Sheng Chen, Jin-Hwan Choi, Hanaa El-Absy, Marian Gofron, Wei-Hsin Gau, Wei-Cheng Hsu, Kuo-Hsing Hwang, Yunn-Lin Hwang, Yehia Khulief, John Kremer, Haichiang Lee, Jalil Rismantab-Sany, and Mohammad Sarwar, and my current students Marcello Berzeri, Marcello Campanelli, Andrew Christensen, Hussien Hussien, and Refaat Yakoub. Their work contributed significantly to the development of the material presented in this book. Special thanks are due to Ms. Denise Burt for the excellent job in typing most of the manuscript. Finally, I thank my family for their patience and encouragement during the time of preparation of this text.

Chicago, Illinois
July 1997

Ahmed Shabana

1 INTRODUCTION

1.1 MULTIBODY SYSTEMS

The primary purpose of this book is to develop methods for the dynamic analysis of *multibody systems* that consist of interconnected *rigid* and *deformable* components. In that sense, the objective may be considered as a generalization of methods of structural and rigid body analysis. Many mechanical and structural systems such as vehicles, space structures, robotics, mechanisms, and aircraft consist of interconnected components that undergo large translational and rotational displacements. Figure 1 shows examples of such systems that can be modeled as multibody systems. In general, a multibody system is defined to be a collection of subsystems called *bodies*, *components*, or *substructures*. The motion of the subsystems is kinematically constrained because of different types of joints, and each subsystem or component may undergo large translations and rotational displacements.

Basic to any presentation of multibody mechanics is the understanding of the motion of subsystems (bodies or components). The motion of material bodies formed the subject of some of the earliest researches pursued in three different fields, namely, *rigid body mechanics*, *structural mechanics*, and *continuum mechanics*. The term *rigid body* implies that the deformation of the body under consideration is assumed small such that the body deformation has no effect on the gross body motion. Hence, for a rigid body, the distance between any two of its particles remains constant at all times and all configurations. The motion of a rigid body in space can be completely described by using six generalized coordinates. However, the resulting mathematical model in general is highly nonlinear because of the large body rotation. On the other hand, the term *structural mechanics* has come into wide use to denote the branch of study in which the deformation is the main concern. Large body rotations are not allowed, thus resulting in inertia-invariant structures. In many applications, however, a large number of elastic coordinates have to be included in the mathematical model in order to accurately describe the body deformation. From the study of these two

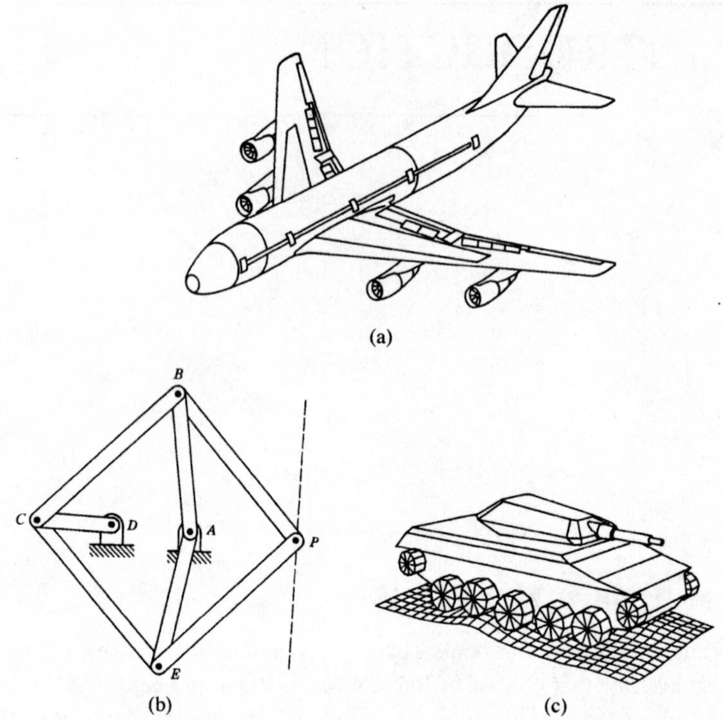

Figure 1.1 Mechanical and structural systems.

subjects, rigid body and structural mechanics, there has evolved the vast field known as *continuum mechanics*, wherein the general body motion is considered, resulting in a mathematical model that has the disadvantages of the previous cases, mainly nonlinearity and large dimensionality. This constitutes many computational problems that will be addressed in subsequent chapters.

The research in the area of multibody dynamics has been motivated by growing interest in the simulation and design of large-scale systems of interconnected bodies that undergo large angular rotations. The analysis and design of such systems require the simultaneous solution of hundreds or thousands of first-order differential equations, a task that could not be accomplished a few decades ago before the development of electronic computers. Most of the work done in this area is based on analyzing rigid multibody systems, and many computer-based techniques that solve complex rigid body systems have been developed.

In recent years, however, greater emphasis has been placed on the design of high-speed, lightweight, precision systems. Generally these systems incorporate various types of driving, sensing, and controlling devices working together to achieve specified performance requirements under different loading conditions. The design and performance analysis of such systems can be greatly enhanced through transient dynamic simulations, provided all significant effects can be incorporated into the mathematical model. The need for a better design, in addition to the fact that many mechanical and structural systems operate in hostile environments, has demanded the

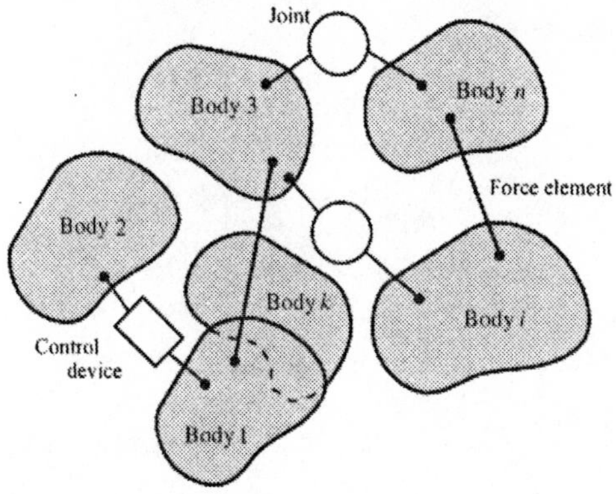

Figure 1.2 Multibody systems.

inclusion of many factors that have been ignored in the past. Systems such as engines, robotics, machine tools, and space structures may operate at high speeds and in very high temperature environments. The neglect of the deformation effect, for example, when these systems are analyzed leads to a mathematical model that poorly represents the actual system.

Consider, for instance, the Peaucellier mechanism shown in Fig. 1(b), which is designed to generate a straight-line path. The geometry of this mechanism is such that $BC = BP = EC = EP$ and $AB = AE$. Points A, C, and P should always lie on a straight line passing through A. The mechanism always satisfies the condition $AC \times AP = c$, where c is a constant called the *inversion constant*. In case $AD = CD$, point C must trace a circular arc and point P should follow an exact straight line. However, this will not be the case when the deformation of the links is considered. If the flexibility of links has to be considered in this specific example, the mechanism can be modeled as a multibody system consisting of interconnected rigid and deformable components, each of which may undergo finite rotations. The connectivity between different components of this mechanism can be described by using revolute joints (turning pairs). This mechanism and other examples shown in Fig. 1, which have different numbers of bodies and different types of mechanical joints, are examples of mechanical and structural systems that can be viewed as a multibody system shown in the abstract drawing in Fig. 2. In this book, computer-based techniques for the dynamic analysis of general multibody systems containing interconnected sets of rigid and deformable bodies will be developed. To this end, methods for the kinematics and dynamics of rigid and deformable bodies that experience large translational and rotational displacements will be presented in the following chapters. In the following sections of this chapter, however, some of the basic concepts that will be subject of detailed analysis in the chapters that follow are briefly discussed.

1.2 REFERENCE FRAMES

The configuration of a multibody system can be described using measurable quantities such as displacements, velocities, and accelerations. These are vector quantities

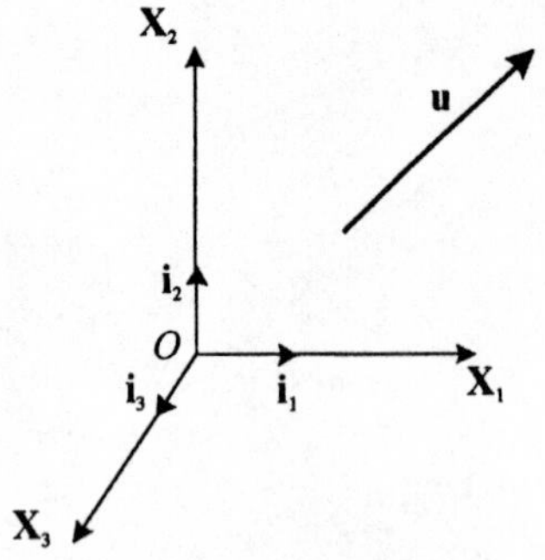

Figure 1.3 Reference frame.

that have to be measured with respect to a proper *frame of reference* or *coordinate system*. In this text, the term *frame of reference*, which can be represented by three orthogonal axes that are rigidly connected at a point called the *origin* of this reference, will be frequently used. Figure 3 shows a frame of reference that consists of the three orthogonal axes $\mathbf{X}_1$, $\mathbf{X}_2$, and $\mathbf{X}_3$. A vector $\mathbf{u}$ in this coordinate system can be defined by three components u_1, u_2, and u_3, along the orthogonal axes $\mathbf{X}_1$, $\mathbf{X}_2$, and $\mathbf{X}_3$, respectively. The vector $\mathbf{u}$ can then be written in terms of its components as

$$\mathbf{u} = [u_1 \quad u_2 \quad u_3]^{\mathrm{T}}$$

or as

$$\mathbf{u} = u_1\mathbf{i}_1 + u_2\mathbf{i}_2 + u_3\mathbf{i}_3$$

where $\mathbf{i}_1$, $\mathbf{i}_2$, and $\mathbf{i}_3$ are unit vectors along the orthogonal axes $\mathbf{X}_1$, $\mathbf{X}_2$, and $\mathbf{X}_3$, respectively.

Generally, in dealing with multibody systems two types of coordinate systems are required. The first is a coordinate system that is fixed in time and represents a unique standard for all bodies in the system. This coordinate system will be referred to as *global*, or *inertial frame* of reference. In addition to this inertial frame of reference, we assign a *body reference* to each component in the system. This body reference translates and rotates with the body; therefore, its location and orientation with respect to the inertial frame change with time. Figure 4 shows a typical body, denoted as body i in the multibody system. The coordinate system $\mathbf{X}_1\mathbf{X}_2\mathbf{X}_3$ is the global inertial frame of reference, and the coordinate system $\mathbf{X}_1^i\mathbf{X}_2^i\mathbf{X}_3^i$ is the body coordinate system. Let $\mathbf{i}_1$, $\mathbf{i}_2$, and $\mathbf{i}_3$ be unit vectors along the axes $\mathbf{X}_1$, $\mathbf{X}_2$, and $\mathbf{X}_3$, respectively, and let $\mathbf{i}_1^i$, $\mathbf{i}_2^i$, and $\mathbf{i}_3^i$ be unit vectors along the body axes $\mathbf{X}_1^i$, $\mathbf{X}_2^i$, and $\mathbf{X}_3^i$, respectively. The unit vectors $\mathbf{i}_1$, $\mathbf{i}_2$, and $\mathbf{i}_3$ are fixed in time; that is, they have constant magnitude and direction, while the unit vectors $\mathbf{i}_1^i$, $\mathbf{i}_2^i$, and $\mathbf{i}_3^i$ have changeable orientations. A vector $\mathbf{u}^i$ defined in the body coordinate system can be written as

$$\mathbf{u}^i = \bar{u}_1^i\mathbf{i}_1^i + \bar{u}_2^i\mathbf{i}_2^i + \bar{u}_3^i\mathbf{i}_3^i$$

where $\bar{u}_1^i$, $\bar{u}_2^i$, and $\bar{u}_3^i$ are the components of the vector $\mathbf{u}^i$ in the local body coordinate system. The same vector $\mathbf{u}^i$ can be expressed in terms of its components in the global

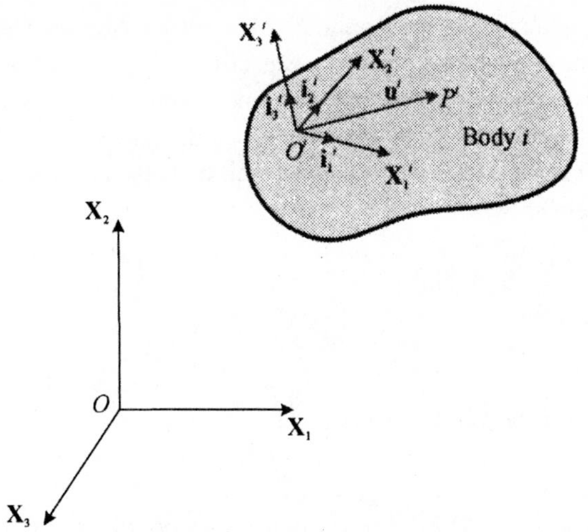

Figure 1.4 Body coordinate system.

coordinate system as

$$\mathbf{u}^i = u_1^i \mathbf{i}_1 + u_2^i \mathbf{i}_2 + u_3^i \mathbf{i}_3$$

where u_1^i, u_2^i, and u_3^i are the components of the vector $\mathbf{u}^i$ in the global coordinate system. We have, therefore, given two different representations for the same vector $\mathbf{u}^i$, one in terms of the body coordinates and the other in terms of global coordinates. Since it is easier to define the vector in terms of the local body coordinates, it is useful to have relationships between the local and global components. Such relationships can be obtained by developing the transformation between the local and global coordinate systems. For instance, consider the *planar motion* of the body shown in Fig. 5. The coordinate system $\mathbf{X}_1\mathbf{X}_2$ represents the inertial frame and $\mathbf{X}_1^i\mathbf{X}_2^i$ is the body coordinate

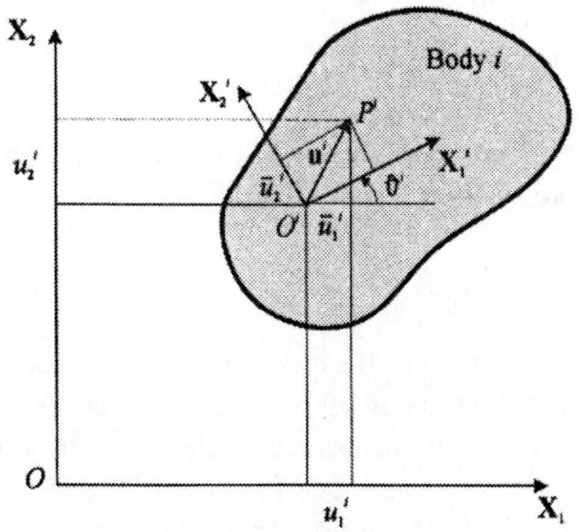

Figure 1.5 Planar motion.

system. Let $\mathbf{i}_1$ and $\mathbf{i}_2$ be unit vectors along the $\mathbf{X}_1$ and $\mathbf{X}_2$ axes, respectively, and let $\mathbf{i}_1^i$ and $\mathbf{i}_2^i$ be unit vectors along the body axes $\mathbf{X}_1^i$ and $\mathbf{X}_2^i$, respectively. The orientation of the body coordinate system with respect to the global frame of reference is defined by the angle θ^i. Since $\mathbf{i}_1^i$ is a unit vector, its component along the $\mathbf{X}_1$ axis is $\cos\theta^i$, while its component along the $\mathbf{X}_2$ axis is $\sin\theta^i$. One can then write the unit vector $\mathbf{i}_1^i$ in the global coordinate system as

$$\mathbf{i}_1^i = \cos\theta^i \mathbf{i}_1 + \sin\theta^i \mathbf{i}_2$$

Similarly, the unit vector $\mathbf{i}_2^i$ is given by

$$\mathbf{i}_2^i = -\sin\theta^i \mathbf{i}_1 + \cos\theta^i \mathbf{i}_2$$

The vector $\mathbf{u}^i$ is defined in the body coordinate system as

$$\mathbf{u}^i = \bar{u}_1^i \mathbf{i}_1^i + \bar{u}_2^i \mathbf{i}_2^i$$

where $\bar{u}_1^i$ and $\bar{u}_2^i$ are the components of the vector $\mathbf{u}^i$ in the body coordinate system. Using the expressions for $\mathbf{i}_1^i$ and $\mathbf{i}_2^i$, one gets

$$\begin{aligned}
\mathbf{u}^i &= \bar{u}_1^i(\cos\theta^i \mathbf{i}_1 + \sin\theta^i \mathbf{i}_2) + \bar{u}_2^i(-\sin\theta^i \mathbf{i}_1 + \cos\theta^i \mathbf{i}_2) \\
&= (\bar{u}_1^i \cos\theta^i - \bar{u}_2^i \sin\theta^i)\mathbf{i}_1 + (\bar{u}_1^i \sin\theta^i + \bar{u}_2^i \cos\theta^i)\mathbf{i}_2 \\
&= u_1^i \mathbf{i}_1 + u_2^i \mathbf{i}_2
\end{aligned}$$

where u_1^i and u_2^i are the components of the vector $\mathbf{u}^i$ defined in the global coordinate system and given by

$$\begin{aligned}
u_1^i &= \bar{u}_1^i \cos\theta^i - \bar{u}_2^i \sin\theta^i \\
u_2^i &= \bar{u}_1^i \sin\theta^i + \bar{u}_2^i \cos\theta^i
\end{aligned}$$

These two equations which provide algebraic relationships between the local and global components in the planar analysis can be expressed in a matrix form as

$$\mathbf{u}^i = \mathbf{A}^i \bar{\mathbf{u}}^i$$

where $\mathbf{u}^i = [u_1^i \quad u_2^i]^{\mathrm{T}}$, $\bar{\mathbf{u}}^i = [\bar{u}_1^i \quad \bar{u}_2^i]^{\mathrm{T}}$, and $\mathbf{A}^i$ is the planar transformation matrix defined as

$$\mathbf{A}^i = \begin{bmatrix} \cos\theta^i & -\sin\theta^i \\ \sin\theta^i & \cos\theta^i \end{bmatrix}$$

In Chapter 2 we will study the spatial kinematics and develop the spatial transformation matrix and study its important properties.

1.3 PARTICLE MECHANICS

Dynamics in general is the science of studying the motion of particles or bodies. The subject of dynamics can be divided into two major branches, *kinematics* and *kinetics*. In kinematic analysis we study the motion regardless of the forces that cause it, while kinetics deals with the motion and forces that produce it. Therefore, in kinematics attention is focused on the geometric aspects of motion. The objective is, then, to determine the positions, velocities, and accelerations of the system under investigation. In order to understand the dynamics of multibody systems containing rigid and deformable bodies, it is important to understand first the body dynamics. We start with a brief discussion on the dynamics of particles that form the rigid and deformable bodies.

Particle Kinematics A *particle* is assumed to have no dimensions and accordingly can be treated as a point in a three-dimensional space. Therefore, in studying the kinematics of particles, we are concerned primarily with the translation of a point with respect to a selected frame of reference. The position of the particle can then be defined using three coordinates. Figure 6 shows a particle p in a three-dimensional space. The position vector of this particle can be written as

$$\mathbf{r} = x_1\mathbf{i}_1 + x_2\mathbf{i}_2 + x_3\mathbf{i}_3 \tag{1.1}$$

where $\mathbf{i}_1$, $\mathbf{i}_2$, and $\mathbf{i}_3$ are unit vectors along the $\mathbf{X}_1$, $\mathbf{X}_2$, and $\mathbf{X}_3$ axes and x_1, x_2, and x_3 are the Cartesian coordinates of the particle.

The velocity of the particle is defined to be the time derivative of the position vector. If we assume that the axes $\mathbf{X}_1$, $\mathbf{X}_2$, and $\mathbf{X}_3$ are fixed in time, the unit vectors $\mathbf{i}_1, \mathbf{i}_2$, and $\mathbf{i}_3$ have a constant magnitude and direction. The velocity vector $\mathbf{v}$ of the particle can be written as

$$\mathbf{v} = \dot{\mathbf{r}} = \frac{d}{dt}(\mathbf{r}) = \dot{x}_1\mathbf{i}_1 + \dot{x}_2\mathbf{i}_2 + \dot{x}_3\mathbf{i}_3 \tag{1.2}$$

where $(\dot{\ })$ denotes differentiation with respect to time and $\dot{x}_1, \dot{x}_2$, and $\dot{x}_3$ are the Cartesian components of the velocity vector.

The acceleration of the particle is defined to be the time derivative of the velocity vector, that is,

$$\mathbf{a} = \frac{d}{dt}(\mathbf{v}) = \ddot{x}_1\mathbf{i}_1 + \ddot{x}_2\mathbf{i}_2 + \ddot{x}_3\mathbf{i}_3 \tag{1.3}$$

where $\mathbf{a}$ is the acceleration vector and $\ddot{x}_1$, $\ddot{x}_2$, and $\ddot{x}_3$ are the Cartesian components of the acceleration vector. Using vector notation, the position vector of the particle in terms of the *Cartesian coordinates* can be written as

$$\mathbf{r} = [x_1 \quad x_2 \quad x_3]^{\mathrm{T}}$$

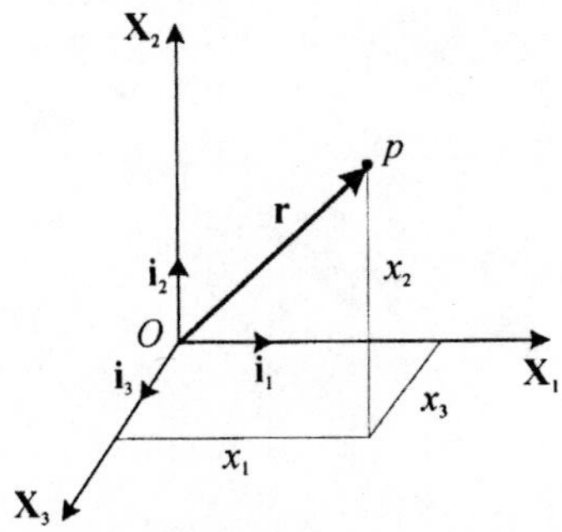

Figure 1.6 Position vector of the particle p.

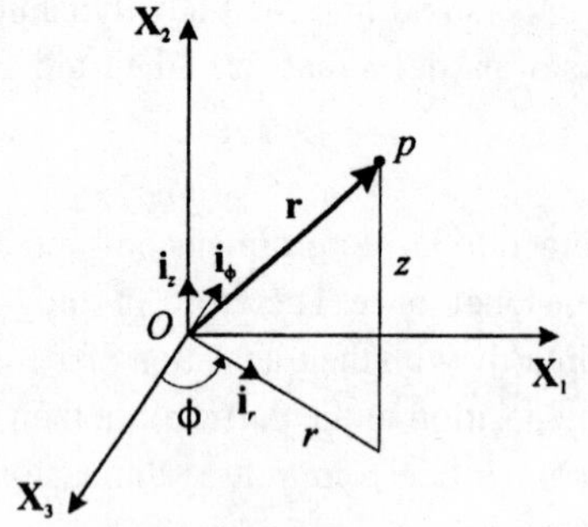

Figure 1.7 Cylindrical coordinates.

while the velocity and acceleration vectors are given by

$$\begin{aligned}\mathbf{v} &= \frac{d\mathbf{r}}{dt} = \begin{bmatrix}\frac{dx_1}{dt} & \frac{dx_2}{dt} & \frac{dx_3}{dt}\end{bmatrix}^{\mathrm{T}}\\ &= [\dot{x}_1 \quad \dot{x}_2 \quad \dot{x}_3]^{\mathrm{T}}\\ \mathbf{a} &= \frac{d\mathbf{v}}{dt} = \frac{d^2\mathbf{r}}{dt^2} = \begin{bmatrix}\frac{d^2x_1}{dt^2} & \frac{d^2x_2}{dt^2} & \frac{d^2x_3}{dt^2}\end{bmatrix}^{\mathrm{T}}\\ &= [\ddot{x}_1 \quad \ddot{x}_2 \quad \ddot{x}_3]^{\mathrm{T}}\end{aligned}$$

The set of coordinates that can be used to define the particle position is not unique. In addition to the Cartesian representation, other sets of coordinates can be used for the same purpose. In Fig. 7, the position of particle p can be defined using the three *cylindrical coordinates*, r, ϕ, and z, while in Fig. 8, the particle position is identified using the *spherical coordinates* r, θ, and ϕ. In many situations, however, it is useful to obtain kinematic relationships between different sets of coordinates. For instance, if we consider the planar motion of a particle p in a circular path as shown in Fig. 9, the position vector of the particle can be written in the fixed coordinate system $\mathbf{X}_1\mathbf{X}_2$ as

$$\mathbf{r} = [x_1 \quad x_2]^{\mathrm{T}} = x_1\mathbf{i}_1 + x_2\mathbf{i}_2$$

where x_1 and x_2 are the coordinates of the particle and $\mathbf{i}_1$ and $\mathbf{i}_2$ are unit vectors along the fixed axes $\mathbf{X}_1$ and $\mathbf{X}_2$, respectively. In terms of the polar coordinates r and θ, the

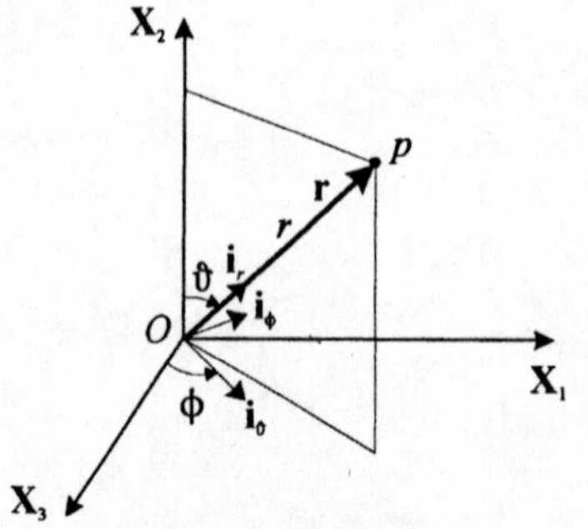

Figure 1.8 Spherical coordinates.

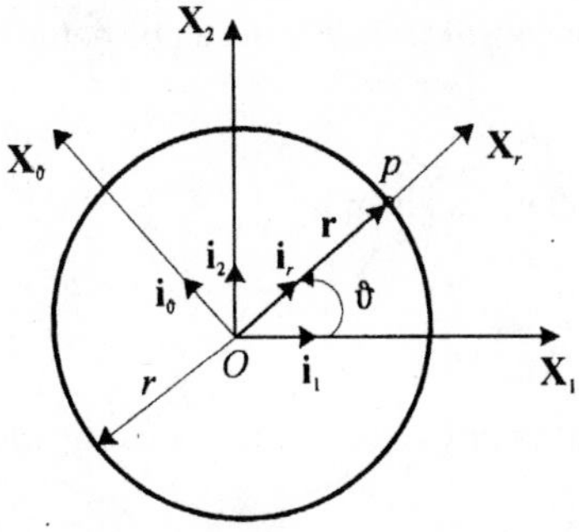

Figure 1.9 Circular motion of a particle.

components x_1 and x_2 are given by

$$x_1 = r\cos\theta, \qquad x_2 = r\sin\theta$$

and the vector **r** can be expressed as

$$\mathbf{r} = r\cos\theta\,\mathbf{i}_1 + r\sin\theta\,\mathbf{i}_2$$

Since r in this example is constant, and $\mathbf{i}_1$ and $\mathbf{i}_2$ are fixed vectors, the velocity of the particle is given by

$$\mathbf{v} = \frac{d\mathbf{r}}{dt} = r\dot{\theta}(-\sin\theta\,\mathbf{i}_1 + \cos\theta\,\mathbf{i}_2)$$

and the acceleration vector **a** is given by

$$\mathbf{a} = \frac{d\mathbf{v}}{dt} = r\ddot{\theta}(-\sin\theta\,\mathbf{i}_1 + \cos\theta\,\mathbf{i}_2) + r(\dot{\theta})^2(-\cos\theta\,\mathbf{i}_1 - \sin\theta\,\mathbf{i}_2)$$

One can verify that this equation can be written in the following compact vector form:

$$\mathbf{a} = \boldsymbol{\alpha} \times \mathbf{r} + \boldsymbol{\omega} \times \mathbf{v}$$

where $\boldsymbol{\omega}$ and $\boldsymbol{\alpha}$ are the vectors

$$\boldsymbol{\omega} = \dot{\theta}\,\mathbf{i}_3, \qquad \boldsymbol{\alpha} = \ddot{\theta}\,\mathbf{i}_3$$

One may also define the position vector of p in the moving coordinate system $\mathbf{X}_r\mathbf{X}_\theta$. Let, as shown in Fig. 9, $\mathbf{i}_r$ and $\mathbf{i}_\theta$ be unit vectors along the axes $\mathbf{X}_r$ and $\mathbf{X}_\theta$, respectively. It can be verified that these two unit vectors can be written in terms of the unit vectors along the fixed axes as

$$\mathbf{i}_r = \cos\theta\,\mathbf{i}_1 + \sin\theta\,\mathbf{i}_2$$
$$\mathbf{i}_\theta = -\sin\theta\,\mathbf{i}_1 + \cos\theta\,\mathbf{i}_2$$

and their time derivatives can be written as

$$\dot{\mathbf{i}}_r = \frac{d\mathbf{i}_r}{dt} = -\dot{\theta}\sin\theta\,\mathbf{i}_1 + \dot{\theta}\cos\theta\,\mathbf{i}_2 = \dot{\theta}\mathbf{i}_\theta$$
$$\dot{\mathbf{i}}_\theta = \frac{d\mathbf{i}_\theta}{dt} = -\dot{\theta}\cos\theta\,\mathbf{i}_1 - \dot{\theta}\sin\theta\,\mathbf{i}_2 = -\dot{\theta}\mathbf{i}_r$$

The position vector of the particle in the moving coordinate system can be defined as

$$\mathbf{r} = r\,\mathbf{i}_r$$

Using this equation, the velocity vector of particle p is given by

$$\mathbf{v} = \frac{d\mathbf{r}}{dt} = \frac{dr}{dt}\mathbf{i}_r + r\frac{d\mathbf{i}_r}{dt}$$

Since the motion of point p is in a circular path, $dr/dt = 0$, and the velocity vector $\mathbf{v}$ reduces to

$$\mathbf{v} = r\frac{d\mathbf{i}_r}{dt} = r\dot{\theta}\mathbf{i}_\theta$$

which shows that the velocity vector of the particle is always tangent to the circular path. The acceleration vector $\mathbf{a}$ is also given by

$$\mathbf{a} = \frac{d\mathbf{v}}{dt} = r\ddot{\theta}\mathbf{i}_\theta + r\dot{\theta}\frac{d\mathbf{i}_\theta}{dt} = r\ddot{\theta}\mathbf{i}_\theta - r(\dot{\theta})^2\mathbf{i}_r$$

The first term, $r\ddot{\theta}$, is called the *tangential component* of the acceleration, while the second term, $-r(\dot{\theta})^2$, is called the *normal component.*

Particle Dynamics The study of *Newtonian mechanics* is based on Newton's three laws, which are used to study particle mechanics. *Newton's first law* states that a particle remains in its state of rest, or of uniform motion in a straight line if there are no forces acting on the particle. This means that the particle can be accelerated if and only if there is a force acting on the particle. *Newton's third law*, which is sometimes called the *law of action and reaction*, states that to every action there is an equal and opposite reaction; that is, when two particles exert forces on one another, these forces will be equal in magnitude and opposite in direction. *Newton's second law*, which is called the *law of motion*, states that the force that acts on a particle and causes its motion is equal to the rate of change of momentum of the particle, that is

$$\mathbf{F} = \dot{\mathbf{P}} \tag{1.4}$$

where $\mathbf{F}$ is the vector of forces acting on the particle and $\mathbf{P}$ is the linear momentum of the particle, which can be written as

$$\mathbf{P} = m\mathbf{v} \tag{1.5}$$

where m is the mass and $\mathbf{v}$ is the velocity vector of the particle. Equations 4 and 5 imply that

$$\mathbf{F} = \frac{d}{dt}(m\mathbf{v}) \tag{1.6}$$

In nonrelativistic mechanics, the mass m is constant and as a consequence, Eq. 6 leads to

$$\mathbf{F} = m\frac{d\mathbf{v}}{dt} = m\mathbf{a} \tag{1.7}$$

where **a** is the acceleration vector of the particle. Equation 7 is a vector equation that has the three scalar components

$$F_1 = ma_1, \quad F_2 = ma_2, \quad F_3 = ma_3$$

where F_1, F_2, and F_3 and a_1, a_2, and a_3 are, respectively, the components of the vectors **F** and **a** defined in the global coordinate system. The vector $m\mathbf{a}$ is sometimes called the *inertia* or the *effective* force vector.

1.4 RIGID BODY MECHANICS

Unlike particles, *rigid bodies* have distributed masses. The configuration of a rigid body in space can be identified by using six coordinates. Three coordinates describe the body translation, and three coordinates define the orientation of the body. Figure 10 shows a rigid body denoted as body i in a three-dimensional space. Let $\mathbf{X}_1\mathbf{X}_2\mathbf{X}_3$ be a coordinate system that is fixed in time, and let $\mathbf{X}_1^i\mathbf{X}_2^i\mathbf{X}_3^i$ be a body coordinate system or body reference whose origin is rigidly attached to a point on the rigid body. The global position of an arbitrary point P^i on the body can be defined as

$$\mathbf{r}^i = \mathbf{R}^i + \mathbf{u}^i \tag{1.8}$$

where $\mathbf{r}^i = [r_1^i \; r_2^i \; r_3^i]^{\mathrm{T}}$ is the global position of point P^i, $\mathbf{R}^i = [R_1^i \; R_2^i \; R_3^i]^{\mathrm{T}}$ is the global position vector of the origin O^i of the body reference, and $\mathbf{u}^i = [u_1^i \; u_2^i \; u_3^i]^{\mathrm{T}}$ is the position vector of point P^i with respect to O^i. Since an assumption is made that the body is rigid, the distance between points P^i and O^i remains constant during the motion of the body; that is, the components of the vector $\mathbf{u}^i$ in the body coordinate system are known and constant. The vectors $\mathbf{r}^i$ and $\mathbf{R}^i$, however, are defined in the global coordinate system; therefore, it is important to be able to express the vector $\mathbf{u}^i$

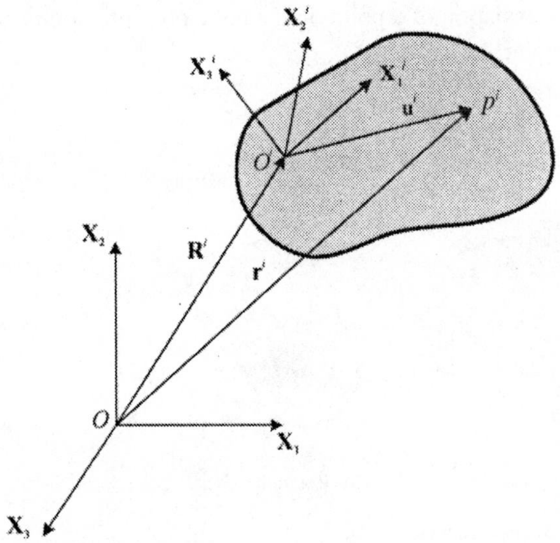

Figure 1.10 Rigid body mechanics.

in terms of its components along the fixed global axes. To this end, one needs to define the orientation of the body coordinate system with respect to the global frame of reference. A transformation between these two coordinate systems can be developed in terms of a set of rotational coordinates. However, this set of rotational coordinates is not unique, and many representations can be found in the literature. In Chapter 2 we develop the transformation matrix that can be used to transform vectors defined in the body coordinate systems to vectors defined in the global coordinate system and vice versa. We also introduce some of the most commonly used orientational coordinates such as *Euler angles*, *Euler parameters*, *Rodriguez parameters*, and the *direction cosines*. In some of these representations more than three orientational coordinates are used. In such cases, the orientational coordinates are not totally independent, since they are related by a set of algebraic equations.

Since Eq. 8 describes the global position of an arbitrary point on the body, the body configuration can be completely defined, provided the components of the vectors in the right-hand side of this equation are known. This equation implies that the general motion of a rigid body is equivalent to a translation of one point, say, O^i, plus a rotation. A rigid body is said to experience pure *translation* if the displacements of any two points on the body are the same. A rigid body is said to experience a pure *rotation* about an axis called the *axis of rotation*, if the particles forming the rigid body move in parallel planes along circles centered on the same axis. Figure 11 shows the translational and rotational motion of a rigid body. It is clear from Fig. 11(b) that in the case of pure rotation, points on the rigid body located on the axis of rotation have zero displacements, velocities, and accelerations. A pure rotation can be obtained if we fix one point on the body, called the *base point*. This will eliminate the translational degrees of freedom of the body. This is, in fact, *Euler's theorem*, which states that the general displacements of a rigid body with one point fixed is a rotation about some axis that passes through that point. If no point is fixed, the general motion of a rigid body is given by *Chasles' theorem*, which states that the most general motion of a rigid body is equivalent to a translation of a point on the body plus a rotation about an axis passing through that point.

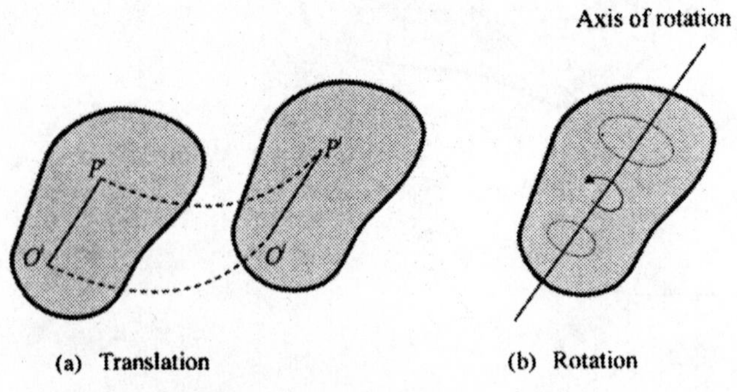

Figure 1.11 Rigid body displacements.

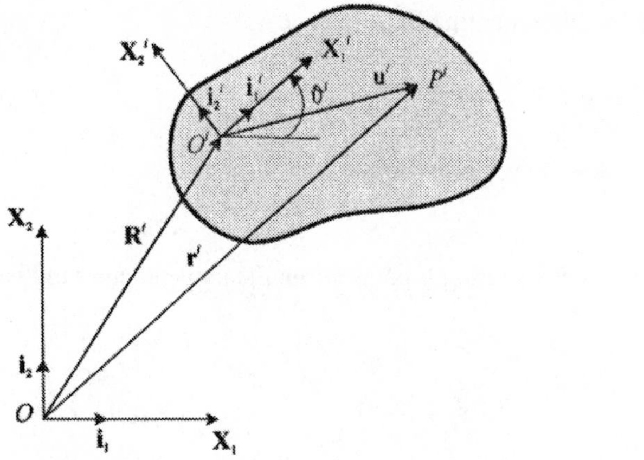

Figure 1.12 Absolute Cartesian coordinates.

Kinematic Equations In two-dimensional analysis, the configuration of the rigid body can be identified by using three coordinates; two coordinates define the translation of a point on the body, and one coordinate defines the orientation of the body with respect to a selected inertial frame of reference. For instance, consider the planar rigid body denoted as body i and shown in Fig. 12. Let $\mathbf{X}_1\mathbf{X}_2$ be the fixed frame of reference and $\mathbf{X}_1^i\mathbf{X}_2^i$ be the body reference whose origin is point O^i, which is rigidly attached to the body. The vector $\mathbf{R}^i = [R_1^i \; R_2^i]^{\mathrm{T}}$ describes the translation of the origin of the body reference, while the angle θ^i describes the body orientation. The set of Cartesian coordinates $\mathbf{q}_r^i$ defined as

$$\mathbf{q}_r^i = \begin{bmatrix} R_1^i & R_2^i & \theta^i \end{bmatrix}^{\mathrm{T}} \tag{1.9}$$

can then be used to define the body configuration; that is, the position, velocity, and acceleration of an arbitrary point on the body can be written in terms of these coordinates. Let P^i be an arbitrary point on the rigid body i and $\mathbf{i}_1^i$ and $\mathbf{i}_2^i$ be unit vectors along the body axes $\mathbf{X}_1^i$ and $\mathbf{X}_2^i$, respectively. The position vector of point P^i can be defined as

$$\mathbf{r}^i = \mathbf{R}^i + \mathbf{u}^i \tag{1.10}$$

where $\mathbf{r}^i = [r_1^i \; r_2^i]^{\mathrm{T}}$ is the global position of the point P^i and $\mathbf{u}^i = [\bar{u}_1^i \; \bar{u}_2^i]^{\mathrm{T}}$ is the position of P^i in the body coordinate system; that is

$$\mathbf{u}^i = \bar{u}_1^i\mathbf{i}_1^i + \bar{u}_2^i\mathbf{i}_2^i \tag{1.11}$$

where $\bar{u}_1^i$ and $\bar{u}_2^i$ are constant because the body is assumed to be rigid. In order to obtain the velocity vector of point P^i we differentiate Eq. 10 with respect to time. This yields

$$\mathbf{v}^i = \frac{d\mathbf{r}^i}{dt} = \dot{\mathbf{R}}^i + \dot{\mathbf{u}}^i \tag{1.12}$$

where $\dot{\mathbf{u}}^i$ can be obtained by differentiating Eq. 11. Then

$$\dot{\mathbf{u}}^i = \bar{u}_1^i \frac{d\mathbf{i}_1^i}{dt} + \bar{u}_2^i \frac{d\mathbf{i}_2^i}{dt} = \bar{u}_1^i \dot{\theta}^i \mathbf{i}_2^i - \bar{u}_2^i \dot{\theta}^i \mathbf{i}_1^i \tag{1.13}$$

We define ω^i, the *angular velocity* vector of body i, as

$$\omega^i = \dot{\theta}^i \mathbf{i}_3^i$$

where $\mathbf{i}_3^i$ is a unit vector that passes through point O^i and is perpendicular to $\mathbf{i}_1^i$ and $\mathbf{i}_2^i$. One can verify that

$$\omega^i \times \mathbf{u}^i = \begin{vmatrix} \mathbf{i}_1^i & \mathbf{i}_2^i & \mathbf{i}_3^i \\ 0 & 0 & \dot{\theta}^i \\ \bar{u}_1^i & \bar{u}_2^i & 0 \end{vmatrix} = -\bar{u}_2^i \dot{\theta}^i \mathbf{i}_1^i + \bar{u}_1^i \dot{\theta}^i \mathbf{i}_2^i \tag{1.14}$$

Comparing Eqs. 13 and 14, we conclude that

$$\dot{\mathbf{u}}^i = \omega^i \times \mathbf{u}^i \tag{1.15}$$

Substituting Eq. 15 into Eq. 12 leads to

$$\mathbf{v}^i = \dot{\mathbf{R}}^i + \omega^i \times \mathbf{u}^i \tag{1.16}$$

which shows that the velocity of an arbitrary point on the rigid body can be written in terms of time derivatives of the coordinates $\mathbf{q}_r^i = [\mathbf{R}^{i\mathrm{T}}\ \theta^i]^{\mathrm{T}}$.

By differentiating Eq. 16 with respect to time, an expression for the acceleration vector can be obtained in terms of the coordinates $\mathbf{q}_r^i$ and their time derivatives as follows

$$\mathbf{a}^i = \frac{d\mathbf{v}^i}{dt} = \ddot{\mathbf{R}}^i + \dot{\omega}^i \times \mathbf{u}^i + \omega^i \times \dot{\mathbf{u}}^i$$

If we define the *angular acceleration* vector α^i of body i as

$$\alpha^i = \ddot{\theta}^i \mathbf{i}_3^i$$

and use Eq. 14, the acceleration vector of point P^i can be written in the familiar vector form as

$$\mathbf{a}^i = \ddot{\mathbf{R}}^i + \alpha^i \times \mathbf{u}^i + \omega^i \times (\omega^i \times \mathbf{u}^i) \tag{1.17}$$

where $\ddot{\mathbf{R}}^i$ is the acceleration of the origin of the body reference. The term $\alpha^i \times \mathbf{u}^i$ is the *tangential component* of the acceleration of point P^i with respect to O^i. This component has magnitude $\ddot{\theta}^i u^i$ and has direction perpendicular to both vectors α^i and $\mathbf{u}^i$. The term $\omega^i \times (\omega^i \times \mathbf{u}^i)$ is called the *normal component* of the acceleration of P^i with respect to O^i. This component has magnitude $(\dot{\theta}^i)^2 u^i$ and is directed from P^i to O^i. In the spatial analysis, similar expressions for the velocity and acceleration of an arbitrary point on the body can be obtained.

Rigid Body Dynamics The dynamic equations that govern the motion of rigid bodies can be systematically obtained from the particle equations by assuming that the rigid body consists of a large number of particles. It can be demonstrated that the unconstrained three-dimensional motion of the rigid body can be described using six equations; three equations are associated with the translation of the rigid body, and three equations are associated with the body rotation. If a centroidal body coordinate system is used, the translational equations are called *Newton equations*, while the rotational equations are called *Euler equations*. Newton–Euler equations which are expressed in terms of the accelerations and forces acting on the body can be used to describe an arbitrary rigid body motion. These general equations are derived in Chapter 3.

In the special case of planar motion, the Newton–Euler equations reduce to three scalar equations that can be written, for body i in the multibody system, as

$$\left.\begin{aligned} m^i\mathbf{a}^i &= \mathbf{F}^i \\ J^i\ddot{\theta}^i &= M^i \end{aligned}\right\} \tag{1.18}$$

where m^i is the total mass of the rigid body, $\mathbf{a}^i$ is a two-dimensional vector that defines the absolute acceleration of the center of mass of the body, $\mathbf{F}^i$ is the vector of forces acting on the body center of mass, J^i is the mass moment of inertia defined with respect to the center of mass, θ^i is the angle that defines the orientation of the body, and M^i is the moment acting on the body. As will be demonstrated in Chapter 3, the choice of the center of mass as the origin of the body coordinate system leads to significant simplifications in the form of the dynamic equations. As the result of such a choice of the body reference, Newton–Euler equations have no inertia coupling between the translational and rotational coordinates of the rigid body. Such a decoupling of the coordinates becomes more difficult when deformable bodies are considered.

1.5 DEFORMABLE BODIES

In rigid body analysis, there is no distinction between the kinematics of the body and the kinematics of its reference. We have seen that the set of coordinates that define the location and orientation of the body reference is sufficient for definition of the location of an arbitrary point on the rigid body. This is mainly because the distance between two points on a rigid body remains invariant. This is not the case, however, when *deformable bodies* are considered. Two arbitrary points on a deformable body move relative to each other, and consequently the reference coordinates are no longer sufficient to describe the kinematics of deformable bodies. In fact, an infinite number of coordinates are required in order to define the exact position of each point on the deformable body. For instance, consider the deformable body i shown in Fig. 13 and let O^i and P^i be two arbitrary points on the body before displacement. After displacement, points O^i and P^i occupy the new positions O_1^i and P_1^i, respectively. In order to be able to measure the relative motion between these two points, we assign to this deformable body a body coordinate system $\mathbf{X}_1^i\mathbf{X}_2^i\mathbf{X}_3^i$ whose origin is rigidly attached to point O^i; that is, the origin of this body reference has the same translational displacements as point O^i, as shown in Fig. 13. Here we employ *body-fixed axes* for simplicity. To determine the change in the distance between points O^i and P^i due to the body displacement, we draw a rigid line element represented by the vector $\mathbf{u}_o^i$ emanating from point O^i that has the same magnitude and direction as the vector

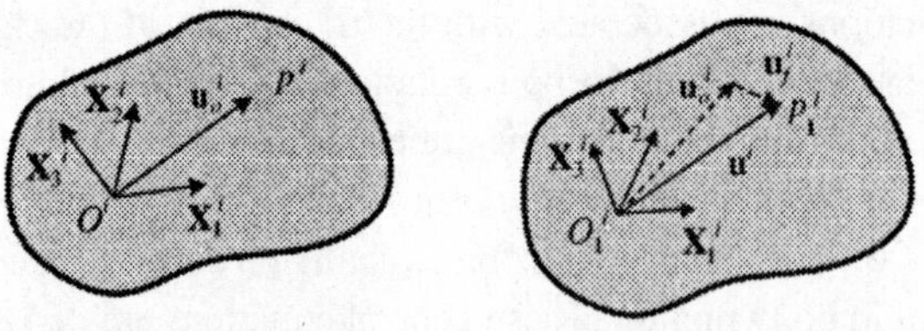

Figure 1.13 Deformation of the deformable body i.

between the two points O^i and P^i in the undeformed state. Furthermore, we assume that this rigid line element $\mathbf{u}_o^i$ has no translational or rotational displacement with respect to the body coordinate system; that is, the components of the vector $\mathbf{u}_o^i$ are constant in the local coordinate system during the motion of the deformable body. Even though the vector $\mathbf{u}_o^i$ is an imaginary line, it represents the position vector of P^i in the body coordinate system in the *undeformed state* and serves as a means for defining the deformation of point P^i as shown in Fig. 14. One can then write an expression for the position vector of point P^i as

$$\mathbf{r}^i = \mathbf{R}^i + \mathbf{u}_o^i + \mathbf{u}_f^i \tag{1.19}$$

where $\mathbf{R}^i = [R_1^i \; R_2^i \; R_3^i]^{\mathrm{T}}$ is the position vector of point O^i, $\mathbf{u}_o^i$ is the undeformed local position of point P^i, and $\mathbf{u}_f^i$ is the deformation vector at this point. While the components of the vector $\mathbf{u}_o^i$ in the body coordinate system are constant, the components of the vector $\mathbf{u}_f^i$ in the body coordinate system are time- and space-dependent. Consequently, the dynamic formulation of such systems leads to a set of partial differential equations that are space- and time-dependent. The exact solutions of these equations require an infinite number of coordinates that can be used to

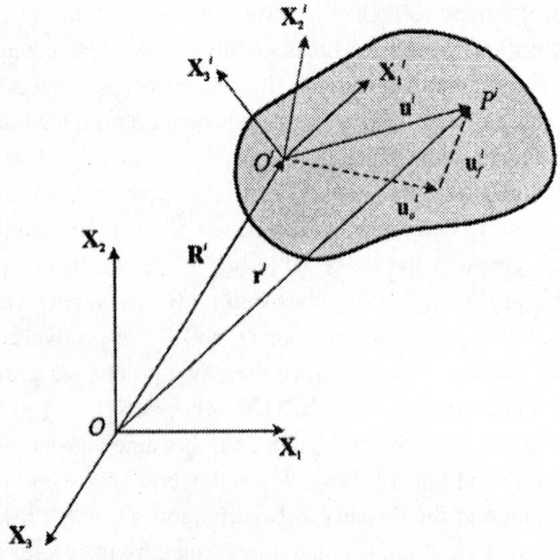

Figure 1.14 Coordinates of deformable bodies.

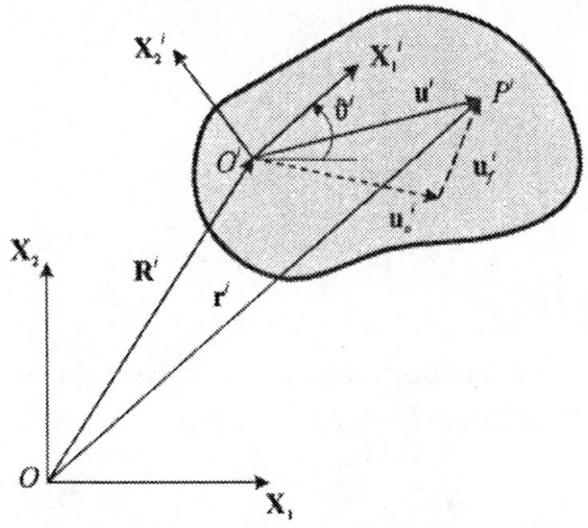

Figure 1.15 Planar motion of deformable bodies.

define the location of each point on the deformable body. To avoid the computational difficulties encountered in dealing with infinite dimensional spaces, approximation techniques such as *Rayleigh–Ritz methods* and the *finite-element methods* are often employed to reduce the number of coordinates to a finite set.

It is clear from Eq. 19 that the position vector of an arbitrary point on a deformable body is different from the rigid body case by the deformation vector $\mathbf{u}^i_f$. It is often convenient to define the deformation vector $\mathbf{u}^i_f$ and the undeformed position vector $\mathbf{u}^i_o$ in the body coordinate system. For instance, consider the planar motion of the deformable body shown in Fig. 15. The location of the origin of the body reference may be defined by the vector $\mathbf{R}^i = [R^i_1 \; R^i_2]^{\mathrm{T}}$, while the orientation of this reference is described by the angle θ^i. Let $\mathbf{i}^i_1$ and $\mathbf{i}^i_2$ be unit vectors along the body axes. The undeformed position vector $\mathbf{u}^i_o$ and the deformation $\mathbf{u}^i_f$ can be written as

$$\mathbf{u}^i_o = \bar{u}^i_{o1}\mathbf{i}^i_1 + \bar{u}^i_{o2}\mathbf{i}^i_2 \tag{1.20}$$

$$\mathbf{u}^i_f = \bar{u}^i_{f1}\mathbf{i}^i_1 + \bar{u}^i_{f2}\mathbf{i}^i_2 \tag{1.21}$$

where $\bar{u}^i_{o1}$ and $\bar{u}^i_{o2}$, and $\bar{u}^i_{f1}$ and $\bar{u}^i_{f2}$ are, respectively, the components of the vectors $\mathbf{u}^i_o$ and $\mathbf{u}^i_f$ defined in the body coordinate system. The components $\bar{u}^i_{o1}$ and $\bar{u}^i_{o2}$ are constant, while the components $\bar{u}^i_{f1}$ and $\bar{u}^i_{f2}$ depend on the location of point P^i as well as time. Keeping this in mind and differentiating Eqs. 20 and 21 with respect to time leads to

$$\dot{\mathbf{u}}^i_o = \bar{u}^i_{o1}\frac{d\mathbf{i}^i_1}{dt} + \bar{u}^i_{o2}\frac{d\mathbf{i}^i_2}{dt} = \bar{u}^i_{o1}\dot{\theta}^i\mathbf{i}^i_2 - \bar{u}^i_{o2}\dot{\theta}^i\mathbf{i}^i_1$$

$$\begin{aligned}\dot{\mathbf{u}}^i_f &= \bar{u}^i_{f1}\frac{d\mathbf{i}^i_1}{dt} + \bar{u}^i_{f2}\frac{d\mathbf{i}^i_2}{dt} + \dot{\bar{u}}^i_{f1}\mathbf{i}^i_1 + \dot{\bar{u}}^i_{f2}\mathbf{i}^i_2 \\ &= \bar{u}^i_{f1}\dot{\theta}^i\mathbf{i}^i_2 - \bar{u}^i_{f2}\dot{\theta}^i\mathbf{i}^i_1 + \dot{\bar{u}}^i_{f1}\mathbf{i}^i_1 + \dot{\bar{u}}^i_{f2}\mathbf{i}^i_2\end{aligned}$$

If we define the angular velocity vector ω^i of the body reference as

$$\omega^i = \dot{\theta}^i\mathbf{i}^i_3$$

we can write $\dot{\mathbf{u}}_o^i$ and $\dot{\mathbf{u}}_f^i$ as

$$\dot{\mathbf{u}}_o^i = \boldsymbol{\omega}^i \times \mathbf{u}_o^i \tag{1.22}$$

$$\dot{\mathbf{u}}_f^i = \boldsymbol{\omega}^i \times \mathbf{u}_f^i + (\dot{\mathbf{u}}_f^i)_r \tag{1.23}$$

where $(\dot{\mathbf{u}}_f^i)_r$ is the vector

$$(\dot{\mathbf{u}}_f^i)_r = \dot{\bar{u}}_{f1}^i \mathbf{i}_1^i + \dot{\bar{u}}_{f2}^i \mathbf{i}_2^i$$

Differentiating Eq. 19 with respect to time and using Eqs. 22 and 23, one obtains an expression for the velocity vector of an arbitrary point on the deformable body as

$$\begin{aligned}\mathbf{v}^i = \dot{\mathbf{r}}^i &= \dot{\mathbf{R}}^i + \dot{\mathbf{u}}_o^i + \dot{\mathbf{u}}_f^i \\ &= \dot{\mathbf{R}}^i + \boldsymbol{\omega}^i \times \mathbf{u}_o^i + \boldsymbol{\omega}^i \times \mathbf{u}_f^i + (\dot{\mathbf{u}}_f^i)_r\end{aligned}$$

or

$$\mathbf{v}^i = \dot{\mathbf{R}}^i + \boldsymbol{\omega}^i \times (\mathbf{u}_o^i + \mathbf{u}_f^i) + (\dot{\mathbf{u}}_f^i)_r \tag{1.24}$$

By comparing Eqs. 16 and 24, we see a clear difference between the velocity expressions for rigid and deformable bodies. The vector $(\dot{\mathbf{u}}_f^i)_r$ represents the rate of change of the deformation vector as seen by an observer stationed on the body.

The acceleration vector of an arbitrary point in the planar analysis can be obtained by differentiating Eq. 24 with respect to time. One can verify that this acceleration vector is given by

$$\mathbf{a}^i = \frac{d\mathbf{v}^i}{dt} = \ddot{\mathbf{R}}^i + \boldsymbol{\omega}^i \times (\boldsymbol{\omega}^i \times \mathbf{u}^i) + \boldsymbol{\alpha}^i \times \mathbf{u}^i + 2\boldsymbol{\omega}^i \times (\dot{\mathbf{u}}_f^i)_r + (\ddot{\mathbf{u}}_f^i)_r \tag{1.25}$$

where $\boldsymbol{\alpha}^i$ is the angular acceleration vector of the body i reference defined as

$$\boldsymbol{\alpha}^i = \ddot{\theta}^i \mathbf{i}_3^i$$

$\mathbf{u}^i$ is the local position of the arbitrary point, that is,

$$\mathbf{u}^i = \mathbf{u}_o^i + \mathbf{u}_f^i$$

and $(\ddot{\mathbf{u}}_f^i)_r$ is the acceleration of the arbitrary point as seen by an observer stationed on the body reference. This component of the acceleration is given by

$$(\ddot{\mathbf{u}}_f^i)_r = \ddot{\bar{u}}_{f1}^i \mathbf{i}_1^i + \ddot{\bar{u}}_{f2}^i \mathbf{i}_2^i \tag{1.26}$$

The first three terms in the right-hand side of Eq. 25 are similar to the ones that appeared in the rigid body analysis. The last two terms, however, are due to the deformation of the body.

Equations similar to Eqs. 24 and 25 can be derived in the spatial case. It is apparent, however, that the position, velocity, and acceleration vectors of an arbitrary point on the deformable body depend on how the deformation vector $\mathbf{u}_f^i$ is defined. This problem is addressed in Chapters 5 and 6 where the floating frame of reference formulation is presented.

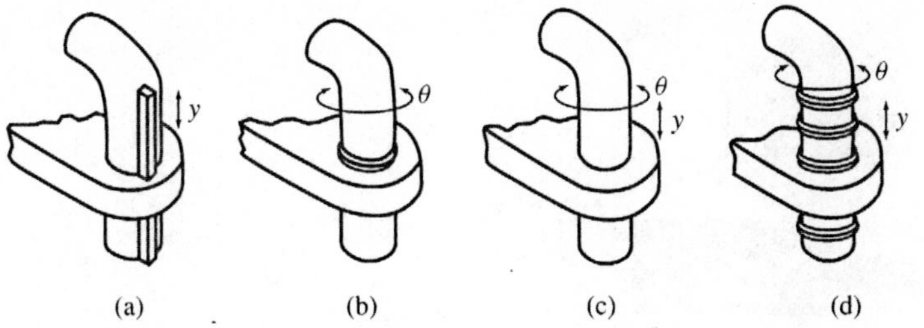

Figure 1.16 Examples of mechanical joints. (a) Prismatic or translational; (b) revolute; (c) cylindrical; (d) screw joint.

1.6 CONSTRAINED MOTION

In multibody systems, the motion of each body is constrained because of the system *mechanical joints* such as revolute, spherical, and prismatic joints or specified trajectories. Since six coordinates are required in order to identify the configuration of a rigid body in space, $6 \times n_b$ coordinates are required to describe the motion of n_b unconstrained bodies. Mechanical joints or specified trajectories reduce the system mobility because the motion of different bodies is no longer independent. Mechanical joints and specified motion trajectories can be described mathematically by using a set of nonlinear algebraic constraint equations. Assuming that these constraint equations are linearly independent, each constraint equation constrains a possible system motion. Therefore, the number of system *degrees of freedom* is defined to be the number of the system coordinates minus the number of independent constraint equations. For an n_b rigid body system with n_c independent constraint equations, the number of system degrees of freedom (DOF) is given by

$$DOF = 6 \times n_b - n_c \tag{1.27}$$

This is sometimes called the *Kutzbach criterion*. Figure 16 shows some of the mechanical joints that appear in many mechanical systems. The *prismatic* or *translational joint* (as it is sometimes called) shown in Fig. 16(a) allows only relative translation between the two bodies common to this joint. This relative translational displacement is along an axis called the *axis of the prismatic joint*. If a set of coordinates in a Cartesian space is used to describe the motion of these two bodies, five kinematic constraints must be imposed in order to allow motion only along the joint axis. These kinematic constraints can be formulated by using a set of algebraic equations that imply that the relative translation between the two bodies along two axes perpendicular to the joint axis as well as the relative rotations between the two bodies must be zero. Similarly, the *revolute joint*, shown in Fig. 16(b), allows only relative rotation between the two bodies about an axis called the *revolute joint axis*. One requires five constraint equations: three equations that constrain the relative translation between the two bodies, and two equations that constrain the relative rotation between the two bodies to be only about the joint axis of rotation. Similar comments apply to the *cylindrical joint*, which allows relative translation and rotation along the joint axis (Fig. 16(c)), and to the *screw joint*, which has one degree of freedom (Fig. 16(d)).

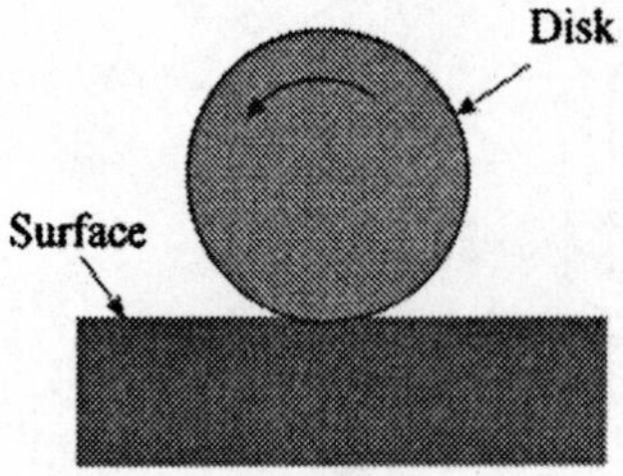

Figure 1.17 Rolling disk.

Another form of constrained motion is the planar motion wherein the body displacements can be represented in a two-dimensional Cartesian space. In this case, as shown in Fig. 12, only three coordinates are required in order to describe the body configuration. Thus the configuration of a set of unconstrained n_b bodies in two-dimensional space is completely defined using $3 \times n_b$ coordinates. Therefore, for a system of constrained n_b bodies in two-dimensional space, the number of degrees of freedom can be determined using the Kutzbach criterion as

$$DOF = 3 \times n_b - n_c \tag{1.28}$$

where n_c is the number of constraint equations that represent mechanical joints in the system as well as specified motion trajectories. One can verify that a revolute joint in plane can be described in Cartesian space using two algebraic constraint equations since the joint has only one degree of freedom that allows relative rotation between the two bodies common to this joint. Similarly, a prismatic joint in planar motion can be described in a mathematical form by using two algebraic constraint equations that allow only relative translation between the two bodies common to this joint along the prismatic joint axis. In the system shown in Fig. 17, a disk rolls on a surface without slipping. In this case the translation of the center of mass and the rotational motion of the disk are not independent. Therefore, the relative motion between the disk and the surface can be described by using only one degree of freedom. If rolling and slipping motions between the disk and the surface occur, then the translation and the rotation of the disk are independent and the relative motion between the disk and the surface can be described by using two degrees of freedom.

Application of the Kutzbach criterion is straightforward. For example, consider the planar slider crank mechanism shown in Fig. 18, which consists of four bodies, the fixed link (ground), the crankshaft OA, the connecting rod AB, and the slider block at B. The system has three revolute joints and one prismatic joint. These joints are the revolute joint at O, which connects the crankshaft to the fixed link; the revolute joint at A between the crankshaft and the connecting rod; the revolute joint at B between the connecting rod and the slider block; and the prismatic joint, which allows translation only between the slider block and the fixed link. Since each revolute and/or prismatic joint eliminates two degrees of freedom, and since three constraints are required in order to eliminate the degrees of freedom of the fixed link, one can verify that $n_c = 11$. Since $n_b = 4$, the mobility of the mechanism can be determined by

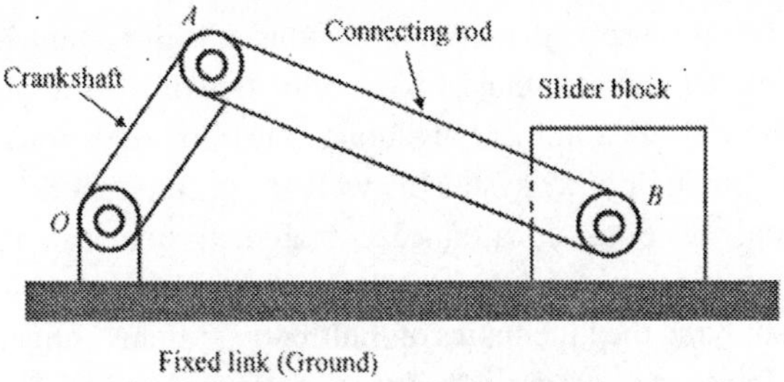

Figure 1.18 Slider crank mechanism.

using the Kutzbach criterion as

$$m = 3 \times n_b - n_c = 3 \times 4 - 11 = 1$$

That is, the planar multibody slider crank mechanism has one degree of freedom. This means that the motion of the mechanism can be controlled by using only one input. In other words, by specifying one variable, say, the angular rotation of the crankshaft or the travel of the slider block, one must be able to completely identify the system configuration.

Another single degree of freedom multibody system is the Peaucellier mechanism shown in Fig. 19. This mechanism has 8 links and 10 revolute joints and is designed such that point P moves in a straight line. The kinematic constraints include the revolute joint constraints as well as the ground (fixed link) constraints. In this case

$$n_b = 8 \quad \text{and} \quad n_c = 23$$

By applying the Kutzbach criterion, one concludes that the system has only one degree of freedom and as a result the configuration of the mechanism can be identified by specifying one variable such as the angular rotation of the crankshaft CD.

In some particular cases in which some geometric restrictions are imposed, the Kutzbach criterion may not give the correct answer. This is not surprising because in developing the Kutzbach criterion, no consideration was given to the dimension or some geometric properties of the multibody system. Nonetheless, the Kutzbach

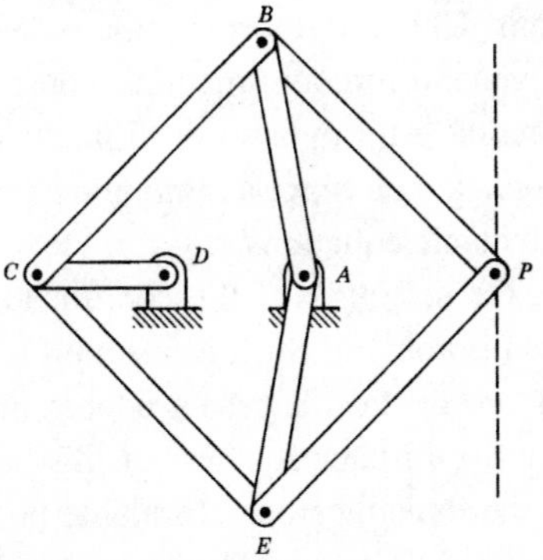

Figure 1.19 Peaucellier–Lipkin mechanism.

criterion is easy to apply and remains useful in most applications. It is important, however, to point out that a complete understanding of the kinematics of the multibody system requires the formulation of the nonlinear algebraic constraint equations that describe mechanical joints in the multibody system as well as specified motion trajectories. By studying the properties of the constraint Jacobian matrix one can obtain useful information about the motion of the multibody system. This is the approach that we will follow in this text in studying the kinematics of multibody systems containing rigid and deformable bodies. In doing so we can introduce the effect of body deformation on the kinematic constraint equations in a systematic and straightforward manner.

1.7 COMPUTER FORMULATION AND COORDINATE SELECTION

Much of the current research in multibody dynamics is devoted to the selection of the system coordinates and system degrees of freedom that can be efficiently used to describe the system configuration. A trade-off must be made between the generality and the efficiency of the dynamic formulation. The methods used in the dynamic analysis of multi-rigid-body systems can, in general, be divided into two main approaches. In the first approach, the configuration of the system is identified by using a set of Cartesian coordinates that describe the locations and orientations of the bodies in the system. This approach has the advantage that the dynamic formulation of the equations that govern the motion of the system is straightforward. Moreover, this approach, in general, allows easy additions of complex force functions and constraint equations. For each spatial rigid body in the system, six coordinates are sufficient to describe the body configuration. The configuration of deformable bodies, however, can be identified by using a coupled set of Cartesian and elastic coordinates, where the Cartesian coordinates define the location and orientation of a selected body reference, while elastic coordinates describe the deformation of the body with respect to the body reference. In this approach, connectivity between different bodies can be introduced to the dynamic formulation by using a set of nonlinear algebraic constraint equations.

In the second approach, relative or joint coordinates are used to formulate a minimum number of dynamic equations that are expressed in terms of the system degrees of freedom. In many applications, this approach leads to a complex recursive formulation based on loop closure equations. Unlike the formulation based on the Cartesian coordinates, the incorporation of general forcing functions, constraint equations, and/or specified trajectories in the recursive formulation is difficult. This approach, however, is more desirable in some applications, since a minimum number of coordinates is employed in formulating the dynamic equations.

As compared with rigid body mechanics, the selection of the coordinates in flexible body dynamics represents a more difficult problem. Such a selection is not limited to the use of Cartesian or relative joint coordinates, but it introduces many conceptual problems that we will attempt to clarify in later chapters of this book. As previously pointed out, exact modeling of the dynamics of deformable bodies requires the use of an infinite number of degrees of freedom. Therefore, the first problem encountered in the computer modeling of deformable bodies is the definition of an acceptable model for the deformable body using a finite set of coordinates. In

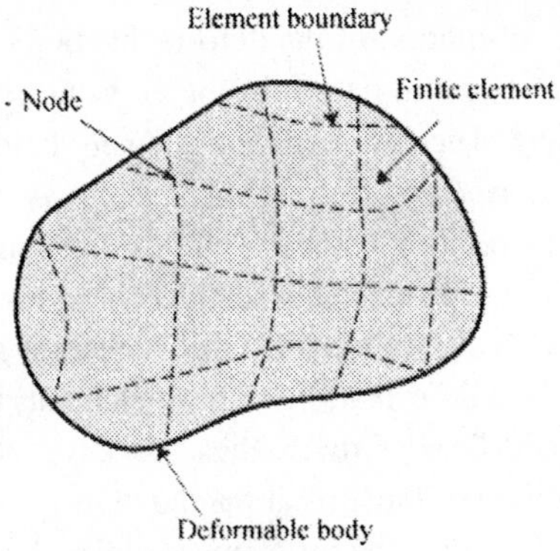

Figure 1.20 Finite element discretization.

the Rayleigh–Ritz method, this problem is solved by assuming that the shape of deformation of the body can be predicted and approximated using a finite set of known functions that define the body deformation with respect to its reference. By so doing, the dynamics of the deformable body can be modeled using a finite set of elastic coordinates as described in Chapter 5. One of the main problems associated with the Rayleigh–Ritz method is the difficulty of determining these approximation functions when the deformable bodies have complex geometrical shapes. This problem can be solved by using the finite element method. In the finite element method, as shown in Fig. 20, deformable bodies are discretized into small regions called *elements* that are connected at points called *nodes*. The coordinates and the spatial derivatives of the coordinates of the nodal points are used as the degrees of freedom. Interpolating polynomials that use the nodal degrees of freedom as coefficients are employed to define the deformation within the element. These interpolating polynomials and the nodal coordinates define the *assumed displacement field* of the finite element in terms of an *element shape function*. There are varieties of finite elements with different geometrical shapes that suit most engineering applications and can be used to represent deformable bodies with very complex geometrical shapes. Examples of these elements are truss, beam, rectangular, and triangular elements used in the planar analysis, and beam, plate, solid, tetrahedral, and shell elements used in the three-dimensional analysis. Some of these finite elements are shown in Fig. 21.

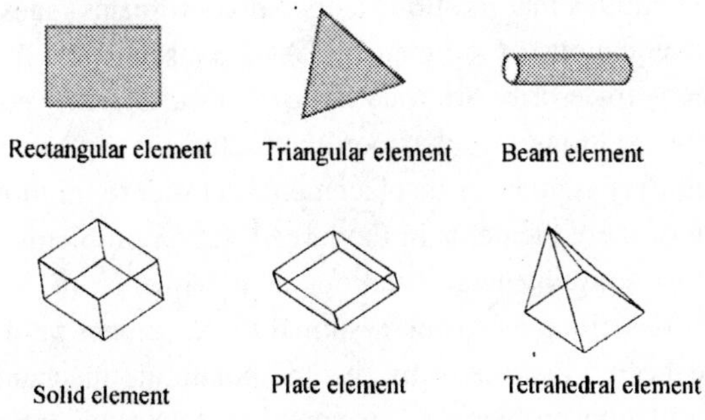

Figure 1.21 Finite elements.

The method of formulating the equations of motion of the deformable body using the finite element depends to a large extent on the nature of the element nodal coordinates and the assumed displacement field. The assumed displacement field of some of the finite elements can be used to describe an arbitrary displacements, and as such, these elements can be used in the large rotation and deformation analysis of flexible bodies. Such elements are not the subject of extensive research. The assumed displacement field of some other elements, as will be discussed in Chapters 6 and 7, cannot be used to describe large rotations and deformations. These elements are the subject of extensive research in the general field of mechanics. Because these elements do not lend themselves easily to solution of large rotation and deformation problems, several methods have been proposed to solve the problems associated with these elements. These methods can be roughly classified into three different basic formulations: the *floating frame of reference formulation*, the *incremental formulation*, and the *large rotation vector formulation*. These three basic methods are briefly discussed below.

Floating Frame of Reference Formulation The kinematic description used in the floating frame of reference formulation is the same as described in Section 5 of this chapter. In this approach, a coordinate system is assigned to each deformable body, which is discretized using a set of rigidly connected finite elements. The large translation and rotation of the deformable body are described using a set of absolute reference coordinates that define the location and the orientation of the selected deformable body coordinate system. The deformation of the body with respect to its coordinate system is defined using the nodal coordinates of the element. It can be demonstrated that the use of the floating frame of reference formulation leads to exact modeling of the rigid body inertia when the deformation is equal to zero. Furthermore, the finite elements defined in the floating frame of reference formulation lead to zero strain under an arbitrary rigid body motion. While the floating frame of reference formulation is the most widely used method in flexible multibody dynamics, its use has been limited to applications in which the deformation of the body with respect to its coordinate system is assumed small. The finite element floating frame of reference formulation is discussed in more detail in Chapters 5 and 6.

Incremental Formulation The nodal coordinates of many important elements such as beams and plates represent nodal displacements and infinitesimal nodal rotations. The use of the infinitesimal rotations as nodal coordinates leads to a linearization of the kinematic equations of the element. As a consequence, these coordinates cannot be used directly to describe arbitrary large rotations. Furthermore, as the result of using infinitesimal rotations as nodal coordinates, the elements do not produce zero strain under an arbitrary rigid body displacement. In order to minimize the error resulting from the use of these elements in the large rotation problems, an incremental procedure is used to represent large rotations as a sequence of small rotations that can be accurately described using the assumed displacement field of the element. This approach has been widely used by the computational mechanics community to solve large deformation problems in structural system applications.

It can be shown, however, that the use of this procedure does not lead to an exact modeling of the rigid body inertia when the element deformation is equal to zero (Shabana 1997b).

Large Rotation Vector Formulation In an attempt to solve the problems resulting from the use of infinitesimal rotations as nodal coordinates, the large rotation vector formulation was proposed. In this formulation, the rotation of the element cross-section is introduced as a field that can be approximated using interpolating polynomials. In this case, finite rotations are used as nodal coordinates. One of the problems associated with this formulation is the need to interpolate finite rotation coordinates. More significantly, this formulation leads to redundancy in describing the large rotation of the cross-section, since, as demonstrated in Chapter 7, the displacement field that describes the locations of the material points on the element can also be used to define the large rotation of the cross-section of the element.

The floating frame of reference formulation has been successfully used in solving many multibody applications. It is also implemented in several commercial and research general purpose flexible multibody computer programs. In comparison to the floating frame of reference formulation, the incremental approach and the large rotation vector formulation are not as widely used in solving flexible multibody applications due to the limitations previously mentioned. However, as previously pointed out, the floating frame of reference formulation was mainly used in solving large reference displacement and small deformation problems. In Chapter 7 of this book, a conceptually different formulation called the *absolute nodal coordinate formulation* is presented. This formulation can be used efficiently for large deformation problems in flexible multibody system applications.

1.8 OBJECTIVES AND SCOPE OF THIS BOOK

This book is designed to introduce the elements that are basic for formulating the dynamic equations of motion of rigid and deformable bodies. Emphasis is placed on the generality of the dynamic formulation developed for the computer-aided analysis of general multibody systems containing rigid and deformable bodies. The materials covered in this book are kept at a level suitable for senior undergraduate and first-year graduate students. Elementary problems and examples are presented in order to demonstrate the basic ideas and concepts presented in this book.

Chapter 2 discusses the kinematics of rigid bodies, or equivalently the kinematics of the rigid frame of reference. In this chapter a rigorous development of the spatial transformation matrix for finite rotations is provided in terms of the four *Euler parameters*. The exponential form of the transformation matrix and some useful identities are also developed and used to study many of the important properties of this spatial transformation such as the orthogonality and the noncommutativity of the finite rotations. The time derivatives of the rotation matrix are then obtained, and the kinematic relationships between the angular velocity vector and the time derivative of the orientational coordinates are established. We conclude this chapter by providing alternate forms of the spatial transformation matrix. Different orientational coordinates such as

Rodriguez parameters, Euler angles, direction cosines, and the 4×4 transformation matrix method are used for this purpose.

Chapter 3 presents some analytical techniques for developing the dynamic equations of motion. The concepts of generalized coordinates and degrees of freedom are first introduced. Examples of some algebraic kinematic constraint equations that describe mechanical joints are presented and computer-oriented techniques are provided in order to determine the system-dependent and -independent coordinates. The generalized coordinate partitioning of the *constraint Jacobian matrix* may be used for this purpose. The concept of virtual work is then introduced and used with D'Alembert's principle to derive Lagrange's equation of motion for mechanical systems subject to *holonomic* and *nonholonomic* constraint equations. The equivalence of Lagrange's equation and Newton's second law, however, is demonstrated by some simple examples. Prior to introducing Hamilton's principle, which represents an alternate approach to developing the dynamic equations of motion, some variational techniques are briefly discussed. This chapter concludes by deriving the dynamic equations of motion of general constrained multibody systems consisting of rigid bodies.

Basic concepts and definitions related to the mechanics of deformable bodies are introduced in Chapter 4. In this chapter the kinematics of deformable bodies are first discussed and the strain displacement relationships are developed. The stress components are then written in terms of the surface forces. This leads to the important *Cauchy stress formula*. The kinematic and stress relationships developed in the first few sections of Chapter 4 are general and can be applied to any kind of material. The stress and strain components, however, are related through the constitutive equations that depend on the material properties. After developing these constitutive equations, we develop the dynamic partial differential equations of equilibrium. We conclude this chapter by developing a general expression for the virtual work of the elastic forces. This expression will be used in chapters that follow. Most of the materials covered in Chapter 4 represent a brief introduction to the classical presentation of the theory of elasticity and continuum mechanics. The concepts and definitions presented in this chapter, however, are essential in the development of the following chapters.

In multibody systems, deformable bodies undergo large translational and rotational displacements. In Chapter 5 approximation techniques such as the Rayleigh–Ritz method are used to reduce the partial differential equations to a set of ordinary differential equations. In this chapter the position and velocity vectors of an arbitrary point on the deformable body are expressed in terms of a finite set of coupled reference and elastic coordinates. The velocity vector is then used to develop the kinetic energy, and the deformable body nonlinear mass matrix is identified in terms of a set of inertia shape integrals that depend on the assumed displacement field. These integrals provide a systematic approach for deriving the inertia properties of the deformable bodies that undergo large translational and rotational displacements. It is also shown that the mass matrices that appear in rigid body and structural analysis are special cases of the mass matrix of the deformable body that undergoes large translational and rotational displacements. The nonlinear terms that represent the inertia coupling between the deformable body reference motion and elastic deformation are identified throughout the development. Virtual work of external and elastic forces as well as kinematic constraint equations that represent mechanical joints and specified trajectories are also expressed in terms of the coupled set of reference and elastic coordinates. The computer implementation of the floating frame of reference formulation is also discussed in Chapter 5.

Two finite element formulations that were introduced and developed by the author are presented in Chapters 6 and 7. In Chapter 6, a finite element Lagrangian formulation for deformable bodies that undergo large translational and rotational displacements is developed. The inertia shape integrals that appear in the mass matrix are developed for each finite element on the deformable body. The body integrals are then developed by summing the integrals of the elements. The use of the formulation presented in this chapter is demonstrated by using two- and three-dimensional beam elements.

In the finite element formulation presented in Chapter 6, infinitesimal rotations can be used as element nodal coordinates. With these types of coordinates, the nonlinear finite element formulation presented in Chapter 6 leads to exact modeling of the rigid body dynamics. In Chapter 7, an *absolute nodal coordinate formulation* is developed for the large deformation and rotation analysis of flexible bodies. This formulation is conceptually different from the finite element formulation presented in Chapter 6 in the sense that all the element nodal coordinates define absolute variables. In the formulation presented in Chapter 7, no infinitesimal or finite rotations are used as the nodal coordinates for the finite elements. The absolute nodal coordinate formulation is also used in Chapter 7 to demonstrate that the floating frame of reference formulations presented in Chapters 5 and 6 do not lead to a separation of the rigid body motion and the elastic deformation of the flexible bodies.

2 REFERENCE KINEMATICS

While a body-fixed coordinate system is commonly employed as a reference for rigid components, a floating coordinate system is suggested for deformable bodies that undergo large rotations. When dealing with rigid body systems, the kinematics of the body is completely described by the kinematics of its coordinate system because the particles of a rigid body do not move with respect to a body-fixed coordinate system. The local position of a particle on the body can then be described in terms of fixed components along the axes of this moving coordinate system. In deformable bodies, however, particles move with respect to the selected body coordinate system, and therefore, we make a distinction between the kinematics of the coordinate system and the body kinematics.

Fundamental to any presentation of kinematics is an understanding of the rotations in space. This chapter, therefore, is devoted mainly to the development of techniques for describing the orientation of the moving body coordinate system in space. A coordinate system, called hereafter a *reference*, is a rigid triad vector whose motion can be described by the translation of the origin of the triad and by the rotation about a line defined in the inertial coordinate system. One may then conclude that if the origin of the body reference is fixed with respect to the inertial frame, the only remaining motion is the rotation of the body reference. Therefore, without loss of generality, we fix the origin of the body reference and develop the transformation matrices that describe the orientation of the reference. Having defined those transformation matrices, we later introduce the translation of the origin of the body reference in order to define the global position of an arbitrary point whose position is defined in terms of components along the axes of the moving reference. In so doing, the configuration of the body reference is described by six independent quantities: three translational and three rotational components. This is consistent with *Chasles' theorem*, which states that the general displacement of a rigid frame may be described by a translation along a line and a rotation about that line.

2.1 ROTATION MATRIX

In multibody systems, the components may undergo large relative translational and rotational displacements. To define the configuration of a body in the multibody system in space, one must be able to determine the location of every point on the body with respect to a selected inertial frame of reference. To this end, it is more convenient to assign for every body in the multibody system a body reference in which the position vectors of the material points can be easily described. The position vectors of these points can then be found in other coordinate systems by defining the relative position and orientation of the body coordinate system with respect to the other coordinate systems. Six variables are sufficient for definition of the position and orientation of one coordinate system $\mathbf{X}_1^i\mathbf{X}_2^i\mathbf{X}_3^i$ with respect to another coordinate system $\mathbf{X}_1\mathbf{X}_2\mathbf{X}_3$. As shown in Fig. 1(a), three variables define the relative translational motion between the two coordinate systems. This relative translational motion can be measured by the position vector of the origin O^i of the coordinate system $\mathbf{X}_1^i\mathbf{X}_2^i\mathbf{X}_3^i$ with respect to the coordinate system $\mathbf{X}_1\mathbf{X}_2\mathbf{X}_3$. The orientation of one coordinate system with respect to another can be defined in terms of three independent variables.

Derivation of the Rotation Matrix To develop the transformation that defines the relative orientation between two coordinate systems $\mathbf{X}_1^i\mathbf{X}_2^i\mathbf{X}_3^i$ and $\mathbf{X}_1\mathbf{X}_2\mathbf{X}_3$, we first – and without loss of generality – assume that the origins of the two coordinate systems coincide as shown in Fig. 1(b). We also assume that the axes of these two coordinate systems are initially parallel. Let the vector $\bar{\mathbf{r}}$ be the position vector of point $\bar{Q}$ whose coordinates are assumed to be fixed in the $\mathbf{X}_1^i\mathbf{X}_2^i\mathbf{X}_3^i$ coordinate system. Therefore, before rotation of the $\mathbf{X}_1^i\mathbf{X}_2^i\mathbf{X}_3^i$ coordinate system relative to the $\mathbf{X}_1\mathbf{X}_2\mathbf{X}_3$ coordinate system, the components of the vector $\bar{\mathbf{r}}$ in both coordinate systems are the same. Let the reference $\mathbf{X}_1^i\mathbf{X}_2^i\mathbf{X}_3^i$ rotate an angle θ about the axis OC as shown in Fig. 2(a). As the result of this rotation, point $\bar{Q}$ is translated to point Q. The position

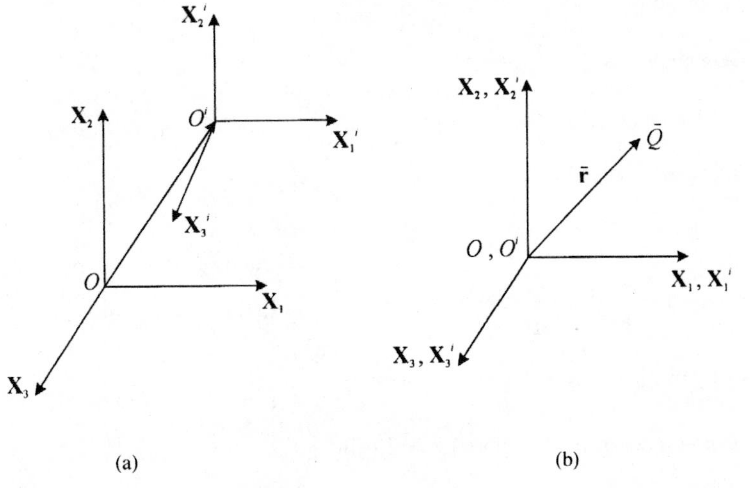

Figure 2.1 Coordinate systems.

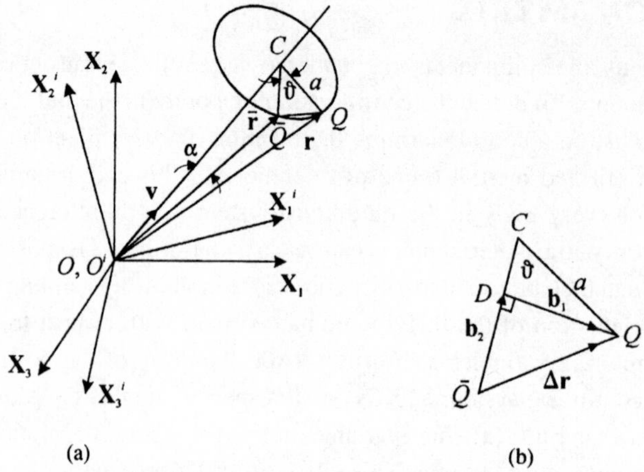

Figure 2.2 Finite rotation.

vector of point Q in the $\mathbf{X}_1\mathbf{X}_2\mathbf{X}_3$ coordinate system is denoted by $\mathbf{r}$. The change in the position vector of point $\bar{Q}$ due to the rotation θ is defined by the vector $\Delta\mathbf{r}$ as shown in Fig. 2(b). It is clear that the vector $\bar{\mathbf{r}}$ as a result of application of the rotation θ about the axis of rotation OC is transformed to the vector $\mathbf{r}$ and the new vector $\mathbf{r}$ can be written as

$$\mathbf{r} = \bar{\mathbf{r}} + \Delta\mathbf{r} \tag{2.1}$$

The vector $\Delta\mathbf{r}$, as shown in Fig. 2(b), can be written as the sum of the two vectors

$$\Delta\mathbf{r} = \mathbf{b}_1 + \mathbf{b}_2 \tag{2.2}$$

where the vector $\mathbf{b}_1$ is drawn perpendicular to the plane $OC\bar{Q}$ and thus has a direction $(\mathbf{v} \times \bar{\mathbf{r}})$, where $\mathbf{v}$ is a unit vector along the axis of rotation OC. The magnitude of the vector $\mathbf{b}_1$ is given by

$$|\mathbf{b}_1| = a \sin\theta$$

From Fig. 2(b), one can see that

$$a = |\bar{\mathbf{r}}| \sin\alpha = |\mathbf{v} \times \bar{\mathbf{r}}|$$

Therefore

$$\mathbf{b}_1 = a \sin\theta \frac{\mathbf{v} \times \bar{\mathbf{r}}}{|\mathbf{v} \times \bar{\mathbf{r}}|} = (\mathbf{v} \times \bar{\mathbf{r}}) \sin\theta \tag{2.3}$$

The vector $\mathbf{b}_2$ in Eq. 2 has a magnitude given by

$$|\mathbf{b}_2| = a - a\cos\theta = (1 - \cos\theta)\, a = 2a \sin^2 \frac{\theta}{2}$$

The vector $\mathbf{b}_2$ is perpendicular to $\mathbf{v}$ and also perpendicular to DQ, whose direction is the same as the unit vecotr $(\mathbf{v} \times \bar{\mathbf{r}})/a$. Therefore, $\mathbf{b}_2$ is the vector

$$\begin{aligned}\mathbf{b}_2 &= 2a \sin^2 \frac{\theta}{2} \cdot \frac{\mathbf{v} \times (\mathbf{v} \times \bar{\mathbf{r}})}{a} \\ &= 2[\mathbf{v} \times (\mathbf{v} \times \bar{\mathbf{r}})]\sin^2 \frac{\theta}{2}\end{aligned} \tag{2.4}$$

Using Eqs. 1–4, one can write

$$\mathbf{r} = \bar{\mathbf{r}} + (\mathbf{v} \times \bar{\mathbf{r}}) \sin\theta + 2[\mathbf{v} \times (\mathbf{v} \times \bar{\mathbf{r}})]\sin^2 \frac{\theta}{2} \tag{2.5}$$

By using the identity

$$\mathbf{v} \times \bar{\mathbf{r}} = \tilde{\mathbf{v}}\bar{\mathbf{r}} = -\tilde{\bar{\mathbf{r}}}\mathbf{v}$$

where $\tilde{\mathbf{v}}$ and $\tilde{\bar{\mathbf{r}}}$ are skew symmetric matrices given by

$$\tilde{\mathbf{v}} = \begin{bmatrix} 0 & -v_3 & v_2 \\ v_3 & 0 & -v_1 \\ -v_2 & v_1 & 0 \end{bmatrix}, \quad \tilde{\bar{\mathbf{r}}} = \begin{bmatrix} 0 & -\bar{r}_3 & \bar{r}_2 \\ \bar{r}_3 & 0 & -\bar{r}_1 \\ -\bar{r}_2 & \bar{r}_1 & 0 \end{bmatrix} \tag{2.6}$$

in which v_1, v_2, and v_3 are the components of the unit vector $\mathbf{v}$ and $\bar{r}_1$, $\bar{r}_2$, and $\bar{r}_3$ are the components of $\bar{\mathbf{r}}$, one can rewrite Eq. 5 in the following form:

$$\mathbf{r} = \bar{\mathbf{r}} + \tilde{\mathbf{v}}\bar{\mathbf{r}} \sin\theta + 2(\tilde{\mathbf{v}})^2\bar{\mathbf{r}} \sin^2 \frac{\theta}{2}$$

This equation can be rewritten as

$$\mathbf{r} = \left[\mathbf{I} + \tilde{\mathbf{v}} \sin\theta + 2(\tilde{\mathbf{v}})^2 \sin^2 \frac{\theta}{2}\right]\bar{\mathbf{r}} \tag{2.7}$$

where $\mathbf{I}$ is a 3×3 identity matrix. Equation 7 can be written as

$$\mathbf{r} = \mathbf{A}\bar{\mathbf{r}} \tag{2.8}$$

where $\mathbf{A} = \mathbf{A}(\theta)$ is the 3×3 *rotation matrix* given by

$$\mathbf{A} = \left[\mathbf{I} + \tilde{\mathbf{v}} \sin\theta + 2(\tilde{\mathbf{v}})^2 \sin^2 \frac{\theta}{2}\right] \tag{2.9}$$

This rotation matrix, referred to as the *Rodriguez formula*, is expressed in terms of the angle of rotation and a unit vector along the axis of rotation.

Euler Parameters Using the trigonometric identity

$$\sin\theta = 2 \sin \frac{\theta}{2} \cos \frac{\theta}{2}$$

one can rewrite the transformation matrix of Eq. 9 as

$$\mathbf{A} = \mathbf{I} + 2\tilde{\mathbf{v}} \sin \frac{\theta}{2}\left(\mathbf{I} \cos \frac{\theta}{2} + \tilde{\mathbf{v}} \sin \frac{\theta}{2}\right) \tag{2.10}$$

The transformation matrix of Eq. 10 can be expressed in terms of the following four *Euler parameters*:

$$\left.\begin{aligned} \theta_0 &= \cos\frac{\theta}{2}, \quad & \theta_1 &= v_1 \sin\frac{\theta}{2} \\ \theta_2 &= v_2 \sin\frac{\theta}{2}, \quad & \theta_3 &= v_3 \sin\frac{\theta}{2} \end{aligned}\right\} \tag{2.11}$$

If $\bar{\boldsymbol{\theta}} = [\theta_1\,\theta_2\,\theta_3]^{\mathrm{T}}$, the transformation matrix **A** can be written as

$$\mathbf{A} = \mathbf{I} + 2\tilde{\bar{\boldsymbol{\theta}}}(\mathbf{I}\theta_0 + \tilde{\bar{\boldsymbol{\theta}}}) \tag{2.12}$$

where the four Euler parameters given by Eq. 11 satisfy the relation

$$\sum_{k=0}^{3}(\theta_k)^2 = \boldsymbol{\theta}^{\mathrm{T}}\boldsymbol{\theta} = 1 \tag{2.13}$$

where $\boldsymbol{\theta}$ is the vector

$$\boldsymbol{\theta} = [\theta_0 \quad \theta_1 \quad \theta_2 \quad \theta_3]^{\mathrm{T}}$$

and $\tilde{\bar{\boldsymbol{\theta}}}$ is a skew symmetric matrix defined as

$$\tilde{\bar{\boldsymbol{\theta}}} = \begin{bmatrix} 0 & -\theta_3 & \theta_2 \\ \theta_3 & 0 & -\theta_1 \\ -\theta_2 & \theta_1 & 0 \end{bmatrix}$$

The transformation matrix **A** can be written explicitly in terms of the four Euler parameters of Eq. 11 as

$$\mathbf{A} = \begin{bmatrix} 1-2(\theta_2)^2-2(\theta_3)^2 & 2(\theta_1\theta_2-\theta_0\theta_3) & 2(\theta_1\theta_3+\theta_0\theta_2) \\ 2(\theta_1\theta_2+\theta_0\theta_3) & 1-2(\theta_1)^2-2(\theta_3)^2 & 2(\theta_2\theta_3-\theta_0\theta_1) \\ 2(\theta_1\theta_3-\theta_0\theta_2) & 2(\theta_2\theta_3+\theta_0\theta_1) & 1-2(\theta_1)^2-2(\theta_2)^2 \end{bmatrix} \tag{2.14a}$$

Using the identity of Eq. 13, an alternative form of the transformation matrix can be obtained as

$$\mathbf{A} = \begin{bmatrix} 2[(\theta_0)^2+(\theta_1)^2]-1 & 2(\theta_1\theta_2-\theta_0\theta_3) & 2(\theta_1\theta_3+\theta_0\theta_2) \\ 2(\theta_1\theta_2+\theta_0\theta_3) & 2[(\theta_0)^2+(\theta_2)^2]-1 & 2(\theta_2\theta_3-\theta_0\theta_1) \\ 2(\theta_1\theta_3-\theta_0\theta_2) & 2(\theta_2\theta_3+\theta_0\theta_1) & 2[(\theta_0)^2+(\theta_3)^2]-1 \end{bmatrix} \tag{2.14b}$$

Note that the vector $\bar{\mathbf{r}}$ in Eq. 8 is the position vector of point $\bar{Q}$, before the rotation, while the vector $\mathbf{r}$ is the position vector of Q after the rotation θ about the axis OC. Equation 8, along with *Euler's theorem*, which states that the general rotation of a rigid frame is equivalent to a rotation about a fixed axis, may give an insight into the significance of the preceding development when body kinematics is considered. We observe that the rotation matrix of Eq. 9 or its explicit form in terms of Euler parameters θ_0, θ_1, θ_2, and θ_3 does not depend on the components of the vector $\bar{\mathbf{r}}$. It depends only on the components of the unit vector $\mathbf{v}$ along the axis of rotation as well as the angle of rotation θ. Therefore, any line element that is rigidly connected to the line $O\bar{Q}$ will be transformed by using the same rotation matrix $\mathbf{A}$ of Eq. 9. One thus concludes that the matrix $\mathbf{A}$ of Eq. 9 can be used to describe the rotation of any line rigidly attached to the rotating frame in which $\bar{\mathbf{r}}$ is defined. Therefore, henceforth, we will denote the matrix $\mathbf{A}$ as the *rotation matrix* or, alternatively, as the *transformation matrix* of the coordinate system $\mathbf{X}_1^i\mathbf{X}_2^i\mathbf{X}_3^i$.

Example 2.1 In the case of a planar motion, one may select the axis of rotation along the unit vector

$$\mathbf{v} = [0 \quad 0 \quad v_3]^{\mathrm{T}} = [0 \quad 0 \quad 1]^{\mathrm{T}}$$

The four Euler parameters of Eq. 11 become

$$\theta_0 = \cos\frac{\theta}{2}, \qquad \theta_1 = \theta_2 = 0, \qquad \theta_3 = v_3 \sin\frac{\theta}{2} = \sin\frac{\theta}{2}$$

Substituting these values in Eq. 14b leads to

$$\mathbf{A} = \begin{bmatrix} 2(\theta_0)^2 - 1 & -2\theta_0\theta_3 & 0 \\ 2\theta_0\theta_3 & 2(\theta_0)^2 - 1 & 0 \\ 0 & 0 & 2[(\theta_0)^2 + (\theta_3)^2] - 1 \end{bmatrix}$$

$$= \begin{bmatrix} 2\cos^2\frac{\theta}{2} - 1 & -2\cos\frac{\theta}{2}\sin\frac{\theta}{2} & 0 \\ 2\cos\frac{\theta}{2}\sin\frac{\theta}{2} & 2\cos^2\frac{\theta}{2} - 1 & 0 \\ 0 & 0 & 2\left(\cos^2\frac{\theta}{2} + \sin^2\frac{\theta}{2}\right) - 1 \end{bmatrix}$$

Using the trigonometric identities

$$2\cos^2\frac{\theta}{2} - 1 = \cos\theta, \qquad 2\cos\frac{\theta}{2}\sin\frac{\theta}{2} = \sin\theta$$

one can write the transformation matrix $\mathbf{A}$ in this special case as

$$\mathbf{A} = \begin{bmatrix} \cos\theta & -\sin\theta & 0 \\ \sin\theta & \cos\theta & 0 \\ 0 & 0 & 1 \end{bmatrix}$$

which is the familiar transformation matrix in the case of planar motion. Since a vector in the plane is defined by two components, we may delete the last row and the last column in the above transformation matrix $\mathbf{A}$ and write $\mathbf{A}$ as a 2×2 matrix as

$$\mathbf{A} = \begin{bmatrix} \cos\theta & -\sin\theta \\ \sin\theta & \cos\theta \end{bmatrix}$$

General Displacement The spatial transformation developed in this section is expressed in terms of the angle of rotation and the three components of a unit vector along the axis of rotation. It is clear that these four variables are not totally independent, since the length of a unit vector along the axis of rotation remains

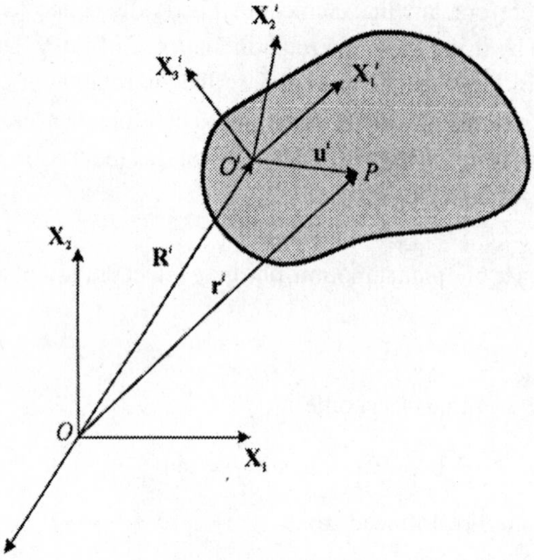

Figure 2.3 Coordinates of rigid bodies.

constant. Similar comments apply also to the rotation matrix written in terms of the four Euler parameters and given by Eq. 12 or 14. The four Euler parameters must satisfy the relationship given by Eq. 13. It is, therefore, clear that the orientation of a rigid frame of reference can be completely defined in terms of three independent variables. The three-variable representation, however, has the disadvantage that the rotation matrix may not be well defined at certain orientations of the rigid frame of reference in space. Some of the most commonly used forms of rotation matrix in terms of three independent parameters will be presented in later sections after we study some of the interesting properties of the spatial transformation.

The *general displacement* of a body i in the multibody system can be described by a rotation plus a translation. The position vector $\bar{\mathbf{u}}^i$ of an arbitrary point P on the rigid body i in the multibody system has a constant component in the body-fixed coordinate system $\mathbf{X}_1^i\mathbf{X}_2^i\mathbf{X}_3^i$. If this body undergoes pure rotation, the position vector of point P in the $\mathbf{X}_1\mathbf{X}_2\mathbf{X}_3$ global frame of reference is defined as shown in Fig. 3 by the vector $\mathbf{u}^i$ according to the equation

$$\mathbf{u}^i = \mathbf{A}^i\bar{\mathbf{u}}^i$$

where the superscript i refers to body i in the multibody system and $\mathbf{A}^i$ is the rotation matrix that defines the orientation of body i with respect to the coordinate system $\mathbf{X}_1\mathbf{X}_2\mathbf{X}_3$. If the body translates in addition to the rotation, the general motion, according to *Chasles' theorem*, can be described by the translation of a point and a rotation along the axis of rotation. The translation of the body can then be described by the position vector of the origin of the body reference. This position vector will be denoted as $\mathbf{R}^i$. Therefore, the global position vector of an arbitrary point on the rigid body can be expressed in terms of the translation and rotation of the body by the vector $\mathbf{r}^i$ given by

$$\mathbf{r}^i = \mathbf{R}^i + \mathbf{A}^i\bar{\mathbf{u}}^i$$

This equation can be used in the position analysis of multibody systems consisting of interconnected rigid bodies as demonstrated by the following planar and spatial multibody system examples.

Example 2.2 Figure 4 shows a slider crank mechanism that consists of four bodies (links). Body 1 is the ground or the fixed link, body 2 is the crankshaft OA, body 3 is the connecting rod AB, and body 4 is the slider block at B. Let l^2 and l^3 denote, respectively, the length of bodies 2 and 3 and θ^2 and θ^3 be the angular orientation of these two bodies. Given $l^2 = 0.15$ m, $l^3 = 0.3$ m, and $\theta^2 = 30^\circ$, determine θ^3 and the location of the slider block at B.

Solution In order to solve this problem we attach at point A a body reference for link 3. Let $\mathbf{R}^3 = [R_1^3 \; R_2^3]^{\mathrm{T}}$ be the global position of point A. Then the global position of point B can be written as

$$\mathbf{r}_B = \mathbf{R}^3 + \mathbf{A}^3\bar{\mathbf{u}}^3$$

where $\bar{\mathbf{u}}^3$ is the local position of point B in the third body coordinate system and $\mathbf{A}^3$ is the transformation matrix from the local coordinate system to the global coordinate system. The vector $\bar{\mathbf{u}}^3$ and the matrix $\mathbf{A}^3$ are given by

$$\bar{\mathbf{u}}^3 = \begin{bmatrix}\bar{u}_1^3 & \bar{u}_2^3\end{bmatrix}^{\mathrm{T}} = [l^3 \quad 0]^{\mathrm{T}}$$

$$\mathbf{A}^3 = \begin{bmatrix}\cos\theta^3 & -\sin\theta^3 \\ \sin\theta^3 & \cos\theta^3\end{bmatrix}$$

Since the block B moves in the horizontal direction, one has

$$\mathbf{r}_B = [x_B \quad 0]^{\mathrm{T}}$$

The global position of point B can then be written as

$$\begin{bmatrix}x_B \\ 0\end{bmatrix} = \begin{bmatrix}R_1^3 \\ R_2^3\end{bmatrix} + \begin{bmatrix}\cos\theta^3 & -\sin\theta^3 \\ \sin\theta^3 & \cos\theta^3\end{bmatrix}\begin{bmatrix}l^3 \\ 0\end{bmatrix}$$

which yields the following two equations:

$$x_B = R_1^3 + l^3\cos\theta^3$$
$$0 = R_2^3 + l^3\sin\theta^3$$

where

$$R_1^3 = l^2\cos\theta^2 \qquad \text{and} \qquad R_2^3 = l^2\sin\theta^2$$

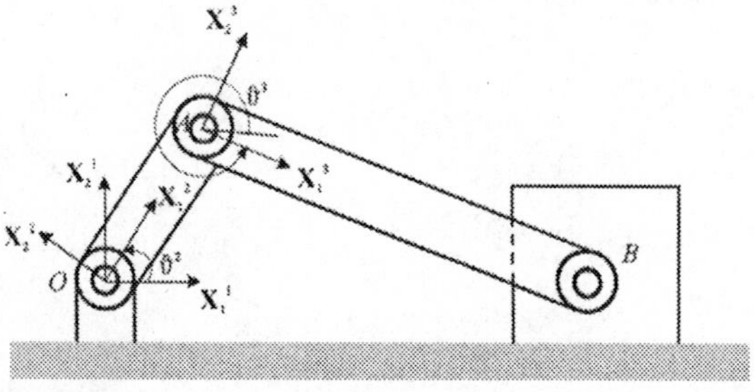

Figure 2.4 Slider crank mechanism.

that is,

$$x_B = l^2 \cos\theta^2 + l^3 \cos\theta^3$$
$$0 = l^2 \sin\theta^2 + l^3 \sin\theta^3$$

The second equation leads to

$$\sin\theta^3 = -\frac{l^2}{l^3}\sin\theta^2 = -\frac{0.15}{0.3}\sin 30^\circ = -0.25$$

which implies, for the mechanism configuration shown, that

$$\theta^3 = 345.522^\circ$$

The position of the slider block can be determined as

$$\begin{aligned} x_B &= l^2 \cos\theta^2 + l^3 \cos\theta^3 \\ &= 0.15 \cos 30^\circ + 0.3 \cos 345.522^\circ = 0.4204 \text{ m} \end{aligned}$$

Example 2.3 Figure 5 shows two robotic arms that are connected by a cylindrical joint that allows relative translational and rotational displacements between the two links. Link 2 rotates and translates relative to link 1 along the axis of the cylindrical joint whose unit vector **v** defined in the link 1 coordinate system is given by

$$\mathbf{v} = [\,v_1 \quad v_2 \quad v_3\,]^{\mathrm{T}} = \frac{1}{\sqrt{3}}[\,1 \quad 1 \quad 1\,]^{\mathrm{T}}$$

If the axes of the coordinate systems of the two links are initially the same and if link 2 translates and rotates with respect to link 1 with a constant speed $\dot{R}^2 = 1$ m/sec and constant angular velocity of $\omega^2 = 0.17453$ rad/sec, respectively, determine the position of point P on link 2 in the first link coordinate system after time $t = 3$ sec, where the local position of point P is given by the vector $\bar{\mathbf{u}} = [0\,1\,0]^{\mathrm{T}}$.

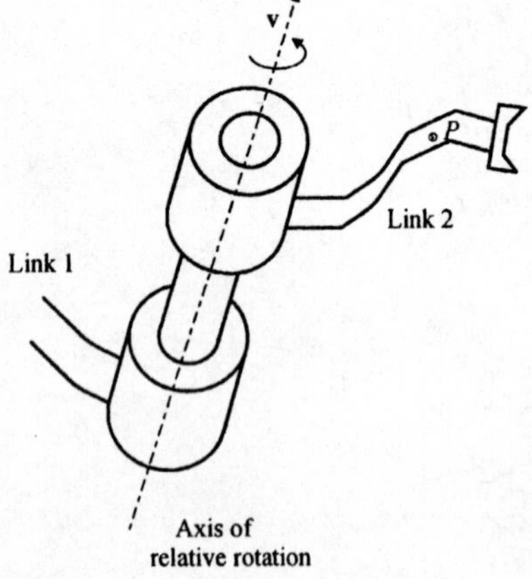

Figure 2.5 Robotic arm.

Solution The distance translated by the origin of the second link coordinate system is given by

$$R^2 = \dot{R}^2 t = (1)(3) = 3 \text{ m}$$

It follows that the position vector of the origin of link 2 in the first link coordinate system is given by the vector $\mathbf{R}^2$ as

$$\mathbf{R}^2 = \begin{bmatrix} R_1^2 & R_2^2 & R_3^2 \end{bmatrix}^{\mathrm{T}} = \sqrt{3}\begin{bmatrix} 1 & 1 & 1 \end{bmatrix}^{\mathrm{T}}$$

The skew symmetric matrix $\tilde{\mathbf{v}}$ is

$$\tilde{\mathbf{v}} = \frac{1}{\sqrt{3}}\begin{bmatrix} 0 & -1 & 1 \\ 1 & 0 & -1 \\ -1 & 1 & 0 \end{bmatrix}$$

and

$$(\tilde{\mathbf{v}})^2 = \frac{1}{3}\begin{bmatrix} -2 & 1 & 1 \\ 1 & -2 & 1 \\ 1 & 1 & -2 \end{bmatrix}$$

$$\sin\theta^2 = \sin 30^\circ = 0.5$$

$$\sin^2\frac{\theta^2}{2} = \sin^2 15^\circ = 0.06699$$

Substituting into Eq. 9, one obtains the transformation matrix $\mathbf{A}^2$ for link 2 as

$$\begin{aligned}\mathbf{A}^2 &= \left[\mathrm{I} + \tilde{\mathbf{v}}\sin\theta^2 + 2(\tilde{\mathbf{v}})^2\sin^2\frac{\theta^2}{2}\right] \\ &= \begin{bmatrix} 0.91068 & -0.24402 & 0.33334 \\ 0.33334 & 0.91068 & -0.24402 \\ -0.24402 & 0.33334 & 0.91068 \end{bmatrix}\end{aligned}$$

The position vector of point P with respect to link 1 is then defined by

$$\begin{aligned}\mathbf{r}_p &= \mathbf{R}^2 + \mathbf{A}^2\bar{\mathbf{u}} \\ &= \sqrt{3}\begin{bmatrix} 1 \\ 1 \\ 1 \end{bmatrix} + \begin{bmatrix} 0.91068 & -0.24402 & 0.33334 \\ 0.33334 & 0.91068 & -0.24402 \\ -0.24402 & 0.33334 & 0.91068 \end{bmatrix}\begin{bmatrix} 0 \\ 1 \\ 0 \end{bmatrix} \\ &= 1.7321\begin{bmatrix} 1 \\ 1 \\ 1 \end{bmatrix} + \begin{bmatrix} -0.24402 \\ 0.91068 \\ 0.33334 \end{bmatrix} = \begin{bmatrix} 1.48808 \\ 2.64278 \\ 2.06544 \end{bmatrix} \text{ m}\end{aligned}$$

2.2 PROPERTIES OF THE ROTATION MATRIX

An important property of the rotation matrix is the *orthogonality*. In the following, a proof of the orthogonality of the rotation matrix is provided and alternate forms of this matrix are presented.

Orthogonality of the Rotation Matrix Note that while $\tilde{\mathbf{v}}$ in Eq. 9 is a skew symmetric matrix (i.e., $\tilde{\mathbf{v}}^{\mathrm{T}} = -\tilde{\mathbf{v}}$), $(\tilde{\mathbf{v}})^2$ is a symmetric matrix. In fact, we have the following identities for the matrix $\tilde{\mathbf{v}}$ (Argyris 1982):

$$(\tilde{\mathbf{v}})^3 = -\tilde{\mathbf{v}}, \quad (\tilde{\mathbf{v}})^4 = -(\tilde{\mathbf{v}})^2, \quad (\tilde{\mathbf{v}})^5 = \tilde{\mathbf{v}}, \quad (\tilde{\mathbf{v}})^6 = (\tilde{\mathbf{v}})^2 \tag{2.15}$$

which leads to the recurrence relations

$$(\tilde{\mathbf{v}})^{2n-1} = (-1)^{n-1}\tilde{\mathbf{v}}, \quad (\tilde{\mathbf{v}})^{2n} = (-1)^{n-1}(\tilde{\mathbf{v}})^2 \tag{2.16}$$

By using Eq. 9, one can write

$$\begin{aligned}\mathbf{A}^{\mathrm{T}}\mathbf{A} &= \left[\mathbf{I} - \tilde{\mathbf{v}}\sin\theta + 2(\tilde{\mathbf{v}})^2\sin^2\frac{\theta}{2}\right]\left[\mathbf{I} + \tilde{\mathbf{v}}\sin\theta + 2(\tilde{\mathbf{v}})^2\sin^2\frac{\theta}{2}\right] \\ &= \mathbf{I} + 4(\tilde{\mathbf{v}})^2\sin^2\frac{\theta}{2} + 4(\tilde{\mathbf{v}})^4\sin^4\frac{\theta}{2} - 4(\tilde{\mathbf{v}})^2\sin^2\frac{\theta}{2}\cos^2\frac{\theta}{2} \\ &= \mathbf{A}\mathbf{A}^{\mathrm{T}}\end{aligned} \tag{2.17}$$

which, on using the identities of Eq. 16, can be written as

$$\mathbf{A}^{\mathrm{T}}\mathbf{A} = \mathbf{I} + 4(\tilde{\mathbf{v}})^4\sin^2\frac{\theta}{2}\left[\left(\sin^2\frac{\theta}{2} + \cos^2\frac{\theta}{2}\right) - 1\right] = \mathbf{I} \tag{2.18}$$

This proves the orthogonality of the transformation matrix $\mathbf{A}$, that is

$$\mathbf{A}^{\mathrm{T}} = \mathbf{A}^{-1} \tag{2.19}$$

This is an important property of the transformation matrix, which can also be checked by reversing the sense of rotation. This is the case in which we transform $\mathbf{r}$ to $\bar{\mathbf{r}}$. In this case, we replace θ by $-\theta$ in Eq. 9, yielding

$$\mathbf{A}^{-1} = \left[\mathbf{I} - \tilde{\mathbf{v}}\sin\theta + 2(\tilde{\mathbf{v}})^2\sin^2\frac{\theta}{2}\right] = \mathbf{A}^{\mathrm{T}}$$

where the identity $\sin(-\theta) = -\sin\theta$ is utilized.

Therefore, $\bar{\mathbf{r}}$ can be written in terms of the vector $\mathbf{r}$ as

$$\bar{\mathbf{r}} = \mathbf{A}^{\mathrm{T}}\mathbf{r}$$

In case of *infinitesimal rotations*, $\sin\theta$ can be approximated as

$$\sin\theta = \theta - \frac{(\theta)^3}{3!} + \frac{(\theta)^5}{5!} + \cdots \approx \theta$$

Equation 9 yields an approximation of $\mathbf{A}$ as

$$\mathbf{A} \approx \mathbf{I} + \tilde{\mathbf{v}}\theta$$

In this special case

$$\mathbf{A}^{\mathrm{T}} = \mathbf{I} - \tilde{\mathbf{v}}\theta$$

where θ, as mentioned earlier, is the angle of rotation about the axis of rotation.

Another Form of the Rotation Matrix The transformation matrix in the form given by Eq. 14 can be written as the product of two matrices, each depending linearly on the four Euler parameters θ_0, θ_1, θ_2, and θ_3. This relation is given by

$$\mathbf{A} = \mathbf{E}\bar{\mathbf{E}}^{\mathrm{T}} \tag{2.20}$$

where $\mathbf{E}$ and $\bar{\mathbf{E}}$ are 3×4 matrices given by

$$\mathbf{E} = \begin{bmatrix} -\theta_1 & \theta_0 & -\theta_3 & \theta_2 \\ -\theta_2 & \theta_3 & \theta_0 & -\theta_1 \\ -\theta_3 & -\theta_2 & \theta_1 & \theta_0 \end{bmatrix} \tag{2.21}$$

$$\bar{\mathbf{E}} = \begin{bmatrix} -\theta_1 & \theta_0 & \theta_3 & -\theta_2 \\ -\theta_2 & -\theta_3 & \theta_0 & \theta_1 \\ -\theta_3 & \theta_2 & -\theta_1 & \theta_0 \end{bmatrix} \tag{2.22}$$

By using the identity of Eq. 13, it can be verified that $\mathbf{E}$ and $\bar{\mathbf{E}}$ satisfy

$$\mathbf{E}\mathbf{E}^{\mathrm{T}} = \bar{\mathbf{E}}\bar{\mathbf{E}}^{\mathrm{T}} = \mathbf{I} \tag{2.23}$$

$$\mathbf{E}^{\mathrm{T}}\mathbf{E} = \bar{\mathbf{E}}^{\mathrm{T}}\bar{\mathbf{E}} = \mathbf{I}_4 - \boldsymbol{\theta}\boldsymbol{\theta}^{\mathrm{T}} \tag{2.24}$$

where $\mathbf{I}_4$ is a 4×4 identity matrix. It follows that

$$\mathbf{A}^{\mathrm{T}}\mathbf{A} = \bar{\mathbf{E}}\mathbf{E}^{\mathrm{T}}\mathbf{E}\bar{\mathbf{E}}^{\mathrm{T}} = \mathbf{I} \tag{2.25}$$

where the fact that $\bar{\mathbf{E}}\boldsymbol{\theta} = \mathbf{0}$ is used. Equation 25 provides another proof for the orthogonality of the transformation matrix.

Occasionally, we will exemplify general developments by presenting two-dimensional cases. A special case of the preceding development is the case of planar analysis in which the axis of rotation is along the unit vector

$$\mathbf{v} = [0 \quad 0 \quad 1]^{\mathrm{T}}$$

In this case, as shown in Example 1, $\theta_1 = \theta_2 = 0$, and the matrices $\mathbf{E}$ and $\bar{\mathbf{E}}$ reduce to

$$\mathbf{E} = \begin{bmatrix} 0 & \theta_0 & -\theta_3 & 0 \\ 0 & \theta_3 & \theta_0 & 0 \\ -\theta_3 & 0 & 0 & \theta_0 \end{bmatrix}, \quad \bar{\mathbf{E}} = \begin{bmatrix} 0 & \theta_0 & \theta_3 & 0 \\ 0 & -\theta_3 & \theta_0 & 0 \\ -\theta_3 & 0 & 0 & \theta_0 \end{bmatrix}$$

The transformation matrix $\mathbf{A}$ of Eq. 14 in the case of planar analysis can then be given as

$$\mathbf{A} = \mathbf{E}\bar{\mathbf{E}}^{\mathrm{T}}$$

Knowing that $\theta_0 = \cos(\theta/2)$ and $\theta_3 = \sin(\theta/2)$ and using the trigonometric identities

$$1 - 2\sin^2\frac{\theta}{2} = \cos\theta \quad \text{and} \quad 2\cos\frac{\theta}{2}\sin\frac{\theta}{2} = \sin\theta$$

one can write the transformation matrix in the two-dimensional case as

$$\mathbf{A} = \begin{bmatrix} \cos\theta & -\sin\theta & 0 \\ \sin\theta & \cos\theta & 0 \\ 0 & 0 & 1 \end{bmatrix}$$

Since the vectors are two-dimensional, the transformation matrix $\mathbf{A}$ in planar analysis reduces to

$$\mathbf{A} = \begin{bmatrix} \cos\theta & -\sin\theta \\ \sin\theta & \cos\theta \end{bmatrix} \tag{2.26}$$

which is the familiar form of the two-dimensional transformation matrix. It is developed here, however, from the general three-dimensional case.

Example 2.4 The position vector of a point on a rigid body is given by

$$\bar{\mathbf{r}} = [2 \quad 3 \quad 4]^{\mathrm{T}}$$

The body rotates an angle $\theta = 45°$ about an axis of rotation whose unit vector is

$$\mathbf{v} = \begin{bmatrix} \frac{1}{\sqrt{3}} & \frac{1}{\sqrt{3}} & \frac{1}{\sqrt{3}} \end{bmatrix}^{\mathrm{T}}$$

Determine the rotation matrix and the transformed vector.

Solution In this case the skew symmetric matrices $\tilde{\mathbf{v}}$ and$(\tilde{\mathbf{v}})^2$ are given by

$$\tilde{\mathbf{v}} = \frac{1}{\sqrt{3}} \begin{bmatrix} 0 & -1 & 1 \\ 1 & 0 & -1 \\ -1 & 1 & 0 \end{bmatrix}, \qquad (\tilde{\mathbf{v}})^2 = \frac{1}{3} \begin{bmatrix} -2 & 1 & 1 \\ 1 & -2 & 1 \\ 1 & 1 & -2 \end{bmatrix}$$

For $\theta = 45°$ one has

$$\sin\theta = \sin 45° = 0.7071$$

$$2\sin^2\frac{\theta}{2} = 2\sin^2\frac{45°}{2} = 0.2929$$

Using Eq. 9, one can write the transformation matrix as

$$\mathbf{A} = \left[\mathbf{I} + \tilde{\mathbf{v}}\sin\theta + 2(\tilde{\mathbf{v}})^2 \sin^2\frac{\theta}{2}\right]$$

$$= \begin{bmatrix} 1 & 0 & 0 \\ 0 & 1 & 0 \\ 0 & 0 & 1 \end{bmatrix} + \frac{0.7071}{\sqrt{3}} \begin{bmatrix} 0 & -1 & 1 \\ 1 & 0 & -1 \\ -1 & 1 & 0 \end{bmatrix}$$

$$+ \frac{0.2929}{3} \begin{bmatrix} -2 & 1 & 1 \\ 1 & -2 & 1 \\ 1 & 1 & -2 \end{bmatrix}$$

$$= \begin{bmatrix} 0.8048 & -0.3106 & 0.5058 \\ 0.5058 & 0.8048 & -0.3106 \\ -0.3106 & 0.5058 & 0.8048 \end{bmatrix}$$

This transformation matrix could also be evaluated by defining the four parameters θ_0, θ_1, θ_2, and θ_3 in Eq. 11 and substituting into Eq. 14. It can also be verified using simple matrix multiplications that the transformation matrix given above is an orthogonal matrix. To define the transformed vector $\mathbf{r}$, we use Eq. 8, which yields

$$\mathbf{r} = \mathbf{A}\bar{\mathbf{r}} = \begin{bmatrix} 0.8048 & -0.3106 & 0.5058 \\ 0.5058 & 0.8048 & -0.3106 \\ -0.3106 & 0.5058 & 0.8048 \end{bmatrix} \begin{bmatrix} 2 \\ 3 \\ 4 \end{bmatrix} = \begin{bmatrix} 2.7010 \\ 2.1836 \\ 4.1154 \end{bmatrix}$$

Therefore, the transformed vector **r** is given by

$$\mathbf{r} = [2.7010 \quad 2.1836 \quad 4.1154]^{\mathrm{T}}$$

Consider an arbitrary vector $\bar{\mathbf{a}}$ defined on the rigid body along the axis of rotation. This vector can be written as

$$\bar{\mathbf{a}} = c\begin{bmatrix} \frac{1}{\sqrt{3}} & \frac{1}{\sqrt{3}} & \frac{1}{\sqrt{3}} \end{bmatrix}^{\mathrm{T}}$$

where c is a constant. If the body rotates 45° about the axis of rotation, the transformed vector **a** is given by

$$\mathbf{a} = \mathbf{A}\bar{\mathbf{a}} = \begin{bmatrix} 0.8048 & -0.3106 & 0.5058 \\ 0.5058 & 0.8048 & -0.3106 \\ -0.3106 & 0.5058 & 0.8048 \end{bmatrix} \begin{bmatrix} \frac{c}{\sqrt{3}} \\ \frac{c}{\sqrt{3}} \\ \frac{c}{\sqrt{3}} \end{bmatrix} = c \begin{bmatrix} \frac{1}{\sqrt{3}} \\ \frac{1}{\sqrt{3}} \\ \frac{1}{\sqrt{3}} \end{bmatrix}$$

which implies that

$$\mathbf{a} = \bar{\mathbf{a}} \quad \text{or} \quad \mathbf{a} = \mathbf{A}\bar{\mathbf{a}}$$

The results of this example show that if the vector **a** is defined along the axis of rotation, then the original and transformed vectors are the same. Furthermore,

$$\mathbf{A}^{\mathrm{T}}\bar{\mathbf{a}} = \bar{\mathbf{a}}$$

The result of the preceding example concerning the transformation of a vector defined along the axis of rotation is not a special situation. Using the definition of the transformation matrix given by Eq. 9, it can be proved that, given a vector $\bar{\mathbf{a}}$ along the axis of rotation, the transformed vector and the original vector are the same. The proof of this statement, although relevant, is very simple and is as follows. Since $\bar{\mathbf{a}}$ is defined along the axis of rotation, $\bar{\mathbf{a}}$ can be written as

$$\bar{\mathbf{a}} = c\mathbf{v}$$

where c is a constant. By direct matrix multiplication or by using the definition of the cross product, one can verify that

$$\tilde{\mathbf{v}}\mathbf{v} = \mathbf{v} \times \mathbf{v} = \mathbf{0}$$

and

$$(\tilde{\mathbf{v}})^2\mathbf{v} = \tilde{\mathbf{v}}(\tilde{\mathbf{v}}\mathbf{v}) = \mathbf{0}$$

Now if the transformation **A** of Eq. 9 is applied to $\bar{\mathbf{a}}$, one gets

$$\mathbf{A}\bar{\mathbf{a}} = \left[\mathbf{I} + \tilde{\mathbf{v}}\sin\theta + 2(\tilde{\mathbf{v}})^2\sin^2\frac{\theta}{2}\right]\bar{\mathbf{a}}$$

$$= c\left[\mathbf{v} + \tilde{\mathbf{v}}\mathbf{v}\sin\theta + 2(\tilde{\mathbf{v}})^2\mathbf{v}\sin^2\frac{\theta}{2}\right]$$

which yields

$$\mathbf{A}\bar{\mathbf{a}} = c\mathbf{v} = \bar{\mathbf{a}}$$

By a similar argument, one can show also that

$$\mathbf{A}^{\mathrm{T}}\bar{\mathbf{a}} = \bar{\mathbf{a}}$$

The preceding two equations indicate that one is an eigenvalue of both $\mathbf{A}$ and its transpose $\mathbf{A}^{\mathrm{T}}$. Associated with this eigenvalue is the eigenvector $\bar{\mathbf{a}}$. Therefore, the direction along the axis of rotation is a principal direction.

2.3 SUCCESSIVE ROTATIONS

In this section, the exponential form of the transformation matrix will be developed and used to provide a simple proof that the product of transformation matrices resulting from two successive rotations about two different axes of rotation is not commutative.

Exponential Form of the Rotation Matrix Equation 9 can be written in terms of the angle of rotation θ as

$$\mathbf{A} = [\mathbf{I} + \tilde{\mathbf{v}}\sin\theta + (1 - \cos\theta)(\tilde{\mathbf{v}})^2]$$

Expanding $\sin\theta$ and $\cos\theta$ using Taylor's series, one obtains

$$\sin\theta = \theta - \frac{(\theta)^3}{3!} + \frac{(\theta)^5}{5!} + \cdots$$

$$\cos\theta = 1 - \frac{(\theta)^2}{2!} + \cdots$$

Substituting these equations in the expression of the transformation matrix given above yields

$$\begin{aligned}\mathbf{A} &= \left[\mathbf{I} + \left(\theta - \frac{(\theta)^3}{3!} + \frac{(\theta)^5}{5!} + \cdots\right)\tilde{\mathbf{v}} + \left(1 - 1 + \frac{(\theta)^2}{2!} - \frac{(\theta)^4}{4!} + \cdots\right)(\tilde{\mathbf{v}})^2\right] \\ &= \left[\mathbf{I} + \left(\theta - \frac{(\theta)^3}{3!} + \frac{(\theta)^5}{5!} + \cdots\right)\tilde{\mathbf{v}} + \left(\frac{(\theta)^2}{2!} - \frac{(\theta)^4}{4!} + \cdots\right)(\tilde{\mathbf{v}})^2\right]\end{aligned}$$

Using the identities of Eq. 16 and rearranging terms, one can write the transformation matrix $\mathbf{A}$ as

$$\mathbf{A} = \left[\mathbf{I} + \theta\tilde{\mathbf{v}} + \frac{(\theta)^2}{2!}(\tilde{\mathbf{v}})^2 + \frac{(\theta)^3}{3!}(\tilde{\mathbf{v}})^3 + \cdots + \frac{(\theta)^n}{n!}(\tilde{\mathbf{v}})^n + \cdots\right]$$

Since

$$\mathbf{e}^{\mathbf{B}} = \mathbf{I} + \mathbf{B} + \frac{(\mathbf{B})^2}{2!} + \frac{(\mathbf{B})^3}{3!} + \cdots$$

where $\mathbf{B}$ is a square matrix, one can write the transformation matrix $\mathbf{A}$ in the following elegant form (Bahar 1970; Argyris 1982):

$$\mathbf{A} = \mathbf{e}^{\theta\tilde{\mathbf{v}}} = \exp(\theta\tilde{\mathbf{v}})$$

Composed finite rotations are in general noncommutative. An exception to this rule occurs only when the axes of rotation are parallel. Consider the case in which two successive rotations θ_1 and θ_2 are performed about two fixed axes. The transformation matrices associated with these two rotations are, respectively, $\mathbf{A}_1$ and $\mathbf{A}_2$. Let $\mathbf{v}_1$ and $\mathbf{v}_2$ be unit vectors in the direction of the two axes of rotations. After the first rotation the vector $\bar{\mathbf{r}}$ occupies a new position denoted by $\mathbf{r}_1$ such that

$$\mathbf{r}_1 = \mathbf{A}_1\bar{\mathbf{r}}$$

where $\mathbf{A}_1$ is the transformation matrix associated with the rotation θ_1. The vector $\mathbf{r}_1$ then rotates about the second axis of rotation whose unit vector is $\mathbf{v}_2$ and as a consequence of this rotation $\mathbf{r}_1$ occupies a new position denoted as $\mathbf{r}_2$ such that

$$\mathbf{r}_2 = \mathbf{A}_2\mathbf{r}_1$$

where $\mathbf{A}_2$ is the transformation matrix resulting from the rotation θ_2. The final position $\mathbf{r}_2$ is then related to the original position defined by $\bar{\mathbf{r}}$ according to

$$\mathbf{r}_2 = \mathbf{A}_2\mathbf{A}_1\bar{\mathbf{r}}$$

Associated with the transformation matrices $\mathbf{A}_1$ and $\mathbf{A}_2$, there are two skew symmetric matrices $\tilde{\mathbf{v}}_1$ and $\tilde{\mathbf{v}}_2$ (Eq. 7). Recall that in general

$$\tilde{\mathbf{v}}_1\tilde{\mathbf{v}}_2 \neq \tilde{\mathbf{v}}_2\tilde{\mathbf{v}}_1$$
$$\mathbf{e}^{\tilde{\mathbf{v}}_2}\mathbf{e}^{\tilde{\mathbf{v}}_1} \neq \mathbf{e}^{(\tilde{\mathbf{v}}_1+\tilde{\mathbf{v}}_2)} \neq \mathbf{e}^{\tilde{\mathbf{v}}_1}\mathbf{e}^{\tilde{\mathbf{v}}_2}$$

unless the product of the matrices $\tilde{\mathbf{v}}_1$ and $\tilde{\mathbf{v}}_2$ is commutative, which is the case when the two axes of rotation are parallel. Therefore, for the two general rotations θ_1 and θ_2, one has

$$\mathbf{A}_2\mathbf{A}_1 = \mathbf{e}^{\theta_2\tilde{\mathbf{v}}_2}\mathbf{e}^{\theta_1\tilde{\mathbf{v}}_1} \neq \mathbf{e}^{(\theta_2\tilde{\mathbf{v}}_2+\theta_1\tilde{\mathbf{v}}_1)}$$

which implies that

$$\mathbf{A}_2\mathbf{A}_1 \neq \mathbf{A}_1\mathbf{A}_2$$

Thus the order of rotation is important. This fact, which implies that the finite rotation is not a vector quantity, can be simply demonstrated by the familiar example shown in Fig. 6. This figure illustrates different sequences of rotations for the same block. In Fig. 6(a) the block is first rotated 90° about the $\mathbf{X}_2$ axis and then 90° about the $\mathbf{X}_3$ axis. In Fig. 6(b) the same rotations in reverse order are employed. That is, the block is first rotated 90° about the $\mathbf{X}_3$ axis and then 90° about the $\mathbf{X}_2$ axis. It is obvious that a change in the sequence of rotations leads to different final positions.

We are now in a position to provide a simple proof for the orthogonality of the rotation matrix $\mathbf{A}$ by using the exponential form developed in this section and the fact that the product of $\tilde{\mathbf{v}}$ and $\tilde{\mathbf{v}}^{\mathrm{T}} = -\tilde{\mathbf{v}}$ is commutative. The matrix product $\mathbf{A}\mathbf{A}^{\mathrm{T}}$ can be written as

$$\mathbf{A}\mathbf{A}^{\mathrm{T}} = \mathbf{e}^{\theta\tilde{\mathbf{v}}}\mathbf{e}^{-\theta\tilde{\mathbf{v}}} = \mathbf{e}^{\mathbf{0}_3} = \mathbf{I} = \mathbf{A}^{\mathrm{T}}\mathbf{A}$$

where $\mathbf{0}_3$ is a 3×3 null matrix.

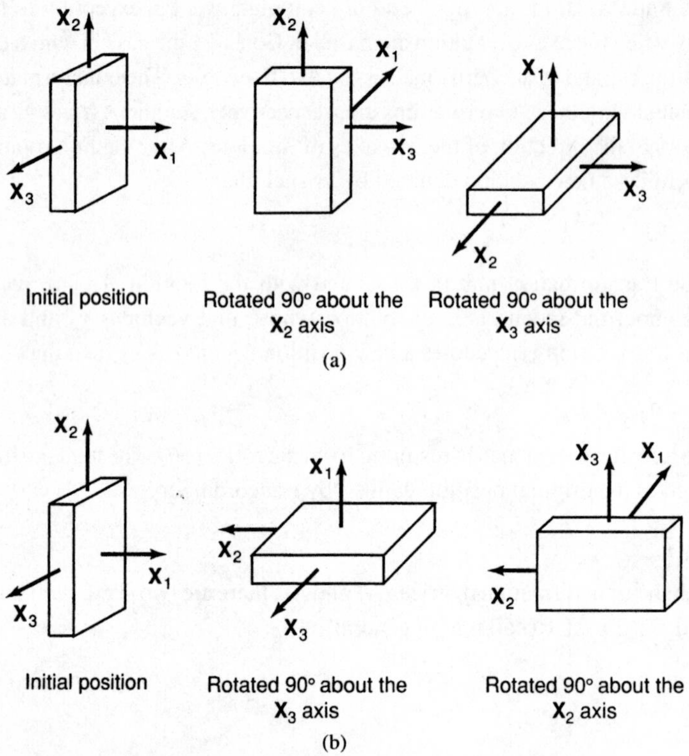

Figure 2.6 Effect of the order of rotation.

There are two methods that can be used to describe successive rotations of rigid bodies. The first is the *single-frame method*, and the second is the *multiframe method*. Both methods are described below and their use is demonstrated by an example.

Single-Frame Method In this method, one fixed coordinate system is used. After each rotation, the new axis of rotation and the vectors on the rigid body are redefined in the fixed coordinate system. Consider the successive rotations $\theta_1, \theta_2, \ldots, \theta_n$ about the vectors $\mathbf{b}_1, \mathbf{b}_2, \ldots, \mathbf{b}_n$ which are fixed in the rigid body. The transformation matrix resulting from the rotation θ_i about the axis $\mathbf{b}_i$ is denoted as $\mathbf{A}_i$. This matrix can be evaluated using the Rodriguez formula with θ equal to θ_i and

$$\mathbf{v} = \mathbf{v}_i = \mathbf{A}^{i-1}\mathbf{A}^{i-2}\cdots\mathbf{A}^2\mathbf{A}^1\mathbf{b}_i$$

with $\mathbf{A}_1$ equal to the identity matrix. An arbitrary vector drawn on the rigid body can be defined in the fixed coordinate system after n successive rotations using the transformation matrix $\mathbf{A}$ defined as

$$\mathbf{A} = \mathbf{A}^n\mathbf{A}^{n-1}\cdots\mathbf{A}^2\mathbf{A}^1$$

The use of the single-frame method is demonstrated by the following example.

Example 2.5 Figure 7 shows a rigid body denoted as body i. Let $\mathbf{X}_1^i\mathbf{X}_2^i\mathbf{X}_3^i$ be the body coordinate system whose origin is rigidly attached to the body and $\mathbf{b}_1$ and $\mathbf{b}_2$ be two lines that are rigidly attached to the body. These two rigid lines are defined in the body coordinate system by the vectors

$$\mathbf{b}_1 = [1 \quad 0 \quad 1]^{\mathrm{T}}, \qquad \mathbf{b}_2 = [0 \quad 1 \quad 0]^{\mathrm{T}}$$

The body is to be subjected to 180° rotation about $\mathbf{b}_1$ and a 90° rotation about $\mathbf{b}_2$. After these successive rotations, determine the global components of the vector $\mathbf{b}_3$ whose components in the body coordinate system are fixed and given by

$$\mathbf{b}_3 = [1 \quad 0 \quad 0]^{\mathrm{T}}$$

Solution A unit vector $\mathbf{v}_1$ along the axis of rotation $\mathbf{b}_1$ is given by

$$\mathbf{v}_1 = \frac{1}{\sqrt{2}}[1 \quad 0 \quad 1]^{\mathrm{T}}$$

Since θ_1 is 180°, the transformation matrix resulting from this finite rotation is given by

$$\begin{aligned}\mathbf{A}_1 &= \mathbf{I} + \tilde{\mathbf{v}}_1 \sin\theta_1 + 2(\tilde{\mathbf{v}}_1)^2 \sin^2\frac{\theta_1}{2}\\ &= \mathbf{I} + \tilde{\mathbf{v}}_1 \sin(180^\circ) + 2(\tilde{\mathbf{v}}_1)^2 \sin^2(90^\circ)\\ &= \mathbf{I} + 2(\tilde{\mathbf{v}}_1)^2\end{aligned}$$

where $(\tilde{\mathbf{v}}_1)^2$ is given by

$$(\tilde{\mathbf{v}}_1)^2 = \begin{bmatrix} 0 & -\frac{1}{\sqrt{2}} & 0 \\ \frac{1}{\sqrt{2}} & 0 & -\frac{1}{\sqrt{2}} \\ 0 & \frac{1}{\sqrt{2}} & 0 \end{bmatrix}^2 = \frac{1}{2}\begin{bmatrix} -1 & 0 & 1 \\ 0 & -2 & 0 \\ 1 & 0 & -1 \end{bmatrix}$$

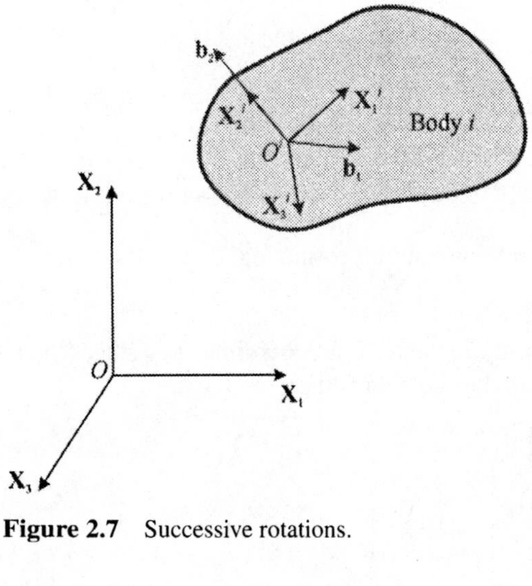

Figure 2.7 Successive rotations.

and the transformation matrix $\mathbf{A}_1$ is given by

$$\mathbf{A}_1 = \mathbf{I} + 2(\tilde{\mathbf{v}}_1)^2 = \begin{bmatrix} 1 & 0 & 0 \\ 0 & 1 & 0 \\ 0 & 0 & 1 \end{bmatrix} + \begin{bmatrix} -1 & 0 & 1 \\ 0 & -2 & 0 \\ 1 & 0 & -1 \end{bmatrix} = \begin{bmatrix} 0 & 0 & 1 \\ 0 & -1 & 0 \\ 1 & 0 & 0 \end{bmatrix}$$

One can verify that $\mathbf{A}_1$ is an orthogonal matrix, that is

$$\mathbf{A}_1^{\mathrm{T}}\mathbf{A}_1 = \mathbf{A}_1\mathbf{A}_1^{\mathrm{T}} = \mathbf{I}$$

As a result of this finite rotation, and assuming that the axes of the coordinate systems $\mathbf{X}_1\mathbf{X}_2\mathbf{X}_3$ and $\mathbf{X}_1^i\mathbf{X}_2^i\mathbf{X}_3^i$ are initially parallel, the components of the vectors $\mathbf{b}_2$ and $\mathbf{b}_3$ in the $\mathbf{X}_1\mathbf{X}_2\mathbf{X}_3$ coordinate system are given, respectively, by

$$\mathbf{b}_{21} = \mathbf{A}_1\mathbf{b}_2 = \begin{bmatrix} 0 & 0 & 1 \\ 0 & -1 & 0 \\ 1 & 0 & 0 \end{bmatrix} \begin{bmatrix} 0 \\ 1 \\ 0 \end{bmatrix} = \begin{bmatrix} 0 \\ -1 \\ 0 \end{bmatrix}$$

$$\mathbf{b}_{31} = \mathbf{A}_1\mathbf{b}_3 = \begin{bmatrix} 0 & 0 & 1 \\ 0 & -1 & 0 \\ 1 & 0 & 0 \end{bmatrix} \begin{bmatrix} 1 \\ 0 \\ 0 \end{bmatrix} = \begin{bmatrix} 0 \\ 0 \\ 1 \end{bmatrix}$$

The second rotation $\theta_2 = 90°$ is performed about the axis of rotation $\mathbf{b}_{21}$. A unit vector $\mathbf{v}_2$ along this axis of rotation is given by

$$\mathbf{v}_2 = [0 \quad -1 \quad 0]^{\mathrm{T}}$$

and the resulting skew symmetric and symmetric matrices $\tilde{\mathbf{v}}_2$ and $(\tilde{\mathbf{v}}_2)^2$ are given, respectively, by

$$\tilde{\mathbf{v}}_2 = \begin{bmatrix} 0 & 0 & -1 \\ 0 & 0 & 0 \\ 1 & 0 & 0 \end{bmatrix}, \quad (\tilde{\mathbf{v}}_2)^2 = \begin{bmatrix} -1 & 0 & 0 \\ 0 & 0 & 0 \\ 0 & 0 & -1 \end{bmatrix}$$

Using Eq. 9, the resulting transformation matrix $\mathbf{A}_2$ is given by

$$\begin{aligned} \mathbf{A}_2 &= \mathbf{I} + \tilde{\mathbf{v}}_2 \sin\theta_2 + 2(\tilde{\mathbf{v}}_2)^2 \sin^2\frac{\theta_2}{2} \\ &= \mathbf{I} + \tilde{\mathbf{v}}_2 \sin(90°) + 2(\tilde{\mathbf{v}}_2)^2 \sin^2(45°) \\ &= \mathbf{I} + \tilde{\mathbf{v}}_2 + (\tilde{\mathbf{v}}_2)^2 \\ &= \begin{bmatrix} 1 & 0 & 0 \\ 0 & 1 & 0 \\ 0 & 0 & 1 \end{bmatrix} + \begin{bmatrix} 0 & 0 & -1 \\ 0 & 0 & 0 \\ 1 & 0 & 0 \end{bmatrix} + \begin{bmatrix} -1 & 0 & 0 \\ 0 & 0 & 0 \\ 0 & 0 & -1 \end{bmatrix} = \begin{bmatrix} 0 & 0 & -1 \\ 0 & 1 & 0 \\ 1 & 0 & 0 \end{bmatrix} \end{aligned}$$

One can verify that $\mathbf{A}_2$ is an orthogonal transformation, that is

$$\mathbf{A}_2^{\mathrm{T}}\mathbf{A}_2 = \mathbf{A}_2\mathbf{A}_2^{\mathrm{T}} = \mathbf{I}$$

The components of the vector $\mathbf{b}_3$, in the $\mathbf{X}_1\mathbf{X}_2\mathbf{X}_3$ coordinate system, as the result of this successive rotation, can thus be obtained as

$$\begin{aligned} \mathbf{b}_{32} &= \mathbf{A}_2\mathbf{b}_{31} = \mathbf{A}_2\mathbf{A}_1\mathbf{b}_3 \\ &= \begin{bmatrix} 0 & 0 & -1 \\ 0 & 1 & 0 \\ 1 & 0 & 0 \end{bmatrix} \begin{bmatrix} 0 \\ 0 \\ 1 \end{bmatrix} = \begin{bmatrix} -1 \\ 0 \\ 0 \end{bmatrix} \end{aligned}$$

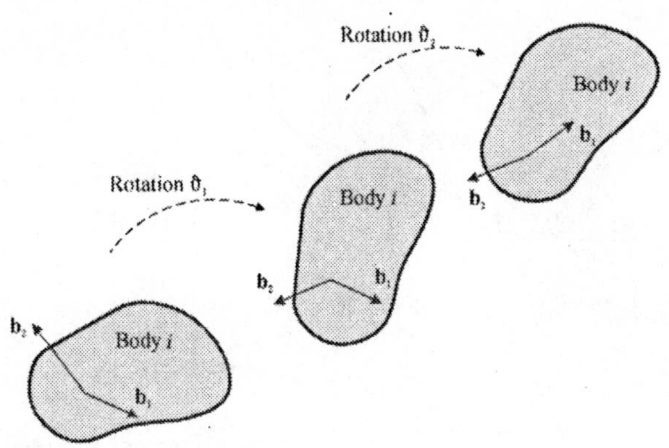

Figure 2.8 Multiframe method.

Multiframe Method In the preceding example, one fixed coordinate system was used. After each rotation, the vectors fixed in the rigid body are defined in the same fixed coordinate system. An alternate approach is to use the concept of the *body fixed coordinate system* previously introduced and define a sequence of configurations of the body that result from the successive rotations $\theta_1, \theta_2, \ldots, \theta_n$ about the vectors $\mathbf{b}_1, \mathbf{b}_2, \ldots, \mathbf{b}_n$ which are fixed in the rigid body. Figure 8 shows three different configurations when $n = 2$. The first configuration shows the body before the two rotations θ_1 and θ_2, the second configuration shows the body after the first rotation θ_1, and the third configuration shows the body after the second rotation θ_2. Let $\mathbf{A}^{i(i-1)}$ denote the transformation matrix that defines the orientation of the body after the $(i-1)$th rotation with respect to the $(i-1)$th configuration. Then, the transformation matrix that defines the orientation of the body after n successive rotations in the fixed coordinate system is

$$\mathbf{A} = \mathbf{A}^{21}\mathbf{A}^{32} \cdots \mathbf{A}^{(n-1)(n-2)}\mathbf{A}^{n(n-1)}$$

In this case, the transformation matrix $\mathbf{A}^{i(i-1)}$ can be evaluated using the Rodriguez formula with the axis of rotation $\mathbf{v}_{i-1} = \mathbf{b}_{i-1}$. That is, the axes of rotations are defined in the body coordinate system and there is no need in this case to define the axes of rotations in the global fixed coordinate system as in the case of the single-frame method. The use of the multiframe method in the case of successive rotations is demonstrated by the following example.

Example 2.6 The problem of the preceding example can be solved using the multiframe method. Before the two rotations, the orientation of the body is as shown in Fig. 9(a). After the first rotation about $\mathbf{b}_1$, the configuration of the body is as shown in Fig. 9(b). Figure 9(c) shows the orientation of the body after the second rotation θ_2 about $\mathbf{b}_2$. The orientation of the coordinate system shown in Fig. 9(c) with respect to the coordinate system shown in Fig. 9(b) can be described by the matrix $\mathbf{A}^{32}$, defined as

$$\mathbf{A}_{32} = \mathbf{I} + \tilde{\mathbf{v}}_2 \sin\theta_2 + 2(\tilde{\mathbf{v}}_2)^2 \sin^2\frac{\theta_2}{2}$$

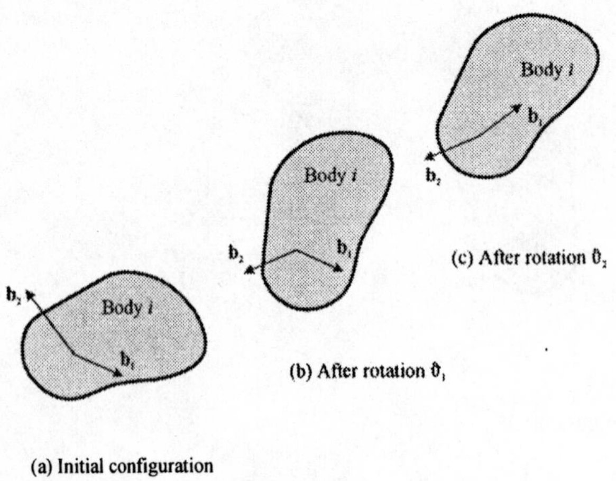

Figure 2.9 Application of the multiframe method.

where $\mathbf{v}_2 = \mathbf{b}_2 = [0\ 1\ 0]^{\mathrm{T}}$, and $\theta_2 = 90°$. It follows that

$$\mathbf{A}^{32} = \begin{bmatrix} 0 & 0 & 1 \\ 0 & 1 & 0 \\ -1 & 0 & 0 \end{bmatrix}$$

The orientation of the coordinate system in Fig. 9(b) can be defined with respect to the coordinate system in Fig. 9(a) as

$$\mathbf{A}^{21} = \mathbf{I} + \tilde{\mathbf{v}}_1 \sin\theta_1 + 2(\tilde{\mathbf{v}}_1)^2 \sin^2\frac{\theta_1}{2}$$

where in this case

$$\mathbf{v}_1 = \frac{\mathbf{b}_1}{|\mathbf{b}_1|} = \frac{1}{\sqrt{2}}[1 \quad 0 \quad 1]^{\mathrm{T}}$$

and $\theta_1 = 180°$. The transformation matrix $\mathbf{A}^{21}$ is then given by

$$\mathbf{A}^{21} = \begin{bmatrix} 0 & 0 & 1 \\ 0 & -1 & 0 \\ 1 & 0 & 0 \end{bmatrix}$$

Therefore, the orientation of the coordinate system in Fig. 9(c) can be defined with respect to the coordinate system in Fig. 9(a) using the matrix

$$\mathbf{A} = \mathbf{A}^{21}\mathbf{A}^{32} = \begin{bmatrix} 0 & 0 & 1 \\ 0 & -1 & 0 \\ 1 & 0 & 0 \end{bmatrix} \begin{bmatrix} 0 & 0 & 1 \\ 0 & 1 & 0 \\ -1 & 0 & 0 \end{bmatrix} = \begin{bmatrix} -1 & 0 & 0 \\ 0 & -1 & 0 \\ 0 & 0 & 1 \end{bmatrix}$$

The vector $\mathbf{b}_3$ can be defined in the coordinate system shown in Fig. 9(a) as

$$\mathbf{b}_{32} = \mathbf{A}\mathbf{b}_3 = \begin{bmatrix} -1 & 0 & 0 \\ 0 & -1 & 0 \\ 0 & 0 & 1 \end{bmatrix} \begin{bmatrix} 1 \\ 0 \\ 0 \end{bmatrix} = \begin{bmatrix} -1 \\ 0 \\ 0 \end{bmatrix}$$

Infinitesimal Rotations We have shown previously that for an infinitesimal rotation, the transformation matrix is given by

$$\mathbf{A} = \mathbf{I} + \tilde{\mathbf{v}}\theta$$

While finite rotation is not a vector quantity, we can use the above equation to prove that the *infinitesimal rotation* is a vector quantity. To this end, we consider two successive infinitesimal rotations θ_1 and θ_2, where θ_1 is a rotation about an axis in the direction of the unit vector $\mathbf{v}_1$, and θ_2 is a rotation about an axis in the direction of the unit vector $\mathbf{v}_2$. Therefore, the transformation matrix associated with the first rotation is

$$\mathbf{A}_1 = \mathbf{I} + \tilde{\mathbf{v}}_1\theta_1$$

and the transformation matrix associated with the second rotation is

$$\mathbf{A}_2 = \mathbf{I} + \tilde{\mathbf{v}}_2\theta_2$$

One can then write

$$\mathbf{A}_1\mathbf{A}_2 = (\mathbf{I} + \tilde{\mathbf{v}}_1\theta_1)(\mathbf{I} + \tilde{\mathbf{v}}_2\theta_2) = \mathbf{I} + \tilde{\mathbf{v}}_1\theta_1 + \tilde{\mathbf{v}}_2\theta_2 + \tilde{\mathbf{v}}_1\tilde{\mathbf{v}}_2\theta_1\theta_2$$

Because the rotations are assumed to be infinitesimal, one can neglect the second order term $\tilde{\mathbf{v}}_1\tilde{\mathbf{v}}_2\theta_1\theta_2$ and write

$$\mathbf{A}_1\mathbf{A}_2 \approx \mathbf{I} + \tilde{\mathbf{v}}_1\theta_1 + \tilde{\mathbf{v}}_2\theta_2 \approx \mathbf{A}_2\mathbf{A}_1$$

which shows that two successive infinitesimal rotations about two different axes of rotation can be added. In fact, for n successive rotations, one can show that

$$\begin{aligned}\mathbf{A}_1\mathbf{A}_2\cdots\mathbf{A}_n &= \prod_{i=1}^{n}\mathbf{A}_i = \mathbf{I} + \tilde{\mathbf{v}}_1\theta_1 + \tilde{\mathbf{v}}_2\theta_2 + \cdots + \tilde{\mathbf{v}}_n\theta_n \\ &= \mathbf{I} + \sum_{i=1}^{n}\tilde{\mathbf{v}}_i\theta_i = \mathbf{A}_n\mathbf{A}_{n-1}\cdots\mathbf{A}_1\end{aligned}$$

Example 2.7 A rigid body experiences only a small rotation about the axis of rotation $\mathbf{b}$. If $\mathbf{b}$ is the vector defined by

$$\mathbf{b} = [2 \quad 2 \quad -1]^{\mathrm{T}}$$

and if the angle of rotation is 3°, determine the transformation matrix for this infinitesimal rotation.

Solution For $\theta = 3^\circ$, one has

$$\theta = 0.05236 \text{ rad}$$

$$\sin\theta = 0.05234, \qquad \tan\theta = 0.05240, \qquad \cos\theta = 0.9986$$

From which

$$\sin\theta \approx \tan\theta \approx \theta, \qquad \cos\theta \approx 1$$

and the angle θ can, indeed, be considered as an infinitesimal rotation. In this case the transformation matrix is given by

$$\mathbf{A} \approx \mathbf{I} + \tilde{\mathbf{v}} \sin\theta \approx \mathbf{I} + \tilde{\mathbf{v}}\theta$$

A unit vector $\mathbf{v}$ along the axis of rotation $\mathbf{b}$ is given by

$$\mathbf{v} = \frac{1}{3}[2 \quad 2 \quad -1]^{\mathrm{T}}$$

and the skew symmetric matrix $\tilde{\mathbf{v}}$ is

$$\tilde{\mathbf{v}} = \frac{1}{3}\begin{bmatrix} 0 & 1 & 2 \\ -1 & 0 & -2 \\ -2 & 2 & 0 \end{bmatrix}$$

Therefore, the infinitesimal transformation matrix is given by

$$\mathbf{A} = \mathbf{I} + \tilde{\mathbf{v}}\theta = \begin{bmatrix} 1 & 0 & 0 \\ 0 & 1 & 0 \\ 0 & 0 & 1 \end{bmatrix} + \frac{0.05236}{3}\begin{bmatrix} 0 & 1 & 2 \\ -1 & 0 & -2 \\ -2 & 2 & 0 \end{bmatrix}$$

$$= \begin{bmatrix} 1 & 0.01745 & 0.03491 \\ -0.01745 & 1 & -0.03491 \\ -0.03491 & 0.03491 & 1 \end{bmatrix}$$

2.4 ANGULAR VELOCITY VECTOR

Differentiating Eq. 8 with respect to time yields

$$\dot{\mathbf{r}} = \mathbf{A}\dot{\bar{\mathbf{r}}} + \dot{\mathbf{A}}\bar{\mathbf{r}} \tag{2.27}$$

where $(\dot{\ })$ denotes the differentiation with respect to time. Equation 27 implies that the time rate of change of a vector is due mainly to two parts. The first part is $\mathbf{A}\dot{\bar{\mathbf{r}}}$. which represents the time rate of change of the length of the vector $\bar{\mathbf{r}}$, while the second term, $\dot{\mathbf{A}}\bar{\mathbf{r}}$, is due to the change of the orientation of the vector $\mathbf{r}$. The discussion that follows will be focused first on the change in the orientation of the vector. Let us consider the product $\dot{\mathbf{A}}\bar{\mathbf{r}}$. As $\mathbf{A}$ depends only on the four parameters θ_0, θ_1, θ_2, and θ_3, a straightforward method for evaluating the product $\dot{\mathbf{A}}\bar{\mathbf{r}}$ is by using the partial derivatives as follows:

$$\dot{\mathbf{A}}\bar{\mathbf{r}} = \sum_{k=0}^{3} (\mathbf{A}_{\theta_k}\bar{\mathbf{r}})\dot{\theta}_k = \mathbf{A}_{\theta_0}\bar{\mathbf{r}}\dot{\theta}_0 + \mathbf{A}_{\theta_1}\bar{\mathbf{r}}\dot{\theta}_1 + \mathbf{A}_{\theta_2}\bar{\mathbf{r}}\dot{\theta}_2 + \mathbf{A}_{\theta_3}\bar{\mathbf{r}}\dot{\theta}_3 \tag{2.28}$$

where $\mathbf{A}_{\theta_k}$ is the partial derivative of $\mathbf{A}$ with respect to the parameter θ_k. Note that these parameters are related by Eq. 13. One may define the following column vectors:

$$\mathbf{B}_0 = \mathbf{A}_{\theta_0}\bar{\mathbf{r}}, \quad \mathbf{B}_1 = \mathbf{A}_{\theta_1}\bar{\mathbf{r}}, \quad \mathbf{B}_2 = \mathbf{A}_{\theta_2}\bar{\mathbf{r}}, \quad \mathbf{B}_3 = \mathbf{A}_{\theta_3}\bar{\mathbf{r}}$$

Therefore, Eq. 28 can be written as

$$\dot{\mathbf{A}}\bar{\mathbf{r}} = \mathbf{B}_0\dot{\theta}_0 + \mathbf{B}_1\dot{\theta}_1 + \mathbf{B}_2\dot{\theta}_2 + \mathbf{B}_3\dot{\theta}_3$$

or in a compact matrix notation as

$$\dot{\mathbf{A}}\bar{\mathbf{r}} = \mathbf{B}\dot{\boldsymbol{\theta}} \tag{2.29}$$

where $\mathbf{B} = \mathbf{B}(\boldsymbol{\theta}, \bar{\mathbf{r}})$, in the case of Euler parameters, is a 3×4 matrix given by

$$\mathbf{B} = [\mathbf{B}_0 \quad \mathbf{B}_1 \quad \mathbf{B}_2 \quad \mathbf{B}_3]$$

Thus, in general, given a transformation matrix $\mathbf{A}$, one should be able, by using the chain rule of differentiation, to isolate the velocity terms. This manipulation will prove useful in the chapters to follow, in particular when writing the kinetic energy expression for rigid and deformable bodies.

An elegant approach to develop Eq. 29 is to use the definition of the rotation matrix given by Eq. 20. It can be verified that the matrices $\mathbf{E}$ and $\bar{\mathbf{E}}$ satisfy the following identities:

$$\mathbf{E}\dot{\bar{\mathbf{E}}}^{\mathrm{T}} = \dot{\mathbf{E}}\bar{\mathbf{E}}^{\mathrm{T}} \tag{2.30}$$

$$\dot{\bar{\mathbf{E}}}^{\mathrm{T}}\bar{\mathbf{r}} = -\tilde{\bar{\mathbf{r}}}_4\dot{\boldsymbol{\theta}} \tag{2.31}$$

where $\tilde{\bar{\mathbf{r}}}_4$ is a 4×4 skew symmetric matrix defined as

$$\tilde{\bar{\mathbf{r}}}_4 = \begin{bmatrix} 0 & \bar{\mathbf{r}}^{\mathrm{T}} \\ -\bar{\mathbf{r}} & \tilde{\bar{\mathbf{r}}} \end{bmatrix} \tag{2.32}$$

and

$$\tilde{\bar{\mathbf{r}}} = \begin{bmatrix} 0 & -\bar{r}_3 & \bar{r}_2 \\ \bar{r}_3 & 0 & -\bar{r}_1 \\ -\bar{r}_2 & \bar{r}_1 & 0 \end{bmatrix} \tag{2.33}$$

Using the identities of Eq. 30, one can write the time derivative of the transformation matrix $\mathbf{A}$ as

$$\dot{\mathbf{A}} = \mathbf{E}\dot{\bar{\mathbf{E}}}^{\mathrm{T}} + \dot{\mathbf{E}}\bar{\mathbf{E}}^{\mathrm{T}} = 2\mathbf{E}\dot{\bar{\mathbf{E}}}^{\mathrm{T}} = 2\dot{\mathbf{E}}\bar{\mathbf{E}}^{\mathrm{T}} \tag{2.34}$$

By doing so and using the identity of Eq. 31, one can write Eq. 29 in the form

$$\dot{\mathbf{A}}\bar{\mathbf{r}} = 2\mathbf{E}\dot{\bar{\mathbf{E}}}^{\mathrm{T}}\bar{\mathbf{r}} = (-2\mathbf{E}\tilde{\bar{\mathbf{r}}}_4 + 2\bar{\mathbf{r}}\boldsymbol{\theta}^{\mathrm{T}})\dot{\boldsymbol{\theta}} \tag{2.35}$$

Thus, defining the matrix $\mathbf{B}$ as

$$\mathbf{B} = -2\mathbf{E}\tilde{\bar{\mathbf{r}}}_4 + 2\bar{\mathbf{r}}\boldsymbol{\theta}^{\mathrm{T}} \tag{2.36}$$

and using the fact that $\boldsymbol{\theta}^{\mathrm{T}}\dot{\boldsymbol{\theta}} = 0$, which is a consequence of Eq. 13, one has

$$\dot{\mathbf{A}}\bar{\mathbf{r}} = -2\mathbf{E}\tilde{\bar{\mathbf{r}}}_4\dot{\boldsymbol{\theta}}$$

It can be noted that, since $\mathbf{E}$ depends linearly on the parameters θ_k, $k = 0, 1, 2, 3$, the matrix $\mathbf{B}$ also depends linearly on these parameters.

Angular Velocity Using the definition of $\dot{\mathbf{A}}$ given in Eq. 34 and premultiplying the second term in the right-hand side of Eq. 27 by the identity matrix $\mathbf{I} = \mathbf{A}\mathbf{A}^{\mathrm{T}}$, one obtains

$$\dot{\mathbf{A}}\bar{\mathbf{r}} = 2\mathbf{A}\mathbf{A}^{\mathrm{T}}\mathbf{E}\dot{\bar{\mathbf{E}}}^{\mathrm{T}}\bar{\mathbf{r}} \tag{2.37}$$

Replacing $\mathbf{A}^{\mathrm{T}}$ by

$$\mathbf{A}^{\mathrm{T}} = \bar{\mathbf{E}}\mathbf{E}^{\mathrm{T}} \tag{2.38}$$

one gets

$$\dot{\mathbf{A}}\bar{\mathbf{r}} = 2\mathbf{A}\bar{\mathbf{E}}\mathbf{E}^{\mathrm{T}}\mathbf{E}\dot{\bar{\mathbf{E}}}^{\mathrm{T}}\bar{\mathbf{r}} \tag{2.39}$$

Substituting Eq. 24 into Eq. 39 yields

$$\dot{\mathbf{A}}\bar{\mathbf{r}} = 2\mathbf{A}\bar{\mathbf{E}}(\mathbf{I}_4 - \boldsymbol{\theta}\boldsymbol{\theta}^{\mathrm{T}})\dot{\bar{\mathbf{E}}}^{\mathrm{T}}\bar{\mathbf{r}} \tag{2.40}$$

By simple matrix multiplication, one can verify the identity

$$\bar{\mathbf{E}}\boldsymbol{\theta} = \mathbf{0} \tag{2.41}$$

which on substituting into Eq. 40 yields

$$\dot{\mathbf{A}}\bar{\mathbf{r}} = 2\mathbf{A}\bar{\mathbf{E}}\dot{\bar{\mathbf{E}}}^{\mathrm{T}}\bar{\mathbf{r}} \tag{2.42}$$

One can show that $2\bar{\mathbf{E}}\dot{\bar{\mathbf{E}}}^{\mathrm{T}}$ is a 3×3 skew symmetric matrix given by

$$2\bar{\mathbf{E}}\dot{\bar{\mathbf{E}}}^{\mathrm{T}} = \tilde{\bar{\omega}} = \begin{bmatrix} 0 & -\bar{\omega}_3 & \bar{\omega}_2 \\ \bar{\omega}_3 & 0 & -\bar{\omega}_1 \\ -\bar{\omega}_2 & \bar{\omega}_1 & 0 \end{bmatrix} \tag{2.43}$$

where the identity of Eq. 13 is used and $\bar{\omega}_1$, $\bar{\omega}_2$, and $\bar{\omega}_3$ are given by

$$\left.\begin{aligned} \bar{\omega}_1 &= 2(\theta_3\dot{\theta}_2 - \theta_2\dot{\theta}_3 - \theta_1\dot{\theta}_0 + \theta_0\dot{\theta}_1) \\ \bar{\omega}_2 &= 2(\theta_1\dot{\theta}_3 + \theta_0\dot{\theta}_2 - \theta_3\dot{\theta}_1 - \theta_2\dot{\theta}_0) \\ \bar{\omega}_3 &= 2(\theta_2\dot{\theta}_1 - \theta_3\dot{\theta}_0 + \theta_0\dot{\theta}_3 - \theta_1\dot{\theta}_2) \end{aligned}\right\} \tag{2.44a}$$

In terms of the angle of rotation and the components of the unit vector $\mathbf{v}$ along the axis of the rotation, the components of Eq. 44a are defined as

$$\left.\begin{aligned} \bar{\omega}_1 &= 2(v_3\dot{v}_2 - v_2\dot{v}_3)\sin^2\frac{\theta}{2} + \dot{v}_1\sin\theta + \dot{\theta}v_1 \\ \bar{\omega}_2 &= 2(v_1\dot{v}_3 - v_3\dot{v}_1)\sin^2\frac{\theta}{2} + \dot{v}_2\sin\theta + \dot{\theta}v_2 \\ \bar{\omega}_3 &= 2(v_2\dot{v}_1 - v_1\dot{v}_2)\sin^2\frac{\theta}{2} + \dot{v}_3\sin\theta + \dot{\theta}v_3 \end{aligned}\right\} \tag{2.44b}$$

which can be written as

$$\bar{\omega} = 2\dot{\mathbf{v}} \times \mathbf{v}\sin^2\frac{\theta}{2} + \dot{\mathbf{v}}\sin\theta + \mathbf{v}\dot{\theta} \tag{2.44c}$$

By using Eq. 43, Eq. 42 can be written as

$$\dot{\mathbf{A}}\bar{\mathbf{r}} = \mathbf{A}\tilde{\bar{\omega}}\bar{\mathbf{r}} \tag{2.45}$$

Since

$$\tilde{\bar{\omega}}\bar{\mathbf{r}} = \bar{\omega} \times \bar{\mathbf{r}} \tag{2.46}$$

where $\bar{\omega}$ is the *angular velocity vector* defined in the body coordinate system and given by

$$\bar{\omega} = [\bar{\omega}_1 \quad \bar{\omega}_2 \quad \bar{\omega}_3]^{\mathrm{T}}$$

one can write Eq. 45 as

$$\dot{\mathbf{A}}\bar{\mathbf{r}} = \mathbf{A}(\bar{\omega} \times \bar{\mathbf{r}}) \tag{2.47}$$

and the velocity vector of Eq. 27 can be written in terms of the angular velocity vector as

$$\dot{\mathbf{r}} = \mathbf{A}[\dot{\bar{\mathbf{r}}} + \bar{\omega} \times \bar{\mathbf{r}}] \tag{2.48}$$

Global Representation The velocity vector $\dot{\mathbf{r}}$ of Eq. 48 can also be expressed in terms of the global components of the angular velocity vector. Using the orthogonality of the transformation matrix **A**, one can write the first term in the right-hand side of Eq. 27 as

$$\dot{\mathbf{A}}\bar{\mathbf{r}} = \dot{\mathbf{A}}\mathbf{A}^{\mathrm{T}}\mathbf{A}\bar{\mathbf{r}} \tag{2.49}$$

Using Eqs. 20 and 34, we may write Eq. 49 in the following form:

$$\dot{\mathbf{A}}\bar{\mathbf{r}} = 2\dot{\mathbf{E}}\bar{\mathbf{E}}^{\mathrm{T}}\bar{\mathbf{E}}\mathbf{E}^{\mathrm{T}}\mathbf{r} \tag{2.50}$$

The identity of Eq. 24 then yields

$$\dot{\mathbf{A}}\bar{\mathbf{r}} = 2\dot{\mathbf{E}}(\mathbf{I}_4 - \boldsymbol{\theta}\boldsymbol{\theta}^{\mathrm{T}})\mathbf{E}^{\mathrm{T}}\mathbf{r} \tag{2.51}$$

By simple matrix multiplication, one can verify that

$$\mathbf{E}\boldsymbol{\theta} = \mathbf{0} \tag{2.52}$$

Substituting Eq. 52 into Eq. 51 yields

$$\dot{\mathbf{A}}\bar{\mathbf{r}} = 2\dot{\mathbf{E}}\mathbf{E}^{\mathrm{T}}\mathbf{r} \tag{2.53}$$

It can be shown that $\dot{\mathbf{E}}\mathbf{E}^{\mathrm{T}}$ is a skew symmetric matrix given by

$$\dot{\mathbf{E}}\mathbf{E}^{\mathrm{T}} = \begin{bmatrix} 0 & \vdots & \dot{\theta}_1\theta_2 + \dot{\theta}_0\theta_3 - \dot{\theta}_3\theta_0 - \dot{\theta}_2\theta_1 & \vdots & \dot{\theta}_1\theta_3 - \dot{\theta}_0\theta_2 - \dot{\theta}_3\theta_1 + \dot{\theta}_2\theta_0 \\ \dot{\theta}_2\theta_1 + \dot{\theta}_3\theta_0 - \dot{\theta}_0\theta_3 - \dot{\theta}_1\theta_2 & \vdots & 0 & \vdots & \dot{\theta}_2\theta_3 - \dot{\theta}_3\theta_2 + \dot{\theta}_0\theta_1 - \dot{\theta}_1\theta_0 \\ \dot{\theta}_3\theta_1 - \dot{\theta}_2\theta_0 - \dot{\theta}_1\theta_3 + \dot{\theta}_0\theta_2 & \vdots & \dot{\theta}_3\theta_2 - \dot{\theta}_2\theta_3 + \dot{\theta}_1\theta_0 - \dot{\theta}_0\theta_1 & \vdots & 0 \end{bmatrix} \tag{2.54}$$

One may define a vector $\boldsymbol{\omega}$ such that

$$\tilde{\omega} = 2\dot{\mathbf{E}}\mathbf{E}^{\mathrm{T}} \tag{2.55}$$

and write Eq. 53 as

$$\dot{\mathbf{A}}\bar{\mathbf{r}} = \tilde{\omega}\mathbf{r} = \tilde{\omega}\mathbf{A}\bar{\mathbf{r}} \tag{2.56}$$

or in the alternative form

$$\dot{\mathbf{A}}\bar{\mathbf{r}} = \boldsymbol{\omega} \times \mathbf{r} \tag{2.57}$$

Therefore, Eq. 27 can be written as

$$\dot{\mathbf{r}} = \mathbf{A}\dot{\bar{\mathbf{r}}} + \boldsymbol{\omega} \times \mathbf{r} \tag{2.58}$$

It can be shown using matrix multiplication that

$$\boldsymbol{\omega} = \mathbf{A}\bar{\boldsymbol{\omega}} \tag{2.59}$$

where $\boldsymbol{\omega}$ is the vector whose components are

$$\left.\begin{aligned} \omega_1 &= 2(\dot{\theta}_3\theta_2 - \dot{\theta}_2\theta_3 + \dot{\theta}_1\theta_0 - \dot{\theta}_0\theta_1) \\ \omega_2 &= 2(\dot{\theta}_1\theta_3 - \dot{\theta}_0\theta_2 - \dot{\theta}_3\theta_1 + \dot{\theta}_2\theta_0) \\ \omega_3 &= 2(\dot{\theta}_2\theta_1 + \dot{\theta}_3\theta_0 - \dot{\theta}_0\theta_3 - \dot{\theta}_1\theta_2) \end{aligned}\right\} \tag{2.60a}$$

In terms of the angle of rotation and the components of the unit vector along the axis of rotation, the components of the angular velocity vector $\boldsymbol{\omega}$ are defined as

$$\left.\begin{aligned} \omega_1 &= 2(\dot{v}_3 v_2 - \dot{v}_2 v_3)\sin^2\frac{\theta}{2} + \dot{v}_1 \sin\theta + \dot{\theta} v_1 \\ \omega_2 &= 2(\dot{v}_1 v_3 - \dot{v}_3 v_1)\sin^2\frac{\theta}{2} + \dot{v}_2 \sin\theta + \dot{\theta} v_2 \\ \omega_3 &= 2(\dot{v}_2 v_1 - \dot{v}_1 v_2)\sin^2\frac{\theta}{2} + \dot{v}_3 \sin\theta + \dot{\theta} v_3 \end{aligned}\right\} \tag{2.60b}$$

which can be written as

$$\boldsymbol{\omega} = 2\mathbf{v} \times \dot{\mathbf{v}} \sin^2\frac{\theta}{2} + \dot{\mathbf{v}}\sin\theta + \mathbf{v}\dot{\theta} \tag{2.60c}$$

One can show by using the definition of $\boldsymbol{\omega}$ and $\bar{\boldsymbol{\omega}}$ that Eq. 59 is valid. To this end, we use Eqs. 44a and 60a to write the following two relations for $\bar{\boldsymbol{\omega}}$ and $\boldsymbol{\omega}$:

$$\bar{\boldsymbol{\omega}} = 2\bar{\mathbf{E}}\dot{\boldsymbol{\theta}} = -2\dot{\bar{\mathbf{E}}}\boldsymbol{\theta} \tag{2.61}$$

$$\boldsymbol{\omega} = 2\mathbf{E}\dot{\boldsymbol{\theta}} = -2\dot{\mathbf{E}}\boldsymbol{\theta} \tag{2.62}$$

By using the definition of $\mathbf{A}$ in Eq. 20, one can write

$$\mathbf{A}\bar{\boldsymbol{\omega}} = 2\mathbf{A}\bar{\mathbf{E}}\dot{\boldsymbol{\theta}} = 2\mathbf{E}\bar{\mathbf{E}}^{\mathrm{T}}\bar{\mathbf{E}}\dot{\boldsymbol{\theta}} \tag{2.63}$$

Using the identity of Eq. 24 one obtains

$$\mathbf{A}\bar{\boldsymbol{\omega}} = 2\mathbf{E}(\mathbf{I}_4 - \boldsymbol{\theta}\boldsymbol{\theta}^{\mathrm{T}})\dot{\boldsymbol{\theta}} \tag{2.64}$$

Since from Eq. 13 $\boldsymbol{\theta}^{\mathrm{T}}\dot{\boldsymbol{\theta}} = 0$, Eq. 64 yields

$$\mathbf{A}\bar{\boldsymbol{\omega}} = 2\mathbf{E}\dot{\boldsymbol{\theta}} = \boldsymbol{\omega}$$

which is the result given by Eq. 59.

Other useful relations that can be easily extracted from the preceding development are

$$\tilde{\bar{\boldsymbol{\omega}}} = \mathbf{A}^{\mathrm{T}}\dot{\mathbf{A}} = -2\dot{\bar{\mathbf{E}}}\bar{\mathbf{E}}^{\mathrm{T}} = 2\bar{\mathbf{E}}\dot{\bar{\mathbf{E}}}^{\mathrm{T}} \tag{2.65}$$

$$\tilde{\boldsymbol{\omega}} = \dot{\mathbf{A}}\mathbf{A}^{\mathrm{T}} = 2\dot{\mathbf{E}}\mathbf{E}^{\mathrm{T}} = -2\mathbf{E}\dot{\mathbf{E}}^{\mathrm{T}} \tag{2.66}$$

It may be interesting to note that the component of the angular velocity vector $\boldsymbol{\omega}$ along the axis of rotation should be $\dot{\theta}$. This can be verified by using the definition of the dot product and the definition of the angular velocity vectors given by Eqs. 44 and 60. It is also clear that if the axis of rotation is fixed in space, then the components of the angular velocity vector $\boldsymbol{\omega}$ in terms of the time derivative of the angle of rotation θ reduce to

$$\boldsymbol{\omega} = [\,\omega_1 \quad \omega_2 \quad \omega_3\,]^{\mathrm{T}} = \dot{\theta}[\,v_1 \quad v_2 \quad v_3\,]^{\mathrm{T}} \tag{2.67}$$

where v_1, v_2, and v_3 are the components of a unit vector along the axis of rotation. The components $\omega_1, \omega_2, \omega_3$ of Eq. 67 are the same as the components of the angular velocity vector as defined with respect to the rotating frame, that is, in this special case also

$$\bar{\boldsymbol{\omega}} = \dot{\theta}[\,v_1 \quad v_2 \quad v_3\,]^{\mathrm{T}} = \boldsymbol{\omega}$$

This is due mainly to the fact that

$$\mathbf{Av} = \mathbf{A}^{\mathrm{T}}\mathbf{v} = \mathbf{v}$$

which has been the subject of earlier discussions.

Example 2.8 If the vector $\bar{\mathbf{r}} = [\,2\ 3\ 4\,]^{\mathrm{T}}$ is defined in a rigid body coordinate system which rotates with a constant angular velocity $\dot{\theta} = 10$ rad/sec about an axis of rotation whose unit vector $\mathbf{v}$ is

$$\mathbf{v} = \left[\frac{1}{\sqrt{3}} \quad \frac{1}{\sqrt{3}} \quad \frac{1}{\sqrt{3}}\right]^{\mathrm{T}}$$

determine, when $t = 0.1$ sec, the angular velocity vector with respect to (1) the fixed frame and (2) the rotating frame.

Solution Since $\mathbf{v}$ is a fixed unit vector, it is obvious that

$$\boldsymbol{\omega} = \bar{\boldsymbol{\omega}} = \dot{\theta}[\,v_1 \quad v_2 \quad v_3\,]^{\mathrm{T}} = 10\left[\frac{1}{\sqrt{3}} \quad \frac{1}{\sqrt{3}} \quad \frac{1}{\sqrt{3}}\right]^{\mathrm{T}}$$
$$= [\,5.773 \quad 5.773 \quad 5.773\,]^{\mathrm{T}}$$

One may try to arrive at this result by using Eqs. 61 and 62. To do this, we use Eq. 11 to evaluate the four parameters $\theta_0, \theta_1, \theta_2$, and θ_3 as follows:

$$\theta_0 = \cos\frac{\theta}{2}$$
$$\theta_1 = \theta_2 = \theta_3 = \frac{1}{\sqrt{3}}\sin\frac{\theta}{2}$$

The time derivatives of these parameters are

$$\dot{\theta}_0 = -\frac{\dot{\theta}}{2}\sin\frac{\theta}{2}$$
$$\dot{\theta}_1 = \dot{\theta}_2 = \dot{\theta}_3 = \frac{\dot{\theta}}{2\sqrt{3}}\cos\frac{\theta}{2}$$

At $t = 0.1$ sec, with the assumption that $\theta(t = 0) = 0$, the angle θ is

$$\theta = \dot{\theta}t = 10(0.1) = 1 \text{ rad} = 57.296^\circ$$

It follows that

$$\theta_0 = \cos\left(\frac{57.296}{2}\right) = 0.8776$$

$$\theta_1 = \theta_2 = \theta_3 = \frac{1}{\sqrt{3}} \sin\left(\frac{57.296}{2}\right) = 0.2768$$

$$\dot{\theta}_0 = -\frac{10}{2} \sin\left(\frac{57.296}{2}\right) = -2.397$$

$$\dot{\theta}_1 = \dot{\theta}_2 = \dot{\theta}_3 = \frac{10}{2(1.732)} \cos\left(\frac{57.296}{2}\right) = 2.533$$

Therefore, by using Eq. 62 and the definition of the matrix $\mathbf{E}$ of Eq. 21, the vector $\boldsymbol{\omega}$ evaluated at $t = 0.1$ sec is given by

$$\boldsymbol{\omega} = 2\mathbf{E}\dot{\boldsymbol{\theta}} = 2\begin{bmatrix} -\theta_1 & \theta_0 & -\theta_3 & \theta_2 \\ -\theta_2 & \theta_3 & \theta_0 & -\theta_1 \\ -\theta_3 & -\theta_2 & \theta_1 & \theta_0 \end{bmatrix}\begin{bmatrix} \dot{\theta}_0 \\ \dot{\theta}_1 \\ \dot{\theta}_2 \\ \dot{\theta}_3 \end{bmatrix}$$

$$= 2\begin{bmatrix} -0.2768 & 0.8776 & -0.2768 & 0.2768 \\ -0.2768 & 0.2768 & 0.8776 & -0.2768 \\ -0.2768 & -0.2768 & 0.2768 & 0.8776 \end{bmatrix}\begin{bmatrix} -2.397 \\ 2.533 \\ 2.533 \\ 2.533 \end{bmatrix}$$

$$= \begin{bmatrix} 5.773 \\ 5.773 \\ 5.773 \end{bmatrix}$$

and by using Eqs. 61 and 22, the vector $\bar{\boldsymbol{\omega}}$ is given by

$$\bar{\boldsymbol{\omega}} = 2\bar{\mathbf{E}}\dot{\boldsymbol{\theta}} = 2\begin{bmatrix} -\theta_1 & \theta_0 & \theta_3 & -\theta_2 \\ -\theta_2 & -\theta_3 & \theta_0 & \theta_1 \\ -\theta_3 & \theta_2 & -\theta_1 & \theta_0 \end{bmatrix}\begin{bmatrix} \dot{\theta}_0 \\ \dot{\theta}_1 \\ \dot{\theta}_2 \\ \dot{\theta}_3 \end{bmatrix}$$

$$= 2\begin{bmatrix} -0.2768 & 0.8776 & 0.2768 & -0.2768 \\ -0.2768 & -0.2768 & 0.8776 & 0.2768 \\ -0.2768 & 0.2768 & -0.2768 & 0.8776 \end{bmatrix}\begin{bmatrix} -2.397 \\ 2.533 \\ 2.533 \\ 2.533 \end{bmatrix}$$

$$= \begin{bmatrix} 5.773 \\ 5.773 \\ 5.773 \end{bmatrix}$$

The vectors $\boldsymbol{\omega}$ and $\bar{\boldsymbol{\omega}}$ could have been evaluated by using their explicit forms given by Eqs. 60 and 44. The reader may try to go a step further and evaluate the transformation matrix at $t = 0.1$ and verify that $\boldsymbol{\omega} = \mathbf{A}\bar{\boldsymbol{\omega}}$. With the result obtained in this example, one can verify that $\mathbf{v}^{\mathrm{T}}\boldsymbol{\omega} = \dot{\theta}$; that is, the component of the angular velocity $\boldsymbol{\omega}$ along the axis of rotation is $\dot{\theta}$.

General Displacement The relationships developed in this section can be used to determine the global velocity of an arbitrary point on a rigid body in the multibody system. It was previously shown that the position vector of an arbitrary point P on the body i in the multibody system can be written as

$$\mathbf{r}^i = \mathbf{R}^i + \mathbf{u}^i$$

where $\mathbf{R}^i$ is the global position vector of the origin of the body reference and $\mathbf{u}^i$ is the local position of point P. The vector $\mathbf{u}^i$ can be written in terms of the local components as

$$\mathbf{u}^i = \mathbf{A}^i \bar{\mathbf{u}}^i$$

in which $\mathbf{A}^i$ is the transformation matrix from the body coordinate system to the global coordinate system, and $\bar{\mathbf{u}}^i$ is the position vector of point P in the body coordinate system. One can, therefore, write the position vector $\mathbf{r}^i$ as

$$\mathbf{r}^i = \mathbf{R}^i + \mathbf{A}^i \bar{\mathbf{u}}^i$$

Differentiating this equation with respect to time leads to

$$\dot{\mathbf{r}}^i = \dot{\mathbf{R}}^i + \dot{\mathbf{A}}^i \bar{\mathbf{u}}^i$$

in which $\dot{\mathbf{r}}^i$ is the absolute velocity vector of point P and $\dot{\mathbf{R}}^i$ is the absolute velocity of the origin of the body reference. It was previously shown that

$$\dot{\mathbf{A}}^i \bar{\mathbf{u}}^i = \boldsymbol{\omega}^i \times \mathbf{u}^i = \mathbf{A}^i (\bar{\boldsymbol{\omega}}^i \times \bar{\mathbf{u}}^i)$$

One can, therefore, write the absolute velocity vector of point P as

$$\dot{\mathbf{r}}^i = \dot{\mathbf{R}}^i + \boldsymbol{\omega}^i \times \mathbf{u}^i$$

or

$$\dot{\mathbf{r}}^i = \dot{\mathbf{R}}^i + \mathbf{A}^i (\bar{\boldsymbol{\omega}}^i \times \bar{\mathbf{u}}^i)$$

Furthermore, the angular velocity vector $\boldsymbol{\omega}^i$ and $\bar{\boldsymbol{\omega}}^i$ can be written in terms of the time derivative of Euler parameters as

$$\boldsymbol{\omega}^i = \mathbf{G}^i \dot{\boldsymbol{\theta}}^i \tag{2.68a}$$

$$\bar{\boldsymbol{\omega}}^i = \bar{\mathbf{G}}^i \dot{\boldsymbol{\theta}}^i \tag{2.68b}$$

where $\mathbf{G}^i$ and $\bar{\mathbf{G}}^i$ are the 3×4 matrices that depend on Euler parameters and given by

$$\mathbf{G}^i = 2\mathbf{E}^i \tag{2.68c}$$

$$\bar{\mathbf{G}}^i = 2\bar{\mathbf{E}}^i \tag{2.68d}$$

Therefore, the vector $\dot{\mathbf{A}}^i \bar{\mathbf{u}}^i$ can be written as

$$\begin{aligned}\dot{\mathbf{A}}^i \bar{\mathbf{u}}^i &= \mathbf{A}^i (\bar{\boldsymbol{\omega}}^i \times \bar{\mathbf{u}}^i) = -\mathbf{A}^i (\bar{\mathbf{u}}^i \times \bar{\boldsymbol{\omega}}^i) \\ &= -\mathbf{A}^i \tilde{\bar{\mathbf{u}}}^i \bar{\mathbf{G}}^i \dot{\boldsymbol{\theta}}^i \end{aligned} \tag{2.68e}$$

in which $\tilde{\bar{\mathbf{u}}}^i$ is the 3×3 skew symmetric matrix

$$\tilde{\bar{\mathbf{u}}}^i = \begin{bmatrix} 0 & -\bar{u}_3^i & \bar{u}_2^i \\ \bar{u}_3^i & 0 & -\bar{u}_1^i \\ -\bar{u}_2^i & \bar{u}_1^i & 0 \end{bmatrix}.$$

As will be seen in sections to follow, equations similar to Eqs. 68a and 68b, in which the time derivatives of the rotational coordinates are isolated, can be obtained when the orientation of the frame of reference is described by using other sets of coordinates such as *Rodriguez parameters* or *Euler angles*. Therefore, the form of Eq. 68e is general and can be developed irrespective of the set of rotational coordinates used. The form of Eq. 68e is convenient and will be used in Chapters 3 and 5 to develop the dynamic equations of rigid and deformable bodies in multibody systems.

Example 2.9 Figure 10 shows a robotic arm that translates and rotates with respect to a vehicle system. Let $\mathbf{X}_1\mathbf{X}_2\mathbf{X}_3$ be the coordinate system of the vehicle and $\mathbf{X}_1^i\mathbf{X}_2^i\mathbf{X}_3^i$ be the coordinate system of the robotic arm as shown in the figure. The robotic arm is assumed to be connected to the vehicle by means of a cylindrical joint; that is, the relative motion between the arm and the vehicle can be described by translational and rotational displacements along and about the joint axis shown in the figure. A unit vector parallel to the joint axis and defined in the vehicle coordinate system is given by

$$\mathbf{v} = \frac{1}{\sqrt{3}}[1 \quad 1 \quad 1]^{\mathrm{T}}$$

Assuming that the arm translates with constant speed 2 m/sec and the angular velocity about the joint axis is 5 rad/sec, determine the velocity of point P on the robotic arm when the arm rotates 30°. The coordinates of P in the arm coordinate system are given by the vector $\bar{\mathbf{u}}_P^i$ where

$$\bar{\mathbf{u}}_P^i = [0 \quad 1 \quad 0]^{\mathrm{T}}\ \mathrm{m}$$

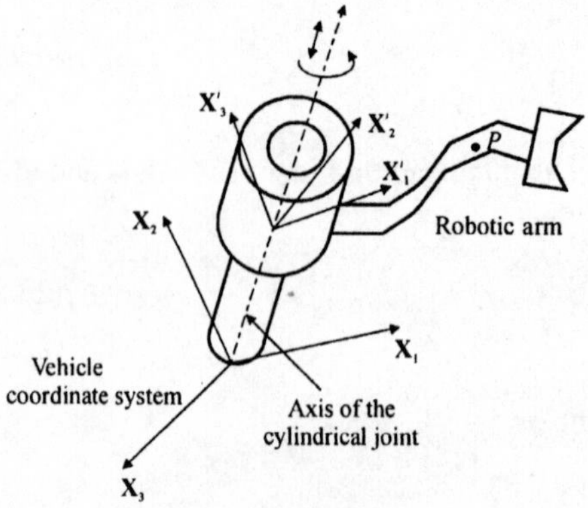

Figure 2.10 Vehicle system.

Assume that the axes of the vehicle and the arm coordinate system are initially the same.

Solution At the specified position, the transformation matrix between the arm coordinate system and the vehicle coordinate system is given by

$$\mathbf{A}^i = \left[\mathbf{I} + \tilde{\mathbf{v}} \sin\theta^i + 2(\tilde{\mathbf{v}})^2 \sin^2\frac{\theta^i}{2}\right]$$

where $\theta^i = 30°$ and $\tilde{\mathbf{v}}$ is the skew symmetric matrix defined as

$$\tilde{\mathbf{v}} = \frac{1}{\sqrt{3}}\begin{bmatrix} 0 & -1 & 1 \\ 1 & 0 & -1 \\ -1 & 1 & 0 \end{bmatrix}$$

After substitution, one can verify that the matrix $\mathbf{A}^i$ is given by

$$\mathbf{A}^i = \begin{bmatrix} 0.91068 & -0.24402 & 0.33334 \\ 0.33334 & 0.91068 & -0.24402 \\ -0.24402 & 0.33334 & 0.91068 \end{bmatrix}$$

The vector $\bar{\mathbf{u}}^i_P$ can be defined in the vehicle coordinate system as

$$\mathbf{u}^i_P = \mathbf{A}^i\bar{\mathbf{u}}^i_P = \begin{bmatrix} 0.91068 & -0.24402 & 0.33334 \\ 0.33334 & 0.91068 & 0.24402 \\ -0.24402 & 0.3334 & 0.91068 \end{bmatrix}\begin{bmatrix} 0 \\ 1 \\ 0 \end{bmatrix}$$

$$= \begin{bmatrix} -0.24402 \\ 0.91068 \\ 0.33334 \end{bmatrix} \text{m}$$

The angular velocity vector $\boldsymbol{\omega}^i$ is defined as

$$\boldsymbol{\omega}^i = \dot{\theta}^i\mathbf{v} = \frac{5}{\sqrt{3}}[1 \quad 1 \quad 1]^{\mathrm{T}} = 2.8868[1 \quad 1 \quad 1]^{\mathrm{T}} \text{ rad/sec}$$

The cross product $\boldsymbol{\omega}^i \times \mathbf{u}^i_P$ is then given by

$$\boldsymbol{\omega}^i \times \mathbf{u}^i_P = \tilde{\boldsymbol{\omega}}^i\mathbf{u}^i_P = 2.8868\begin{bmatrix} 0 & -1 & 1 \\ 1 & 0 & -1 \\ -1 & 1 & 0 \end{bmatrix}\begin{bmatrix} -0.24402 \\ 0.91068 \\ 0.33334 \end{bmatrix}$$

$$= \begin{bmatrix} -1.6667 \\ -1.6667 \\ 3.3334 \end{bmatrix} \text{m/sec}$$

and the velocity of point P in the vehicle coordinate system is given by

$$\dot{\mathbf{r}}_P = \dot{\mathbf{R}}^i + \boldsymbol{\omega}^i \times \mathbf{u}^i_P$$

where $\dot{\mathbf{R}}^i$ is the absolute velocity of the origin of the arm coordinate system, which is given by

$$\dot{\mathbf{R}}^i = \frac{2}{\sqrt{3}}[1 \quad 1 \quad 1]^{\mathrm{T}} = 1.1547[1 \quad 1 \quad 1]^{\mathrm{T}} \text{ m/sec}$$

Therefore, the velocity of P is given by

$$\dot{\mathbf{r}}_P = 1.1547\begin{bmatrix}1\\1\\1\end{bmatrix} + \begin{bmatrix}-1.6667\\-1.6667\\3.3334\end{bmatrix} = \begin{bmatrix}-0.512\\-0.512\\4.4881\end{bmatrix} \text{ m/sec}$$

Relative Angular Velocity The transformation matrix that defines the orientation of an arbitrary body i can be expressed in terms of the transformation matrix that defines the orientation of another body j as

$$\mathbf{A}^i = \mathbf{A}^j\mathbf{A}^{ij}$$

It follows that

$$\begin{aligned}\tilde{\omega}^i &= \dot{\mathbf{A}}^i\mathbf{A}^{i\mathrm{T}} = (\dot{\mathbf{A}}^j\mathbf{A}^{ij} + \mathbf{A}^j\dot{\mathbf{A}}^{ij})(\mathbf{A}^j\mathbf{A}^{ij})^{\mathrm{T}}\\ &= (\tilde{\omega}^j\mathbf{A}^j\mathbf{A}^{ij} + \mathbf{A}^j(\tilde{\omega}^{ij})_i\mathbf{A}^{ij})(\mathbf{A}^j\mathbf{A}^{ij})^{\mathrm{T}}\end{aligned}$$

where $(\tilde{\omega}^{ij})_i$ is the skew symmetric matrix associated with the angular velocity of body i with respect to body j defined in the coordinate system of body j. Since the following identity

$$\tilde{\omega}^{ij} = \mathbf{A}^j(\tilde{\omega}^{ij})_i\mathbf{A}^{j\mathrm{T}}$$

holds, where $\tilde{\omega}^{ij}$ is the skew symmetric matrix associated with the angular velocity of body i with respect to body j defined in the global coordinate system, the preceding equation yields

$$\tilde{\omega}^i = \tilde{\omega}^j + \tilde{\omega}^{ij}$$

which implies that

$$\omega^i = \omega^j + \omega^{ij}$$

This equation shows that the absolute angular velocity of body i is equal to the absolute angular velocity of body j plus the angular velocity of body i with respect to body j.

2.5 ACCELERATION EQUATIONS

In the preceding section, the time rate of change of a vector was defined. In this section, the second derivative of a vector with respect to time is evaluated. For convenience, we reproduce Eq. 56 here:

$$\dot{\mathbf{A}}\bar{\mathbf{r}} = \tilde{\omega}\mathbf{r} \tag{2.69a}$$

where the vector $\mathbf{r}$ is equal to $\mathbf{A}\bar{\mathbf{r}}$. Therefore, $\dot{\mathbf{A}}\bar{\mathbf{r}}$ can also be written as

$$\dot{\mathbf{A}}\bar{\mathbf{r}} = \tilde{\omega}\mathbf{A}\bar{\mathbf{r}} \tag{2.69b}$$

and the velocity vector of Eq. 27 can be written as

$$\dot{\mathbf{r}} = \mathbf{A}\dot{\bar{\mathbf{r}}} + \tilde{\omega}\mathbf{A}\bar{\mathbf{r}} \tag{2.70}$$

Differentiating Eq. 70 with respect to time yields

$$\ddot{\mathbf{r}} = \dot{\mathbf{A}}\dot{\bar{\mathbf{r}}} + \mathbf{A}\ddot{\bar{\mathbf{r}}} + \dot{\tilde{\omega}}\mathbf{A}\bar{\mathbf{r}} + \tilde{\omega}\dot{\mathbf{A}}\bar{\mathbf{r}} + \tilde{\omega}\mathbf{A}\dot{\bar{\mathbf{r}}} \tag{2.71}$$

In a manner similar to that for Eq. 69a, one can write

$$\dot{\mathbf{A}}\dot{\bar{\mathbf{r}}} = \tilde{\omega}\mathbf{v}_g$$

where $\mathbf{v}_g = \mathbf{A}\dot{\bar{\mathbf{r}}}$ is the time derivative of the vector $\bar{\mathbf{r}}$ defined in the global coordinates system. Equation 71, after rearranging terms, yields

$$\ddot{\mathbf{r}} = \mathbf{A}\ddot{\bar{\mathbf{r}}} + 2\tilde{\omega}\mathbf{v}_g + \dot{\tilde{\omega}}\mathbf{r} + \tilde{\omega}\dot{\mathbf{A}}\bar{\mathbf{r}} \tag{2.72}$$

One may substitute Eq. 69a into Eq. 72 to get

$$\ddot{\mathbf{r}} = \mathbf{A}\ddot{\bar{\mathbf{r}}} + 2\tilde{\omega}\mathbf{v}_g + \dot{\tilde{\omega}}\mathbf{r} + \tilde{\omega}\tilde{\omega}\mathbf{r} \tag{2.73}$$

Using the identities

$$\tilde{\omega}\mathbf{v}_g = \omega \times \mathbf{v}_g, \qquad \dot{\tilde{\omega}}\mathbf{r} = \dot{\omega} \times \mathbf{r}$$
$$\tilde{\omega}\tilde{\omega}\mathbf{r} = \omega \times (\omega \times \mathbf{r}),$$

one obtains

$$\ddot{\mathbf{r}} = \mathbf{A}\ddot{\bar{\mathbf{r}}} + 2\omega \times \mathbf{v}_g + \dot{\omega} \times \mathbf{r} + \omega \times (\omega \times \mathbf{r}) \tag{2.74}$$

Angular Acceleration Vector Using the notation

$$\alpha = \dot{\omega}$$

where α is the *angular acceleration vector*, Eq. 74 reduces to

$$\ddot{\mathbf{r}} = \mathbf{A}\ddot{\bar{\mathbf{r}}} + 2\omega \times \mathbf{v}_g + \alpha \times \mathbf{r} + \omega \times (\omega \times \mathbf{r}) \tag{2.75}$$

Using the definition of ω given by Eq. 60, one can show that the vector α can be expressed in terms of Euler parameters as

$$\alpha = 2\begin{bmatrix} \ddot{\theta}_3\theta_2 - \ddot{\theta}_2\theta_3 + \ddot{\theta}_1\theta_0 - \ddot{\theta}_0\theta_1 \\ \ddot{\theta}_1\theta_3 - \ddot{\theta}_0\theta_2 - \ddot{\theta}_3\theta_1 + \ddot{\theta}_2\theta_0 \\ \ddot{\theta}_2\theta_1 + \ddot{\theta}_3\theta_0 - \ddot{\theta}_0\theta_3 - \ddot{\theta}_1\theta_2 \end{bmatrix}$$

which can be written in matrix form as

$$\alpha = 2\mathbf{E}\ddot{\theta}$$

where the matrix $\mathbf{E}$ is given by Eq. 21.

Equation 75 can also be written as

$$\ddot{\mathbf{r}} = \mathbf{a}_l + \mathbf{a}_c + \mathbf{a}_t + \mathbf{a}_n$$

where

$$\mathbf{a}_l = \mathbf{A}\ddot{\bar{\mathbf{r}}}, \qquad \mathbf{a}_c = 2\omega \times \mathbf{v}_g$$
$$\mathbf{a}_t = \alpha \times \mathbf{r}, \qquad \mathbf{a}_n = \omega \times (\omega \times \mathbf{r})$$

In these equations, $\ddot{\bar{\mathbf{r}}}$ is the acceleration of the point whose position is defined by the vector $\mathbf{r}$ as seen by an observer stationed on the rotating frame; thus $\mathbf{a}_l$ represents the local acceleration defined in the global coordinate system. This component of the acceleration is zero if the vector $\bar{\mathbf{r}}$ has a constant length, that is, if the vector $\bar{\mathbf{r}}$ has fixed components in the rotating coordinate system. The other three components

$\mathbf{a}_c$, $\mathbf{a}_t$, and $\mathbf{a}_n$ are, respectively, the *Coriolis*, *tangential*, and *normal* components of the acceleration vector. One may observe from the definition of these components that the normal component $\mathbf{a}_n$ has a magnitude equal to $(\dot{\theta})^2 r$, where r is the length of the vector $\mathbf{r}$. By using the definition of the cross product, it can be shown that the direction of this component is along the vector $(-\mathbf{r})$. The tangential component $\mathbf{a}_t$ has a magnitude $\ddot{\theta} r$, and its direction is perpendicular to both $\boldsymbol{\alpha}$ and $\mathbf{r}$. The Coriolis component $\mathbf{a}_c$ has a magnitude equal to $2\dot{\theta} v_g$, where v_g is the norm of the velocity vector $\mathbf{v}_g$. The Coriolis component has a direction that is perpendicular to both $\boldsymbol{\omega}$ and $\mathbf{v}_g$.

If the axis of rotation is fixed in space, the unit vector $\mathbf{v}$ is a constant vector, and since

$$\boldsymbol{\theta} = [\theta_0 \quad \theta_1 \quad \theta_2 \quad \theta_3]^{\mathrm{T}} = \left[\cos\frac{\theta}{2} \quad v_1 \sin\frac{\theta}{2} \quad v_2 \sin\frac{\theta}{2} \quad v_3 \sin\frac{\theta}{2}\right]^{\mathrm{T}}$$

one has

$$\dot{\boldsymbol{\theta}} = \frac{\dot{\theta}}{2}\left[-\sin\frac{\theta}{2} \quad v_1 \cos\frac{\theta}{2} \quad v_2 \cos\frac{\theta}{2} \quad v_3 \cos\frac{\theta}{2}\right]^{\mathrm{T}}$$

and

$$\ddot{\boldsymbol{\theta}} = \frac{\ddot{\theta}}{2}\left[-\sin\frac{\theta}{2} \quad v_1 \cos\frac{\theta}{2} \quad v_2 \cos\frac{\theta}{2} \quad v_3 \cos\frac{\theta}{2}\right]^{\mathrm{T}} - \frac{(\dot{\theta})^2}{4}\boldsymbol{\theta}$$

Since $\boldsymbol{\omega} = \dot{\theta}\mathbf{v}$, as a result of the assumption that the axis of rotation is fixed in space, one can verify that the angular acceleration vector is

$$\boldsymbol{\alpha} = \ddot{\theta}[v_1 \quad v_2 \quad v_3]^{\mathrm{T}} = \ddot{\theta}\mathbf{v}$$

If we consider the planar case in which $\mathbf{X}_3$ is the axis of rotation, it is obvious that

$$\boldsymbol{\alpha} = \ddot{\theta}[0 \quad 0 \quad 1]^{\mathrm{T}}$$

as expected.

General Displacement The kinematic relationships developed in this section can be used to determine the absolute acceleration of an arbitrary point on a rigid body in multibody systems. It has been shown in the previous section that the absolute velocity of a point on the rigid body i can be written as

$$\dot{\mathbf{r}}^i = \dot{\mathbf{R}}^i + \dot{\mathbf{A}}^i\bar{\mathbf{u}}^i = \dot{\mathbf{R}}^i + \mathbf{A}^i(\bar{\boldsymbol{\omega}}^i \times \bar{\mathbf{u}}^i)$$

where $\dot{\mathbf{R}}^i$ is the absolute velocity of the origin of the body reference, $\mathbf{A}^i$ is the transformation matrix, $\bar{\boldsymbol{\omega}}^i$ is the angular velocity vector defined in the body coordinate system, and $\bar{\mathbf{u}}^i$ is the local position of the arbitrary point. Differentiating the preceding equation with respect to time and using the relationships developed in this section, one can verify that the absolute acceleration of an arbitrary point on the rigid body can be written as

$$\ddot{\mathbf{r}}^i = \ddot{\mathbf{R}}^i + \boldsymbol{\omega}^i \times (\boldsymbol{\omega}^i \times \mathbf{u}^i) + \boldsymbol{\alpha}^i \times \mathbf{u}^i$$

where $\ddot{\mathbf{R}}^i$ is the absolute acceleration of the origin of the body reference, $\boldsymbol{\omega}^i$ and $\boldsymbol{\alpha}^i$ are, respectively, the angular velocity and angular acceleration vectors defined in the global coordinate system, and $\mathbf{u}^i$ is the vector

$$\mathbf{u}^i = \mathbf{A}^i \bar{\mathbf{u}}^i$$

The acceleration vector $\ddot{\mathbf{r}}^i$ can also be written in terms of vectors defined in the body coordinate system as

$$\ddot{\mathbf{r}}^i = \ddot{\mathbf{R}}^i + \mathbf{A}^i[\bar{\omega}^i \times (\bar{\omega}^i \times \bar{\mathbf{u}}^i)] + \mathbf{A}^i(\bar{\alpha}^i \times \bar{\mathbf{u}}^i)$$

where $\bar{\omega}^i$ and $\bar{\alpha}^i$ are, respectively, the angular velocity vector and angular acceleration vector defined in the body coordinate system.

Alternate forms of the rotation matrix are developed in the following three sections. These forms are, respectively, in terms of *Rodriguez parameters*, *Euler angles*, and the *direction cosines*.

2.6 RODRIGUEZ PARAMETERS

As pointed out earlier, only three independent variables are required to describe the orientation of the body reference. The transformation matrix developed in the preceding sections is expressed in terms of four parameters, that is, one more than the number of degrees of freedom. In this section, an alternative representation, which uses three parameters called *Rodriguez parameters*, is developed.

For convenience, we reproduce the transformation matrix of Eq. 9:

$$\mathbf{A} = \mathbf{I} + \tilde{\mathbf{v}} \sin\theta + 2(\tilde{\mathbf{v}})^2 \sin^2\frac{\theta}{2} \tag{2.76}$$

We now define the vector γ of Rodriguez parameters as

$$\gamma = \mathbf{v} \tan\frac{\theta}{2} \tag{2.77}$$

that is,

$$\gamma_1 = v_1 \tan\frac{\theta}{2}, \qquad \gamma_2 = v_2 \tan\frac{\theta}{2}, \qquad \gamma_3 = v_3 \tan\frac{\theta}{2} \tag{2.78}$$

where θ is the angle of rotation about the axis of rotation and $\mathbf{v}$ is a unit vector along the axis of rotation. Note that the Rodriguez parameter representation has the disadvantage of becoming infinite when the angle of rotation θ is equal to π.

Using the trigonometric identities

$$\sin\theta = 2\sin\frac{\theta}{2}\cos\frac{\theta}{2}, \qquad \sin\frac{\theta}{2} = \tan\frac{\theta}{2}\cos\frac{\theta}{2}$$

$$\sec^2\frac{\theta}{2} = \frac{1}{\cos^2(\theta/2)} = 1 + \tan^2\frac{\theta}{2}$$

one can write $\sin\theta$ as

$$\sin\theta = 2\sin\frac{\theta}{2}\cos\frac{\theta}{2} = 2\tan\frac{\theta}{2}\cos^2\frac{\theta}{2} = \frac{2\tan(\theta/2)}{\sec^2(\theta/2)} = \frac{2\tan(\theta/2)}{1+\tan^2(\theta/2)} \tag{2.79}$$

Since $\mathbf{v}$ is a unit vector, one has

$$\gamma^{\mathrm{T}}\gamma = \tan^2\frac{\theta}{2} = (\gamma)^2 \tag{2.80}$$

Therefore, Eq. 79 leads to

$$\sin\theta = \frac{2\tan(\theta/2)}{1+(\gamma)^2} \tag{2.81}$$

Similarly

$$\sin^2\frac{\theta}{2} = \frac{\tan^2(\theta/2)}{1+\tan^2(\theta/2)} = \frac{\tan^2(\theta/2)}{1+(\gamma)^2} \tag{2.82}$$

Substituting Eqs. 81 and 82 into Eq. 76 yields

$$\mathbf{A} = \mathbf{I} + \frac{2}{1+(\gamma)^2}\left(\tilde{\mathbf{v}}\tan\frac{\theta}{2} + (\tilde{\mathbf{v}})^2\tan^2\frac{\theta}{2}\right) \tag{2.83}$$

which, on using Eq. 77, yields

$$\mathbf{A} = \mathbf{I} + \frac{2}{1+(\gamma)^2}(\tilde{\gamma} + (\tilde{\gamma})^2) \tag{2.84}$$

where $\tilde{\gamma}$ is the skew symmetric matrix

$$\tilde{\gamma} = \begin{bmatrix} 0 & -\gamma_3 & \gamma_2 \\ \gamma_3 & 0 & -\gamma_1 \\ -\gamma_2 & \gamma_1 & 0 \end{bmatrix} \tag{2.85}$$

In a more explicit form, the transformation matrix $\mathbf{A}$ can be written in terms of the three parameters γ_1, γ_2, and γ_3 as

$$\mathbf{A} = \frac{1}{1+(\gamma)^2} \times \begin{bmatrix} 1+(\gamma_1)^2-(\gamma_2)^2-(\gamma_3)^2 & 2(\gamma_1\gamma_2-\gamma_3) & 2(\gamma_1\gamma_3+\gamma_2) \\ 2(\gamma_1\gamma_2+\gamma_3) & 1-(\gamma_1)^2+(\gamma_2)^2-(\gamma_3)^2 & 2(\gamma_2\gamma_3-\gamma_1) \\ 2(\gamma_1\gamma_3-\gamma_2) & 2(\gamma_2\gamma_3+\gamma_1) & 1-(\gamma_1)^2-(\gamma_2)^2+(\gamma_3)^2 \end{bmatrix} \tag{2.86}$$

Relationships with Euler Parameters Using the definition of Euler parameters of Eq. 11, one can show that Rodriguez parameters can be written in terms of Euler parameters as follows:

$$\left.\begin{aligned} \gamma_1 &= v_1\frac{\sin(\theta/2)}{\cos(\theta/2)} = \frac{\theta_1}{\theta_0} \\ \gamma_2 &= \frac{\theta_2}{\theta_0}, \quad \gamma_3 = \frac{\theta_3}{\theta_0} \end{aligned}\right\} \tag{2.87}$$

That is,

$$\gamma_i = \frac{\theta_i}{\theta_0}, \quad i = 1, 2, 3 \tag{2.88}$$

From Eq. 82 it is obvious that

$$\cos^2\frac{\theta}{2} = \frac{1}{1 + (\gamma)^2}$$

that is,

$$\theta_0 = \frac{1}{\sqrt{1 + (\gamma)^2}}$$

Using this equation, one can then write the remaining Euler parameters in terms of Rodriguez parameters as follows

$$\left.\begin{aligned} \theta_1 &= \frac{\gamma_1}{\sqrt{1 + (\gamma)^2}} \\ \theta_2 &= \frac{\gamma_2}{\sqrt{1 + (\gamma)^2}} \\ \theta_3 &= \frac{\gamma_3}{\sqrt{1 + (\gamma)^2}} \end{aligned}\right\} \tag{2.89}$$

that is,

$$\theta_i = \frac{\gamma_i}{\sqrt{1 + (\gamma)^2}}, \quad i = 1, 2, 3$$

It follows that the relation between the time derivatives of Rodriguez parameters and Euler parameters are given by

$$\dot{\gamma}_i = \frac{\dot{\theta}_i\theta_0 - \dot{\theta}_0\theta_i}{(\theta_0)^2}, \quad i = 1, 2, 3$$

which can be written in a matrix form as

$$\dot{\boldsymbol{\gamma}} = \mathbf{C}\dot{\boldsymbol{\theta}}$$

where $\boldsymbol{\theta} = [\theta_0\ \theta_1\ \theta_2\ \theta_3]^{\mathrm{T}}$ are the four Euler parameters and $\mathbf{C}$ is a 3×4 matrix defined as

$$\mathbf{C} = \frac{1}{(\theta_0)^2}\begin{bmatrix} -\theta_1 & \theta_0 & 0 & 0 \\ -\theta_2 & 0 & \theta_0 & 0 \\ -\theta_3 & 0 & 0 & \theta_0 \end{bmatrix}$$

The inverse relation is given by

$$\dot{\boldsymbol{\theta}} = \mathbf{D}\dot{\boldsymbol{\gamma}}$$

where $\mathbf{D}$ is a 4×3 matrix given by

$$\mathbf{D} = \frac{1}{\sqrt{1+(\gamma)^2}}\begin{bmatrix} 0 & 0 & 0 \\ 1 & 0 & 0 \\ 0 & 1 & 0 \\ 0 & 0 & 1 \end{bmatrix} - \frac{1}{[1+(\gamma)^2]^{3/2}}\begin{bmatrix} \gamma_1 & \gamma_2 & \gamma_3 \\ (\gamma_1)^2 & \gamma_1\gamma_2 & \gamma_1\gamma_3 \\ \gamma_2\gamma_1 & (\gamma_2)^2 & \gamma_2\gamma_3 \\ \gamma_3\gamma_1 & \gamma_3\gamma_2 & (\gamma_3)^2 \end{bmatrix}$$

Angular Velocity Vector It can be also verified that the transformation matrix expressed in terms of Rodriguez parameters can be written as the product of two semitransformation matrices $\mathbf{E}$ and $\bar{\mathbf{E}}$ as follows:

$$\mathbf{A} = \mathbf{E}\bar{\mathbf{E}}^{\mathrm{T}}$$

where the 3×4 matrices $\mathbf{E}$ and $\bar{\mathbf{E}}$ are defined as follows:

$$\mathbf{E} = \frac{1}{\sqrt{1+(\gamma)^2}}\begin{bmatrix} -\gamma_1 & 1 & -\gamma_3 & \gamma_2 \\ -\gamma_2 & \gamma_3 & 1 & -\gamma_1 \\ -\gamma_3 & -\gamma_2 & \gamma_1 & 1 \end{bmatrix}$$

$$\bar{\mathbf{E}} = \frac{1}{\sqrt{1+(\gamma)^2}}\begin{bmatrix} -\gamma_1 & 1 & \gamma_3 & -\gamma_2 \\ -\gamma_2 & -\gamma_3 & 1 & \gamma_1 \\ -\gamma_3 & \gamma_2 & -\gamma_1 & 1 \end{bmatrix}$$

Using these matrices, one can verify that the angular velocity vector $\bar{\omega}$ in the local coordinate system can be written in terms of Rodriguez parameters as follows:

$$\bar{\omega} = \bar{\mathbf{G}}\dot{\gamma} \tag{2.90}$$

where

$$\bar{\mathbf{G}} = 2\bar{\mathbf{E}}\mathbf{D}$$

By carrying out this matrix multiplication, it can be shown that the matrix $\bar{\mathbf{G}}$ is given by

$$\bar{\mathbf{G}} = \frac{2}{1+(\gamma)^2}\begin{bmatrix} 1 & \gamma_3 & -\gamma_2 \\ -\gamma_3 & 1 & \gamma_1 \\ \gamma_2 & -\gamma_1 & 1 \end{bmatrix}$$

The global angular velocity vector ω is also given by

$$\omega = \mathbf{G}\dot{\gamma} \tag{2.91}$$

where

$$\mathbf{G} = 2\mathbf{E}\mathbf{D}$$

By carrying out this matrix multiplication, one can verify that the matrix $\mathbf{G}$ is given by

$$\mathbf{G} = \frac{2}{1+(\gamma)^2}\begin{bmatrix} 1 & -\gamma_3 & \gamma_2 \\ \gamma_3 & 1 & -\gamma_1 \\ -\gamma_2 & \gamma_1 & 1 \end{bmatrix}$$

2.7 EULER ANGLES

One of the most common and widely used parameters in describing reference orientations are the three independent *Euler angles*. In this section, we define these angles and develop the transformation matrix in terms of them. Euler angles involve three successive rotations about three axes that are not orthogonal in general. Euler angles, however, are not unique; therefore, we follow the most widely used set given by Goldstein (1950). To this end we carry out the transformation between two coordinate systems by means of three successive rotations, called *Euler angles*, performed in a given sequence. For instance, we consider the coordinate systems $\mathbf{X}_1\mathbf{X}_2\mathbf{X}_3$ and $\xi_1\xi_2\xi_3$, which initially coincide. The sequence starts by rotating the system $\xi_1\xi_2\xi_3$ an angle ϕ about the $\mathbf{X}_3$ axis. The result of this rotation is shown in Fig. 11(a). Since ϕ is the angle of rotation in the plane $\mathbf{X}_1\mathbf{X}_2$, we have

$$\xi = \mathbf{D}_1\mathbf{x} \tag{2.92}$$

where $\mathbf{D}_1$ is the transformation matrix

$$\mathbf{D}_1 = \begin{bmatrix} \cos\phi & \sin\phi & 0 \\ -\sin\phi & \cos\phi & 0 \\ 0 & 0 & 1 \end{bmatrix}$$

Next we consider the coordinate system $\eta_1\eta_2\eta_3$, which coincides with the system $\xi_1\xi_2\xi_3$ and rotate this system an angle θ about the axis ξ_1, which at the current position is called the *line of nodes*. The result of this rotation is shown in Fig. 11(b). Since the rotation θ is in the plane $\xi_2\xi_3$, we have

$$\eta = \mathbf{D}_2\xi \tag{2.93}$$

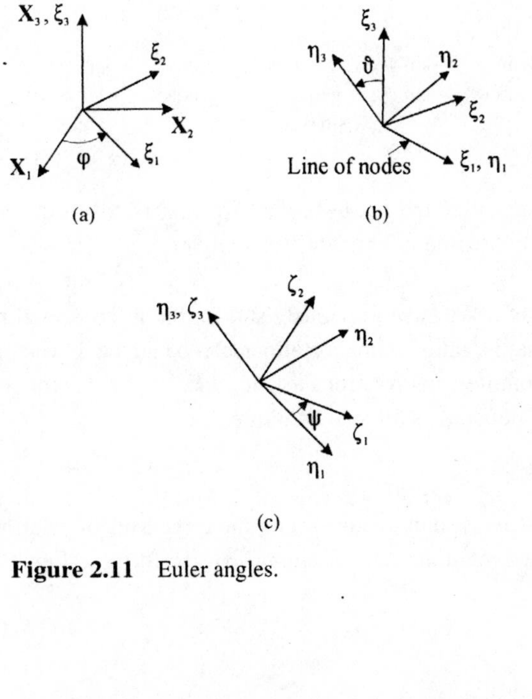

Figure 2.11 Euler angles.

where $\mathbf{D}_2$ is the transformation matrix defined as

$$\mathbf{D}_2 = \begin{bmatrix} 1 & 0 & 0 \\ 0 & \cos\theta & \sin\theta \\ 0 & -\sin\theta & \cos\theta \end{bmatrix}$$

Finally, we consider the coordinate system $\zeta_1\zeta_2\zeta_3$, which coincides with the coordinate system $\eta_1\eta_2\eta_3$. We rotate the coordinate system $\zeta_1\zeta_2\zeta_3$ an angle ψ about the η_3 axis as shown in Fig. 11(c). In this case, we can write

$$\boldsymbol{\zeta} = \mathbf{D}_3\boldsymbol{\eta} \tag{2.94}$$

where $\mathbf{D}_3$ is the transformation matrix

$$\mathbf{D}_3 = \begin{bmatrix} \cos\psi & \sin\psi & 0 \\ -\sin\psi & \cos\psi & 0 \\ 0 & 0 & 1 \end{bmatrix}$$

Using Eqs. 92–94, one can write the transformation between the initial coordinate system $\mathbf{X}_1\mathbf{X}_2\mathbf{X}_3$ and the final coordinate system $\zeta_1\zeta_2\zeta_3$ as follows:

$$\boldsymbol{\zeta} = \mathbf{D}_3\mathbf{D}_2\mathbf{D}_1\mathbf{x}$$

which can be written as

$$\boldsymbol{\zeta} = \mathbf{A}^{\mathrm{T}}\mathbf{x}$$

where $\mathbf{A}^{\mathrm{T}} = \mathbf{D}_3\mathbf{D}_2\mathbf{D}_1$. Also one can write

$$\mathbf{x} = \mathbf{A}\boldsymbol{\zeta} \tag{2.95}$$

where $\mathbf{A}$ is the transformation matrix

$$\mathbf{A} = \begin{bmatrix} \cos\psi\cos\phi - \cos\theta\sin\phi\sin\psi & -\sin\psi\cos\phi - \cos\theta\sin\phi\cos\psi & \sin\theta\sin\phi \\ \cos\psi\sin\phi + \cos\theta\cos\phi\sin\psi & -\sin\psi\sin\phi + \cos\theta\cos\phi\cos\psi & -\sin\theta\cos\phi \\ \sin\theta\sin\psi & \sin\theta\cos\psi & \cos\theta \end{bmatrix} \tag{2.96}$$

The three angles ϕ, θ, and ψ are called the *Euler angles*. The matrix $\mathbf{A}$ of Eq. 96 is then the transformation matrix expressed in terms of Euler angles.

Angular Velocity Vector We have previously shown that infinitesimal rotations are vector quantities; that is, infinitesimal rotations can be added as vectors. Therefore, if we consider the infinitesimal rotations $\delta\phi$, $\delta\theta$, and $\delta\psi$, the vector sum of these infinitesimal rotations, denoted as $\delta\boldsymbol{\theta}$, can be written as

$$\delta\boldsymbol{\theta} = \mathbf{v}_1\delta\phi + \mathbf{v}_2\delta\theta + \mathbf{v}_3\delta\psi$$

where $\mathbf{v}_1$, $\mathbf{v}_2$, and $\mathbf{v}_3$ are, respectively, unit vectors along the three axes of rotations about which the three successive rotations $\delta\phi$, $\delta\theta$, and $\delta\psi$ are performed (Fig. 12).

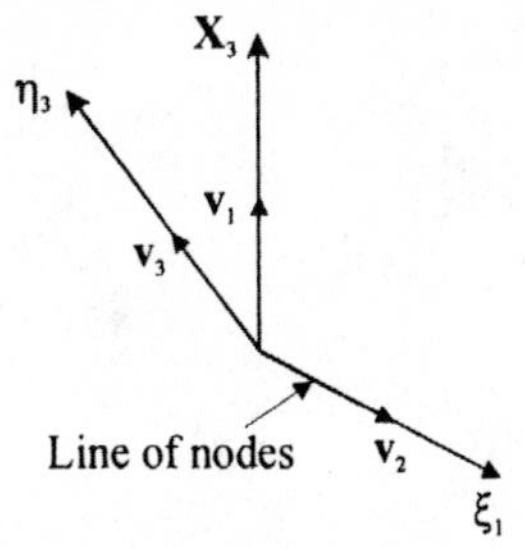

Figure 2.12 Axes of rotation.

The vectors $\mathbf{v}_1$, $\mathbf{v}_2$, and $\mathbf{v}_3$ in the $\mathbf{X}_1\mathbf{X}_2\mathbf{X}_3$ coordinate system are given by

$$\mathbf{v}_1 = [0 \quad 0 \quad 1]^{\mathrm{T}}$$
$$\mathbf{v}_2 = [\cos\phi \quad \sin\phi \quad 0]^{\mathrm{T}}$$
$$\mathbf{v}_3 = [\sin\theta\sin\phi \quad -\sin\theta\cos\phi \quad \cos\theta]^{\mathrm{T}}$$

Therefore, the vector $\delta\boldsymbol{\theta}$ in the $\mathbf{X}_1\mathbf{X}_2\mathbf{X}_3$ coordinate system can be written in matrix form as

$$\delta\boldsymbol{\theta} = \begin{bmatrix} 0 & \cos\phi & \sin\theta\sin\phi \\ 0 & \sin\phi & -\sin\theta\cos\phi \\ 1 & 0 & \cos\theta \end{bmatrix} \begin{bmatrix} \delta\phi \\ \delta\theta \\ \delta\psi \end{bmatrix}$$

By definition, the angular velocity vector $\boldsymbol{\omega}$ is

$$\boldsymbol{\omega} = \frac{d\boldsymbol{\theta}}{dt}$$

One can then write the angular velocity in terms of Euler angles as

$$\boldsymbol{\omega} = \mathbf{G}\dot{\boldsymbol{\nu}}$$

where $\boldsymbol{\nu} = (\phi, \theta, \psi)$ and $\mathbf{G}$ is the matrix

$$\mathbf{G} = \begin{bmatrix} 0 & \cos\phi & \sin\theta\sin\phi \\ 0 & \sin\phi & -\sin\theta\cos\phi \\ 1 & 0 & \cos\theta \end{bmatrix}$$

That is,

$$\omega_1 = \dot{\theta}\cos\phi + \dot{\psi}\sin\theta\sin\phi$$
$$\omega_2 = \dot{\theta}\sin\phi - \dot{\psi}\sin\theta\cos\phi$$
$$\omega_3 = \dot{\phi} + \dot{\psi}\cos\theta$$

In the $\zeta_1\zeta_2\zeta_3$ coordinate system the angular velocity vector is given by

$$\bar{\omega}_1 = \dot{\phi}\sin\theta\sin\psi + \dot{\theta}\cos\psi$$
$$\bar{\omega}_2 = \dot{\phi}\sin\theta\cos\psi - \dot{\theta}\sin\psi$$
$$\bar{\omega}_3 = \dot{\phi}\cos\theta + \dot{\psi}$$

which can be written in a matrix form as

$$\bar{\omega} = \bar{\mathbf{G}}\dot{\nu}$$

where $\bar{\mathbf{G}}$ is the matrix

$$\bar{\mathbf{G}} = \begin{bmatrix} \sin\theta\sin\psi & \cos\psi & 0 \\ \sin\theta\cos\psi & -\sin\psi & 0 \\ \cos\theta & 0 & 1 \end{bmatrix}$$

Relationship with Euler Parameters The relationship between Euler angles and Euler parameters is given by

$$\theta_0 = \cos\frac{\theta}{2}\cos\frac{\psi+\phi}{2}, \qquad \theta_1 = \sin\frac{\theta}{2}\cos\frac{\phi-\psi}{2}$$

$$\theta_2 = \sin\frac{\theta}{2}\sin\frac{\phi-\psi}{2}, \qquad \theta_3 = \cos\frac{\theta}{2}\sin\frac{\phi+\psi}{2}$$

Euler angles can be also written in terms of Euler parameters as follows:

$$\theta = \cos^{-1}\left[2\left[(\theta_0)^2 + (\theta_3)^2\right] - 1\right]$$

$$\phi = \cos^{-1}\left\{\frac{-2(\theta_2\theta_3 - \theta_0\theta_1)}{\sin\theta}\right\}$$

$$\psi = \cos^{-1}\left\{\frac{2(\theta_2\theta_3 + \theta_0\theta_1)}{\sin\theta}\right\}$$

The rotations ϕ, θ, and ψ were chosen by Euler to study the motion of the *gyroscope*. The term *gyroscope* refers to any rotating rigid body whose axis of rotation changes its orientation. The gyroscope is shown in Fig. 13 where the circular disk spins about its axis of rotational symmetry $O\zeta$. This axis is mounted in a ring called

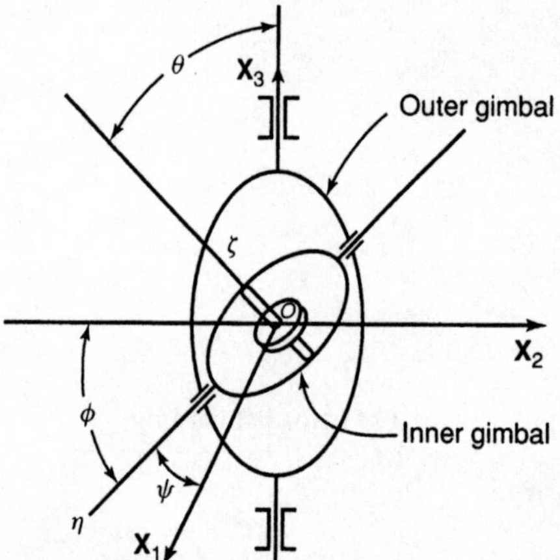

Figure 2.13 Gyroscope.

the *inner gimbal*, and the rotation of the disk about its axis of symmetry $O\zeta$ relative to the inner gimbal is allowed. The inner gimbal rotates freely about the axis $O\eta$ in its own plane. The axis $O\eta$ is perpendicular to the axis of the disk and mounted on a second gimbal, called the *outer gimbal*, which is free to rotate about the axis OX_3 in its own plane. The axis $\mathbf{X}_3$ is fixed in the inertial coordinate system $\mathbf{X}_1\mathbf{X}_2\mathbf{X}_3$. The disk whose center of gravity remains stationary may then attain any arbitrary position by the following three successive rotations: a rotation ϕ of the outer gimbal about the axis $\mathbf{X}_3$, a rotation θ of the inner gimbal about the axis $O\eta$, and a rotation ψ of the disk about its own axis $O\zeta$. These three Euler angles are called the *precession*, the *nutation*, and the *spin*, and the type of mounting used in the gyroscope is called a *cardan suspension*.

Example 2.10 The orientation of a rigid body is defined by the four Euler parameters

$$\theta_0 = 0.9239, \quad \theta_1 = \theta_2 = \theta_3 = 0.2209$$

At the given configuration, the body has an instantaneous absolute angular velocity defined by the vector

$$\boldsymbol{\omega} = [\,120.72 \quad 75.87 \quad -46.59\,]^{\mathrm{T}} \text{ rad/sec}$$

Find the time derivatives of Euler angles.

Solution Euler angles are

$$\theta = \cos^{-1}\{2[(\theta_0)^2 + (\theta_3)^2] - 1\} = 36.41^\circ$$

$$\phi = \cos^{-1}\left\{\frac{-2(\theta_2\theta_3 - \theta_0\theta_1)}{\sin\theta}\right\} = \cos^{-1}(0.5232) = 58.4496^\circ$$

$$\psi = \cos^{-1}\left\{\frac{2(\theta_2\theta_3 + \theta_0\theta_1)}{\sin\theta}\right\} = \cos^{-1}(0.8521) = -31.5604^\circ$$

The negative sign of ψ was selected to ensure that all the elements of the transformation matrix evaluated using Euler angles are the same as the elements of the transformation matrix evaluated using Euler parameters. Recall that

$$\boldsymbol{\omega} = \mathbf{G}\dot{\boldsymbol{\gamma}}$$

where

$$\mathbf{G} = \begin{bmatrix} 0 & \cos\phi & \sin\theta\sin\phi \\ 0 & \sin\phi & -\sin\theta\cos\phi \\ 1 & 0 & \cos\theta \end{bmatrix}$$

It follows that

$$\dot{\boldsymbol{\gamma}} = \mathbf{G}^{-1}\boldsymbol{\omega}$$

where

$$\mathbf{G}^{-1} = \frac{1}{\sin\theta}\begin{bmatrix} -\sin\phi\cos\theta & \cos\phi\cos\theta & \sin\theta \\ \sin\theta\cos\phi & \sin\theta\sin\phi & 0 \\ \sin\phi & -\cos\phi & 0 \end{bmatrix}$$

Note the singularity that occurs when θ is equal to zero or π. Using the values of Euler angles, one has

$$\mathbf{G}^{-1} = \begin{bmatrix} -1.1554 & 0.7093 & 1 \\ 0.5232 & 0.8522 & 0 \\ 1.4356 & -0.8814 & 0 \end{bmatrix}$$

The time derivatives of Euler angles can then be evaluated as

$$\begin{bmatrix} \dot{\phi} \\ \dot{\theta} \\ \dot{\psi} \end{bmatrix} = \mathbf{G}^{-1}\omega = \begin{bmatrix} -1.1554 & 0.7093 & 1 \\ 0.5232 & 0.8522 & 0 \\ 1.4356 & -0.8814 & 0 \end{bmatrix} \begin{bmatrix} 120.72 \\ 75.87 \\ -46.59 \end{bmatrix}$$

$$= \begin{bmatrix} -132.2553 \\ 127.8171 \\ 106.4338 \end{bmatrix}$$

2.8 DIRECTION COSINES

Even though the *direction cosines* are rarely used in describing the three dimensional rotations in multibody system dynamics, we discuss this method in this section for the completeness of our presentation. To this end, we consider the two coordinate systems $\mathbf{X}_1\mathbf{X}_2\mathbf{X}_3$ and $\mathbf{X}_1^i\mathbf{X}_2^i\mathbf{X}_3^i$ shown in Fig. 14. Let $\mathbf{i}_1^i$, $\mathbf{i}_2^i$, and $\mathbf{i}_3^i$ be unit vectors along the $\mathbf{X}_1^i$, $\mathbf{X}_2^i$, and $\mathbf{X}_3^i$ axes, respectively, and $\mathbf{i}_1$, $\mathbf{i}_2$, and $\mathbf{i}_3$ be unit vectors along the $\mathbf{X}_1$, $\mathbf{X}_2$, and $\mathbf{X}_3$ axes, respectively. Let β_1 be the angle between $\mathbf{X}_1^i$ and $\mathbf{X}_1$, β_2 be the angle between $\mathbf{X}_1^i$ and $\mathbf{X}_2$, and β_3 be the angle between $\mathbf{X}_1^i$ and $\mathbf{X}_3$. Then the components of the unit vector $\mathbf{i}_1^i$ along the $\mathbf{X}_1$, $\mathbf{X}_2$, and $\mathbf{X}_3$ axes are given by

$$\begin{aligned} \alpha_{11} &= \cos\beta_1 = \mathbf{i}_1^i \cdot \mathbf{i}_1 \\ \alpha_{12} &= \cos\beta_2 = \mathbf{i}_1^i \cdot \mathbf{i}_2 \\ \alpha_{13} &= \cos\beta_3 = \mathbf{i}_1^i \cdot \mathbf{i}_3 \end{aligned}$$

where α_{11}, α_{12}, and α_{13} are the direction cosines of the $\mathbf{X}_1^i$ axis with respect to $\mathbf{X}_1$, $\mathbf{X}_2$, and $\mathbf{X}_3$, respectively. In a similar manner, we denote the direction cosines of the $\mathbf{X}_2^i$ axis with respect to the $\mathbf{X}_1$, $\mathbf{X}_2$, and $\mathbf{X}_3$ axes by α_{21}, α_{22}, and α_{23}, respectively, and the direction cosines of the $\mathbf{X}_3^i$ axis by α_{31}, α_{32}, and α_{33}. One then can write these

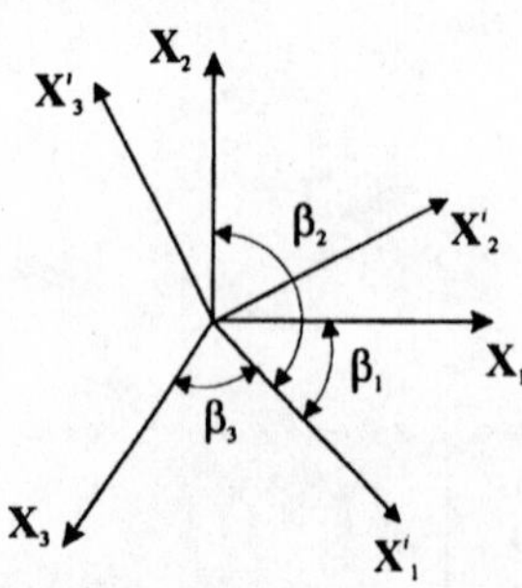

Figure 2.14 Direction cosines.

direction cosines using the dot product between unit vectors as

$$\alpha_{21} = \mathbf{i}_2^i \cdot \mathbf{i}_1, \quad \alpha_{22} = \mathbf{i}_2^i \cdot \mathbf{i}_2, \quad \alpha_{23} = \mathbf{i}_2^i \cdot \mathbf{i}_3$$

and

$$\alpha_{31} = \mathbf{i}_3^i \cdot \mathbf{i}_1, \quad \alpha_{32} = \mathbf{i}_3^i \cdot \mathbf{i}_2, \quad \alpha_{33} = \mathbf{i}_3^i \cdot \mathbf{i}_3$$

which can also be written as

$$\alpha_{jk} = \mathbf{i}_j^i \cdot \mathbf{i}_k, \quad j, k = 1, 2, 3$$

Using these relations, one can express the unit vectors $\mathbf{i}_1^i$, $\mathbf{i}_2^i$, and $\mathbf{i}_3^i$, in terms of the unit vectors $\mathbf{i}_1$, $\mathbf{i}_2$, and $\mathbf{i}_3$ as

$$\left.\begin{aligned} \mathbf{i}_1^i &= \alpha_{11}\mathbf{i}_1 + \alpha_{12}\mathbf{i}_2 + \alpha_{13}\mathbf{i}_3 \\ \mathbf{i}_2^i &= \alpha_{21}\mathbf{i}_1 + \alpha_{22}\mathbf{i}_2 + \alpha_{23}\mathbf{i}_3 \\ \mathbf{i}_3^i &= \alpha_{31}\mathbf{i}_1 + \alpha_{32}\mathbf{i}_2 + \alpha_{33}\mathbf{i}_3 \end{aligned}\right\} \tag{2.97}$$

which can be written in a more abbreviated form using the *summation convention* as

$$\mathbf{i}_k^i = \alpha_{kl}\mathbf{i}_l, \quad k, l = 1, 2, 3$$

In a similar manner, the unit vectors $\mathbf{i}_1$, $\mathbf{i}_2$, and $\mathbf{i}_3$ can be written in terms of the unit vectors $\mathbf{i}_1^i$, $\mathbf{i}_2^i$, and $\mathbf{i}_3^i$ as follows:

$$\begin{aligned} \mathbf{i}_1 &= \alpha_{11}\mathbf{i}_1^i + \alpha_{21}\mathbf{i}_2^i + \alpha_{31}\mathbf{i}_3^i \\ \mathbf{i}_2 &= \alpha_{12}\mathbf{i}_1^i + \alpha_{22}\mathbf{i}_2^i + \alpha_{32}\mathbf{i}_3^i \\ \mathbf{i}_3 &= \alpha_{13}\mathbf{i}_1^i + \alpha_{23}\mathbf{i}_2^i + \alpha_{33}\mathbf{i}_3^i \end{aligned}$$

which can be written in an abbreviated form using the *summation convention* as

$$\mathbf{i}_k = \alpha_{lk}\mathbf{i}_l^i, \quad k, l = 1, 2, 3$$

We now consider the three-dimensional vector $\mathbf{x}$ whose components in the coordinate systems $\mathbf{X}_1\mathbf{X}_2\mathbf{X}_3$ and $\mathbf{X}_1^i\mathbf{X}_2^i\mathbf{X}_3^i$ are denoted, respectively, as x_1, x_2, and x_3 and $\bar{x}_1$, $\bar{x}_2$, and $\bar{x}_3$. That is,

$$\begin{aligned} \mathbf{r} &= x_1\mathbf{i}_1 + x_2\mathbf{i}_2 + x_3\mathbf{i}_3 \\ &= \bar{x}_1\mathbf{i}_1^i + \bar{x}_2\mathbf{i}_2^i + \bar{x}_3\mathbf{i}_3^i \end{aligned}$$

Since the unit vectors $\mathbf{i}_1$, $\mathbf{i}_2$, $\mathbf{i}_3$ and $\mathbf{i}_1^i$, $\mathbf{i}_2^i$, and $\mathbf{i}_3^i$ satisfy the relations

$$\left.\begin{aligned} \mathbf{i}_k \cdot \mathbf{i}_l = \mathbf{i}_k^i \cdot \mathbf{i}_l^i = 1 \quad \text{for } k = l \\ \mathbf{i}_k \cdot \mathbf{i}_l = \mathbf{i}_k^i \cdot \mathbf{i}_l^i = 0 \quad \text{for } k \neq l \end{aligned}\right\} \tag{2.98}$$

one concludes that

$$\begin{aligned} \bar{x}_1 &= \mathbf{r} \cdot \mathbf{i}_1^i = \alpha_{11}x_1 + \alpha_{12}x_2 + \alpha_{13}x_3 \\ \bar{x}_2 &= \mathbf{r} \cdot \mathbf{i}_2^i = \alpha_{21}x_1 + \alpha_{22}x_2 + \alpha_{23}x_3 \\ \bar{x}_3 &= \mathbf{r} \cdot \mathbf{i}_3^i = \alpha_{31}x_1 + \alpha_{32}x_2 + \alpha_{33}x_3 \end{aligned}$$

that is,

$$\bar{\mathbf{x}} = \mathbf{A}\mathbf{x} \tag{2.99}$$

where

$$\bar{\mathbf{x}} = [\bar{x}_1 \quad \bar{x}_2 \quad \bar{x}_3]^{\mathrm{T}}, \qquad \mathbf{x} = [x_1 \quad x_2 \quad x_3]^{\mathrm{T}}$$

and $\mathbf{A}$ is the transformation matrix given by

$$\mathbf{A} = \begin{bmatrix} \alpha_{11} & \alpha_{12} & \alpha_{13} \\ \alpha_{21} & \alpha_{22} & \alpha_{23} \\ \alpha_{31} & \alpha_{32} & \alpha_{33} \end{bmatrix}$$

In a similar manner, one can show that

$$\mathbf{x} = \mathbf{A}^{\mathrm{T}}\bar{\mathbf{x}} \tag{2.100}$$

in which the transformation matrix $\mathbf{A}$ between the two coordinate systems is expressed in terms of the direction cosines α_{ij}, $(i, j = 1, 2, 3)$. Since only three variables are required in order to describe the orientation of a rigid frame in space, the nine quantities α_{ij} are not independent. In fact, those quantities satisfy six equations that are the result of the orthonormality of the vectors $\mathbf{i}_1$, $\mathbf{i}_2$, and $\mathbf{i}_3$ (or $\mathbf{i}_1^i$, $\mathbf{i}_2^i$, and $\mathbf{i}_3^i$) given by Eq. 98. These equations can be obtained by substituting Eq. 97 into Eq. 98, resulting in

$$\alpha_{1k}\alpha_{1l} + \alpha_{2k}\alpha_{2l} + \alpha_{3k}\alpha_{3l} = \delta_{kl}, \quad k, l = 1, 2, 3 \tag{2.101}$$

where δ_{kl} is the Kronecker delta, that is,

$$\delta_{kl} = \begin{cases} 1 & \text{if } k = l \\ 0 & \text{if } k \neq l \end{cases}$$

Equation 101 gives six relations that are satisfied by the direction cosines α_{ij}.

Even though we have previously proved the orthogonality of the transformation matrix, Eqs. 99 and 100 can be used to provide another proof. Substituting Eq. 99 into Eq. 100 yields

$$\mathbf{x} = \mathbf{A}^{\mathrm{T}}\mathbf{A}\mathbf{x}$$

that is,

$$\begin{aligned} x_l &= \sum_j \sum_k \alpha_{kl}\alpha_{kj}x_j \\ &= \sum_j \left(\sum_k \alpha_{kl}\alpha_{kj} \right) x_j \end{aligned}$$

To satisfy this equation, the coefficient of x_j must be zero for $j \neq l$ and must be one for $j = l$; that is,

$$\sum_k \alpha_{kl}\alpha_{kj} = \delta_{lj} \tag{2.102}$$

where δ_{lj} is the Kronecker delta. In a matrix form, Eq. 102 can be written as

$$\mathbf{A}^{\mathrm{T}}\mathbf{A} = \mathbf{I} \tag{2.103}$$

where $\mathbf{I}$ is a 3×3 identity matrix. In a similar manner, by substituting Eq. 100 into

Eq. 99 it can be shown that

$$\mathbf{A}\mathbf{A}^{\mathrm{T}} = \mathbf{I} \tag{2.104}$$

In view of Eqs. 103 and 104, it follows that

$$\mathbf{A}^{\mathrm{T}} = \mathbf{A}^{-1}$$

This completes the proof.

Example 2.11 At the initial configuration, the axes $\mathbf{X}_1^i$, $\mathbf{X}_2^i$, and $\mathbf{X}_3^i$ of the coordinate system of body i are defined in the global coordinate system $\mathbf{X}_1\mathbf{X}_2\mathbf{X}_3$ by the vectors $[0.5\ 0.0\ 0.5]^{\mathrm{T}}$, $[0.25\ 0.25\ -0.25]^{\mathrm{T}}$, and $[-2.0\ 4.0\ 2.0]^{\mathrm{T}}$, respectively. Starting with this initial configuration, the body rotates an angle $\theta_1 = 45°$ about its $\mathbf{X}_3^i$ axis followed by another rotation $\theta_2 = 60°$ about its $\mathbf{X}_1^i$ axis. Determine the transformation matrix that defines the orientation of the body coordinate system with respect to the global coordinate system.

Solution Before the rotations, the direction cosines that define the axes of the body coordinate system are

$$[\alpha_{11}\quad \alpha_{12}\quad \alpha_{13}]^{\mathrm{T}} = [0.7071\quad 0.0\quad 0.7071]^{\mathrm{T}}$$
$$[\alpha_{21}\quad \alpha_{22}\quad \alpha_{23}]^{\mathrm{T}} = [0.5774\quad 0.5774\quad -0.5774]^{\mathrm{T}}$$
$$[\alpha_{31}\quad \alpha_{32}\quad \alpha_{33}]^{\mathrm{T}} = [-0.4082\quad 0.8165\quad 0.4082]^{\mathrm{T}}$$

Therefore the transformation matrix that defines the body orientation before the rotations is

$$\mathbf{A}_0^i = \begin{bmatrix} \alpha_{11} & \alpha_{21} & \alpha_{31} \\ \alpha_{12} & \alpha_{22} & \alpha_{32} \\ \alpha_{13} & \alpha_{23} & \alpha_{33} \end{bmatrix} = \begin{bmatrix} 0.7071 & 0.5774 & -0.4082 \\ 0.0 & 0.5774 & 0.8165 \\ 0.7071 & -0.5774 & 0.4082 \end{bmatrix}$$

Since θ_1 is about the $\mathbf{X}_3^i$ axis, the transformation matrix due to this simple rotation is defined as

$$\mathbf{A}_1^i = \begin{bmatrix} \cos\theta_1 & -\sin\theta_1 & 0 \\ \sin\theta_1 & \cos\theta_1 & 0 \\ 0 & 0 & 1 \end{bmatrix} = \begin{bmatrix} 0.7071 & -0.7071 & 0 \\ 0.7071 & 0.7071 & 0 \\ 0 & 0 & 1 \end{bmatrix}$$

Since θ_2 is about the $\mathbf{X}_1^i$ axis, the transformation matrix due to this simple rotation is

$$\mathbf{A}_2^i = \begin{bmatrix} 1 & 0 & 0 \\ 0 & \cos\theta_2 & -\sin\theta_2 \\ 0 & \sin\theta_2 & \cos\theta_2 \end{bmatrix} = \begin{bmatrix} 1 & 0 & 0 \\ 0 & 0.5 & -0.8660 \\ 0 & 0.8660 & 0.5 \end{bmatrix}$$

The transformation matrix that defines the orientation of body i in the global coordinate system can then be written as

$$\mathbf{A}^i = \mathbf{A}_0^i\mathbf{A}_1^i\mathbf{A}_2^i = \begin{bmatrix} 0.9082 & -0.3994 & -0.1247 \\ 0.4082 & 0.9112 & 0.0547 \\ 0.0917 & -0.1007 & 0.9906 \end{bmatrix}$$

2.9 THE 4 × 4 TRANSFORMATION MATRIX

It was previously shown that the position of an arbitrary point P on the rigid body i can be written in the $\mathbf{X}_1\mathbf{X}_2\mathbf{X}_3$ coordinate system as

$$\mathbf{r}^i = \mathbf{R}^i + \mathbf{A}^i\bar{\mathbf{u}}^i \tag{2.105}$$

where $\mathbf{R}^i$ is the position of the origin of the rigid body reference in the $\mathbf{X}_1\mathbf{X}_2\mathbf{X}_3$ coordinate system as shown in Fig. 3, $\mathbf{A}^i$ is the rotation matrix, and $\bar{\mathbf{u}}^i$ is the position of the arbitrary point P in the $\mathbf{X}_1^i\mathbf{X}_2^i\mathbf{X}_3^i$ coordinate system. In Eq. 105, the vector $\mathbf{R}^i$ describes the rigid body translation. The second term on the right-hand side of Eq. 105, however, represents the contribution from the rotation of the body. An alternative for writing Eq. 105, is to use the 4 × 4 transformation. Equation 105 represents a transformation of a vector $\bar{\mathbf{u}}^i$ defined in the $\mathbf{X}_1^i\mathbf{X}_2^i\mathbf{X}_3^i$ coordinate system to another coordinate system $\mathbf{X}_1\mathbf{X}_2\mathbf{X}_3$. This transformation mapping can be written in another form as

$$\mathbf{r}_4^i = \mathbf{A}_4^i\bar{\mathbf{u}}_4^i \tag{2.106}$$

where $\mathbf{r}_4^i$ and $\bar{\mathbf{u}}_4^i$ are the four-dimensional vectors defined as

$$\mathbf{r}_4^i = \begin{bmatrix} r_1^i & r_2^i & r_3^i & 1 \end{bmatrix}^{\mathrm{T}} \tag{2.107}$$

$$\bar{\mathbf{u}}_4^i = \begin{bmatrix} \bar{u}_1^i & \bar{u}_2^i & \bar{u}_3^i & 1 \end{bmatrix}^{\mathrm{T}} \tag{2.108}$$

and $\mathbf{A}_4^i$ is the 4 × 4 transformation matrix defined by

$$\mathbf{A}_4^i = \left[\begin{array}{c|c} \mathbf{A}^i & \mathbf{R}^i \\ \hline \mathbf{0}_3^{\mathrm{T}} & 1 \end{array}\right] \tag{2.109}$$

where $\mathbf{0}_3$ is the null vector, that is,

$$\mathbf{0}_3 = [0 \quad 0 \quad 0]^{\mathrm{T}}$$

The 4 × 4 transformation matrix of Eq. 109 is sometimes called the *homogeneous transform*. The advantage of using this notation is that the translation and rotation of the body can be described by one matrix. It is important, however, to realize that the 4 × 4 transformation matrix $\mathbf{A}_4^i$ is not orthogonal, and, as a result, the inverse of $\mathbf{A}_4^i$ is not equal to its transpose. One can verify, however, that the inverse of the 4 × 4 transformation matrix $\mathbf{A}_4^i$ is given by

$$(\mathbf{A}_4^i)^{-1} = \left[\begin{array}{c|c} \mathbf{A}^{i\mathrm{T}} & -\mathbf{A}^{i\mathrm{T}}\mathbf{R}^i \\ \hline \mathbf{0}_3^{\mathrm{T}} & 1 \end{array}\right] \tag{2.110}$$

Therefore, in order to find the inverse of the 4 × 4 transformation $\mathbf{A}_4^i$, one has to find only the transpose of the rotation matrix $\mathbf{A}^i$.

Example 2.12 A rigid body has a coordinate system $\mathbf{X}_1^i\mathbf{X}_2^i\mathbf{X}_3^i$. The position vector of the origin O^i of this coordinate system is defined by the vector

$$\mathbf{R}^i = [1 \quad 1 \quad -5]^{\mathrm{T}}$$

The rigid body rotates an angle $\theta^i = 30°$ about the $\mathbf{X}_3$ axis. Define the new position of point P that has coordinates $\bar{\mathbf{u}}^i = [0\ 1\ 0]^{\mathrm{T}}$ as the result of this rotation. Assume that the axes of the coordinate systems $\mathbf{X}_1\mathbf{X}_2\mathbf{X}_3$ and $\mathbf{X}_1^i\mathbf{X}_2^i\mathbf{X}_3^i$ are initially parallel.

Solution In this case, the rotation matrix $\mathbf{A}^i$ is given by

$$\mathbf{A}^i = \begin{bmatrix} \cos\theta^i & -\sin\theta^i & 0 \\ \sin\theta^i & \cos\theta^i & 0 \\ 0 & 0 & 1 \end{bmatrix} = \begin{bmatrix} \cos 30 & -\sin 30 & 0 \\ \sin 30 & \cos 30 & 0 \\ 0 & 0 & 1 \end{bmatrix}$$

$$= \begin{bmatrix} 0.8660 & -0.5000 & 0 \\ 0.5000 & 0.8660 & 0 \\ 0 & 0 & 1 \end{bmatrix}$$

The 4 × 4 transformation matrix of Eq. 109 is then given by

$$\mathbf{A}_4^i = \left[\begin{array}{c|c} \mathbf{A}^i & \mathbf{R}^i \\ \hline \mathbf{0}_3^{\mathrm{T}} & 1 \end{array}\right] = \begin{bmatrix} 0.8660 & -0.5000 & 0 & 1 \\ 0.5000 & 0.8660 & 0 & 1 \\ 0 & 0 & 1 & -5 \\ 0 & 0 & 0 & 1 \end{bmatrix}$$

The four-dimensional vector $\bar{\mathbf{u}}_4^i$ of Eq. 108 is given by

$$\bar{\mathbf{u}}_4^i = [0 \quad 1 \quad 0 \quad 1]^{\mathrm{T}}$$

and the four-dimensional vector $\mathbf{r}_4^i$ of Eq. 107 can be obtained by using Eq. 106 as

$$\begin{bmatrix} r_1^i \\ r_2^i \\ r_3^i \\ 1 \end{bmatrix} = \begin{bmatrix} 0.8660 & -0.5000 & 0 & 1 \\ 0.5000 & 0.8660 & 0 & 1 \\ 0 & 0 & 1 & -5 \\ 0 & 0 & 0 & 1 \end{bmatrix} \begin{bmatrix} 0 \\ 1 \\ 0 \\ 1 \end{bmatrix} = \begin{bmatrix} 0.5000 \\ 1.8660 \\ -5.000 \\ 1 \end{bmatrix}$$

The inverse of the 4 × 4 transformation matrix $\mathbf{A}_4^i$ is given by

$$[\mathbf{A}_4^i]^{-1} = \begin{bmatrix} 0.8660 & 0.5000 & 0 & -1.3660 \\ -0.5000 & 0.8660 & 0 & -0.3660 \\ 0 & 0 & 1 & 5.000 \\ 0 & 0 & 0 & 1 \end{bmatrix}$$

Relative Motion Many multibody systems can be modeled as a set of bodies connected in a kinematic chain by sets of mechanical joints such as *revolute*, *prismatic*, and/or *cylindrical* joints. Examples of these systems are *robotic manipulators* such as the one shown in Fig. 15. The bodies in such systems are sometimes called *links*. The relative motion between two neighboring links connected by a revolute joint can be

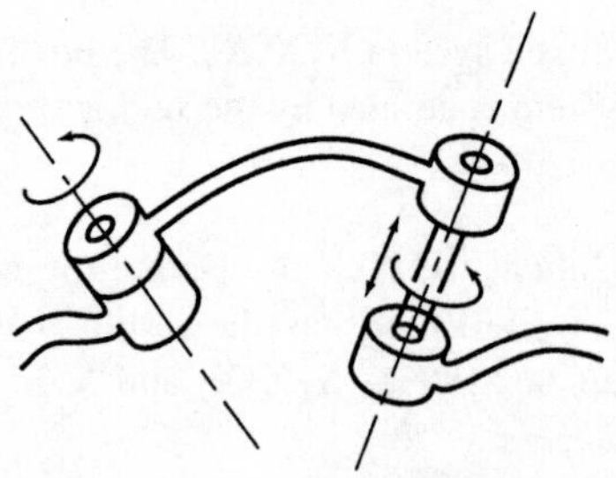

Figure 2.15 Robotic manipulators.

described by a rotation about the joint axis. In the case of a prismatic joint, the relative motion is described by a translation along the joint axis. A more general motion is the case of two neighboring links connected by a cylindrical joint. In this case, the relative motion is represented by a translation along and a rotation about the joint axis. It is obvious that the revolute and prismatic joints can be obtained as special cases of the cylindrical joint by fixing one of the cylindrical joint degrees of freedom. A relative translation and rotation between neighboring links can be represented in terms of two variables or two joint degrees of freedom. As a consequence, the 4×4 transformation matrix will be a function of only two parameters; one parameter represents the relative translation, while the other represents the relative rotation between the two links. To illustrate this we consider the two links i and $i-1$ shown in Fig. 16. The two links i and $i-1$ are assumed to be connected by a cylindrical joint; that is, link i translates and rotates with respect to link $i-1$ along the joint axis i. Let $\mathbf{X}_1^{i-1}\mathbf{X}_2^{i-1}\mathbf{X}_3^{i-1}$ be a coordinate system whose origin is rigidly attached to point O^{i-1} on link $i-1$, and let $\mathbf{X}_1^i\mathbf{X}_2^i\mathbf{X}_3^i$ be a coordinate system whose origin is rigidly attached to point O^i on link i. A unit vector along the axis of rotation i is denoted by the vector $\mathbf{v}$. This unit vector can be represented by a rigid line emanating from point O^{i-1}, and accordingly the

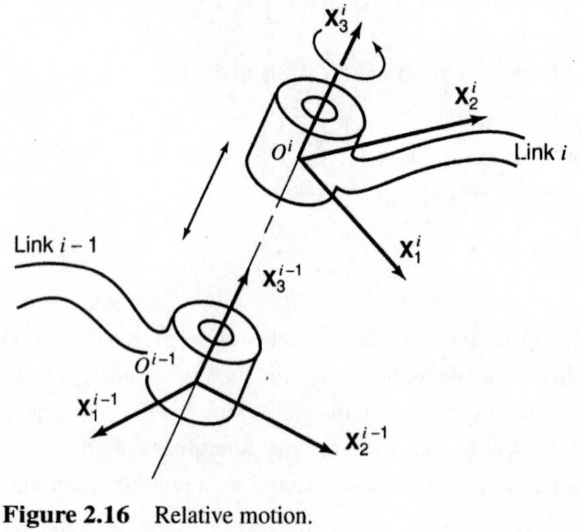

Figure 2.16 Relative motion.

components of this unit vector in the coordinate system $X_1^{i-1}X_2^{i-1}X_3^{i-1}$ are constant. The relative translation between the two frames $X_1^{i-1}X_2^{i-1}X_3^{i-1}$ and $X_1^iX_2^iX_3^i$ can be described by the vector $\mathbf{R}^{i,i-1}$. Since the unit vector $\mathbf{v}$ along the joint axis has constant components in the coordinate system $X_1^{i-1}X_2^{i-1}X_3^{i-1}$, in this coordinate system the vector $\mathbf{R}^{i,i-1}$ can be written in terms of one variable only as

$$\mathbf{R}^{i,i-1} = \mathbf{v}d^i \tag{2.111}$$

where d^i is the distance between the origins O^{i-1} and O^i. The variable d^i is a measure of the relative translational motion between link i and link $i-1$. The relative rotation between the two neighboring links can also be described in terms of one variable θ^i. Using Eq. 9, and assuming that the axes of the two coordinate systems $X_1^{i-1}X_2^{i-1}X_3^{i-1}$ and $X_1^iX_2^iX_3^i$ initially parallel, the rotation matrix $\mathbf{A}^{i,i-1}$ that defines the orientation of link i with respect to link $i-1$ can be written as

$$\mathbf{A}^{i,i-1} = \mathbf{I} + \tilde{\mathbf{v}}\sin\theta^i + 2(\tilde{\mathbf{v}})^2\sin^2\frac{\theta^i}{2} \tag{2.112}$$

where $\tilde{\mathbf{v}}$ is the skew symmetric matrix and $\mathbf{I}$ is the identity matrix. The rotation matrix of Eq. 112 depends on only one variable θ^i, since the components of the unit vector $\mathbf{v}$ are constant in the $X_1^{i-1}X_2^{i-1}X_3^{i-1}$ coordinate system. Using the identities of Eq. 15 or Eq. 16, the following interesting identities for the rotation matrix can be obtained:

$$(\mathbf{A}^{i,i-1})^{\mathrm{T}}\mathbf{A}_\theta^{i,i-1} = \tilde{\mathbf{v}} \tag{2.113}$$

$$(\mathbf{A}_\theta^{i,i-1})^{\mathrm{T}}\mathbf{A}_\theta^{i,i-1} = -(\tilde{\mathbf{v}})^2 \tag{2.114}$$

where

$$\begin{aligned}\mathbf{A}_\theta^{i,i-1} &= \frac{\partial}{\partial\theta^i}\mathbf{A}^{i,i-1} = \tilde{\mathbf{v}}\cos\theta^i + 2(\tilde{\mathbf{v}})^2\sin\frac{\theta^i}{2}\cos\frac{\theta^i}{2}\\ &= \tilde{\mathbf{v}}\cos\theta^i + (\tilde{\mathbf{v}})^2\sin\theta^i = \tilde{\mathbf{v}}\mathbf{A}^{i,i-1} = \mathbf{A}^{i,i-1}\tilde{\mathbf{v}}\end{aligned} \tag{2.115}$$

since the multiplication of the rotation matrix $\mathbf{A}^{i,i-1}$ with the skew symmetric matrix $\tilde{\mathbf{v}}$ is commutative. Furthermore, the angular velocity vector $\boldsymbol{\omega}^{i,i-1}$ of link i with respect to link $i-1$ is simply defined by the equation

$$\boldsymbol{\omega}^{i,i-1} = \dot{\theta}^i\mathbf{v} \tag{2.116}$$

The time derivative of the rotation matrix also takes the following simple form:

$$\begin{aligned}\dot{\mathbf{A}}^{i,i-1} &= \mathbf{A}_\theta^{i,i-1}\dot{\theta}^i = [\tilde{\mathbf{v}}\cos\theta^i + (\tilde{\mathbf{v}})^2\sin\theta^i]\dot{\theta}^i = \dot{\theta}^i\tilde{\mathbf{v}}\mathbf{A}^{i,i-1}\\ &= \mathbf{A}^{i,i-1}\tilde{\mathbf{v}}\dot{\theta}^i\end{aligned} \tag{2.117}$$

Using Eqs. 111 and 112, the position vector $\mathbf{r}^{i,i-1}$ of an arbitrary point P on link i in the $X_1^{i-1}X_2^{i-1}X_3^{i-1}$ coordinate system can be written as

$$\begin{aligned}\mathbf{r}^{i,i-1} &= \mathbf{R}^{i,i-1} + \mathbf{A}^{i,i-1}\bar{\mathbf{u}}^i\\ &= \mathbf{v}d^i + \mathbf{A}^{i,i-1}\bar{\mathbf{u}}^i\end{aligned} \tag{2.118}$$

where $\bar{\mathbf{u}}^i$ is the local position vector of point P defined in the $X_1^iX_2^iX_3^i$ coordinate system. One can also use the 4 × 4 transformation matrix to write Eq. 118 in the form of Eq. 106, where in this case the 4×4 transformation matrix of Eq. 109 is defined as

$$\mathbf{A}_4^{i,i-1} = \left[\begin{array}{c|c} \mathbf{A}^{i,i-1} & \mathbf{v}d^i \\ \hline \mathbf{0}_3^{\mathrm{T}} & 1 \end{array}\right] \tag{2.119}$$

Because $\mathbf{A}^{i,i-1}$ depends only on the joint variable θ^i, the 4×4 transformation matrix $\mathbf{A}_4^{i,i-1}$ is a function of the two joint variables θ^i and d^i.

The velocity vector of the arbitrary point P on link i with respect to the frame $\mathbf{X}_1^{i-1}\mathbf{X}_2^{i-1}\mathbf{X}_3^{i-1}$, which is rigidly attached to link $i-1$, can be obtained by differentiating Eq. 118 with respect to time, that is,

$$\dot{\mathbf{r}}^{i,i-1} = \mathbf{v}\dot{d}^i + \dot{\mathbf{A}}^{i,i-1}\bar{\mathbf{u}}^i = \mathbf{v}\dot{d}^i + \mathbf{A}_\theta^{i,i-1}\bar{\mathbf{u}}^i\dot{\theta}^i \tag{2.120}$$

Using the 4×4 matrix notation, Eq. 120 can be written as

$$\dot{\mathbf{r}}_4^{i,i-1} = \dot{\mathbf{A}}_4^{i,i-1}\bar{\mathbf{u}}_4^i \tag{2.121}$$

where $\dot{\mathbf{A}}_4^{i,i-1}$ is given by

$$\dot{\mathbf{A}}_4^{i,i-1} = \left[\begin{array}{c|c} \dot{\theta}^i\mathbf{A}_\theta^{i,i-1} & \mathbf{v}\dot{d}^i \\ \hline \mathbf{0}_3^{\mathrm{T}} & 0 \end{array}\right] = \left[\begin{array}{c|c} \dot{\theta}^i\tilde{\mathbf{v}}\mathbf{A}^{i,i-1} & \mathbf{v}\dot{d}^i \\ \hline \mathbf{0}_3^{\mathrm{T}} & 0 \end{array}\right] \tag{2.122}$$

and the vector $\dot{\mathbf{r}}_4^{i,i-1}$ is the time derivative of the vector $\mathbf{r}_4^{i,i-1}$, that is,

$$\dot{\mathbf{r}}_4^{i,i-1} = [\dot{r}_1^{i,i-1} \quad \dot{r}_2^{i,i-1} \quad \dot{r}_3^{i,i-1} \quad 0]^{\mathrm{T}}$$

The acceleration of point P can be obtained by differentiating Eq. 120 with respect to time, that is,

$$\ddot{\mathbf{r}}^{i,i-1} = \mathbf{v}\ddot{d}^i + \mathbf{A}_\theta^{i,i-1}\bar{\mathbf{u}}^i\ddot{\theta}^i + \dot{\mathbf{A}}_\theta^{i,i-1}\bar{\mathbf{u}}^i\dot{\theta}^i \tag{2.123}$$

in which $\dot{\mathbf{A}}_\theta^{i,i-1}$ is given by

$$\begin{aligned} \dot{\mathbf{A}}_\theta^{i,i-1} &= [-\tilde{\mathbf{v}}\sin\theta^i + (\tilde{\mathbf{v}})^2\cos\theta^i]\dot{\theta}^i \\ &= (\tilde{\mathbf{v}})^2\mathbf{A}^{i,i-1}\dot{\theta}^i = \mathbf{A}^{i,i-1}(\tilde{\mathbf{v}})^2\dot{\theta}^i \end{aligned} \tag{2.124}$$

Using Eqs. 115 and 124, we can write Eq. 123 as

$$\begin{aligned} \ddot{\mathbf{r}}^{i,i-1} &= \mathbf{v}\ddot{d}^i + \mathbf{A}^{i,i-1}[\tilde{\mathbf{v}}\bar{\mathbf{u}}^i\ddot{\theta}^i + (\tilde{\mathbf{v}})^2\bar{\mathbf{u}}^i(\dot{\theta}^i)^2] \\ &= \mathbf{v}\ddot{d}^i + \mathbf{A}^{i,i-1}[\bar{\alpha}^i \times \bar{\mathbf{u}}^i + \bar{\omega}^i \times (\bar{\omega}^i \times \bar{\mathbf{u}}^i)] \end{aligned} \tag{2.125}$$

where $\bar{\omega}^i$ and $\bar{\alpha}^i$ are defined as

$$\bar{\omega}^i = \mathbf{v}\dot{\theta}^i, \quad \bar{\alpha}^i = \mathbf{v}\ddot{\theta}^i$$

Equation 125 can be written by using the 4×4 matrix notation as

$$\ddot{\mathbf{r}}_4^{i,i-1} = \ddot{\mathbf{A}}_4^{i,i-1}\bar{\mathbf{u}}_4^i \tag{2.126}$$

where

$$\ddot{\mathbf{r}}_4^{i,i-1} = \frac{d}{dt}\dot{\mathbf{r}}_4^{i,i-1}$$

and $\ddot{\mathbf{A}}_4^{i,i-1}$ is the 4 × 4 matrix defined as

$$\ddot{\mathbf{A}}_4^{i,i-1} = \left[\begin{array}{c|c} \mathbf{A}^{i,i-1}[\tilde{\mathbf{v}}\ddot{\theta}^i + (\tilde{\mathbf{v}})^2(\dot{\theta}^i)^2] & \mathbf{v}\ddot{d}^i \\ \hline \mathbf{0}_3^{\mathrm{T}} & 0 \end{array}\right] \tag{2.127}$$

In developing these kinematic equations, the fact that the product of the rotation matrix $\mathbf{A}^{i,i-1}$ and the skew symmetric matrix $\tilde{\mathbf{v}}$ is commutative is used. In fact, for any rotation matrix $\mathbf{A}$ that is the result of a finite rotation θ, a more general expression than the one given by Eq. 115 can be obtained as (Shabana 1989)

$$\frac{\partial^n \mathbf{A}}{\partial \theta^n} = (\tilde{\mathbf{v}})^n \mathbf{A} = \mathbf{A}(\tilde{\mathbf{v}})^n \tag{2.128}$$

Equations 118, 120, and 123 are, respectively, the kinematic *position*, *velocity*, and *acceleration* equations for a general *relative motion*. The special case of a revolute joint can be obtained from the above kinematic equations by assuming d^i to be constant, while the case of the prismatic joint is obtained by assuming θ^i to be constant. Furthermore, many of the properties obtained in the previous sections for the planar rotation matrix can be obtained from the equations developed in this section by assuming that the unit vector $\mathbf{v}$ is along the $\mathbf{X}_3^{i-1}$ axis.

Denavit–Hartenberg Transformation Another 4 × 4 transformation matrix method for describing the relative translational and rotational motion is based on *Denavit–Hartenberg* notation (Denavit and Hartenberg 1955). This method of describing the motion is popular among researchers in the field of robotics and mechanisms. The 4 × 4 Denavit–Hartenberg transformation matrix is a function of four parameters: two constant parameters that depend on the geometry of the rigid links and two variable parameters that are sufficient for description of the relative motion. Figure 17 shows link $i-1$ in a kinematic chain. Joint $i-1$ is assumed to be at the proximal end of the link $i-1$, while joint i is located at the distal end of the link. For the joint axes $i-1$ and i, there exists a well-defined distance a^{i-1} between them.

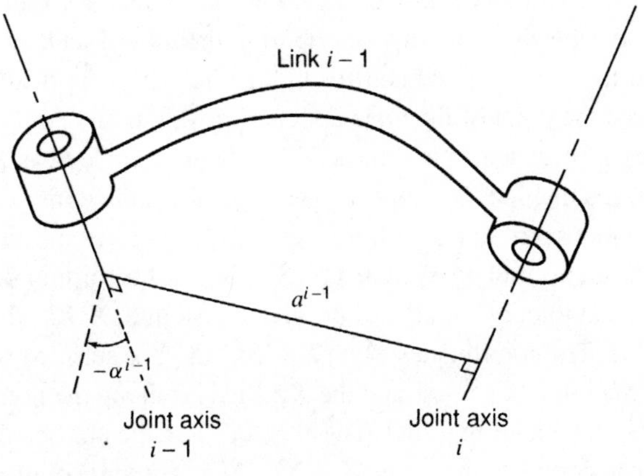

Figure 2.17 Link parameters.

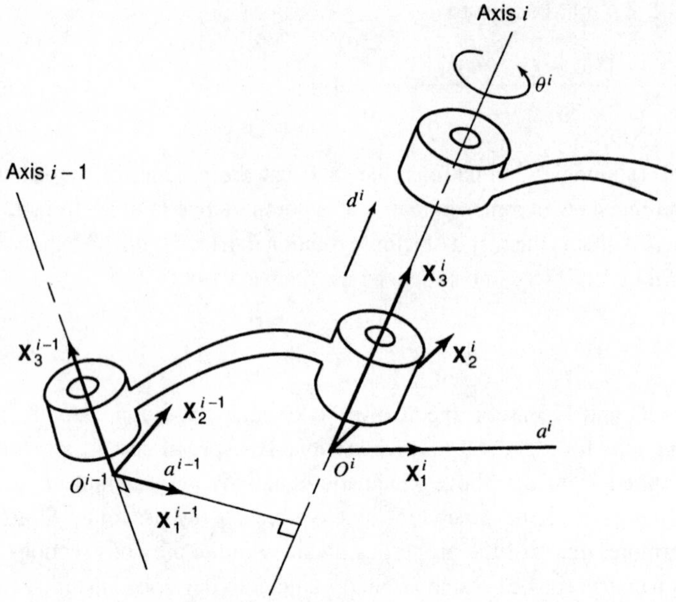

Figure 2.18 Joint degrees of freedom.

The distance a^{i-1} is measured along the line that is perpendicular to both axes $i-1$ and i. This perpendicular line is unique except in the special case in which the joint axes are parallel. The distance a^{i-1} is the first constant link parameter and is called the *link length*. The second constant link parameter is called the *link twist*. The link twist denoted as α^{i-1} is the angle between the axes in a plane perpendicular to a^{i-1}. This angle is measured from the joint axis $i-1$ to axis i in the right-hand sense about a^{i-1}. The two other variable parameters used in the 4×4 Denavit–Hartenberg transformation matrix are the *link offset* d^i and the *joint angle* θ^i. The link offset d^i describes the relative translation between link $i-1$ and link i, while the joint angle θ^i is used as a measure of the change in the orientation of link i with respect to link $i-1$. As shown in Fig. 18, link $i-1$ and link i are connected at joint i, and accordingly axis i is the common joint axis between the two neighboring links $i-1$ and i. If a^i is the line perpendicular to the joint axis defined for link i, then d^i is the distance along the joint axis i between the point of the intersection of a^{i-1} with the joint axis i and the point of intersection of a^i with the joint axis i. The joint angle θ^i is also defined to be the angle between the lines a^{i-1} and a^i measured about the joint axis i.

To describe the relative motion of link i with respect to link $i-1$, we introduce two joint coordinate systems: the coordiante system $\mathbf{X}_1^{i-1}\mathbf{X}_2^{i-1}\mathbf{X}_3^{i-1}$ whose origin O^{i-1} is rigidly attached to link $i-1$ at joint $i-1$ and the coordinate system $\mathbf{X}_1^i\mathbf{X}_2^i\mathbf{X}_3^i$ whose origin is O^i on the joint axis i. The coordinate system $\mathbf{X}_1^{i-1}\mathbf{X}_2^{i-1}\mathbf{X}_3^{i-1}$ is selected such that the $\mathbf{X}_3^{i-1}$ axis is along the joint axis $i-1$ and the $\mathbf{X}_1^{i-1}$ axis is along the normal a^{i-1} and in the direction from joint $i-1$ to joint i. The axis $\mathbf{X}_2^{i-1}$ can then be obtained by using the right-hand rule to complete the frame $\mathbf{X}_1^{i-1}\mathbf{X}_2^{i-1}\mathbf{X}_3^{i-1}$. Similar comments apply for the coordinate system $\mathbf{X}_1^i\mathbf{X}_2^i\mathbf{X}_3^i$ as shown in Fig. 18. It is, therefore, clear that the link length a^{i-1} is the distance between $\mathbf{X}_3^{i-1}$ and $\mathbf{X}_3^i$ measured along $\mathbf{X}_1^{i-1}$,

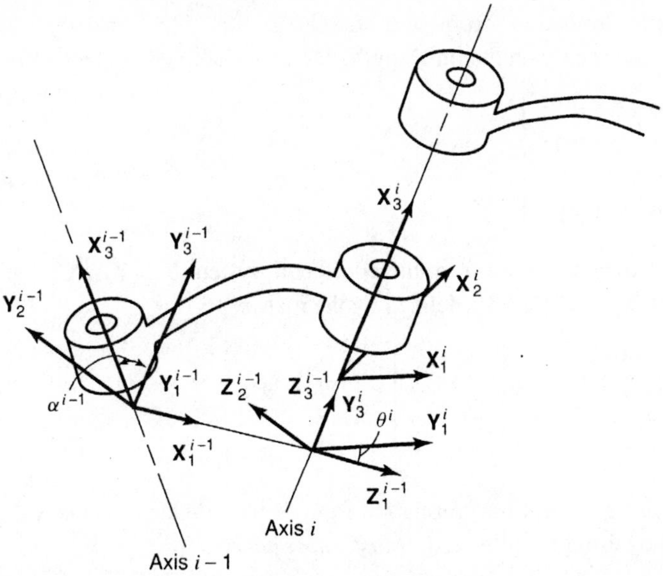

Figure 2.19 Intermediate coordinate systems.

the link twist α^{i-1} is the angle between $\mathbf{X}_3^{i-1}$ and $\mathbf{X}_3^i$ measured about the axis $\mathbf{X}_1^{i-1}$, the link offset d^i is the distance from $\mathbf{X}_1^{i-1}$ to $\mathbf{X}_1^i$ measured along the $\mathbf{X}_3^i$ axis, and the joint angle θ^i is the angle between the axis $\mathbf{X}_1^{i-1}$ and the axis $\mathbf{X}_1^i$ measured about $\mathbf{X}_3^i$.

To determine the position and orientation of the coordinate system $\mathbf{X}_1^i\mathbf{X}_2^i\mathbf{X}_3^i$ with respect to the coordinate system $\mathbf{X}_1^{i-1}\mathbf{X}_2^{i-1}\mathbf{X}_3^{i-1}$, three intermediate coordinate systems are introduced as shown in Fig. 19. A coordinate system $\mathbf{Y}_1^{i-1}\mathbf{Y}_2^{i-1}\mathbf{Y}_3^{i-1}$ is obtained by rotating the coordinate system $\mathbf{X}_1^{i-1}\mathbf{X}_2^{i-1}\mathbf{X}_3^{i-1}$ by an angle α^{i-1} about the $\mathbf{X}_1^{i-1}$ axis. The coordinate system $\mathbf{Z}_1^{i-1}\mathbf{Z}_2^{i-1}\mathbf{Z}_3^{i-1}$ is obtained from translating the coordinate system $\mathbf{Y}_1^{i-1}\mathbf{Y}_2^{i-1}\mathbf{Y}_3^{i-1}$ by a translation a^{i-1} along the $\mathbf{Y}_1^{i-1}$ axis. The coordinate system $\mathbf{Y}_1^i\mathbf{Y}_2^i\mathbf{Y}_3^i$ is obtained by rotating the coordinate system $\mathbf{Z}_1^{i-1}\mathbf{Z}_2^{i-1}\mathbf{Z}_3^{i-1}$ an angle θ^i about the $\mathbf{Z}_3^i$ axis. It is then clear that the coordinate system $\mathbf{X}_1^i\mathbf{X}_2^i\mathbf{X}_3^i$ can be obtained by translating the coordinate system $\mathbf{Y}_1^i\mathbf{Y}_2^i\mathbf{Y}_3^i$ a distance d^i along the $\mathbf{X}_3^i$ axis. The 4 × 4 transformation matrix from the frame $\mathbf{X}_1^i\mathbf{X}_2^i\mathbf{X}_3^i$ to the frame $\mathbf{Y}_1^i\mathbf{Y}_2^i\mathbf{Y}_3^i$ depends only on the link offset d^i and is given by

$$\mathbf{A}_1 = \begin{bmatrix} 1 & 0 & 0 & 0 \\ 0 & 1 & 0 & 0 \\ 0 & 0 & 1 & d^i \\ 0 & 0 & 0 & 1 \end{bmatrix}$$

The 4 × 4 transformation from the frame $\mathbf{Y}_1^i\mathbf{Y}_2^i\mathbf{Y}_3^i$ to $\mathbf{Z}_1^{i-1}\mathbf{Z}_2^{i-1}\mathbf{Z}_3^{i-1}$ depends only on the joint angle θ^i and is defined by the matrix

$$\mathbf{A}_2 = \begin{bmatrix} \cos\theta^i & -\sin\theta^i & 0 & 0 \\ \sin\theta^i & \cos\theta^i & 0 & 0 \\ 0 & 0 & 1 & 0 \\ 0 & 0 & 0 & 1 \end{bmatrix}$$

The 4×4 transformation from the frame $\mathbf{Z}_1^{i-1}\mathbf{Z}_2^{i-1}\mathbf{Z}_3^{i-1}$ to the frame $\mathbf{Y}_1^{i-1}\mathbf{Y}_2^{i-1}\mathbf{Y}_3^{i-1}$ depends only on the link length a^{i-1} and is defined by the matrix

$$\mathbf{A}_3 = \begin{bmatrix} 1 & 0 & 0 & a^{i-1} \\ 0 & 1 & 0 & 0 \\ 0 & 0 & 1 & 0 \\ 0 & 0 & 0 & 1 \end{bmatrix}$$

Finally, the 4×4 transformation from the coordinate system $\mathbf{Y}_1^{i-1}\mathbf{Y}_2^{i-1}\mathbf{Y}_3^{i-1}$ to the coordinate system $\mathbf{X}_1^{i-1}\mathbf{X}_2^{i-1}\mathbf{X}_3^{i-1}$ is defined by the matrix

$$\mathbf{A}_4 = \begin{bmatrix} 1 & 0 & 0 & 0 \\ 0 & \cos\alpha^{i-1} & -\sin\alpha^{i-1} & 0 \\ 0 & \sin\alpha^{i-1} & \cos\alpha^{i-1} & 0 \\ 0 & 0 & 0 & 1 \end{bmatrix}$$

One can then obtain the 4×4 transformation from the coordinate systems $\mathbf{X}_1^i\mathbf{X}_2^i\mathbf{X}_3^i$ to $\mathbf{X}_1^{i-1}\mathbf{X}_2^{i-1}\mathbf{X}_3^{i-1}$ by using the following *transform equation*:

$$\mathbf{A}^{i,i-1} = \mathbf{A}_4\mathbf{A}_3\mathbf{A}_2\mathbf{A}_1$$

where the 4×4 *Denavit–Hartenberg transformation matrix* $\mathbf{A}^{i,i-1}$ is defined as

$$\mathbf{A}^{i,i-1} = \begin{bmatrix} \cos\theta^i & -\sin\theta^i & 0 & a^{i-1} \\ \sin\theta^i\cos\alpha^{i-1} & \cos\theta^i\cos\alpha^{i-1} & -\sin\alpha^{i-1} & -d^i\sin\alpha^{i-1} \\ \sin\theta^i\sin\alpha^{i-1} & \cos\theta^i\sin\alpha^{i-1} & \cos\alpha^{i-1} & d^i\cos\alpha^{i-1} \\ 0 & 0 & 0 & 1 \end{bmatrix} \quad (2.129)$$

2.10 RELATIONSHIP BETWEEN DIFFERENT ORIENTATION COORDINATES

In this chapter, the rotation matrix for both planar and spatial motion was developed. Different forms for the 3×3 orthogonal rotation matrix were presented in terms of different orientational coordinates such as Euler parameters, Euler angles. Rodriguez parameters, and the direction cosines. In the forms that employ Euler angles and Rodriguez parameters, only three variables are required to identify the body orientation in space. The representation using Euler angles and Rodriguez parameters, however, has the disadvantage that *singularities* may occur at certain orientations of the body in space. Therefore, the use of Euler parameters or the *Rodriguez formula* of Eq. 9 has been recommended. Even though the four Euler parameters are not independent, by using these parameters one can avoid the problem of singularities of the rotation matrix. Furthermore, the rotation matrix is only a quadratic function of Euler parameters, and many interesting identities that can be used to simplify the dynamic formulation can be developed. Different forms of the rotation matrix, however, are equivalent. In fact, any set of coordinates such as Euler parameters, Euler angles, Rodriguez parameters, and so on can be extracted from a given rotation matrix by solving a set of transcendental equations. By equating two 3×3 rotation matrices, one obtains nine equations. Nonetheless, a set of independent equations equal to the number of the unknowns can be identified and solved for the orientational coordinates. For instance, given the following rotation matrix:

$$\mathbf{A} = \begin{bmatrix} a_{11} & a_{12} & a_{13} \\ a_{21} & a_{22} & a_{23} \\ a_{31} & a_{32} & a_{33} \end{bmatrix}$$

one can equate this matrix, for example, with Eq. 9, which is the form of the rotation matrix in terms of the angle of rotation θ and the components of the unit vector $\mathbf{v} = [v_1\ v_2\ v_3]^{\mathrm{T}}$ along the axis of rotation, to obtain

$$\theta = \cos^{-1}\left\{\frac{a_{11} + a_{22} + a_{33} - 1}{2}\right\}$$

$$\mathbf{v} = \begin{bmatrix} v_1 \\ v_2 \\ v_3 \end{bmatrix} = \frac{1}{2\sin\theta}\begin{bmatrix} a_{32} - a_{23} \\ a_{13} - a_{31} \\ a_{21} - a_{12} \end{bmatrix}$$

Clearly, when the angle of rotation is very small, the unit vector $\mathbf{v}$ along the axis of rotation is not well defined. More detailed discussions on this subject and also methods for extracting the orientational coordinates from a given rotation matrix can be found in the literature (Klumpp 1976; Paul 1981; Craig 1986).

In the preceding section we also discussed the 4×4 transformation matrix in which the translation and rotation of the body are described in one matrix. Even though the 4×4 transformation matrix is not an orthogonal matrix, it was shown that its inverse can be computed in a straightforward manner from the transpose of the orthogonal 3×3 rotation matrix. Methods for describing the relative motion between two bodies were also discussed. The angle and axis of rotation between the two bodies were first used to formulate the 4×4 transformation matrix. Kinematic relationships for the position, velocity, and acceleration were developed and more interesting identities were presented in order to simplify these kinematic equations. Finally, the 4×4 transformation matrix was formulated by use of Denavit–Hartenberg notations. The transformation was developed in terms of the four parameters: the link length, the link twist, the link offset, and the joint angle. The link length and link offset are constant and depend on the geometry of the rigid link. The link offset and joint angle are both variable in the case of a cylindrical joint. In the case of a revolute joint the link offset is constant, while in the case of a prismatic joint the joint angle is constant. The Denavit-Hartenberg 4×4 transformation matrix has been used extensively in applications related to robotic manipulators and spatial mechanisms (Paul 1976; Craig 1986).

The concept of the angular velocity vector was introduced in Section 4 by using many of the identities obtained using Euler parameters. This route was chosen because it gives an explicit definition of the angular velocity in terms of the angle of rotation and a unit vector along the axis of rotation. A simpler and more general definition of the skew symmetric matrix associated with the angular velocity vector can be obtained without using a particular set of orientation coordinates. This can be demonstrated by using the orthogonality property of the transformation matrix, which can be expressed as

$$\mathbf{A}^{\mathrm{T}}\mathbf{A} = \mathbf{I}$$

Differentiating this equation with respect to time, one obtains

$$\dot{\mathbf{A}}^{\mathrm{T}}\mathbf{A} + \mathbf{A}^{\mathrm{T}}\dot{\mathbf{A}} = \mathbf{0}$$

which implies that

$$\mathbf{A}^{\mathrm{T}}\dot{\mathbf{A}} = -(\mathbf{A}^{\mathrm{T}}\dot{\mathbf{A}})^{\mathrm{T}}$$

A matrix which is equal to the negative of its transpose is a skew symmetric matrix. One, therefore, concludes that

$$\mathbf{A}^{\mathrm{T}}\dot{\mathbf{A}} = \tilde{\bar{\omega}}$$

which leads to the known equation

$$\dot{\mathbf{A}} = \mathbf{A}\tilde{\bar{\omega}}$$

The identity for the skew symmetric matrix associated with the global angular velocity vector can be obtained by differentiating $\mathbf{A}\mathbf{A}^{\mathrm{T}} = \mathbf{I}$ instead of $\mathbf{A}^{\mathrm{T}}\mathbf{A} = \mathbf{I}$.

Problems

1. If $\mathbf{v} = [\,v_1\; v_2\; v_3\,]^{\mathrm{T}}$ is a unit vector, and $\tilde{\mathbf{v}}$ is a skew symmetric matrix such that

$$\tilde{\mathbf{v}} = \begin{bmatrix} 0 & -v_3 & v_2 \\ v_3 & 0 & -v_1 \\ -v_2 & v_1 & 0 \end{bmatrix}$$

show that $(\tilde{\mathbf{v}})^2$ is a symmetric matrix

2. Using mathematical induction, or by direct matrix multiplication, verify that

$$(\tilde{\mathbf{v}})^{2n-1} = (-1)^{n-1}\tilde{\mathbf{v}}$$
$$(\tilde{\mathbf{v}})^{2n} = (-1)^{n-1}(\tilde{\mathbf{v}})^2$$

where $\tilde{\mathbf{v}}$ is given in Problem 1.

3. Given a vector $\bar{\mathbf{r}} = [\,-2\;1\;5\,]^{\mathrm{T}}$ defined on a rigid body that rotates an angle $\theta = 30°$ about an axis of rotation along the vector $\mathbf{a} = [\,3\;-1\;7\,]^{\mathrm{T}}$, derive an expression for the transformation matrix $\mathbf{A}$ that defines the orientation of the body. Evaluate also the transformed vector $\mathbf{r}$.

4. Find the transformation matrix $\mathbf{A}$ that results from a rotation $\theta = 20°$ of a vector $\bar{\mathbf{r}} = [\,0\;2\;-6\,]^{\mathrm{T}}$ about another vector $\mathbf{a} = [\,-2\;1\;3\,]^{\mathrm{T}}$. Evaluate the transformed vector $\mathbf{r}$.

5. Use the principle of mathematical induction to prove the identity

$$\frac{\partial^n \mathbf{A}}{\partial \theta^n} = (\tilde{\mathbf{v}})^n \mathbf{A} = \mathbf{A}(\tilde{\mathbf{v}})^n$$

where $\mathbf{A}$ is the three-dimensional transformation matrix expressed in terms of the angle of rotation θ and a unit vector $\mathbf{v}$ along the axis of rotation.

6. The angular velocity vector defined with respect to the fixed frame is given by

$$\omega_1 = 2(\dot\theta_3\theta_2 - \dot\theta_2\theta_3 + \dot\theta_1\theta_0 - \dot\theta_0\theta_1)$$
$$\omega_2 = 2(\dot\theta_1\theta_3 - \dot\theta_0\theta_2 - \dot\theta_3\theta_1 + \dot\theta_2\theta_0)$$
$$\omega_3 = 2(\dot\theta_2\theta_1 + \dot\theta_3\theta_0 - \dot\theta_0\theta_3 - \dot\theta_1\theta_2)$$

Assuming that **v** is a unit vector along the axis of rotation, show that the component of the angular velocity vector along this axis is $\dot\theta$, where θ is the angle of rotation.

7. The vector $\bar{\mathbf{r}}$ has components defined in a rigid body coordinate system by the vector $\bar{\mathbf{r}} = [\,0\ 1\ 5\,]^{\mathrm{T}}$. The rigid body rotates with a constant angular velocity $\dot\theta = 20$ rad/sec about an axis of rotation defined by the vector $\mathbf{v} = [\,1\ 0\ 3\,]^{\mathrm{T}}$. Determine the angular velocity vector and the transformation matrix at $t = 0.1$ sec.

8. In the preceding problem, determine the global velocity vector at time $t = 0.2$ sec.

9. The vector $\bar{\mathbf{r}}$ has components defined in a rigid body coordinate system by the vector $\bar{\mathbf{r}} = [\,t\ t^2\ 0\,]^{\mathrm{T}}$, where t is time. The rigid body rotates with a constant angular velocity $\dot\theta = 10$ rad/sec about an axis defined by the unit vector $\mathbf{v} = [\,\frac{1}{\sqrt{3}}\ \frac{1}{\sqrt{3}}\ \frac{1}{\sqrt{3}}\,]^{\mathrm{T}}$. Determine the global velocity vector at time $t = 0.1$ sec.

10. Show that the angular velocity vector in the global and moving coordinate systems are defined, respectively, by

$$\omega = 2\mathbf{v} \times \dot{\mathbf{v}} \sin^2\frac{\theta}{2} + \dot{\mathbf{v}}\sin\theta + \mathbf{v}\dot\theta$$
$$\bar\omega = 2\dot{\mathbf{v}} \times \mathbf{v} \sin^2\frac{\theta}{2} + \dot{\mathbf{v}}\sin\theta + \mathbf{v}\dot\theta$$

where **v** is a unit vector along the instantaneous axis of rotation and θ is the angle of rotation.

11. Show that the rotation matrix **A** satisfies the following identity

$$\widetilde{(\mathbf{A}\mathbf{u})} = \mathbf{A}\tilde{\mathbf{u}}\mathbf{A}^{\mathrm{T}}$$

where **u** is an arbitrary three-dimensional vector.

12. The orientation of a rigid body is defined by the four Euler parameters

$$\theta_0 = 0.9239, \qquad \theta_1 = \theta_2 = \theta_3 = 0.2209$$

At the given configuration, the body has an instantaneous angular velocity defined in the global coordinate system by the vector

$$\omega = [\,120.72 \quad 75.87 \quad -46.59\,]^{\mathrm{T}} \text{ rad/sec}$$

Find the time derivatives of Euler parameters.

13. The orientation of a rigid body is defined by the four Euler parameters

$$\theta_0 = 0.8660, \qquad \theta_1 = \theta_2 = \theta_3 = 0.2887$$

At this given orientation, the body has an instantaneous absolute angular velocity defined as

$$\omega = [\,110.0 \quad 90.0 \quad 0.0\,]^{\mathrm{T}}$$

Determine the time derivatives of Euler parameters.

14. In the preceding problem, determine the time derivatives of Rodriguez parameters and the time derivatives of Euler angles.

15. The orientation of a rigid body is defined by the four Euler parameters

$$\theta_0 = 0.9239, \qquad \theta_1 = \theta_2 = \theta_3 = 0.2209$$

At the given configuration, the body has an instantaneous absolute angular velocity defined by the vector

$$\omega = [\,120.72 \quad 75.87 \quad -46.59\,]^{\mathrm{T}} \text{ rad/sec}$$

Find the time derivatives of Rodriguez parameters and the time derivatives of Euler angles.

16. Discuss the singularity associated with the transformation matrix when Euler angles are used.

17. Discuss the singularity problem associated with the use of Rodriguez parameters.

18. In Section 7, the transformation matrix in terms of Euler angles was derived using the multiframe method. Show that the single-frame method and the Rodriguez formula can also be used to obtain the same transformation matrix.

19. At the initial configuration, the axes $\mathbf{X}_1^i$, $\mathbf{X}_2^i$, and $\mathbf{X}_3^i$ of the coordinate system of body i are defined in the coordinate system $\mathbf{X}_1\mathbf{X}_2\mathbf{X}_3$ by the vectors $[\,0.0\ 1.0\ 1.0\,]^{\mathrm{T}}$, $[-1.0\ 1.0\ -1.0]^{\mathrm{T}}$, and $[\,-2.0\ -1.0\ 1.0\,]^{\mathrm{T}}$, respectively. Starting with this initial configuration, the body rotates an angle $\theta_1 = 60°$ about its $\mathbf{X}_3^i$ axis followed by another rotation $\theta_2 = 45°$ about its $\mathbf{X}_1^i$ axis. Determine the transformation matrix that defines the orientation of the body coordinate system with respect to the global coordinate system.

20. Derive the relationships between Euler angles and Euler parameters.

21. For the transformation matrix determined in Problem 7, evaluate the eigenvalues and the determinant. Verify that a unit vector along the axis of rotation is an eigenvector of this transformation matrix.

22. Given the following rotation matrix

$$\mathbf{A} = \begin{bmatrix} 0.91068 & -0.24402 & 0.33334 \\ 0.33334 & 0.91068 & -0.24402 \\ -0.24402 & 0.33334 & 0.91068 \end{bmatrix}$$

extract the following:

(i) the angle of rotation θ and a unit vector along the axis of rotation
(ii) The four Euler parameters
(iii) Rodriguez parameters
(iv) Euler angles

3 ANALYTICAL TECHNIQUES

In the preceding chapter, methods for the kinematic analysis of moving frames of reference were presented. The kinematic analysis presented in the preceding chapter is of a preliminary nature and is fundamental for understanding the dynamic motion of moving rigid bodies or coordinate systems. In this chapter, techniques for developing the dynamic equations of motion of *multibody systems* consisting of interconnected rigid bodies are introduced. The analysis of multibody systems consisting of deformable bodies that undergo large translational and rotational displacements will be deferred until we discuss in later chapters some concepts related to the deformation of the bodies. In the first three sections, a few basic concepts and definitions to be used repeatedly in this book are introduced. In these sections, the important concepts of the system *generalized coordinates*, *holonomic* and *nonholonomic* constraints, *degrees of freedom*, *virtual work*, and the system *generalized forces* are discussed. Although the reader previously may very well have met some, or even all, of these concepts and definitions, they are so fundamental for our purposes that it seems desirable to present them here in some detail. Since the direct application of *Newton's second law* becomes difficult when large-scale multibody systems are considered, in Section 4, *D'Alembert's principle* is used to derive *Lagrange's equation*, which circumvents to some extent some of the difficulties found in applying Newton's second law. In contrast to Newton's second law, the application of Lagrange's equation requires scalar quantities such as the kinetic energy, potential energy, and virtual work. In Sections 5 and 6 the variational principles of dynamics, including *Hamilton's principle*, are presented. Hamilton's principle can also be used to derive the dynamic equations of motion of multibody systems from scalar quantities. We conclude this chapter by developing the equations of motion of multibody systems consisting of interconnected rigid components.

To maintain the generality of the formulations presented in this chapter, the *Cartesian coordinates* are used to describe the motion of the multibody system. To this end, a reference coordinate system, henceforth called the *body reference* or the body coordinate system, is assigned for each body in the multibody system. The configuration of the rigid body in the system can then be identified by defining the location of the origin and the orientation of the body coordinate system with respect to an inertial global frame of reference.

3.1 GENERALIZED COORDINATES AND KINEMATIC CONSTRAINTS

The configuration of a multibody system is identified by a set of variables called *coordinates* or *generalized coordinates* that completely define the location and orientation of each body in the system. The configuration of a particle in space is defined by using three coordinates that describe the translation of this particle with respect to the three axes of the inertial frame. No rotational coordinates are required for a description of the motion of the particle, and, therefore, the three translational coordinates completely define the particle position. This simplified description of the particle kinematics is the result of the assumption that the particle has such small dimensions that a point in the three-dimensional space can be used to define the position of the particle. This assumption is not valid, however, when rigid bodies are considered. The configuration of a rigid body can be completely described by using six independent coordinates: three coordinates describing the location of the origin of the body axes and three rotational coordinates describing the orientation of the body with respect to the fixed frame. Once this set of coordinates is identified, the global position of an arbitrary point on the body can be expressed in terms of these coordinates. For instance, the global position of an arbitrary point P on a body, denoted as body i in the multibody system, can be written as (Fig. 1)

$$\mathbf{r}^i_P = \mathbf{R}^i + \mathbf{A}^i\bar{\mathbf{u}}^i$$

where $\mathbf{R}^i$ is the position of the origin of a selected body reference $\mathbf{X}^i_1\mathbf{X}^i_2\mathbf{X}^i_3$, $\mathbf{A}^i$ is the transformation matrix from the body $\mathbf{X}^i_1\mathbf{X}^i_2\mathbf{X}^i_3$ coordinate system to the global $\mathbf{X}_1\mathbf{X}_2\mathbf{X}_3$ coordinate system, and $\bar{\mathbf{u}}^i$ is the local position of the point P measured with respect to the $\mathbf{X}^i_1\mathbf{X}^i_2\mathbf{X}^i_3$ coordinate system. Thus, by defining the position vector $\mathbf{R}^i$ and the transformation matrix $\mathbf{A}^i$, we can identify the position of an arbitrary point P on the body i. It has been shown in the preceding chapter that the transformation matrix $\mathbf{A}^i$ is a function of the set of rotational coordinates $\boldsymbol{\theta}^i$, where the set $\boldsymbol{\theta}^i$ has one element in the *planar analysis* and three or four elements in the *spatial analysis* depending on whether Euler angles, Rodriguez parameters, or Euler parameters are employed. Therefore, through definition of the translational and rotational coordinates $\mathbf{R}^i$ and $\boldsymbol{\theta}^i$, respectively, of the body reference, the configuration of the rigid body is completely identified. This is not the case when deformable bodies are considered because $\bar{\mathbf{u}}^i$ is no longer a constant vector.

Reference Coordinates For convenience, we will use the notation $\mathbf{q}^i_r$ to denote the generalized coordinates of the body reference, that is,

$$\mathbf{q}^i_r = [\mathbf{R}^{i\mathrm{T}} \quad \boldsymbol{\theta}^{i\mathrm{T}}]^{\mathrm{T}} \tag{3.1}$$

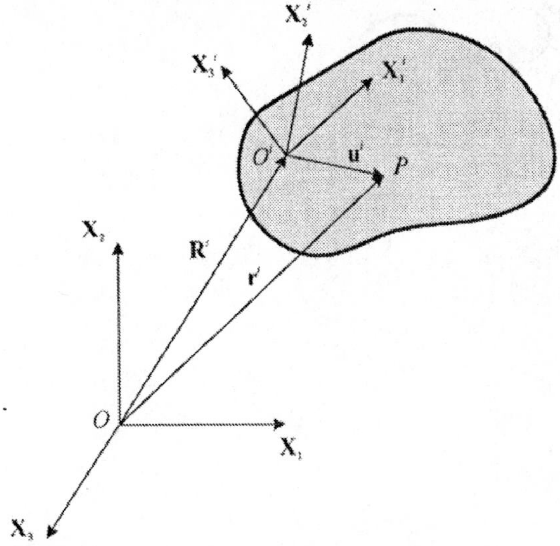

Figure 3.1 Reference coordinates of the rigid body.

Consequently, in planar analysis the vector $\mathbf{q}_r^i$ is given by

$$\mathbf{q}_r^i = \begin{bmatrix} R_1^i & R_2^i & \theta^i \end{bmatrix}^{\mathrm{T}}$$

where R_1^i and R_2^i are the coordinates of the origin of the body reference and θ^i is the rotation of the body reference about the $\mathbf{X}_3$ axis. In the three-dimensional analysis, one may write $\mathbf{q}_r^i$ as

$$\mathbf{q}_r^i = \begin{bmatrix} R_1^i & R_2^i & R_3^i & \boldsymbol{\theta}^{i\mathrm{T}} \end{bmatrix}^{\mathrm{T}}$$

where R_1^i, R_2^i, and R_3^i define the global position of the origin of the body axes and $\boldsymbol{\theta}^i$ is the set of rotational coordinates that can be used to formulate the rotation matrix. This set of coordinates can be Euler angles, Rodriguez parameters, or Euler parameters. In the case of Euler angles and Rodriguez parameters the set $\boldsymbol{\theta}^i$ contains three variables, while in the case of Euler parameters the set $\boldsymbol{\theta}^i$ contains four variables that are not totally independent. Therefore, in the spatial analysis, if three independent orientational coordinates are used, the vector $\mathbf{q}_r^i$ is a six-dimensional vector and if four parameters are used, $\mathbf{q}_r^i$ is a seven-dimensional vector.

A multibody system as shown in Fig. 2, which consists of n_b interconnected rigid bodies, requires $6n_b$ coordinates in order to describe the system configuration in space. These generalized coordinates, however, are not totally independent because of the mechanical joints between adjacent bodies. The motion of each component in the system is influenced by the motion of the others through the kinematic constraints that relate the generalized coordinates and velocities. To understand and control the motion of the multibody system it is important to identify a set of independent generalized coordinates called *degrees of freedom*. Consider, for example, the Peaucellier mechanism shown in Fig. 3, which has several bodies whose motions are constrained by a set of revolute joints. As mentioned in Chapter 1, the purpose of this mechanism

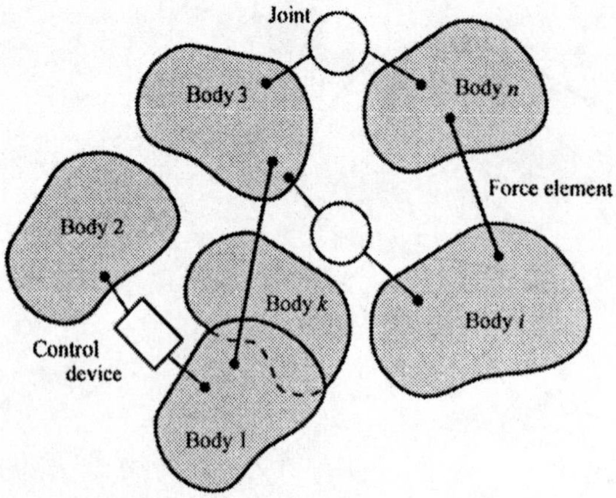

Figure 3.2 Multibody systems.

is to generate a straight-line motion at point P. The motion of point P is completely controlled by the rotation of the crankshaft CD. This mechanism has eight bodies, and yet the number of degrees of freedom is one.

Kinematic Constraints Henceforth, the set of generalized coordinates of the multibody system will be denoted by the vector $\mathbf{q} = [q_1\ q_2\ q_3\ \cdots\ q_n]^{\mathrm{T}}$, where n is the number of coordinates. In a multibody system, these n generalized coordinates are related by n_c constraint equations where $n_c \leq n$. If these n_c constraint equations can be written in the following vector form:

$$\mathbf{C}(q_1, q_2, \ldots, q_n, t) = \mathbf{C}(\mathbf{q}, t) = \mathbf{0} \tag{3.2}$$

where $\mathbf{C} = [C_1(\mathbf{q}, t)\ C_2(\mathbf{q}, t)\ \cdots\ C_{n_c}(\mathbf{q}, t)]^{\mathrm{T}}$ is the set of independent constraint equations, then the constraints are said to be *holonomic*. If the time t does not appear explicitly in Eq. 2, then the system is said to be *scleronomic*. Otherwise, if the system is holonomic and t appears explicitly in Eq. 2, the system is said to be *rheonomic*. A

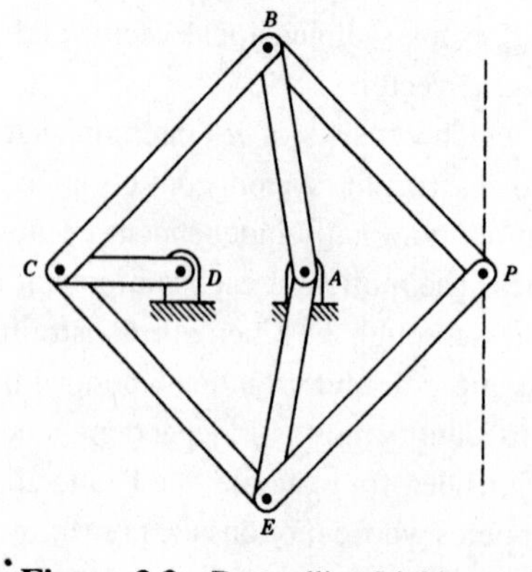

Figure 3.3 Peaucellier–Lipkin mechanism.

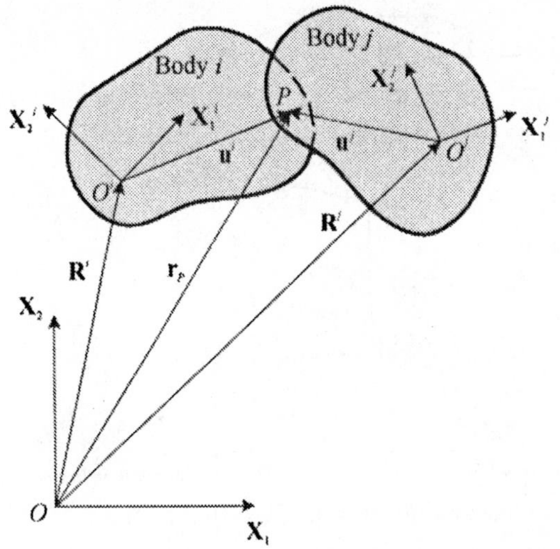

Figure 3.4 Revolute (pin) joint between two rigid bodies.

simple example of scleronomic constraint equations is the revolute joint between any two bodies in the Peaucellier mechanism shown in Fig. 3.

Figure 4 depicts a revolute joint between two arbitrary planar bodies i and j in the system. The constraint equations in this case, which allow only relative angular rotations between the two bodies, require that the global position of point P defined by the set of coordinates of body i be equal to the global position of point P defined by the set of coordinates of body j. This condition gives two constraint equations that can be written as

$$\mathbf{r}_P^i = \mathbf{r}_P^j \quad \text{or} \quad \mathbf{R}^i + \mathbf{A}^i\bar{\mathbf{u}}^i = \mathbf{R}^j + \mathbf{A}^j\bar{\mathbf{u}}^j \tag{3.3}$$

or in a more explicit form as

$$\begin{bmatrix} R_1^i \\ R_2^i \end{bmatrix} + \begin{bmatrix} \cos\theta^i & -\sin\theta^i \\ \sin\theta^i & \cos\theta^i \end{bmatrix} \begin{bmatrix} \bar{u}_1^i \\ \bar{u}_2^i \end{bmatrix} = \begin{bmatrix} R_1^j \\ R_2^j \end{bmatrix} + \begin{bmatrix} \cos\theta^j & -\sin\theta^j \\ \sin\theta^j & \cos\theta^j \end{bmatrix} \begin{bmatrix} \bar{u}_1^j \\ \bar{u}_2^j \end{bmatrix}$$

where $\mathbf{R}^k = [R_1^k \ R_2^k]^{\mathrm{T}}$ and $\bar{\mathbf{u}}^k = [\bar{u}_1^k \ \bar{u}_2^k]^{\mathrm{T}}$. A set of equations similar to Eq. 3 can be written for the spherical joint in the three-dimensional analysis.

An example of rheonomic constraints can be given if we consider the motion of the manipulator shown in Fig. 5. In many applications the motion of the end effector (hand) of the manipulator has to follow a specified path. Robotic manipulators are examples of an open-loop multibody system. We may then denote the end effector as body i and write the specified trajectory of a point P on the end effector as

$$\mathbf{r}_P^i = \mathbf{R}^i + \mathbf{A}^i\bar{\mathbf{u}}^i = \mathbf{f}(t) \tag{3.4}$$

where $\mathbf{f}(t) = [f_1(t) \ f_2(t) \ f_3(t)]^{\mathrm{T}}$ is a time-dependent function and $\mathbf{A}^i$ is the 3×3 rotation matrix given in the preceding chapter.

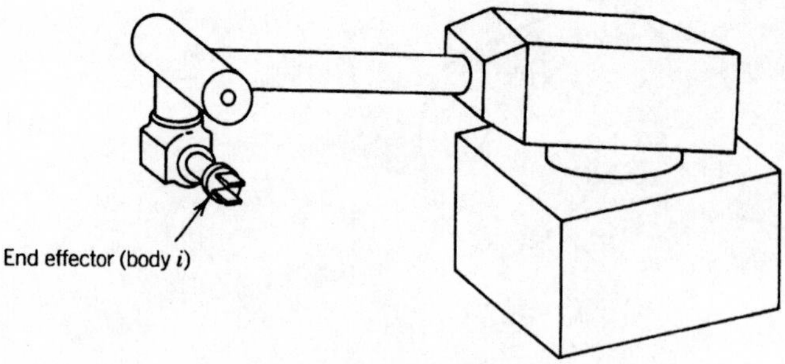

Figure 3.5 Three-dimensional manipulator.

Constraints that cannot be written in the form of Eq. 2 are called *nonholonomic* constrainsts. Simple nonholonomic constraints may be given in the form

$$\mathbf{a}_0 + \mathbf{B}\dot{\mathbf{q}} = \mathbf{0} \tag{3.5}$$

where $\mathbf{a}_0 = \mathbf{a}_0(\mathbf{q}, t) = [a_{01}\ a_{02}\ \cdots\ a_{0n_c}]^{\mathrm{T}}$, $\dot{\mathbf{q}} = [\dot{q}_1\ \dot{q}_2\ \cdots\ \dot{q}_n]^{\mathrm{T}}$ is the vector of the system generalized velocities, and $\mathbf{B}$ is an $n_c \times n$ coefficient matrix having the form

$$\mathbf{B} = \begin{bmatrix} b_{11} & b_{12} & \cdots & b_{1n} \\ b_{21} & b_{22} & \cdots & b_{2n} \\ \vdots & \vdots & \ddots & \vdots \\ b_{n_c1} & b_{n_c2} & \cdots & b_{n_cn} \end{bmatrix} = \mathbf{B}(\mathbf{q}, t)$$

One should not be able to integrate Eq. 5 and write it in terms of the generalized coordinates only; otherwise Eq. 2 will follow. Nonholonomic constraints arise in many applications. For instance, the spinning top shown in Fig. 6 that rotates about its $\mathbf{X}_3^i$ axis with an arbitrary angular velocity, or the rotation of the propeller of a ship engine or aircraft that rotates with an arbitrary angular velocity about nonfixed axes, are examples of nonholonomic systems. One may recall that the angular velocity vector defined with respect to the body reference and in terms of the four Euler

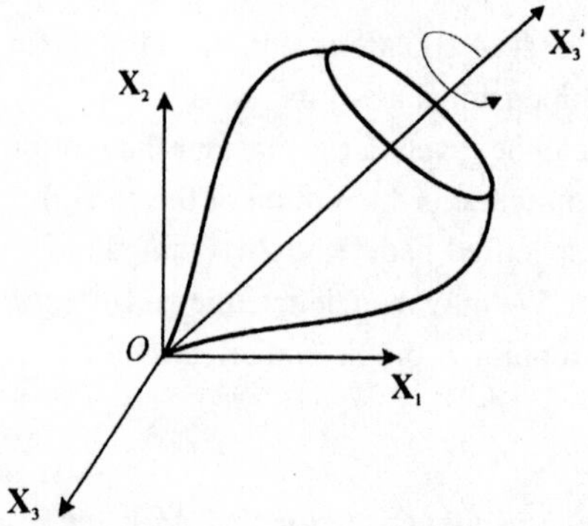

Figure 3.6 Spinning top.

parameters $\theta_0^i, \theta_1^i, \theta_2^i, \theta_3^i$ can be written as

$$\bar{\omega}^i = 2\bar{\mathbf{E}}^i\dot{\boldsymbol{\theta}}^i$$

where $\bar{\mathbf{E}}^i$ is the matrix

$$\bar{\mathbf{E}}^i = \begin{bmatrix} -\theta_1 & \theta_0 & \theta_3 & -\theta_2 \\ -\theta_2 & -\theta_3 & \theta_0 & \theta_1 \\ -\theta_3 & \theta_2 & -\theta_1 & \theta_0 \end{bmatrix}^i$$

If body i rotates about a follower axis with a specified angular velocity, which is the case of the top example shown in Fig. 6, the constraint equations in this case can be written as

$$\bar{\omega}^i = 2\bar{\mathbf{E}}^i\dot{\boldsymbol{\theta}}^i = \mathbf{f}(\mathbf{q}, t) \tag{3.6}$$

where in this case, according to Eq. 5, $\mathbf{a}_0 = -\mathbf{f}(\mathbf{q}, t)$, $\mathbf{B} = 2\bar{\mathbf{E}}^i$, and $\mathbf{f}(\mathbf{q}, t)$ is a specified function that depends on the system coordinates and time. Equation 6 can be written in a more explicit form as

$$\begin{aligned} 2(\theta_3\dot{\theta}_2 - \theta_2\dot{\theta}_3 - \theta_1\dot{\theta}_0 + \theta_0\dot{\theta}_1)^i &= f_1(\mathbf{q}, t) \\ 2(\theta_1\dot{\theta}_3 + \theta_0\dot{\theta}_2 - \theta_3\dot{\theta}_1 - \theta_2\dot{\theta}_0)^i &= f_2(\mathbf{q}, t) \\ 2(\theta_2\dot{\theta}_1 - \theta_3\dot{\theta}_0 + \theta_0\dot{\theta}_3 - \theta_1\dot{\theta}_2)^i &= f_3(\mathbf{q}, t) \end{aligned}$$

where $f_1(\mathbf{q}, t)$, $f_2(\mathbf{q}, t)$, and $f_3(\mathbf{q}, t)$ are the components of the vector $\mathbf{f}(\mathbf{q}, t)$.

Other forms of constraints are inequality relationships between the system coordinates, which can be written in the vector form

$$\mathbf{C}(\mathbf{q}, t) \geq \mathbf{0} \tag{3.7}$$

For example, the motion of a particle P placed on the surface of a sphere has to satisfy the relation

$$\mathbf{r}_P^{\mathrm{T}}\mathbf{r}_P - (a)^2 \geq 0$$

where $\mathbf{r}_P$ is the position vector of point P measured from the center of the sphere and a is the radius of the sphere.

The nonholonomic inequality constraints may be expressed in a form that depends on the system coordinates as well as velocities as

$$\mathbf{C}(\mathbf{q}, \dot{\mathbf{q}}, t) \geq \mathbf{0}$$

These constraints are called *one-sided* and *nonrestrictive* or *nonlimiting*. If the equality holds, that is

$$\mathbf{C}(\mathbf{q}, \dot{\mathbf{q}}, t) = \mathbf{0}$$

the constraints are said to be *two-sided* and *restrictive* or *limiting*. In some textbooks the difference between holonomic and nonholonomic systems is made by classification of restrictive constraints as *geometric* or *kinematic*. The constraints are said to be geometric if they are expressed in the form of Eq. 2, that is,

$$\mathbf{C}(\mathbf{q}, t) = \mathbf{0}$$

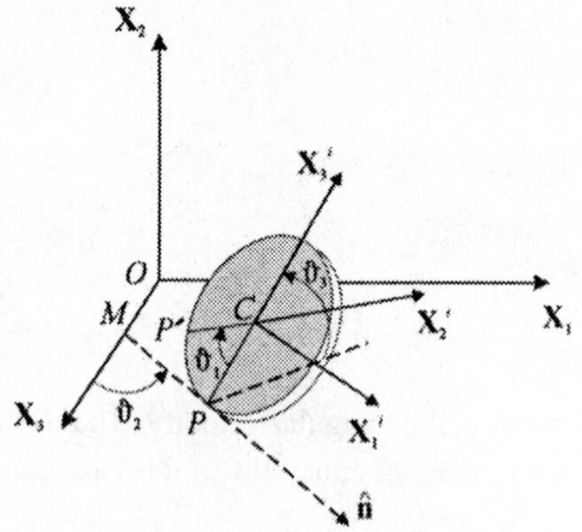

Figure 3.7 Rolling disk.

The constraints are said to be kinematic if they contain velocities, that is,

$$\mathbf{C}(\mathbf{q}, \dot{\mathbf{q}}, t) = \mathbf{0}$$

Integrable kinematic constraints are, essentially, geometric constraints. The converse is not generally true; that is, nonintegrable kinematic constraints are not generally equivalent to geometric constraints. Therefore, we may define a *nonholonomic multibody system* as a system with nonintegrable kinematic constraints that cannot be reduced to geometric constraints. In this chapter and the following chapters, the term *kinematic constraints* stands for both holonomic and nonholonomic constraints. One may observe, however, that holonomic constraints impose restrictions on the possible motion of the individual bodies in the mechanical system, while nonholonomic constraints restrict the kinematically possible values of the velocities of the bodies in the system. It is clear that every holonomic constraint, at the same time, gives rise to certain kinematic constraints on velocities. The converse, however, is not true; that is, nonintegrable constraints on the system velocities do not necessarily imply restrictions on the system coordinates. Therefore, in a nonholonomic system, some coordinates may be independent, but their variations are dependent.

Simple Nonholonomic Systems We consider the disk shown in Fig. 7. which has a sharp rim and rolls without sliding on the $\mathbf{X}_1\mathbf{X}_2$ plane. Assuming that $\mathbf{X}_1\mathbf{X}_2\mathbf{X}_3$ is a fixed frame of reference, the configuration of the disk at any instant of time can be identified using the parameters x_1, x_2, x_3, which define the coordinates of point C with respect to the global coordinate system, and θ_1, θ_2, and θ_3, where θ_1 is the angle between the two lines CP and CP' where P is a chosen point of the rim, θ_2 is the angle between the tangent to the disk at the contact point P and the $\mathbf{X}_1$ axis, and θ_3 is the angle of inclination between the $\mathbf{X}_1\mathbf{X}_2$ plane and the disk. Therefore, the vector of generalized coordinates $\mathbf{q}$ can be written as

$$\mathbf{q} = [q_1 \quad q_2 \quad q_3 \quad q_4 \quad q_5 \quad q_6]^{\mathrm{T}} = [x_1 \quad x_2 \quad x_3 \quad \theta_1 \quad \theta_2 \quad \theta_3]^{\mathrm{T}}$$

That is, $q_1 = x_1$, $q_2 = x_2$, $q_3 = x_3$, $q_4 = \theta_1$, $q_5 = \theta_2$, and $q_6 = \theta_3$. A small variation in these coordinates is denoted by

$$\delta\mathbf{q} = [\delta q_1 \quad \delta q_2 \quad \delta q_3 \quad \delta q_4 \quad \delta q_5 \quad \delta q_6]^{\mathrm{T}}$$
$$= [\delta x_1 \quad \delta x_2 \quad \delta x_3 \quad \delta\theta_1 \quad \delta\theta_2 \quad \delta\theta_3]^{\mathrm{T}}$$

The condition of rolling without sliding on the plane $\mathbf{X}_1\mathbf{X}_2$ implies that the instantaneous velocity of the point of contact on the disk is equal to zero. The position vector of point C can be written in terms of the coordinates x_1, x_2, and x_3 as

$$\mathbf{R}_C = \begin{bmatrix} x_1 \\ x_2 \\ x_3 \end{bmatrix}$$

Clearly

$$\mathbf{R}_C = \mathbf{R}_P + \mathbf{R}_{PC}$$

where $\mathbf{R}_P$ is the position vector of the contact point P and $\mathbf{R}_{PC}$ is the position of point C with respect to point P. Differentiating the vector $\mathbf{R}_C$ with respect to time yields

$$\dot{\mathbf{R}}_C = \dot{\mathbf{R}}_P + \dot{\mathbf{R}}_{PC}$$

Since the contact point P has zero velocity, that is, $\dot{\mathbf{R}}_P = \mathbf{0}$, one has

$$\dot{\mathbf{R}}_C = \dot{\mathbf{R}}_{PC}$$

The vector $\dot{\mathbf{R}}_{PC}$ can be written as

$$\dot{\mathbf{R}}_{PC} = \boldsymbol{\omega} \times \mathbf{R}_{PC}$$

where the angular velocity vector $\boldsymbol{\omega}$ is defined as

$$\boldsymbol{\omega} = \dot{\theta}_1\mathbf{i}_2' + \dot{\theta}_3\hat{\mathbf{n}}$$

where $\mathbf{i}_2'$ is a unit vector along the disk $\mathbf{X}_2^i$ axis and $\hat{\mathbf{n}}$ is a unit vector along the line MP. The velocity vector $\dot{\mathbf{R}}_C$ can then be written as

$$\dot{\mathbf{R}}_C = \dot{\mathbf{R}}_{PC} = (\dot{\theta}_1\mathbf{i}_2' + \dot{\theta}_3\hat{\mathbf{n}}) \times \mathbf{R}_{PC}$$

Clearly

$$\mathbf{i}_2' \times \mathbf{R}_{PC} = a\,\hat{\mathbf{n}}$$
$$\hat{\mathbf{n}} \times \mathbf{R}_{PC} = -a\,\mathbf{i}_2'$$

where a is the radius of the disk. Since the contact point P has zero velocity, that is, $\dot{\mathbf{R}}_P = \mathbf{0}$, one has

$$\hat{\mathbf{n}} = \cos\theta_2\mathbf{i}_1 + \sin\theta_2\mathbf{i}_2$$
$$\mathbf{i}_2' = -\sin\theta_3\sin\theta_2\,\mathbf{i}_1 + \sin\theta_3\cos\theta_2\,\mathbf{i}_2 - \cos\theta_3\,\mathbf{i}_3$$

where $\mathbf{i}_1$, $\mathbf{i}_2$, and $\mathbf{i}_3$ are unit vectors along the fixed axes $\mathbf{X}_1$, $\mathbf{X}_2$, and $\mathbf{X}_3$, respectively, Using the above two equations, the velocity vector of point C can be written as

$$\dot{\mathbf{R}}_C = \begin{bmatrix} \dot{x}_1 \\ \dot{x}_2 \\ \dot{x}_3 \end{bmatrix} = \begin{bmatrix} \dot{\theta}_1 a\cos\theta_2 + \dot{\theta}_3 a\sin\theta_2\sin\theta_3 \\ \dot{\theta}_1 a\sin\theta_2 - \dot{\theta}_3 a\cos\theta_2\sin\theta_3 \\ \dot{\theta}_3 a\cos\theta_3 \end{bmatrix}$$

The last equation, $\dot{x}_3 = \dot{\theta}_3 a \cos\theta_3$, is simply the time derivative of the $\mathbf{X}_3$ coordinate of the center of the disk, which is given by

$$R_{C3} = x_3 = a \sin\theta_3$$

Therefore, the condition

$$\dot{x}_3 = \dot{\theta}_3 a \cos\theta_3$$

can be integrated and accordingly represents a holonomic constraint equation. The two remaining conditions,

$$\dot{x}_1 = \dot{\theta}_3 a \sin\theta_2 \sin\theta_3 + \dot{\theta}_1 a \cos\theta_2$$
$$\dot{x}_2 = \dot{\theta}_1 a \sin\theta_2 - \dot{\theta}_3 a \cos\theta_2 \sin\theta_3$$

are in a form that cannot be integrated and, therefore, are nonholonomic constraint equations. For a small change in the system coordinates, the above equations yield

$$\begin{bmatrix} \delta x_1 \\ \delta x_2 \\ \delta x_3 \end{bmatrix} = \begin{bmatrix} a \cos\theta_2 \delta\theta_1 + a \sin\theta_2 \sin\theta_3 \delta\theta_3 \\ a \sin\theta_2 \delta\theta_1 - a \cos\theta_2 \sin\theta_3 \delta\theta_3 \\ a \cos\theta_3 \delta\theta_3 \end{bmatrix}$$

which can be written in a matrix form as

$$\begin{bmatrix} \delta x_1 \\ \delta x_2 \\ \delta x_3 \end{bmatrix} = \begin{bmatrix} a \cos\theta_2 & 0 & a \sin\theta_2 \sin\theta_3 \\ a \sin\theta_2 & 0 & -a \cos\theta_2 \sin\theta_3 \\ 0 & 0 & a \cos\theta_3 \end{bmatrix} \begin{bmatrix} \delta\theta_1 \\ \delta\theta_2 \\ \delta\theta_3 \end{bmatrix}$$

The small change in the system coordinates $\delta\mathbf{q}$ can then be written in terms of the independent variations $\delta\theta_1$, $\delta\theta_2$, and $\delta\theta_3$ as

$$\begin{bmatrix} \delta x_1 \\ \delta x_2 \\ \delta x_3 \\ \delta\theta_1 \\ \delta\theta_2 \\ \delta\theta_3 \end{bmatrix} = \begin{bmatrix} a \cos\theta_2 & 0 & a \sin\theta_2 \sin\theta_3 \\ a \sin\theta_2 & 0 & -a \cos\theta_2 \sin\theta_3 \\ 0 & 0 & a \cos\theta_3 \\ 1 & 0 & 0 \\ 0 & 1 & 0 \\ 0 & 0 & 1 \end{bmatrix} \begin{bmatrix} \delta\theta_1 \\ \delta\theta_2 \\ \delta\theta_3 \end{bmatrix}$$

Clearly, there are only three independent variations. There are, however, five independent coordinates, $x_1, x_2, \theta_1, \theta_2$, and θ_3; that is, the configuration of the disk is defined in terms of five independent coordinates, while there are only three independent velocities because of the nonholonomic constraint equations. Although these nonholonomic kinematic constraints must be satisfied throughout the motion of the disk, the coordinates $x_1, x_2, \theta_1, \theta_2$, and θ_3 may take any values as the disk rolls without sliding. For instance, the disk can be brought from a given position $x_{1o}, x_{2o}, \theta_{1o}, \theta_{2o}$, and θ_{3o} to any other position $x_1, x_2, \theta_1, \theta_2$, and θ_3 by first rolling the disk from the contact point P_0 to point P along any curve of length $a(\theta_1 - \theta_{1o} + 2\pi k)$ where k is an integer number. The disk is then rotated about a line that connects point C to the contact point until the angle θ_2 takes the desired value. The disk is finally tilted into the position θ_3 (Neimark and Fufaev 1972).

3.2 DEGREES OF FREEDOM AND GENERALIZED COORDINATE PARTITIONING

Because of the constraints imposed on the multibody system, the system coordinates are not independent. They are, in general, related by a set of nonlinear constraint equations that represent mechanical joints as well as specified motion trajectories. For holonomic systems, each constraint equation can be used to eliminate one coordinate by writing this coordinate in terms of the others, provided the constraint equations are linearly independent. Therefore, a system with n coordinates and n_c constraint equations has $n - n_c$ independent coordinates. The independent coordinates are also called the system *degrees of freedom*. For example, in the planar two-body system shown in Fig. 4, where the two bodies are connected by a revolute joint whose constraint equations are given by Eq. 3, one may rewrite this equation in the following form

$$\mathbf{R}^j = \mathbf{R}^i + \mathbf{A}^i\bar{\mathbf{u}}^i - \mathbf{A}^j\bar{\mathbf{u}}^j$$

where $\mathbf{R}^j$, the set of dependent coordinates in this system, is written in terms of the set of independent coordinates $\mathbf{q}_r^i = [\mathbf{R}^{i\mathrm{T}}\ \theta^i]^\mathrm{T}$ and θ^j. The vectors $\mathbf{R}^i = [R_1^i\ R_2^i]^\mathrm{T}$ and $\mathbf{R}^j = [R_1^j\ R_2^j]^\mathrm{T}$ define the location of the origin of the reference of body i and body j, respectively; $\bar{\mathbf{u}}^i = [\bar{u}_1^i\ \bar{u}_2^i]^\mathrm{T}$ and $\bar{\mathbf{u}}^j = [\bar{u}_1^j\ \bar{u}_2^j]^\mathrm{T}$ are the local positions of the joint definition point defined, respectively, in the coordinate system of body i and body j; and $\mathbf{A}^i$ and $\mathbf{A}^j$ are the planar transformations given by

$$\mathbf{A}^i = \begin{bmatrix} \cos\theta^i & -\sin\theta^i \\ \sin\theta^i & \cos\theta^i \end{bmatrix}, \qquad \mathbf{A}^j = \begin{bmatrix} \cos\theta^j & -\sin\theta^j \\ \sin\theta^j & \cos\theta^j \end{bmatrix}$$

One can write the dependent coordinates $\mathbf{R}^j$ in terms of the independent ones in a more explicit form as

$$\mathbf{R}^j = \begin{bmatrix} R_1^j \\ R_2^j \end{bmatrix} = \begin{bmatrix} R_1^i \\ R_2^i \end{bmatrix} + \begin{bmatrix} \cos\theta^i & -\sin\theta^i \\ \sin\theta^i & \cos\theta^i \end{bmatrix}\begin{bmatrix} \bar{u}_1^i \\ \bar{u}_2^i \end{bmatrix} - \begin{bmatrix} \cos\theta^j & -\sin\theta^j \\ \sin\theta^j & \cos\theta^j \end{bmatrix}\begin{bmatrix} \bar{u}_1^j \\ \bar{u}_2^j \end{bmatrix}$$

that is,

$$R_1^j = R_1^i + \bar{u}_1^i\cos\theta^i - \bar{u}_2^i\sin\theta^i - \bar{u}_1^j\cos\theta^j + \bar{u}_2^j\sin\theta^j$$
$$R_2^j = R_2^i + \bar{u}_1^i\sin\theta^i + \bar{u}_2^i\cos\theta^i - \bar{u}_1^j\sin\theta^j - \bar{u}_2^j\cos\theta^j$$

Alternatively, one may also select R_1^i and R_2^i as dependent coordinates and write them in terms of the other coordinates as

$$R_1^i = R_1^j + \bar{u}_1^j\cos\theta^j - \bar{u}_2^j\sin\theta^j - \bar{u}_1^i\cos\theta^i + \bar{u}_2^i\sin\theta^i$$
$$R_2^i = R_2^j + \bar{u}_1^j\sin\theta^j + \bar{u}_2^j\cos\theta^j - \bar{u}_1^i\sin\theta^i - \bar{u}_2^i\cos\theta^i$$

Therefore, the set of independent coordinates is not unique. It is important, however, to realize that the number of dependent coordinates is equal to the number of linearly independent constraint equations. In that sense we may define the degrees of freedom as the minimum number of independent variables required to describe the system configuration. For the two-body system shown in Fig. 4, the total number of system coordinates is six and yet the number of independent coordinates or system degrees of freedom is four because of the kinematic constraints of the revolute joint. The Peaucellier mechanism shown in Fig. 3 consists of eight links, including the fixed link (ground). If three Cartesian coordinates are used to describe the configuration of each link, the mechanism will have 24 coordinates. These coordinates, however, are not independent because of kinematic constraints, and it can be shown that the mechanism has only one degree of freedom; that is, the motion of the mechanism can be controlled by specifying only one variable, say, the rotation of the crankshaft.

Generalized Coordinate Partitioning In the following, we make use of the concept of the *virtual displacement*, which refers to a change in the configuration of the system as the result of any arbitrary infinitesimal change of the coordinates $\mathbf{q}$, consistent with the forces and constraints imposed on the system at the given instant t. The displacement is called "virtual" to distinguish it from an actual displacement of the system occurring in a time interval dt, during which the forces and constraints may be changing.

As the result of a virtual change in the system coordinates and using Taylor's expansion, the constraints given by Eq. 2 yield

$$\mathbf{C}_{q_1}\delta q_1 + \mathbf{C}_{q_2}\delta q_2 + \cdots + \mathbf{C}_{q_n}\delta q_n = \mathbf{0} \tag{3.8}$$

where $\mathbf{C}_{q_i} = \partial \mathbf{C}/\partial q_i = [\partial C_1/\partial q_i \; \partial C_2/\partial q_i \; \cdots \; \partial C_{n_c}/\partial q_i]^{\mathrm{T}}$. We may write Eq. 8 in matrix form as

$$\mathbf{C}_{\mathbf{q}}\delta\mathbf{q} = \mathbf{0} \tag{3.9}$$

where

$$\mathbf{C}_{\mathbf{q}} = \begin{bmatrix} C_{11} & C_{12} & \cdots & C_{1n} \\ C_{21} & C_{22} & \cdots & C_{2n} \\ \vdots & \vdots & \ddots & \vdots \\ C_{n_c1} & C_{n_c2} & \cdots & C_{n_cn} \end{bmatrix}$$

is an $n_c \times n$ matrix called the *system Jacobian* and $C_{ij} = \partial C_i/\partial q_j$. If the constraint equations are linearly independent, $\mathbf{C}_{\mathbf{q}}$ has a full row rank. In this case, one may partition the system generalized coordinates as

$$\mathbf{q} = [\mathbf{q}_i^{\mathrm{T}} \;\; \mathbf{q}_d^{\mathrm{T}}]^{\mathrm{T}} \tag{3.10}$$

where $\mathbf{q}_i$ and $\mathbf{q}_d$ are two vectors having $n - n_c$ and n_c components, respectively. According to the coordinate partitioning of Eq. 10, Eq. 9 can be written in the form

$$\mathbf{C}_{\mathbf{q}_i}\delta\mathbf{q}_i + \mathbf{C}_{\mathbf{q}_d}\delta\mathbf{q}_d = \mathbf{0} \tag{3.11}$$

where $\mathbf{C}_{\mathbf{q}_d}$ is selected to be a nonsingular $n_c \times n_c$ matrix and $\mathbf{C}_{\mathbf{q}_i}$ is an $n_c \times (n - n_c)$ matrix. Equation 11 yields

$$\mathbf{C}_{\mathbf{q}_d}\delta\mathbf{q}_d = -\mathbf{C}_{\mathbf{q}_i}\delta\mathbf{q}_i \tag{3.12}$$

Since $\mathbf{C}_{\mathbf{q}_d}$ is nonsingular, and thus invertible, Eq. 12 may be rewritten as

$$\delta \mathbf{q}_d = \mathbf{C}_{di}\delta \mathbf{q}_i \tag{3.13}$$

where

$$\mathbf{C}_{di} = -\mathbf{C}_{\mathbf{q}_d}^{-1}\mathbf{C}_{\mathbf{q}_i} \tag{3.14}$$

is an $n_c \times (n - n_c)$ matrix. Therefore, using the coordinate partitioning of Eq. 10, one can write the change in a set of coordinates $\mathbf{q}_d$ in terms of the change in the other set $\mathbf{q}_i$. The set $\mathbf{q}_d$ is called the set of *dependent coordinates*, while the set $\mathbf{q}_i$ is called the set of *independent coordinates* or the system *degrees of freedom.*

Illustrative Example An illustrative example for the preceding development is the planar revolute joint between the two bodies shown in Fig. 4. By considering the variation of Eq. 3, we obtain

$$\delta \mathbf{R}^i + \delta(\mathbf{A}^i\bar{\mathbf{u}}^i) - \delta \mathbf{R}^j - \delta(\mathbf{A}^j\bar{\mathbf{u}}^j) = \mathbf{0} \tag{3.15}$$

Since $\bar{\mathbf{u}}^i$ is constant in the case of rigid body analysis and $\delta \mathbf{A}^i = (\partial \mathbf{A}^i/\partial\theta^i)\delta\theta^i$, we can write $\delta(\mathbf{A}^i\bar{\mathbf{u}}^i)$ as

$$\delta(\mathbf{A}^i\bar{\mathbf{u}}^i) = \left(\mathbf{A}_\theta^i\bar{\mathbf{u}}^i\right)\delta\theta^i$$

where $\mathbf{A}_\theta^i = (\partial \mathbf{A}^i/\partial\theta^i)$ is given by

$$\mathbf{A}_\theta^i = \begin{bmatrix} -\sin\theta^i & -\cos\theta^i \\ \cos\theta^i & -\sin\theta^i \end{bmatrix}$$

A similar comment applies to body j, and Eq. 15 can be written as

$$\delta \mathbf{R}^i + \mathbf{A}_\theta^i\bar{\mathbf{u}}^i\delta\theta^i - \delta \mathbf{R}^j - \mathbf{A}_\theta^j\bar{\mathbf{u}}^j\delta\theta^j = \mathbf{0}$$

We may select $\mathbf{R}^j$ as dependent coordinates and $\mathbf{R}^i$, θ^i, and θ^j as independent coordinates and write

$$\delta \mathbf{R}^j = \delta \mathbf{R}^i + \mathbf{A}_\theta^i\bar{\mathbf{u}}^i\delta\theta^i - \mathbf{A}_\theta^j\bar{\mathbf{u}}^j\delta\theta^j$$

or equivalently as

$$\delta \mathbf{R}^j = \begin{bmatrix} \mathbf{I}_2 & \mathbf{A}_\theta^i\bar{\mathbf{u}}^i & -\mathbf{A}_\theta^j\bar{\mathbf{u}}^j \end{bmatrix} \begin{bmatrix} \delta \mathbf{R}^i \\ \delta\theta^i \\ \delta\theta^j \end{bmatrix} \tag{3.16}$$

where $\mathbf{I}_2$ is a 2×2 identity matrix. Comparing Eqs. 12 and 16, we recognize $\mathbf{C}_{\mathbf{q}_d}$ as an identity matrix and

$$-\mathbf{C}_{\mathbf{q}_i} = \begin{bmatrix} \mathbf{I}_2 & \mathbf{A}_\theta^i\bar{\mathbf{u}}^i & -\mathbf{A}_\theta^j\bar{\mathbf{u}}^j \end{bmatrix} \tag{3.17}$$

It is obvious in this simple example that the matrix $\mathbf{C}_{di}$ of Eq. 14 is equal to the matrix $-\mathbf{C}_{\mathbf{q}_i}$ because of the fact that $\mathbf{C}_{\mathbf{q}_d}$ is an identity matrix; that is, $\mathbf{C}_{di}$ is a 2×4 matrix defined as

$$\mathbf{C}_{di} = -\mathbf{C}_{\mathbf{q}_i} = \begin{bmatrix} \mathbf{I}_2 & \mathbf{A}_\theta^i\bar{\mathbf{u}}^i & -\mathbf{A}_\theta^j\bar{\mathbf{u}}^j \end{bmatrix}$$

where

$$\mathbf{A}_\theta^i \bar{\mathbf{u}}^i = \begin{bmatrix} -\sin\theta^i & -\cos\theta^i \\ \cos\theta^i & -\sin\theta^i \end{bmatrix} \begin{bmatrix} \bar{u}_1^i \\ \bar{u}_2^i \end{bmatrix} = \begin{bmatrix} -\bar{u}_1^i \sin\theta^i - \bar{u}_2^i \cos\theta^i \\ \bar{u}_1^i \cos\theta^i - \bar{u}_2^i \sin\theta^i \end{bmatrix}$$

$$\mathbf{A}_\theta^j \bar{\mathbf{u}}^j = \begin{bmatrix} -\sin\theta^j & -\cos\theta^j \\ \cos\theta^j & -\sin\theta^j \end{bmatrix} \begin{bmatrix} \bar{u}_1^j \\ \bar{u}_2^j \end{bmatrix} = \begin{bmatrix} -\bar{u}_1^j \sin\theta^j - \bar{u}_2^j \cos\theta^j \\ \bar{u}_1^j \cos\theta^j - \bar{u}_2^j \sin\theta^j \end{bmatrix}$$

Therefore, the matrix $\mathbf{C}_{di}$ can be written as

$$\mathbf{C}_{di} = \begin{bmatrix} 1 & 0 & c_{13} & c_{14} \\ 0 & 1 & c_{23} & c_{24} \end{bmatrix}$$

where the coefficients c_{13}, c_{14}, c_{23}, and c_{24} are defined as

$$c_{13} = -\bar{u}_1^i \sin\theta^i - \bar{u}_2^i \cos\theta^i$$

$$c_{23} = \bar{u}_1^i \cos\theta^i - \bar{u}_2^i \sin\theta^i$$

$$c_{14} = \bar{u}_1^j \sin\theta^j + \bar{u}_2^j \cos\theta^j$$

$$c_{24} = -\bar{u}_1^j \cos\theta^j + \bar{u}_2^j \sin\theta^j$$

In large-scale multibody systems, identifying the dependent or the independent coordinates and accordingly identifying the nonsingular matrix $\mathbf{C}_{\mathbf{q}_d}$ may be difficult because of the complexity of the system. In such cases, numerical methods can be employed to determine a nonsingular sub-Jacobian matrix, thus identifying the independent and dependent coordinates. This subject will be discussed in more detail in Chapter 5 after introducing the dynamics of flexible multibody systems.

Example 3.1 The multibody slider crank mechanism shown in Fig. 8 consists of four rigid bodies. Body 1 is the fixed link or the ground, body 2 is the crankshaft OA, body 3 is the connecting rod AB, and body 4 is the slider block whose center is located at B. By rotating the crankshaft (body 2) with a specified angular velocity, the slider block (body 4) will produce a straight-line motion. To study the motion of this mechanism using Cartesian coordinates, we select a coordinate system for each body. The origins of these coordinate systems are assumed to be rigidly attached to the geometric center of the respective bodies.

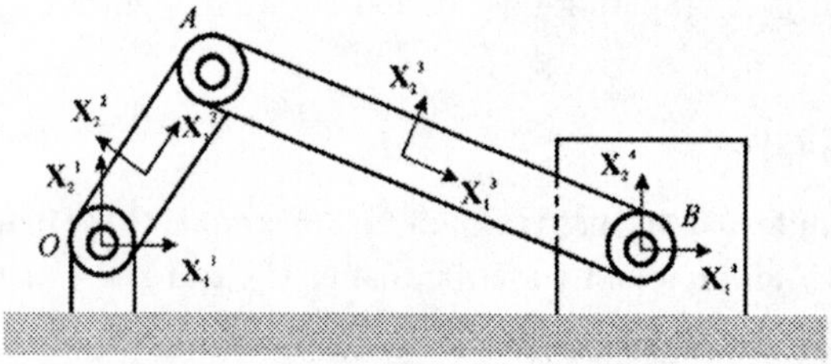

Figure 3.8 Multibody slider crank mechanism.

Therefore, we define the Cartesian coordinates of the bodies (links) as follows:

$$\mathbf{q}_r^1 = \begin{bmatrix} R_1^1 & R_2^1 & \theta^1 \end{bmatrix}^{\mathrm{T}}, \qquad \mathbf{q}_r^2 = \begin{bmatrix} R_1^2 & R_2^2 & \theta^2 \end{bmatrix}^{\mathrm{T}}$$

$$\mathbf{q}_r^3 = \begin{bmatrix} R_1^3 & R_2^3 & \theta^3 \end{bmatrix}^{\mathrm{T}}, \qquad \mathbf{q}_r^4 = \begin{bmatrix} R_1^4 & R_2^4 & \theta^4 \end{bmatrix}^{\mathrm{T}}$$

where R_1^i and R_2^i are the Cartesian coordinates of the origin of the ith body coordinate system $\mathbf{X}_1^i\mathbf{X}_2^i$ defined with respect to the global coordinate system and θ^i is the angular orientation of the ith body. Thus, the vector $\mathbf{q}$ of the system Cartesian coordinates is defined as

$$\begin{aligned} \mathbf{q} &= [q_1 \quad q_2 \quad q_3 \quad \cdots \quad q_{12}]^{\mathrm{T}} = \begin{bmatrix} \mathbf{q}_r^{1\mathrm{T}} & \mathbf{q}_r^{2\mathrm{T}} & \mathbf{q}_r^{3\mathrm{T}} & \mathbf{q}_r^{4\mathrm{T}} \end{bmatrix}^{\mathrm{T}} \\ &= \begin{bmatrix} R_1^1 & R_2^1 & \theta^1 & R_1^2 & R_2^2 & \theta^2 & R_1^3 & R_2^3 & \theta^3 & R_1^4 & R_2^4 & \theta^4 \end{bmatrix}^{\mathrm{T}} \end{aligned}$$

These coordinates, however, are not independent because of the kinematic constraints imposed on the motion of the mechanism members. These constraints can be recognized as follows. Body 1 is the fixed link, that is,

$$R_1^1 = 0, \qquad R_2^1 = 0, \qquad \theta^1 = 0$$

We will call these constraints *ground constraints*. The motion of the crankshaft can be considered as a pure rotation about point O. This implies that point O has zero coordinates with respect to the global coordinate system $\mathbf{X}_1^1\mathbf{X}_2^1$. This can be expressed mathematically as

$$\mathbf{R}^2 + \mathbf{A}^2\bar{\mathbf{u}}_o^2 = \mathbf{0}$$

where $\mathbf{R}^2 = [R_1^2 \; R_2^2]^{\mathrm{T}}$, $\mathbf{A}^2$ is the transformation matrix from the coordinate system of the crankshaft (body 2) to the global inertial frame, and $\bar{\mathbf{u}}_o^2$ is the position vector of point O defined in the coordinate system of the crankshaft, that is,

$$\bar{\mathbf{u}}_o^2 = \begin{bmatrix} -\dfrac{l^2}{2} & 0 \end{bmatrix}^{\mathrm{T}}$$

where l^2 is the length of the crankshaft.

The crankshaft (body 2) is connected to the connecting rod (body 3) by a revolute joint at point A. Let l^3 denote the length of the connecting rod (body 3). Then the revolute joint constraint equations can be written in terms of the Cartesian coordinates of the two bodies as

$$\mathbf{R}^2 + \mathbf{A}^2\bar{\mathbf{u}}_A^2 - \mathbf{R}^3 - \mathbf{A}^3\bar{\mathbf{u}}_A^3 = \mathbf{0}$$

where $\mathbf{R}^i = [R_1^i \; R_2^i]^{\mathrm{T}}$, $\mathbf{A}^i$ is the planar transformation matrix from the body i coordinate system to the global frame of reference, and $\bar{\mathbf{u}}_A^i$ $(i = 2, 3)$ is the local coordinates of the joint definition point, that is

$$\bar{\mathbf{u}}_A^2 = \begin{bmatrix} \dfrac{l^2}{2} & 0 \end{bmatrix}^{\mathrm{T}}, \qquad \bar{\mathbf{u}}_A^3 = \begin{bmatrix} -\dfrac{l^3}{2} & 0 \end{bmatrix}^{\mathrm{T}}$$

Bodies 3 and 4 are also connected by a revolute joint at point B in a manner similar to the revolute joint at A; therefore, we may write the following matrix equation, which describes the connectivity between body 3 and body 4:

$$\mathbf{R}^3 + \mathbf{A}^3\bar{\mathbf{u}}_B^3 - \mathbf{R}^4 - \mathbf{A}^4\bar{\mathbf{u}}_B^4 = \mathbf{0}$$

in which

$$\bar{\mathbf{u}}_B^3 = \begin{bmatrix} \frac{l^3}{2} & 0 \end{bmatrix}^{\mathrm{T}}, \qquad \bar{\mathbf{u}}_B^4 = [0 \quad 0]^{\mathrm{T}}$$

Finally, the motion of the slider block (body 4) must satisfy the following kinematic constraints:

$$R_2^4 = 0, \qquad \theta^4 = 0$$

It is clear that the slider crank mechanism discussed in this example has 12 Cartesian coordinates and 11 algebraic constraint equations that can be summarized as follows: 3 ground constraints, 2 constraints that fix the coordinates of point O on the crankshaft, 4 constraints that describe the revolute (pin) joints at A and B, and 2 constraints that restrict the motion of the slider block (body 4). Thus, the number of degrees of freedom of the mechanism is 1. By taking a virtual change in the system generalized coordinates, the ground constraints lead to

$$\delta R_1^1 = 0, \qquad \delta R_2^1 = 0, \qquad \delta\theta^1 = 0$$

which can be written in a matrix form as

$$\begin{bmatrix} 1 & 0 & 0 \\ 0 & 1 & 0 \\ 0 & 0 & 1 \end{bmatrix} \begin{bmatrix} \delta R_1^1 \\ \delta R_2^1 \\ \delta\theta^1 \end{bmatrix} = \begin{bmatrix} 0 \\ 0 \\ 0 \end{bmatrix}$$

The constraints on the global position of point O lead to

$$\delta\mathbf{R}^2 + \mathbf{A}_\theta^2\bar{\mathbf{u}}_O^2\delta\theta^2 = \mathbf{0}$$

where $\mathbf{A}_\theta^2$ is the partial derivative of the planar transformation $\mathbf{A}^2$ with respect to θ^2. Using the definition of $\bar{\mathbf{u}}_O^2$, we can write the above equation in a more explicit form as

$$\begin{bmatrix} 1 & 0 \\ 0 & 1 \end{bmatrix} \begin{bmatrix} \delta R_1^2 \\ \delta R_2^2 \end{bmatrix} + \begin{bmatrix} \frac{l^2}{2}\sin\theta^2 \\ -\frac{l^2}{2}\cos\theta^2 \end{bmatrix} \delta\theta^2 = \begin{bmatrix} 0 \\ 0 \end{bmatrix}$$

or alternatively

$$\begin{bmatrix} 1 & 0 & \frac{l^2}{2}\sin\theta^2 \\ 0 & 1 & -\frac{l^2}{2}\cos\theta^2 \end{bmatrix} \begin{bmatrix} \delta R_1^2 \\ \delta R_2^2 \\ \delta\theta^2 \end{bmatrix} = \begin{bmatrix} 0 \\ 0 \end{bmatrix}$$

The revolute joint constraint at point A leads to

$$\begin{bmatrix} 1 & 0 & -\frac{l^2}{2}\sin\theta^2 \\ 0 & 1 & \frac{l^2}{2}\cos\theta^2 \end{bmatrix} \begin{bmatrix} \delta R_1^2 \\ \delta R_2^2 \\ \delta\theta^2 \end{bmatrix} - \begin{bmatrix} 1 & 0 & \frac{l^3}{2}\sin\theta^3 \\ 0 & 1 & -\frac{l^3}{2}\cos\theta^3 \end{bmatrix} \begin{bmatrix} \delta R_1^3 \\ \delta R_2^3 \\ \delta\theta^3 \end{bmatrix} = \begin{bmatrix} 0 \\ 0 \end{bmatrix}$$

or alternatively

$$\begin{bmatrix} 1 & 0 & -\frac{l^2}{2}\sin\theta^2 & -1 & 0 & -\frac{l^3}{2}\sin\theta^3 \\ 0 & 1 & \frac{l^2}{2}\cos\theta^2 & 0 & -1 & \frac{l^3}{2}\cos\theta^3 \end{bmatrix} \begin{bmatrix} \delta R_1^2 \\ \delta R_2^2 \\ \delta\theta^2 \\ \delta R_1^3 \\ \delta R_2^3 \\ \delta\theta^3 \end{bmatrix} = \begin{bmatrix} 0 \\ 0 \end{bmatrix}$$

For the revolute joint at B, we have

$$\begin{bmatrix} 1 & 0 & -\frac{l^3}{2}\sin\theta^3 & -1 & 0 & 0 \\ 0 & 1 & \frac{l^3}{2}\cos\theta^3 & 0 & -1 & 0 \end{bmatrix} \begin{bmatrix} \delta R_1^3 \\ \delta R_2^3 \\ \delta\theta^3 \\ \delta R_1^4 \\ \delta R_2^4 \\ \delta\theta^4 \end{bmatrix} = \begin{bmatrix} 0 \\ 0 \end{bmatrix}$$

Finally, the constraints on the motion of the slider block at B provide

$$\begin{bmatrix} 1 & 0 \\ 0 & 1 \end{bmatrix} \begin{bmatrix} \delta R_2^4 \\ \delta\theta^4 \end{bmatrix} = \begin{bmatrix} 0 \\ 0 \end{bmatrix}$$

or

$$\begin{bmatrix} 0 & 1 & 0 \\ 0 & 0 & 1 \end{bmatrix} \begin{bmatrix} \delta R_1^4 \\ \delta R_2^4 \\ \delta\theta^4 \end{bmatrix} = \begin{bmatrix} 0 \\ 0 \end{bmatrix}$$

Combining the above equations, one obtains

$$\mathbf{C_q}\,\delta\mathbf{q} = \mathbf{0}$$

where $\mathbf{q} = [R_1^1 \ R_2^1 \ \theta^1 \ R_1^2 \ R_2^2 \ \theta^2 \ R_1^3 \ R_2^3 \ \theta^3 \ R_1^4 \ R_2^4 \ \theta^4]^{\mathrm{T}}$ and $\mathbf{C_q}$ is an 11×12 system Jacobian matrix, which can be written as

$$\mathbf{C_q} = [C_{i,j}]$$

where the nonzero elements $C_{i,j}$ are defined as

$$C_{1,1} = C_{2,2} = C_{3,3} = C_{4,4} = C_{5,5} = C_{6,4} = C_{7,5} = C_{8,7}$$
$$= C_{9,8} = C_{10,11} = C_{11,12} = 1$$
$$C_{6,7} = C_{7,8} = C_{8,10} = C_{9,11} = -1$$
$$C_{4,6} = \frac{l^2}{2}\sin\theta^2, \qquad C_{5,6} = -\frac{l^2}{2}\cos\theta^2$$
$$C_{6,6} = -\frac{l^2}{2}\sin\theta^2, \qquad C_{7,6} = \frac{l^2}{2}\cos\theta^2$$

$$C_{6,9} = -\frac{l^3}{2}\sin\theta^3, \qquad C_{7,9} = \frac{l^3}{2}\cos\theta^3$$

$$C_{8,9} = -\frac{l^3}{2}\sin\theta^3, \qquad C_{9,9} = \frac{l^3}{2}\cos\theta^3$$

One may select θ^2 as an independent coordinate or the system degree of freedom. In this case, the Jacobian matrix can be partitioned according to

$$\mathbf{C}_{\mathbf{q}_d}\delta\mathbf{q}_d + \mathbf{C}_{\mathbf{q}_i}\delta\mathbf{q}_i = \mathbf{0}$$

where $\mathbf{C}_{\mathbf{q}_d}$ is the Jacobian matrix associated with the dependent coordinates. It is an 11×11 square matrix. $\mathbf{C}_{\mathbf{q}_i}$ is the Jacobian matrix associated with the independent coordinate θ^2. In this case, $\mathbf{C}_{\mathbf{q}_i}$ is an 11-dimensional vector. The vectors of dependent and independent coordinates are defined as

$$\mathbf{q}_d = \begin{bmatrix} R_1^1 & R_2^1 & \theta^1 & R_1^2 & R_2^2 & R_1^3 & R_2^3 & \theta^3 & R_1^4 & R_2^4 & \theta^4 \end{bmatrix}^{\mathrm{T}}$$

$$\mathbf{q}_i = \theta^2$$

The vector $\mathbf{C}_{\mathbf{q}_i}$ is defined as

$$\begin{aligned}\mathbf{C}_{\mathbf{q}_i} &= [0 \quad 0 \quad 0 \quad C_{4,6} \quad C_{5,6} \quad C_{6,6} \quad C_{7,6} \quad 0 \quad 0 \quad 0 \quad 0]^{\mathrm{T}} \\ &= \begin{bmatrix} 0 & 0 & 0 & \frac{l^2}{2}\sin\theta^2 & -\frac{l^2}{2}\cos\theta^2 & -\frac{l^2}{2}\sin\theta^2 & \frac{l^2}{2}\cos\theta^2 & 0 & 0 & 0 & 0 \end{bmatrix}^{\mathrm{T}}\end{aligned}$$

and the matrix $\mathbf{C}_{\mathbf{q}_d}$ is defined as

$$\mathbf{C}_{\mathbf{q}_d} = \begin{bmatrix} 1 & 0 & 0 & 0 & 0 & 0 & 0 & 0 & 0 & 0 & 0 \\ 0 & 1 & 0 & 0 & 0 & 0 & 0 & 0 & 0 & 0 & 0 \\ 0 & 0 & 1 & 0 & 0 & 0 & 0 & 0 & 0 & 0 & 0 \\ 0 & 0 & 0 & 1 & 0 & 0 & 0 & 0 & 0 & 0 & 0 \\ 0 & 0 & 0 & 0 & 1 & 0 & 0 & 0 & 0 & 0 & 0 \\ 0 & 0 & 0 & 1 & 0 & -1 & 0 & C_{6,9} & 0 & 0 & 0 \\ 0 & 0 & 0 & 0 & 1 & 0 & -1 & C_{7,9} & 0 & 0 & 0 \\ 0 & 0 & 0 & 0 & 0 & 1 & 0 & C_{8,9} & -1 & 0 & 0 \\ 0 & 0 & 0 & 0 & 0 & 0 & 1 & C_{9,9} & 0 & -1 & 0 \\ 0 & 0 & 0 & 0 & 0 & 0 & 0 & 0 & 0 & 1 & 0 \\ 0 & 0 & 0 & 0 & 0 & 0 & 0 & 0 & 0 & 0 & 1 \end{bmatrix}$$

One can see that $\mathbf{C}_{\mathbf{q}_d}$ is a nonsingular matrix that can be inverted to write the vector $\delta\mathbf{q}_d$ in terms of the variation in the system degree of freedom $\delta\theta^2$ as

$$\delta\mathbf{q}_d = -\mathbf{C}_{\mathbf{q}_d}^{-1}\mathbf{C}_{\mathbf{q}_i}\delta\theta^2$$

As it was pointed out earlier, the set of independent coordinates is not unique. In this multibody slider crank mechanism, one may also select R_1^4, which describes the translation of the slider block in the horizontal direction, as an independent coordinate, that is

$$\mathbf{q}_i = R_1^4$$

$$\mathbf{q}_d = \begin{bmatrix} R_1^1 & R_2^1 & \theta^1 & R_1^2 & R_2^2 & \theta^2 & R_1^3 & R_2^3 & \theta^3 & R_2^4 & \theta^4 \end{bmatrix}^{\mathrm{T}}$$

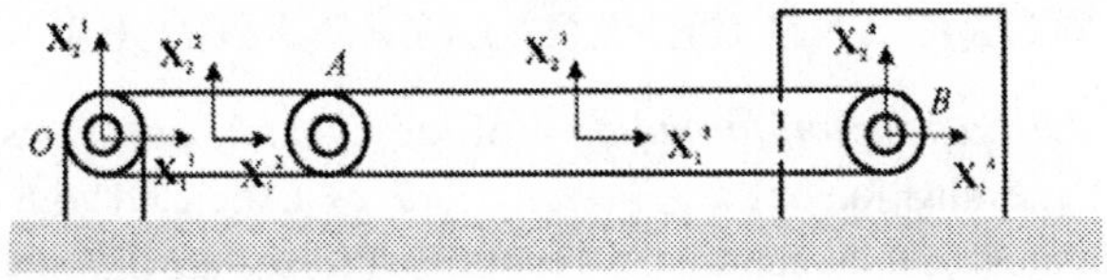

Figure 3.9 Special configuration.

It can be shown that the vector $\mathbf{C}_{\mathbf{q}_i}$ is defined in this case as

$$\mathbf{C}_{\mathbf{q}_i} = [0 \quad 0 \quad 0 \quad 0 \quad 0 \quad 0 \quad 0 \quad -1 \quad 0 \quad 0 \quad 0]^{\mathrm{T}}$$

and the matrix $\mathbf{C}_{\mathbf{q}_d}$ as

$$\mathbf{C}_{\mathbf{q}_d} = \begin{bmatrix} 1 & 0 & 0 & 0 & 0 & 0 & 0 & 0 & 0 & 0 & 0 \\ 0 & 1 & 0 & 0 & 0 & 0 & 0 & 0 & 0 & 0 & 0 \\ 0 & 0 & 1 & 0 & 0 & 0 & 0 & 0 & 0 & 0 & 0 \\ 0 & 0 & 0 & 1 & 0 & C_{4,6} & 0 & 0 & 0 & 0 & 0 \\ 0 & 0 & 0 & 0 & 1 & C_{5,6} & 0 & 0 & 0 & 0 & 0 \\ 0 & 0 & 0 & 1 & 0 & C_{6,6} & -1 & 0 & C_{6,9} & 0 & 0 \\ 0 & 0 & 0 & 0 & 1 & C_{7,6} & 0 & -1 & C_{7,9} & 0 & 0 \\ 0 & 0 & 0 & 0 & 0 & 0 & 1 & 0 & C_{8,9} & 0 & 0 \\ 0 & 0 & 0 & 0 & 0 & 0 & 0 & 1 & C_{9,9} & -1 & 0 \\ 0 & 0 & 0 & 0 & 0 & 0 & 0 & 0 & 0 & 1 & 0 \\ 0 & 0 & 0 & 0 & 0 & 0 & 0 & 0 & 0 & 0 & 1 \end{bmatrix}$$

Therefore, the system-dependent coordinates $\mathbf{q}_d$ can be written in terms of the independent ones, provided the matrix $\mathbf{C}_{\mathbf{q}_d}$ is nonsingular.

Consider, now, the special configuration of the mechanism shown in Fig. 9 in which $\theta^2 = \theta^3 = 0$; one can verify that in this special configuration

$$C_{4,6} = C_{6,6} = C_{6,9} = C_{8,9} = 0$$

By substituting these values in the preceding matrix $\mathbf{C}_{\mathbf{q}_d}$, one can verify that this matrix is singular because, for example, adding the sixth and eighth rows will produce the fourth row; that is, the fourth row is a linear combination of the sixth and eighth rows and $\mathbf{C}_{\mathbf{q}_d}$ at this configuration does not have a full row rank, and as a consequence, it is a singular matrix. This implies that at this special configuration, the selected dependent coordinates cannot be written in terms of the variation δR_1^4. Physically, this means that at this configuration the mechanism cannot be controlled by specifying the motion of the slider block (body 4) in the horizontal direction. This special configuration is called the *singular configuration*. This situation, however, will not occur at other configurations where $\mathbf{C}_{\mathbf{q}_d}$ is nonsingular.

The slider crank mechanism presented in the preceding example is a simple multibody system, and yet identifying the dependent and independent coordinates may not be an easy task because of the size of the Jacobian matrix. When large-scale multibody systems are considered, the use of numerical techniques that use the numerical structure of the Jacobian matrix is recommended.

3.3 VIRTUAL WORK AND GENERALIZED FORCES

An essential step in the *Lagrangian formulation* of the dynamic equations of the multibody systems is the evaluation of the *generalized forces* associated with the system generalized coordinates. In this section, the generalized forces are introduced by application of the principle of *virtual work* in both cases of static and dynamic analysis. In the development presented in this section, a system of particles is employed. By assuming that rigid bodies consist of a large number of particles, similar expressions for the body generalized forces can be developed.

Static Equilibrium Consider a system of n_p particles in a three-dimensional space as shown in Fig. 10. An arbitrary particle i in the system is acted on by a system of forces whose resultant is the vector $\mathbf{F}^i$. If particle i is in static equilibrium, we have

$$\mathbf{F}^i = \mathbf{0} \tag{3.18}$$

where $\mathbf{F}^i = [F_1^i \; F_2^i \; F_3^i]^{\mathrm{T}}$. If Eq. 18 holds, it is clear that

$$\mathbf{F}^i \cdot \delta \mathbf{r}^i = 0 \tag{3.19}$$

for any arbitrary virtual displacement $\delta \mathbf{r}^i$ for particle i. If the system of particles is in equilibrium, it follows that

$$\sum_{i=1}^{n_p} \mathbf{F}^i \cdot \delta \mathbf{r}^i = 0 \tag{3.20}$$

If the system configuration has to satisfy a set of constraint equations, we may write the resultant force $\mathbf{F}^i$ acting on the particle i as

$$\mathbf{F}^i = \mathbf{F}_e^i + \mathbf{F}_c^i \tag{3.21}$$

where $\mathbf{F}_e^i$ is the vector of externally applied forces and $\mathbf{F}_c^i$ is the vector of constraint forces that arise because of the existence of connections between the individual particles of the system. Substitution of Eq. 21 into Eq. 20 yields

$$\sum_{i=1}^{n_p} \mathbf{F}^i \cdot \delta \mathbf{r}^i = \sum_{i=1}^{n_p} (\mathbf{F}_e^i + \mathbf{F}_c^i) \cdot \delta \mathbf{r}^i = 0 \tag{3.22}$$

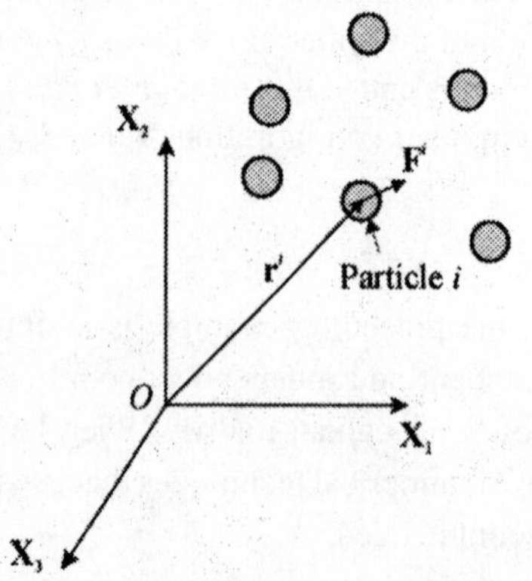

Figure 3.10 System of particles.

Since the dot product is distributive, we have

$$\sum_{i=1}^{n_p} \mathbf{F}^i \cdot \delta\mathbf{r}^i = \sum_{i=1}^{n_p} \mathbf{F}_e^i \cdot \delta\mathbf{r}^i + \sum_{i=1}^{n_p} \mathbf{F}_c^i \cdot \delta\mathbf{r}^i = 0 \tag{3.23}$$

Using the following notations:

$$\delta W = \sum_{i=1}^{n_p} \mathbf{F}^i \cdot \delta\mathbf{r}^i, \qquad \delta W_e = \sum_{i=1}^{n_p} \mathbf{F}_e^i \cdot \delta\mathbf{r}^i, \qquad \delta W_c = \sum_{i=1}^{n_p} \mathbf{F}_c^i \cdot \delta\mathbf{r}^i$$

where δW is defined as the virtual work of all the forces acting on the system, δW_e is the virtual work of externally applied forces, and δW_c is the virtual work of constraint forces. Equation 23 can then be written as

$$\delta W = \delta W_e + \delta W_c = 0 \tag{3.24}$$

If we consider constraints that do no work, denoted henceforth as *workless constraints*, the virtual work of the constraint forces is zero, that is,

$$\delta W_c = \sum_{i=1}^{n_p} \mathbf{F}_c^i \cdot \delta\mathbf{r}^i = 0 \tag{3.25}$$

Examples of workless constraints are the frictionless revolute and prismatic joints wherein the constraint forces act in a direction perpendicular to the direction of the displacement. In this case, Eq. 24 reduces to

$$\delta W = \delta W_e = \sum_{i=1}^{n_p} \mathbf{F}_e^i \cdot \delta\mathbf{r}^i = 0 \tag{3.26}$$

Equation 26 is the *principle of virtual work for static equilibrium*, which states that the virtual work of the externally applied forces of a system of particles in equilibrium with workless constraints is equal to zero. The condition of Eq. 26, however, does not imply that $\mathbf{F}_e^i = \mathbf{0}$ for all i values, since $\mathbf{r}^i$, $(i = 1, 2, \ldots, n_p)$, are not linearly independent in a constrained system of particles.

Previously, it was mentioned that the system configuration can be identified by using a set of generalized coordinates $\mathbf{q} = [q_1 \; q_2 \; \cdots \; q_n]^{\mathrm{T}}$. In this case $\mathbf{r}^i$ can be written as

$$\mathbf{r}^i = \mathbf{r}^i(q_1, q_2, \ldots, q_n) \tag{3.27}$$

and the virtual displacement can be written as

$$\begin{aligned} \delta\mathbf{r}^i &= \frac{\partial \mathbf{r}^i}{\partial q_1}\delta q_1 + \frac{\partial \mathbf{r}^i}{\partial q_2}\delta q_2 + \cdots + \frac{\partial \mathbf{r}^i}{\partial q_n}\delta q_n \\ &= \sum_{j=1}^{n} \frac{\partial \mathbf{r}^i}{\partial q_j}\delta q_j \end{aligned} \tag{3.28}$$

Substitution of this equation into Eq. 26 yields

$$\delta W = \delta W_e = \sum_{i=1}^{n_p} \mathbf{F}_e^i \cdot \sum_{j=1}^{n} \frac{\partial \mathbf{r}^i}{\partial q_j}\delta q_j = 0 \tag{3.29}$$

which can be written as

$$\delta W = \delta W_e = \sum_{j=1}^{n} \sum_{i=1}^{n_p} \mathbf{F}_e^i \cdot \frac{\partial \mathbf{r}^i}{\partial q_j} \delta q_j = 0 \tag{3.30}$$

One may define Q_j such that

$$Q_j = \sum_{i=1}^{n_p} \mathbf{F}_e^i \cdot \frac{\partial \mathbf{r}^i}{\partial q_j} = \sum_{i=1}^{n_p} \mathbf{F}_e^{i\mathrm{T}} \mathbf{r}_{q_j}^i$$

where $\mathbf{r}_{q_j}^i$ is the vector

$$\mathbf{r}_{q_j}^i = \frac{\partial \mathbf{r}^i}{\partial q_j}$$

With the definition of Q_j, Eq. 30 reduces to

$$\delta W = \delta W_e = \sum_{j=1}^{n} Q_j \delta q_j = \mathbf{Q}^{\mathrm{T}} \delta \mathbf{q} = 0 \tag{3.31}$$

where $\mathbf{Q} = [Q_1 \; Q_2 \; \cdots \; Q_n]^{\mathrm{T}}$ is called the *vector of generalized forces*. The element Q_j in this vector is denoted as the generalized force associated with the generalized coordinate q_j. If the components of the generalized coordinates are independent, then the equilibrium condition of Eq. 31 yields

$$Q_j = 0, \quad j = 1, 2, \ldots, n \tag{3.32}$$

These are n algebraic equations which can be nonlinear functions in the system generalized coordinates $q_1, q_2, \ldots, q_n$. These equations can be solved for the n coordinates. The position of the particles in the system can be obtained by using the kinematic relationships of Eq. 27.

Example 3.2 Figure 11 shows a system of two particles in the $\mathbf{X}_1\mathbf{X}_2$ plane. The two particles, whose masses are denoted as m^1 and m^2, are constrained to move along the rod shown in the figure using friction-free prismatic joints. The

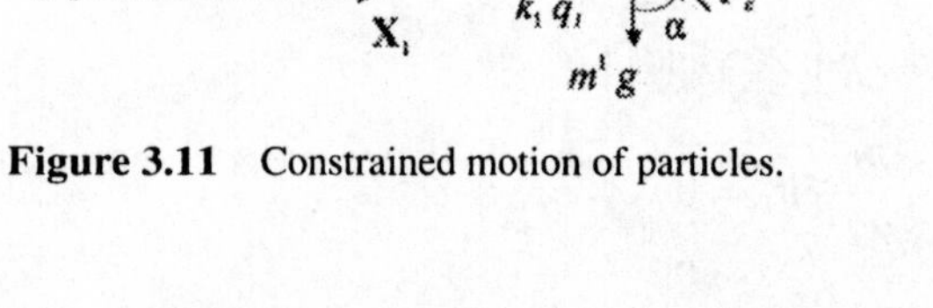

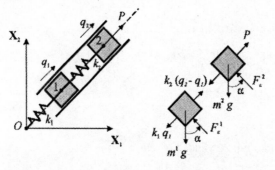

Figure 3.11 Constrained motion of particles.

particles are supported by two springs with stiffness coefficients k_1 and k_2. Given a constant force of magnitude P acting on particle 2, determine the equilibrium conditions of Eq. 32.

Solution Let $\mathbf{r}^1 = [R_1^1 \; R_2^1 \; 0]^{\mathrm{T}}$ denote the displacement vector of particle 1 in the Cartesian coordinate system, and let $\mathbf{r}^2 = [R_1^2 \; R_2^2 \; 0]^{\mathrm{T}}$ denote the displacement vector of particle 2. It is clear, however, that the system has only two independent coordinates q_1 and q_2. The virtual changes in the vectors of coordinates of the two particles can be expressed in terms of the virtual changes in the coordinates q_1 and q_2 as

$$\delta \mathbf{r}^1 = \begin{bmatrix} \cos\alpha \\ \sin\alpha \\ 0 \end{bmatrix} \delta q_1, \qquad \delta \mathbf{r}^2 = \begin{bmatrix} \cos\alpha \\ \sin\alpha \\ 0 \end{bmatrix} \delta q_2$$

where α is a constant angle. As shown in Fig. 11, the vectors of constraint forces that act on the two particles are defined in the Cartesian coordinate system as

$$\mathbf{F}_c^1 = \begin{bmatrix} -F_c^1 \sin\alpha \\ F_c^1 \cos\alpha \\ 0 \end{bmatrix}, \qquad \mathbf{F}_c^2 = \begin{bmatrix} -F_c^2 \sin\alpha \\ F_c^2 \cos\alpha \\ 0 \end{bmatrix}$$

The virtual work of the constraint forces is defined as

$$\begin{aligned} \delta W_c &= \sum_{i=1}^{n_p} \mathbf{F}_c^{i\,\mathrm{T}} \delta \mathbf{r}^i = \mathbf{F}_c^{1\mathrm{T}} \delta \mathbf{r}^1 + \mathbf{F}_c^{2\mathrm{T}} \delta \mathbf{r}^2 \\ &= \begin{bmatrix} -F_c^1 \sin\alpha & F_c^1 \cos\alpha & 0 \end{bmatrix} \begin{bmatrix} \cos\alpha \\ \sin\alpha \\ 0 \end{bmatrix} \delta q_1 \\ &\quad + \begin{bmatrix} -F_c^2 \sin\alpha & F_c^2 \cos\alpha & 0 \end{bmatrix} \begin{bmatrix} \cos\alpha \\ \sin\alpha \\ 0 \end{bmatrix} \delta q_2 = 0 \end{aligned}$$

That is, the constraints are workless, since the reaction forces act in a direction perpendicular to the direction of the displacement. From the force diagram shown in Fig. 11, it is clear that the vectors of external forces in the Cartesian coordinate system are

$$\mathbf{F}_e^1 = \begin{bmatrix} \{k_2(q_2 - q_1) - k_1 q_1\} \cos\alpha \\ \{k_2(q_2 - q_1) - k_1 q_1\} \sin\alpha - m^1 g \\ 0 \end{bmatrix}$$

$$\mathbf{F}_e^2 = \begin{bmatrix} \{P - k_2(q_2 - q_1)\} \cos\alpha \\ \{P - k_2(q_2 - q_1)\} \sin\alpha - m^2 g \\ 0 \end{bmatrix}$$

where g is the gravitational constant.

The virtual work of the external forces can then be written as

$$\delta W_e = \delta W = \sum_{i=1}^{n_p} \mathbf{F}_e^{i\mathrm{T}} \delta \mathbf{r}^i = \mathbf{F}_e^{1\mathrm{T}} \delta \mathbf{r}^1 + \mathbf{F}_e^{2\mathrm{T}} \delta \mathbf{r}^2$$

$$= [\{k_2(q_2 - q_1) - k_1 q_1\}(\cos^2\alpha + \sin^2\alpha) - m^1 g \sin\alpha]\, \delta q_1$$
$$+ [\{P - k_2(q_2 - q_1)\}(\cos^2\alpha + \sin^2\alpha) - m^2 g \sin\alpha]\, \delta q_2$$

Since $\cos^2\alpha + \sin^2\alpha = 1$ and since q_1 and q_2 are assumed to be linearly independent, their coefficients in the above equations can be set equal to zero, that is,

$$Q_1 = k_2(q_2 - q_1) - k_1 q_1 - m^1 g \sin\alpha = 0$$
$$Q_2 = P - k_2(q_2 - q_1) - m^2 g \sin\alpha = 0$$

These equations can be written in the following matrix form:

$$\begin{bmatrix} k_1 + k_2 & -k_2 \\ -k_2 & k_2 \end{bmatrix} \begin{bmatrix} q_1 \\ q_2 \end{bmatrix} = \begin{bmatrix} -m^1 g \sin\alpha \\ P - m^2 g \sin\alpha \end{bmatrix}$$

These are the equilibrium equations that can be solved for q_1 and q_2 as

$$\begin{bmatrix} q_1 \\ q_2 \end{bmatrix} = \frac{1}{k_1 k_2} \begin{bmatrix} k_2 & k_2 \\ k_2 & k_1 + k_2 \end{bmatrix} \begin{bmatrix} -m^1 g \sin\alpha \\ P - m^2 g \sin\alpha \end{bmatrix}$$

As can be seen from the formulation and example presented in this section, connection forces can be considered as auxiliary quantities that we are forced to introduce when we study the equilibrium of each particle separately. These forces can then be eliminated by considering the equilibrium of the entire system of particles. Such constraints are sometimes called *ideal* since their connection forces do not do work. Clearly, the internal reaction forces between material points that form a rigid body are connection forces of this type. The distance between two particles i and j on the rigid body must remain constant. This condition can be expressed mathematically as

$$(\mathbf{r}^i - \mathbf{r}^j)^{\mathrm{T}}(\mathbf{r}^i - \mathbf{r}^j) = c$$

where $\mathbf{r}^i$ and $\mathbf{r}^j$ are, respectively, the position vectors of particles i and j and c is a constant. By assuming a virtual change in the position vectors, the constraint equation yields

$$(\mathbf{r}^i - \mathbf{r}^j)^{\mathrm{T}}(\delta\mathbf{r}^i - \delta\mathbf{r}^j) = 0$$

If $\mathbf{F}_c^{ij}$ is the connection force acting on particle i as the result of this constraint, then according to Newton's third law, $\mathbf{F}_c^{ji} = -\mathbf{F}_c^{ij}$ is the reaction force that acts on particle j. Clearly, the two reactions $\mathbf{F}_c^{ji}$ and $\mathbf{F}_c^{ij}$ are equal in magnitude and opposite in direction and must be directed along a straight line joining the two particles i and j, that is

$$\mathbf{F}_c^{ij} = k(\mathbf{r}^i - \mathbf{r}^j)$$

where k is a constant. The virtual work of the constraint forces in this case can then be written as

$$\begin{aligned}\delta W_c &= \mathbf{F}_c^{ij\mathrm{T}}\delta\mathbf{r}^i + \mathbf{F}_c^{ji\mathrm{T}}\delta\mathbf{r}^j \\ &= \mathbf{F}_c^{ij\mathrm{T}}\delta\mathbf{r}^i - \mathbf{F}_c^{ij\mathrm{T}}\delta\mathbf{r}^j = \mathbf{F}_c^{ij\mathrm{T}}(\delta\mathbf{r}^i - \delta\mathbf{r}^j) \\ &= k(\mathbf{r}^i - \mathbf{r}^j)^{\mathrm{T}}(\delta\mathbf{r}^i - \delta\mathbf{r}^j) = 0\end{aligned}$$

That is, the virtual work of the connection forces resulting from constraints between the particles forming the rigid body is equal to zero.

Dynamic Equilibrium In a similar manner, the principle of virtual work in the dynamic case can be developed. Newton's second law states that the resultant of the forces acting on a particle is equal to the rate of change of momentum of this particle, that is,

$$\mathbf{F}^i = \dot{\mathbf{P}}^i$$

or equivalently

$$\mathbf{F}^i - \dot{\mathbf{P}}^i = \mathbf{0} \tag{3.33}$$

where $\mathbf{P}^i$ is the momentum of the particle i. If condition 33 is satisfied, we say that particle i is in *dynamic equilibrium*. The dynamic equilibrium condition implies that

$$(\mathbf{F}^i - \dot{\mathbf{P}}^i)\cdot\delta\mathbf{r}^i = 0 \tag{3.34}$$

If the system of particles is in dynamic equilibrium, we can then write

$$\sum_{i=1}^{n_p}(\mathbf{F}^i - \dot{\mathbf{P}}^i)\cdot\delta\mathbf{r}^i = 0 \tag{3.35}$$

According to Eq. 21, $\mathbf{F}^i$ can be written as the sum of the external and constraint forces, yielding

$$\sum_{i=1}^{n_p}(\mathbf{F}_e^i + \mathbf{F}_c^i - \dot{\mathbf{P}}^i)\cdot\delta\mathbf{r}^i = 0$$

or

$$\sum_{i=1}^{n_p}(\mathbf{F}_e^i - \dot{\mathbf{P}}^i)\cdot\delta\mathbf{r}^i + \sum_{i=1}^{n_p}\mathbf{F}_c^i\cdot\delta\mathbf{r}^i = 0$$

If the constraints are workless, we have

$$\sum_{i=1}^{n_p}\mathbf{F}_c^i\cdot\delta\mathbf{r}^i = 0$$

which yields

$$\sum_{i=1}^{n_p}(\mathbf{F}_e^i - \dot{\mathbf{P}}^i)\cdot\delta\mathbf{r}^i = 0 \tag{3.36}$$

The result of this equation is often called *D'Alembert's principle*. Using Eq. 28, we may write Eq. 36 in terms of the system generalized coordinates as

$$\sum_{i=1}^{n_p} (\mathbf{F}_e^i - \dot{\mathbf{P}}^i) \cdot \sum_{j=1}^{n} \frac{\partial \mathbf{r}^i}{\partial q_j} \delta q_j = 0$$

or equivalently

$$\sum_{j=1}^{n} \sum_{i=1}^{n_p} (\mathbf{F}_e^i - \dot{\mathbf{P}}^i) \cdot \frac{\partial \mathbf{r}^i}{\partial q_j} \delta q_j = 0 \tag{3.37}$$

Define $\bar{Q}_j$ such that

$$\bar{Q}_j = \sum_{i=1}^{n_p} (\mathbf{F}_e^i - \dot{\mathbf{P}}^i) \cdot \frac{\partial \mathbf{r}^i}{\partial q_j}, \qquad j = 1, 2, \ldots, n$$

We may then write Eq. 37 as

$$\sum_{j=1}^{n} \sum_{i=1}^{n_p} (\mathbf{F}_e^i - \dot{\mathbf{P}}^i) \cdot \frac{\partial \mathbf{r}^i}{\partial q_j} \delta q_j = \sum_{j=1}^{n} \bar{Q}_j \delta q_j = \bar{\mathbf{Q}}^{\mathrm{T}} \delta \mathbf{q} = 0 \tag{3.38}$$

where $\bar{\mathbf{Q}} = [\bar{Q}_1 \; \bar{Q}_2 \; \cdots \; \bar{Q}_n]^{\mathrm{T}}$. If the components of the vector of generalized coordinates are independent, Eq. 38 yields

$$\bar{\mathbf{Q}} = \begin{bmatrix} \bar{Q}_1 & \bar{Q}_2 & \cdots & \bar{Q}_n \end{bmatrix}^{\mathrm{T}} = \mathbf{0} \tag{3.39}$$

that is,

$$\bar{Q}_j = 0, \qquad j = 1, 2, \ldots, n$$

Equation 39 is a set of n second-order ordinary differential equations of motion that describe the dynamics of the system. These equations are expressed in terms of the independent coordinates, and as a consequence, the constraint forces are automatically eliminated (Shabana 1994a). Equation 39 can be integrated in order to determine the generalized coordinates and velocities. The position of the particles can then be determined by using the kinematic relationships of Eq. 27.

Example 3.3 Figure 12 shows a particle of mass m that slides freely in the $\mathbf{X}_1\mathbf{X}_2$ plane on a slender massless rod that rotates with angular velocity $\dot{\theta}$ and angular acceleration $\ddot{\theta}$ about the $\mathbf{X}_3$ axis. Determine the dynamic equilibrium equations for this particle.

Solution The configuration of the system shown in Fig. 12 can be identified by using the independent coordinates q and θ. In Fig. 12, the force components F_{c1}^1 and F_{c2}^1 are the reactions of the workless pin joint constraints. The displacement and velocity of the particle in the Cartesian coordinate system can be written in terms of the independent coordinates as

$$\mathbf{r} = \begin{bmatrix} \cos\theta \\ \sin\theta \\ 0 \end{bmatrix} q, \qquad \dot{\mathbf{r}} = \dot{\theta} \begin{bmatrix} -\sin\theta \\ \cos\theta \\ 0 \end{bmatrix} q + \begin{bmatrix} \cos\theta \\ \sin\theta \\ 0 \end{bmatrix} \dot{q}$$

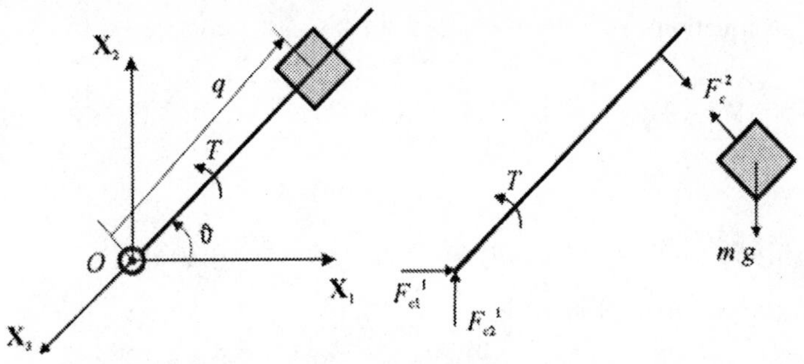

Figure 3.12 Dynamic equilibrium of particles.

where θ is the angular position of the particle and q is the displacement of the particle with respect to point O. A virtual change in the system coordinates leads to

$$\delta\mathbf{r} = \begin{bmatrix} -\sin\theta \\ \cos\theta \\ 0 \end{bmatrix} q\,\delta\theta + \begin{bmatrix} \cos\theta \\ \sin\theta \\ 0 \end{bmatrix} \delta q$$

Since the reaction forces acting on the rod and the particle are equal in magnitude and opposite in direction, the virtual work of these forces is equal to zero.

The vector of external forces acting on the particle is defined in the Cartesian coordinate system as

$$\mathbf{F}_e = \begin{bmatrix} 0 \\ -mg \\ 0 \end{bmatrix}$$

where m is the mass of the particle and g is the gravitational constant. The virtual work of the external forces and moments that act on the system can then be written as

$$\delta W_e = \mathbf{F}_e^{\mathrm{T}}\delta\mathbf{r} + T\delta\theta = -mg\sin\theta\,\delta q + (T - mgq\cos\theta)\,\delta\theta$$

where T is the external moment that acts on the rod. The momentum of the particle is defined in the Cartesian coordinate system as

$$\mathbf{P} = m\dot{\mathbf{r}}$$

and the rate of change of momentum $\dot{\mathbf{P}}$ is given by

$$\dot{\mathbf{P}} = m\ddot{\mathbf{r}} = m\ddot{\theta}q \begin{bmatrix} -\sin\theta \\ \cos\theta \\ 0 \end{bmatrix} + 2m\dot{\theta}\dot{q} \begin{bmatrix} -\sin\theta \\ \cos\theta \\ 0 \end{bmatrix}$$

$$+ m\ddot{q} \begin{bmatrix} \cos\theta \\ \sin\theta \\ 0 \end{bmatrix} + m(\dot{\theta})^2 q \begin{bmatrix} -\cos\theta \\ -\sin\theta \\ 0 \end{bmatrix}$$

One can then verify that the virtual work of this inertia force is given by

$$\delta W_i = \dot{\mathbf{P}}^{\mathrm{T}}\delta\mathbf{r} = (2mq\dot{q}\dot{\theta} + m(q)^2\ddot{\theta})\delta\theta + (m\ddot{q} - m(\dot{\theta})^2 q)\,\delta q$$

Applying the equation

$$\sum_{i=1}^{n_p} (\mathbf{F}_c^i + \mathbf{F}_e^i - \dot{\mathbf{P}}^i)^{\mathrm{T}} \delta \mathbf{r}^i = 0$$

we obtain

$$-mg \sin\theta \delta q + (T - mgq\cos\theta)\,\delta\theta - (2mq\dot{q}\dot{\theta} + m(q)^2\ddot{\theta})\,\delta\theta$$
$$-(m\ddot{q} - m(\dot{\theta})^2 q)\,\delta q = 0$$

or

$$\left(T - mgq\cos\theta - 2mq\dot{q}\dot{\theta} - m(q)^2\ddot{\theta}\right)\delta\theta$$
$$+(-mg\sin\theta - m\ddot{q} + m(\dot{\theta})^2 q)\delta q = 0$$

Since θ and q are assumed to be independent, the coefficients of $\delta\theta$ and δq in the preceding equation can be set equal to zero. This leads to the following two nonlinear second-order differential equations of motion:

$$m\ddot{q} - m(\dot{\theta})^2 q + mg\sin\theta = 0$$
$$m(q)^2\ddot{\theta} + 2mq\dot{q}\dot{\theta} + mgq\cos\theta - T = 0$$

These equations can be integrated numerically to determine the independent coordinates and velocities. The displacements of the particle in the Cartesian coordinate system can then be determined from the kinematic equations in which these displacements are written in terms of the independent coordinates.

Generalized Forces of Rigid Bodies We have seen that the virtual work of a force is defined to be the dot product of the force with the virtual change in the vector of displacements of the point of application of the force. Even though in the preceding development we considered a system of particles, the definition of the virtual work can be extended to the case of rigid bodies, as demonstrated by the following examples.

Example 3.4 Consider the planar motion of the rigid body i shown in Fig. 13, where $\mathbf{F}^i = [F_1^i \; F_2^i]^{\mathrm{T}}$ is an arbitrary forcing function whose components are defined with respect to the inertial frame. The force $\mathbf{F}^i$ is acting at point P on the body whose global position vector is $\mathbf{r}_P^i$. The virtual work of this force can be written as

$$\delta W^i = \mathbf{F}^{i\mathrm{T}} \delta \mathbf{r}_P^i$$

where $\mathbf{r}_P^i$ can be written in terms of the generalized coordinates of body i as

$$\mathbf{r}_P^i = \mathbf{R}^i + \mathbf{A}^i \bar{\mathbf{u}}_P^i$$

in which $\mathbf{A}^i$ is the transformation matrix given by

$$\mathbf{A}^i = \begin{bmatrix} \cos\theta^i & -\sin\theta^i \\ \sin\theta^i & \cos\theta^i \end{bmatrix}$$

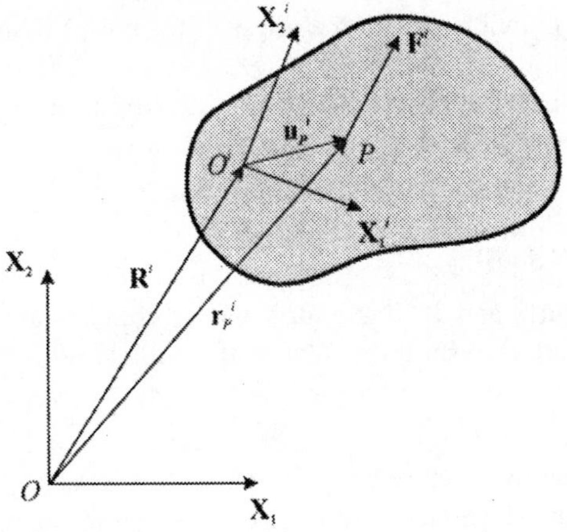

Figure 3.13 Planar rigid body.

and $\bar{\mathbf{u}}^i_P$ is the local position of point P. It follows that

$$\delta\mathbf{r}^i_P = \delta\mathbf{R}^i + \mathbf{A}^i_\theta\bar{\mathbf{u}}^i_P\delta\theta^i$$

where $\mathbf{A}^i_\theta$ is the derivative of $\mathbf{A}^i$ with respect to θ^i and is given by

$$\mathbf{A}^i_\theta = \begin{bmatrix} -\sin\theta^i & -\cos\theta^i \\ \cos\theta^i & -\sin\theta^i \end{bmatrix}$$

$\delta\mathbf{r}^i_P$ can then be written in a partitioned form as

$$\delta\mathbf{r}^i_P = \begin{bmatrix}\mathbf{I}_2 & \mathbf{A}^i_\theta\bar{\mathbf{u}}^i_P\end{bmatrix}\begin{bmatrix}\delta\mathbf{R}^i \\ \delta\theta^i\end{bmatrix}$$

where $\mathbf{I}_2$ is a 2×2 identity matrix. The virtual work δW^i due to the application of the force $\mathbf{F}^i$ is given by

$$\begin{aligned}\delta W^i &= \mathbf{F}^{i\mathrm{T}}\delta\mathbf{r}^i_P \\ &= \mathbf{F}^{i\mathrm{T}}\begin{bmatrix}\mathbf{I}_2 & \mathbf{A}^i_\theta\bar{\mathbf{u}}^i_P\end{bmatrix}\begin{bmatrix}\delta\mathbf{R}^i \\ \delta\theta^i\end{bmatrix} = \begin{bmatrix}\mathbf{F}^{i\mathrm{T}} & \mathbf{F}^{i\mathrm{T}}\mathbf{A}^i_\theta\bar{\mathbf{u}}^i_P\end{bmatrix}\begin{bmatrix}\delta\mathbf{R}^i \\ \delta\theta^i\end{bmatrix}\end{aligned}$$

One may write δW^i in a more simplified form as

$$\delta W^i = \begin{bmatrix}\mathbf{Q}^{i\mathrm{T}}_r & Q^i_\theta\end{bmatrix}\begin{bmatrix}\delta\mathbf{R}^i \\ \delta\theta^i\end{bmatrix}$$

where

$$\mathbf{Q}^i_r = \mathbf{F}^i$$

and

$$Q^i_\theta = \mathbf{F}^{i\mathrm{T}}\mathbf{A}^i_\theta\bar{\mathbf{u}}^i_P = \pm\left|(\mathbf{A}^i\bar{\mathbf{u}}^i_P) \times \mathbf{F}^i\right| = \pm\left|\bar{\mathbf{u}}^i_P \times (\mathbf{A}^{i\mathrm{T}}\mathbf{F}^i)\right|$$

are the generalized forces associated with the generalized coordinates $\mathbf{R}^i$ and θ^i, respectively. This implies that a force acting on an arbitrary point P is equivalent to a force that has the same magnitude acting at the origin of the body reference and a moment acting on this body.

The generalized forces in the spatial analysis can be derived in a similar manner. In this case $\mathbf{F}^i = [F_1^i\ F_2^i\ F_3^i]^{\mathrm{T}}$ and

$$\delta \mathbf{r}_P^i = [\mathbf{I}_3 \quad \mathbf{B}^i] \begin{bmatrix} \delta \mathbf{R}^i \\ \delta \boldsymbol{\theta}^i \end{bmatrix}$$

where $\mathbf{I}_3$ is a 3×3 identity matrix and $\mathbf{B}^i$ is a matrix whose columns are the result of differentiating $\mathbf{A}^i \bar{\mathbf{u}}_P^i$ with respect to the rotational coordinates

Example 3.5 Figure 14 shows two bodies, body i and body j, connected by a spring–damper–actuator element. The attachment points of the spring–damper–actuator element on body i and body j are, respectively P^i and P^j. The spring constant is k, the damping coefficient is c, and the actuator force acting along a line connecting points P^i and P^j is f_a. The undeformed length of the spring is denoted as l_o. The component of the spring-damper-actuator force along a line connecting points P^i and P^j can then be written as

$$F_s = k(l - l_o) + c\dot{l} + f_a$$

where l is the spring length and $\dot{l}$ is the time derivative of l. The first term in this equation is the spring force, the second term represents the damping force, and the third term is the actuator force. Realizing that the spring force acts in a direction opposite to the direction of the increase in length, we may write the virtual work of the force F_s as

$$\delta W = -F_s \delta l$$

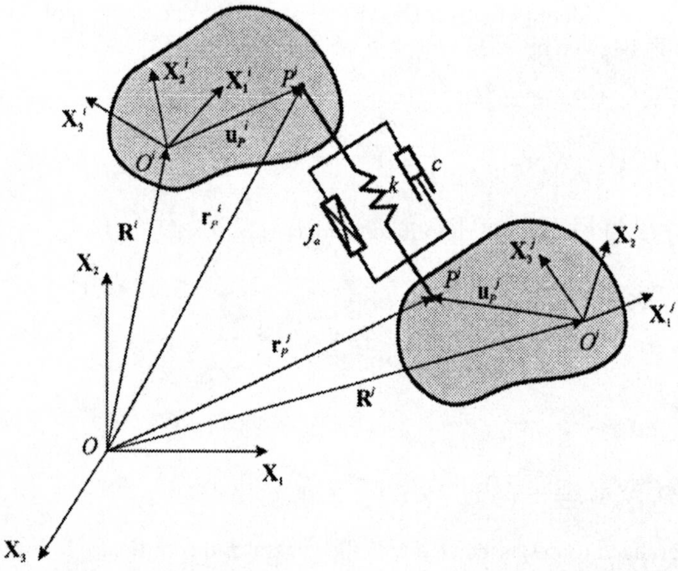

Figure 3.14 Spring–damper–actuator force element.

where δl is the virtual change in the spring length. Denoting the vector $P^i P^j$ as $\mathbf{l}_s$ whose components are

$$\mathbf{l}_s = [l_1 \quad l_2 \quad l_3]^{\mathrm{T}}$$

the spring length l can be evaluated from the relation

$$l = (\mathbf{l}_s^{\mathrm{T}}\mathbf{l}_s)^{1/2} = \left[(l_1)^2 + (l_2)^2 + (l_3)^2\right]^{1/2}$$

in which

$$\mathbf{l}_s = \mathbf{r}_P^i - \mathbf{r}_P^j = \mathbf{R}^i + \mathbf{A}^i\bar{\mathbf{u}}_P^i - \mathbf{R}^j - \mathbf{A}^j\bar{\mathbf{u}}_P^j$$

where $\bar{\mathbf{u}}_P^i$ and $\bar{\mathbf{u}}_P^j$ are the local positions of P^i and P^j, $\mathbf{R}^i$ and $\mathbf{R}^j$ are the global positions of the origins of the body axes of body i and body j, respectively, and $\mathbf{A}^i$ and $\mathbf{A}^j$ are the transformation matrices from the local to the global coordinate systems. One can show that the virtual change in the length δl can be written as

$$\begin{aligned}\delta l &= \frac{\partial l}{\partial l_1}\delta l_1 + \frac{\partial l}{\partial l_2}\delta l_2 + \frac{\partial l}{\partial l_3}\delta l_3 \\ &= \frac{1}{l}[l_1\delta l_1 + l_2\delta l_2 + l_3\delta l_3]\end{aligned}$$

which in vector notation can be written as

$$\delta l = \frac{1}{l}\mathbf{l}_s^{\mathrm{T}}\delta\mathbf{l}_s = \hat{\mathbf{l}}_s^{\mathrm{T}}\delta\mathbf{l}_s$$

where $\hat{\mathbf{l}}_s$ is a unit vector along $\mathbf{l}_s$ and $\delta\mathbf{l}_s$ is given by

$$\delta\mathbf{l}_s = \delta\mathbf{R}^i + \mathbf{B}^i\delta\boldsymbol{\theta}^i - \delta\mathbf{R}^j - \mathbf{B}^j\delta\boldsymbol{\theta}^j$$

where $\mathbf{q}_r^i = [\mathbf{R}^{i\mathrm{T}} \; \boldsymbol{\theta}^{i\mathrm{T}}]^{\mathrm{T}}$ and $\mathbf{q}_r^j = [\mathbf{R}^{j\mathrm{T}} \; \boldsymbol{\theta}^{j\mathrm{T}}]^{\mathrm{T}}$ are the generalized coordinates of bodies i and j, respectively, and $\mathbf{B}^k$ is the partial derivative of $\mathbf{A}^k\bar{\mathbf{u}}_P^k$ with respect to the rotational coordinates $\boldsymbol{\theta}^k$ of body $k (k = i, j)$. In matrix notation $\delta\mathbf{l}_s$ can be written as

$$\delta\mathbf{l}_s = [\mathbf{I}_3 \quad \mathbf{B}^i]\begin{bmatrix}\delta\mathbf{R}^i \\ \delta\boldsymbol{\theta}^i\end{bmatrix} - [\mathbf{I}_3 \quad \mathbf{B}^j]\begin{bmatrix}\delta\mathbf{R}^j \\ \delta\boldsymbol{\theta}^j\end{bmatrix}$$

It follows that the virtual work δW can be written as

$$\begin{aligned}\delta W &= -F_s\delta l = -F_s\hat{\mathbf{l}}_s^{\mathrm{T}}\delta\mathbf{l}_s \\ &= -F_s\hat{\mathbf{l}}_s^{\mathrm{T}}[\mathbf{I}_3 \quad \mathbf{B}^i]\begin{bmatrix}\delta\mathbf{R}^i \\ \delta\boldsymbol{\theta}^i\end{bmatrix} + F_s\hat{\mathbf{l}}_s^{\mathrm{T}}[\mathbf{I}_3 \quad \mathbf{B}^j]\begin{bmatrix}\delta\mathbf{R}^j \\ \delta\boldsymbol{\theta}^j\end{bmatrix} \\ &= [\mathbf{Q}_R^{i\mathrm{T}} \quad \mathbf{Q}_\theta^{i\mathrm{T}}]\begin{bmatrix}\delta\mathbf{R}^i \\ \delta\boldsymbol{\theta}^i\end{bmatrix} + [\mathbf{Q}_R^{j\mathrm{T}} \quad \mathbf{Q}_\theta^{j\mathrm{T}}]\begin{bmatrix}\delta\mathbf{R}^j \\ \delta\boldsymbol{\theta}^j\end{bmatrix}\end{aligned}$$

where $\mathbf{I}_3$ is a 3×3 identity matrix and $\mathbf{Q}_R^i$, $\mathbf{Q}_\theta^i$, $\mathbf{Q}_R^j$, and $\mathbf{Q}_\theta^j$ are the vectors of generalized forces associated with the generalized coordinates $\mathbf{R}^i$, $\boldsymbol{\theta}^i$, $\mathbf{R}^j$, and $\boldsymbol{\theta}^j$ and given by

$$\begin{aligned}\mathbf{Q}_R^{i\mathrm{T}} &= -F_s\hat{\mathbf{l}}_s^{\mathrm{T}}, & \mathbf{Q}_\theta^{i\mathrm{T}} &= -F_s\hat{\mathbf{l}}_s^{\mathrm{T}}\mathbf{B}^i \\ \mathbf{Q}_R^{j\mathrm{T}} &= F_s\hat{\mathbf{l}}_s^{\mathrm{T}}, & \mathbf{Q}_\theta^{j\mathrm{T}} &= -F_s\hat{\mathbf{l}}_s^{\mathrm{T}}\mathbf{B}^j\end{aligned}$$

It is left to the reader as an exercise to exemplify the preceding formulation in the two-dimensional case and also in the three-dimensional case by using different sets of orientational coordinates.

Constrained Motion In the preceding examples we derived the virtual work expression and generalized forces for unconstrained rigid bodies. The virtual work and generalized forces can also be derived for rigid body systems with constraints. This can be achieved by identifying the system-independent coordinates, and, by inspection, one may try to determine the generalized forces associated with these coordinates. For large-scale constrained mechanical systems, however, this approach may be difficult to follow. An alternate and systematic approach is to develop first the virtual work in terms of the system Cartesian coordinates that can be written in a vector form as

$$\delta W = \mathbf{Q}^{\mathrm{T}} \delta \mathbf{q}$$

where $\mathbf{Q}$ is the generalized force vector and $\delta \mathbf{q}$ is the virtual change in the vector of system coordinates. One can then, as described in the previous section, use the constraint Jacobian matrix to identify a set of independent coordinates. In this case, the system coordinates can be written in terms of the independent coordinates as

$$\delta \mathbf{q} = \mathbf{B}_{di} \delta \mathbf{q}_i$$

where $\mathbf{q}_i$ is the vector of system independent coordinates or degrees of freedom and $\mathbf{B}_{di}$ is an appropriate transformation matrix. In terms of these independent coordinates, the virtual work can be written as

$$\delta W = \mathbf{Q}^{\mathrm{T}} \mathbf{B}_{di} \delta \mathbf{q}_i = \mathbf{Q}_i^{\mathrm{T}} \delta \mathbf{q}_i$$

where

$$\mathbf{Q}_i^{\mathrm{T}} = \mathbf{Q}^{\mathrm{T}} \mathbf{B}_{di}$$

is the vector of generalized forces associated with the independent coordinates or the system degrees of freedom.

3.4 LAGRANGIAN DYNAMICS

In this section, D'Alembert's principle, discussed in the preceding section, will be used to derive Lagrange's equation. The development will be exemplified by using a system of n_p particles. The displacement $\mathbf{r}^i$ of the ith particle is assumed to depend on a set of system generalized coordinates q_j, where $j = 1, 2, \ldots, n$. Hence

$$\mathbf{r}^i = \mathbf{r}^i(q_1, q_2, \ldots, q_n, t) \tag{3.40}$$

where t is the time. Differentiating Eq. 40 with respect to time using the chain rule of differentiation yields

$$\dot{\mathbf{r}}^i = \frac{\partial \mathbf{r}^i}{\partial q_1}\dot{q}_1 + \frac{\partial \mathbf{r}^i}{\partial q_2}\dot{q}_2 + \cdots + \frac{\partial \mathbf{r}^i}{\partial q_n}\dot{q}_n + \frac{\partial \mathbf{r}^i}{\partial t} = \sum_{j=1}^{n} \frac{\partial \mathbf{r}^i}{\partial q_j}\dot{q}_j + \frac{\partial \mathbf{r}^i}{\partial t} \tag{3.41}$$

The virtual displacement $\delta\mathbf{r}^i$ can be written in terms of the coordinates q_j as

$$\delta\mathbf{r}^i = \sum_{j=1}^{n} \frac{\partial\mathbf{r}^i}{\partial q_j}\delta q_j \tag{3.42}$$

Using this expression for the virtual displacement, one may write the virtual work of the force $\mathbf{F}^i$ acting on the ith particle as

$$\mathbf{F}^{i\mathrm{T}}\delta\mathbf{r}^i = \sum_{j=1}^{n} \mathbf{F}^{i\mathrm{T}}\frac{\partial\mathbf{r}^i}{\partial q_j}\delta q_j \tag{3.43}$$

Equation 43 can be written for every particle in the system. By summing up these expressions, one gets

$$\begin{aligned}\sum_{i=1}^{n_p} \mathbf{F}^{i\mathrm{T}}\delta\mathbf{r}^i &= \sum_{i=1}^{n_p}\sum_{j=1}^{n} \mathbf{F}^{i\mathrm{T}}\frac{\partial\mathbf{r}^i}{\partial q_j}\delta q_j \\ &= \sum_{j=1}^{n}\sum_{i=1}^{n_p} \mathbf{F}^{i\mathrm{T}}\frac{\partial\mathbf{r}^i}{\partial q_j}\delta q_j = \sum_{j=1}^{n} Q_j\delta q_j\end{aligned} \tag{3.44}$$

where Q_j is called the *component of the generalized force associated with the coordinate* q_j, that is

$$Q_j = \sum_{i=1}^{n_p} \mathbf{F}^{i\mathrm{T}}\frac{\partial\mathbf{r}^i}{\partial q_j} \tag{3.45}$$

The virtual work of the inertia force of the ith particle can be written as

$$\delta W_i^i = m^i\ddot{\mathbf{r}}^i\cdot\delta\mathbf{r}^i$$

where m^i and $\ddot{\mathbf{r}}^i$ are, respectively, the mass and acceleration vector of particle i. The virtual work due to all inertia forces in the system can then be written as

$$\delta W_i = \sum_{i=1}^{n_p} m^i\ddot{\mathbf{r}}^i\cdot\delta\mathbf{r}^i \tag{3.46}$$

Using Eq. 42, we may write Eq. 46 in the following form:

$$\delta W_i = \sum_{i=1}^{n_p}\sum_{j=1}^{n} m^i\ddot{\mathbf{r}}^i\cdot\frac{\partial\mathbf{r}^i}{\partial q_j}\delta q_j \tag{3.47}$$

The following identity can be verified:

$$\sum_{i=1}^{n_p}\frac{d}{dt}\left(m^i\dot{\mathbf{r}}^i\cdot\frac{\partial\mathbf{r}^i}{\partial q_j}\right) = \sum_{i=1}^{n_p} m^i\ddot{\mathbf{r}}^i\cdot\frac{\partial\mathbf{r}^i}{\partial q_j} + \sum_{i=1}^{n_p} m^i\dot{\mathbf{r}}^i\cdot\frac{d}{dt}\left(\frac{\partial\mathbf{r}^i}{\partial q_j}\right)$$

which yields

$$\sum_{i=1}^{n_p}\left(m^i\ddot{\mathbf{r}}^i\cdot\frac{\partial\mathbf{r}^i}{\partial q_j}\right) = \sum_{i=1}^{n_p}\left[\frac{d}{dt}\left(m^i\dot{\mathbf{r}}^i\cdot\frac{\partial\mathbf{r}^i}{\partial q_j}\right) - m^i\dot{\mathbf{r}}^i\cdot\frac{d}{dt}\left(\frac{\partial\mathbf{r}^i}{\partial q_j}\right)\right] \tag{3.48}$$

By using Eq. 41 and by interchanging the differentiation with respect to t and q_j, one gets

$$\frac{d}{dt}\left(\frac{\partial \mathbf{r}^i}{\partial q_j}\right) = \sum_{k=1}^{n} \frac{\partial^2 \mathbf{r}^i}{\partial q_j \partial q_k}\dot{q}_k + \frac{\partial^2 \mathbf{r}^i}{\partial q_j \partial t} = \frac{\partial \dot{\mathbf{r}}^i}{\partial q_j} \tag{3.49}$$

By taking the partial derivative of $\dot{\mathbf{r}}^i$ in Eq. 41 with respect to $\dot{q}_j$ we obtain

$$\frac{\partial \dot{\mathbf{r}}^i}{\partial \dot{q}_j} = \frac{\partial \mathbf{r}^i}{\partial q_j} \tag{3.50}$$

It follows then from Eq. 48 that

$$\begin{aligned}\sum_{i=1}^{n_p} m^i \ddot{\mathbf{r}}^i \cdot \frac{\partial \mathbf{r}^i}{\partial q_j} &= \sum_{i=1}^{n_p}\left[\frac{d}{dt}\left(m^i \dot{\mathbf{r}}^i \cdot \frac{\partial \mathbf{r}^i}{\partial q_j}\right) - m^i \dot{\mathbf{r}}^i \cdot \frac{d}{dt}\left(\frac{\partial \mathbf{r}^i}{\partial q_j}\right)\right] \\ &= \sum_{i=1}^{n_p}\left\{\frac{d}{dt}\left[\frac{\partial}{\partial \dot{q}_j}\left(\frac{1}{2}m^i \dot{\mathbf{r}}^{i\mathrm{T}}\dot{\mathbf{r}}^i\right)\right] - \frac{\partial}{\partial q_j}\left(\frac{1}{2}m^i \dot{\mathbf{r}}^{i\mathrm{T}}\dot{\mathbf{r}}^i\right)\right\}\end{aligned} \tag{3.51}$$

One may denote the ith particle kinetic energy as T^i, that is

$$T^i = \frac{1}{2}m^i \dot{\mathbf{r}}^{i\mathrm{T}}\dot{\mathbf{r}}^i,$$

and write Eq. 51 in a more simplified form as

$$\sum_{i=1}^{n_p} m^i \ddot{\mathbf{r}}^i \cdot \frac{\partial \mathbf{r}^i}{\partial q_j} = \sum_{i=1}^{n_p}\left\{\frac{d}{dt}\left[\frac{\partial}{\partial \dot{q}_j}(T^i)\right] - \frac{\partial T^i}{\partial q_j}\right\}$$

or alternatively

$$\sum_{i=1}^{n_p} m^i \ddot{\mathbf{r}}^i \cdot \frac{\partial \mathbf{r}^i}{\partial q_j} = \frac{d}{dt}\left(\frac{\partial T}{\partial \dot{q}_j}\right) - \frac{\partial T}{\partial q_j} \tag{3.52}$$

where T is the total system kinetic energy given by

$$T = \sum_{i=1}^{n_p} T^i = \sum_{i=1}^{n_p} \frac{1}{2}m^i \dot{\mathbf{r}}^{i\mathrm{T}}\dot{\mathbf{r}}^i$$

Substituting Eq. 52 into Eq. 47 and using D'Alembert's principle of Eq. 35 yields

$$\sum_j \left[\frac{d}{dt}\left(\frac{\partial T}{\partial \dot{q}_j}\right) - \frac{\partial T}{\partial q_j} - Q_j\right]\delta q_j = 0 \tag{3.53}$$

This equation is sometimes called *D'Alembert–Lagrange's* equation. If the set of generalized coordinates q_j is linearly independent, Eq. 53 leads to *Lagrange's equation*, which is given by

$$\frac{d}{dt}\left(\frac{\partial T}{\partial \dot{q}_j}\right) - \frac{\partial T}{\partial q_j} - Q_j = 0, \quad j = 1, 2, \ldots, n \tag{3.54}$$

It is sometimes convenient to write D'Alembert–Lagrange's equation in a matrix form. To this end, we write Eq. 53 in a more explicit form as

$$\left[\frac{d}{dt}\left(\frac{\partial T}{\partial \dot{q}_1}\right) - \frac{\partial T}{\partial q_1} - Q_1\right]\delta q_1 + \left[\frac{d}{dt}\left(\frac{\partial T}{\partial \dot{q}_2}\right) - \frac{\partial T}{\partial q_2} - Q_2\right]\delta q_2 + \cdots + \left[\frac{d}{dt}\left(\frac{\partial T}{\partial \dot{q}_n}\right) - \frac{\partial T}{\partial q_n} - Q_n\right]\delta q_n = 0$$

which can be rewritten as

$$\left[\frac{d}{dt}\left(\frac{\partial T}{\partial \dot{q}_1}\right) \quad \frac{d}{dt}\left(\frac{\partial T}{\partial \dot{q}_2}\right) \quad \cdots \quad \frac{d}{dt}\left(\frac{\partial T}{\partial \dot{q}_n}\right)\right]\begin{bmatrix}\delta q_1\\ \delta q_2\\ \vdots\\ \delta q_n\end{bmatrix}$$

$$-\left[\frac{\partial T}{\partial q_1} \quad \frac{\partial T}{\partial q_2} \quad \cdots \quad \frac{\partial T}{\partial q_n}\right]\begin{bmatrix}\delta q_1\\ \delta q_2\\ \vdots\\ \delta q_n\end{bmatrix} - [Q_1 \quad Q_2 \quad \cdots \quad Q_n]\begin{bmatrix}\delta q_1\\ \delta q_2\\ \vdots\\ \delta q_n\end{bmatrix} = 0$$

That is,

$$\left[\frac{d}{dt}\left(\frac{\partial T}{\partial \dot{\mathbf{q}}}\right) - \frac{\partial T}{\partial \mathbf{q}} - \mathbf{Q}^{\mathrm{T}}\right]\delta\mathbf{q} = 0$$

where

$$\frac{d}{dt}\left(\frac{\partial T}{\partial \dot{\mathbf{q}}}\right) = \left[\frac{d}{dt}\left(\frac{\partial T}{\partial \dot{q}_1}\right) \quad \frac{d}{dt}\left(\frac{\partial T}{\partial \dot{q}_2}\right) \quad \cdots \quad \frac{d}{dt}\left(\frac{\partial T}{\partial \dot{q}_n}\right)\right]$$

$$\frac{\partial T}{\partial \mathbf{q}} = \left[\frac{\partial T}{\partial q_1} \quad \frac{\partial T}{\partial q_2} \quad \cdots \quad \frac{\partial T}{\partial q_n}\right], \qquad \mathbf{Q}^{\mathrm{T}} = [Q_1 \quad Q_2 \quad \cdots \quad Q_n]$$

Example 3.6 Derive the differential equations of motion of the system given in Example 3 using Lagrange's equation.

Solution It was shown in Example 3 that the velocity of the particle can be written in terms of the independent coordinates and their time derivatives as

$$\dot{\mathbf{r}} = \dot{\theta}\begin{bmatrix}-\sin\theta\\ \cos\theta\\ 0\end{bmatrix} q + \begin{bmatrix}\cos\theta\\ \sin\theta\\ 0\end{bmatrix}\dot{q} = \begin{bmatrix}-q\sin\theta & \cos\theta\\ q\cos\theta & \sin\theta\\ 0 & 0\end{bmatrix}\begin{bmatrix}\dot{\theta}\\ \dot{q}\end{bmatrix} = \mathbf{B}\dot{\mathbf{q}}$$

where

$$\mathbf{q} = [\theta \quad q]^{\mathrm{T}}$$

$$\mathbf{B} = \begin{bmatrix}-q\sin\theta & \cos\theta\\ q\cos\theta & \sin\theta\\ 0 & 0\end{bmatrix}$$

Since the rod is assumed to be massless, the kinetic energy of the system is given by

$$T = \frac{1}{2}m\dot{\mathbf{r}}^{\mathrm{T}}\dot{\mathbf{r}} = \frac{1}{2}m\dot{\mathbf{q}}^{\mathrm{T}}\mathbf{B}^{\mathrm{T}}\mathbf{B}\dot{\mathbf{q}}$$

in which $\mathbf{B}^{\mathrm{T}}\mathbf{B}$ is the 2×2 matrix given by

$$\mathbf{B}^{\mathrm{T}}\mathbf{B} = \begin{bmatrix} -q\sin\theta & q\cos\theta & 0 \\ \cos\theta & \sin\theta & 0 \end{bmatrix} \begin{bmatrix} -q\sin\theta & \cos\theta \\ q\cos\theta & \sin\theta \\ 0 & 0 \end{bmatrix} = \begin{bmatrix} (q)^2 & 0 \\ 0 & 1 \end{bmatrix}$$

Therefore, the kinetic energy T is given by

$$T = \frac{1}{2}m(\dot{\theta})^2(q)^2 + \frac{1}{2}m(\dot{q})^2$$

It can be shown that

$$\frac{\partial T}{\partial \dot{\theta}} = m(q)^2\dot{\theta}, \qquad \frac{d}{dt}\left(\frac{\partial T}{\partial \dot{\theta}}\right) = m(q)^2\ddot{\theta} + 2mq\dot{q}\dot{\theta}, \qquad \frac{\partial T}{\partial \theta} = 0$$

$$\frac{\partial T}{\partial \dot{q}} = m\dot{q}, \qquad \frac{d}{dt}\left(\frac{\partial T}{\partial \dot{q}}\right) = m\ddot{q}, \qquad \frac{\partial T}{\partial q} = m(\dot{\theta})^2 q$$

The virtual work of the external forces is given by

$$\delta W = T\delta\theta - mg\delta(q\sin\theta) = (T - mgq\cos\theta)\delta\theta - mg\sin\theta\,\delta q$$

That is, the generalized forces Q_θ and Q_q associated, respectively, with the generalized coordinates θ and q are given by

$$Q_\theta = T - mgq\cos\theta, \qquad Q_q = -mg\sin\theta$$

Since we have two independent coordinates, θ and q, we have the following two Lagrange's equations:

$$\frac{d}{dt}\left(\frac{\partial T}{\partial \dot{\theta}}\right) - \frac{\partial T}{\partial \theta} = Q_\theta$$

$$\frac{d}{dt}\left(\frac{\partial T}{\partial \dot{q}}\right) - \frac{\partial T}{\partial q} = Q_q$$

which lead to the following two differential equations of motion:

$$m(q)^2\ddot{\theta} + 2mq\dot{q}\dot{\theta} = T - mgq\cos\theta$$

$$m\ddot{q} - m(\dot{\theta})^2 q = -mg\sin\theta$$

which are the same differential equations obtained in Example 3.

To arrive at Eq. 54 from Eq. 53, it was assumed that the virtual changes in the vector of system coordinates $\mathbf{q}$ are independent. In multibody systems, kinematic constraint equations may exist because of mechanical joints or specified motion trajectories. In this case, two procedures can be followed to formulate the dynamic equations of constrained multibody systems. These procedures are the *embedding technique* and the *augmented formulation*. In the embedding technique, the system dynamic equations are formulated in terms of the degrees of freedom. This technique leads to a minimum set of dynamic equations that do not contain any constraint forces. In the augmented formulation, on the other hand, the dynamic equations are formulated in terms of a set of redundant coordinates. As a consequence, the resulting equations are expressed in terms of dependent coordinates as well as the constraint forces. The numerical solution of the equations obtained using the embedding technique requires only the integration of a system of differential equations, while the solution of the equations obtained using the augmented formulation requires the solution of a system of differential and algebraic equations.

Embedding Technique The constraint equations of the multibody system can be written as

$$\mathbf{C}(\mathbf{q}, t) = \mathbf{0} \tag{3.55}$$

where $\mathbf{C} = [C_1(\mathbf{q}, t)\ C_2(\mathbf{q}, t) \cdots C_{n_c}(\mathbf{q}, t)]^{\mathrm{T}}$ is the vector of constraint functions and n_c is the number of constraint equations. For a virtual displacement $\delta\mathbf{q}$, Eq. 55 yields

$$\mathbf{C}_{\mathbf{q}}\, \delta\mathbf{q} = \mathbf{0} \tag{3.56}$$

where $\mathbf{C}_{\mathbf{q}}$ is the constraint Jacobian matrix.

For holonomic systems, one should be able to identify a set of independent coordinates (degrees of freedom) and write the system coordinates in terms of these independent ones. Let $\mathbf{q}_d$ denote the set of dependent coordinates and $\mathbf{q}_i$ the set of independent ones. The vector $\mathbf{q}$ of generalized coordinates can then be written in partitioned form as

$$\mathbf{q} = \begin{bmatrix}\mathbf{q}_d^{\mathrm{T}} & \mathbf{q}_i^{\mathrm{T}}\end{bmatrix}^{\mathrm{T}}$$

It follows that

$$\delta\mathbf{q} = \begin{bmatrix}\delta\mathbf{q}_d^{\mathrm{T}} & \delta\mathbf{q}_i^{\mathrm{T}}\end{bmatrix}^{\mathrm{T}}$$

Equation 56 can then be written according to this coordinate partitioning as

$$\mathbf{C}_{\mathbf{q}_d}\delta\mathbf{q}_d + \mathbf{C}_{\mathbf{q}_i}\delta\mathbf{q}_i = \mathbf{0} \tag{3.57}$$

where $\mathbf{C}_{\mathbf{q}_d}$ and $\mathbf{C}_{\mathbf{q}_i}$ are the constraint Jacobian matrices associated with the dependent and independent coordinates, respectively. If the constraints of Eq. 55 are linearly independent, one should be able to identify the coordinates $\mathbf{q}_i$ such that $\mathbf{C}_{\mathbf{q}_d}$ has a full row rank and, hence, nonsingular. If $\mathbf{C}_{\mathbf{q}_d}$ is nonsingular, the inverse of $\mathbf{C}_{\mathbf{q}_d}$, denoted as $\mathbf{C}_{\mathbf{q}_d}^{-1}$, exists and Eq. 57 yields

$$\delta\mathbf{q}_d = -\mathbf{C}_{\mathbf{q}_d}^{-1}\mathbf{C}_{\mathbf{q}_i}\delta\mathbf{q}_i$$

By doing this, the virtual displacement of the dependent coordinates is written in terms of the virtual displacement of the independent ones. In a more compact form, this relation can be stated as

$$\delta\mathbf{q}_d = \mathbf{C}_{di}\delta\mathbf{q}_i$$

where

$$\mathbf{C}_{di} = -\mathbf{C}_{\mathbf{q}_d}^{-1}\mathbf{C}_{\mathbf{q}_i}$$

Therefore, one can write the vector $\delta\mathbf{q}$ as

$$\delta\mathbf{q} = \begin{bmatrix} \delta\mathbf{q}_i \\ \delta\mathbf{q}_d \end{bmatrix} = \begin{bmatrix} \delta\mathbf{q}_i \\ \mathbf{C}_{di}\delta\mathbf{q}_i \end{bmatrix} = \begin{bmatrix} \mathbf{I} \\ \mathbf{C}_{di} \end{bmatrix} \delta\mathbf{q}_i$$

which can also be written as

$$\delta\mathbf{q} = \mathbf{B}_{di}\delta\mathbf{q}_i \tag{3.58}$$

where the matrix $\mathbf{B}_{di}$ is given by

$$\mathbf{B}_{di} = \begin{bmatrix} \mathbf{I} \\ \mathbf{C}_{di} \end{bmatrix}$$

in which $\mathbf{I}$ is an identity matrix, with dimension $n - n_c$.

Using vector notation, it was shown that Eq. 53 can be written as

$$\left[\frac{d}{dt}(T_{\dot{\mathbf{q}}}) - T_{\mathbf{q}} - \mathbf{Q}^{\mathrm{T}}\right]\delta\mathbf{q} = 0 \tag{3.59}$$

where $\mathbf{Q}$ is the vector of the system generalized forces and the subscript vector denotes differentiation with respect to this vector.

Equations 58 and 59 yield

$$\left[\frac{d}{dt}(T_{\dot{\mathbf{q}}}) - T_{\mathbf{q}} - \mathbf{Q}^{\mathrm{T}}\right]\mathbf{B}_{di}\delta\mathbf{q}_i = 0 \tag{3.60}$$

Since δq_i, $i = 1, 2, \ldots, n - n_c$, are linearly independent, by using Eq. 60, one arrives at

$$\left[\frac{d}{dt}(T_{\dot{\mathbf{q}}}) - T_{\mathbf{q}} - \mathbf{Q}^{\mathrm{T}}\right]\mathbf{B}_{di} = \mathbf{0}^{\mathrm{T}} \tag{3.61}$$

where Eq. 61 contains $(n - n_c)$ differential equations. Furthermore, in these equations the constraint forces are automatically eliminated since only independent coordinates are used.

Augmented Formulation In the augmented formulation, the method of Lagrange multipliers that can be applied to both holonomic and nonholonomic systems is used. If Eq. 56 holds and/or the constraint relationships are velocity-dependent and nonintegrable, then it is also true that

$$\boldsymbol{\lambda}^{\mathrm{T}}\mathbf{C}_{\mathbf{q}}\delta\mathbf{q} = 0 \tag{3.62}$$

where $\boldsymbol{\lambda} = [\lambda_1 \; \lambda_2 \; \cdots \; \lambda_{n_c}]^{\mathrm{T}}$ is the vector of *Lagrange multipliers* (Shabana 1994a). Equations 59 and 62 can be combined to yield

$$\delta\mathbf{q}^{\mathrm{T}}\left[\frac{d}{dt}\left(\frac{\partial T}{\partial \dot{\mathbf{q}}}\right)^{\mathrm{T}} - \left(\frac{\partial T}{\partial \mathbf{q}}\right)^{\mathrm{T}} - \mathbf{Q} + \mathbf{C}_{\mathbf{q}}^{\mathrm{T}}\boldsymbol{\lambda}\right] = 0 \tag{3.63}$$

The components of the virtual displacement vector $\delta\mathbf{q}$ are still not independent because of the holonomic or nonholonomic constraint equations. Suppose that we select $\lambda_k, k = 1, 2, \ldots, n_c$ such that

$$\frac{d}{dt}\left(\frac{\partial T}{\partial \dot{\mathbf{q}}_d}\right)^{\mathrm{T}} - \left(\frac{\partial T}{\partial \mathbf{q}_d}\right)^{\mathrm{T}} - \mathbf{Q}_d + \mathbf{C}_{\mathbf{q}_d}^{\mathrm{T}}\boldsymbol{\lambda} = \mathbf{0} \tag{3.64}$$

where $\mathbf{q}_d = [q_1\, q_2 \cdots q_{n_c}]^{\mathrm{T}}$ are selected to be the dependent variables. Using Eq. 64, we can write Eq. 63 for the independent variables as

$$\delta\mathbf{q}_i^{\mathrm{T}}\left[\frac{d}{dt}\left(\frac{\partial T}{\partial \dot{\mathbf{q}}_i}\right)^{\mathrm{T}} - \left(\frac{\partial T}{\partial \mathbf{q}_i}\right)^{\mathrm{T}} - \mathbf{Q}_i + \mathbf{C}_{\mathbf{q}_i}^{\mathrm{T}}\boldsymbol{\lambda}\right] = 0$$

where $\mathbf{Q}_d$ and $\mathbf{Q}_i$ are, respectively, the vectors of generalized forces associated with the vectors of dependent and independent coordinates. Since the elements of the vector $\delta\mathbf{q}_i$ in this equation are independent, the following equations hold:

$$\frac{d}{dt}\left(\frac{\partial T}{\partial \dot{\mathbf{q}}_i}\right)^{\mathrm{T}} - \left(\frac{\partial T}{\partial \mathbf{q}_i}\right)^{\mathrm{T}} - \mathbf{Q}_i + \mathbf{C}_{\mathbf{q}_i}^{\mathrm{T}}\boldsymbol{\lambda} = \mathbf{0} \tag{3.65}$$

Since $\mathbf{q}_d$ and $\mathbf{q}_i$ are the partitions of $\mathbf{q}$, we may combine Eqs. 64 and 65 in one vector equation as follows:

$$\frac{d}{dt}\left(\frac{\partial T}{\partial \dot{\mathbf{q}}}\right)^{\mathrm{T}} - \left(\frac{\partial T}{\partial \mathbf{q}}\right)^{\mathrm{T}} + \mathbf{C}_{\mathbf{q}}^{\mathrm{T}}\boldsymbol{\lambda} = \mathbf{Q} \tag{3.66}$$

Equation 66 is a system of differential equations of motion that along with the constraint equations can be solved for the vector of system generalized coordinates $\mathbf{q}$ and the vector of Lagrange multipliers $\boldsymbol{\lambda}$. This equation is used as a basis for developing many general computational algorithms for the dynamic analysis of multibody systems subject to both holonomic and nonholonomic constraints.

Application to Rigid Body Dynamics Thus far, we have used a system of particles to derive the principle of virtual work in dynamics and Lagrange's equation of motion. By considering the rigid body to consist of a large number of particles, one expects that these two approaches are applicable to rigid bodies as well. This can be demonstrated by solving a simple example using Newton's second law and then attempting to arrive at the same results by using the principle of virtual work in dynamics and Lagrange's equation. To this end, the pendulum shown in Fig. 15 is considered. Applying Newton's second law, the equations of motion of this pendulum can be written as

$$m\ddot{R}_1 = F_1, \qquad m\ddot{R}_2 = F_2, \qquad I_c\ddot{\theta} = M_c$$

where m and I_c are, respectively, the mass of the rod and the mass moment of inertia of the rod about its center of mass, that is,

$$I_c = \frac{m(l)^2}{12}$$

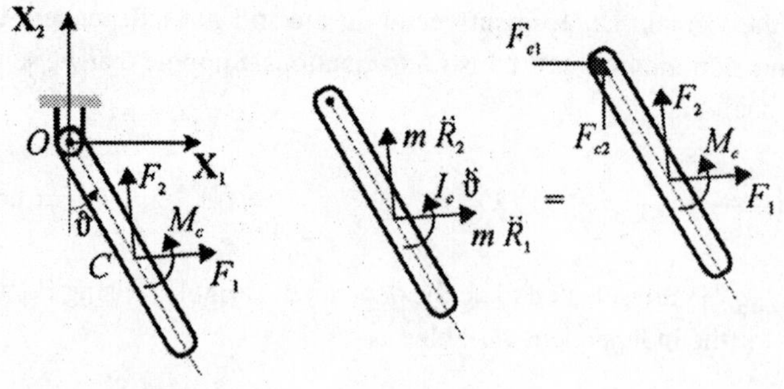

Figure 3.15 Planar pendulum.

l is the length of the rod; R_1 and R_2 are, respectively, the horizontal and vertical displacements of the center of mass of the rod; θ is the angular displacement; F_1 and F_2 are, respectively, the resultant forces in the horizontal and vertical directions; and M_c is the applied moment. The above-stated equations of motion imply that the inertia (effective) forces and moments should be equal, respectively, to the applied forces and moments. This is shown diagrammatically in Fig. 15, where F_{c1} and F_{c2} are the reaction forces at the pin joint. By taking the moment of the two systems of forces about point O, one obtains the scalar equation

$$m\ddot{R}_1\frac{l}{2}\cos\theta + m\ddot{R}_2\frac{l}{2}\sin\theta + I_c\ddot{\theta} = F_1\frac{l}{2}\cos\theta + F_2\frac{l}{2}\sin\theta + M_c$$

Since

$$R_1 = \frac{l}{2}\sin\theta, \qquad R_2 = -\frac{l}{2}\cos\theta$$

one has

$$\dot{R}_1 = \frac{l}{2}\dot{\theta}\cos\theta, \qquad \ddot{R}_1 = \frac{l}{2}\ddot{\theta}\cos\theta - \frac{l}{2}(\dot{\theta})^2\sin\theta$$

$$\dot{R}_2 = \frac{l}{2}\dot{\theta}\sin\theta, \qquad \ddot{R}_2 = \frac{l}{2}\ddot{\theta}\sin\theta + \frac{l}{2}(\dot{\theta})^2\cos\theta$$

The preceding equations lead to

$$\left(m\frac{(l)^2}{4} + I_c\right)\ddot{\theta} = M_o$$

where M_o is the external moment about O given by

$$M_o = F_1\frac{l}{2}\cos\theta + F_2\frac{l}{2}\sin\theta + M_c$$

Since $I_c = m(l)^2/12$, we conclude that

$$\frac{m(l)^2}{4} + I_c = \frac{m(l)^2}{3} = I_o$$

that is,

$$I_o\ddot{\theta} = M_o$$

This is the equation of motion of the single degree of freedom pendulum derived by applying Newton's second law of motion. The same equation can be derived by using Lagrange's equation. To this end, we write the kinetic energy of the rod as

$$T = \frac{1}{2}m(\dot{R}_1)^2 + \frac{1}{2}m(\dot{R}_2)^2 + \frac{1}{2}I_c(\dot{\theta})^2$$

Substituting the values of $\dot{R}_1$ and $\dot{R}_2$ in the kinetic energy expression leads to

$$T = \frac{1}{2}\left(m\frac{(l)^2}{4} + I_c\right)(\dot{\theta})^2 = \frac{1}{2}I_o(\dot{\theta})^2$$

The virtual work of external forces and moments is given by

$$\begin{aligned}
\delta W &= F_1\delta R_1 + F_2\delta R_2 + M_c\delta\theta \\
&= F_1\frac{l}{2}\cos\theta\delta\theta + F_2\frac{l}{2}\sin\theta\delta\theta + M_c\delta\theta \\
&= \left(F_1\frac{l}{2}\cos\theta + F_2\frac{l}{2}\sin\theta + M_c\right)\delta\theta \\
&= M_o\delta\theta
\end{aligned}$$

Using Lagrange's equation and keeping in mind that we have only one independent coordinate θ, we obtain the single equation

$$\frac{d}{dt}\left(\frac{\partial T}{\partial\dot{\theta}}\right) - \frac{\partial T}{\partial\theta} = M_o$$

which leads to

$$I_o\ddot{\theta} = M_o$$

which is the same equation obtained by applying Newton's second law.

The differential equation of the simple pendulum obtained in this section using Newton's second law and Lagrange's equation can also be derived using the principle of virtual work in dynamics. In this case, one defines the virtual displacements

$$\delta R_1 = \frac{l}{2}\cos\theta\delta\theta, \qquad \delta R_2 = \frac{l}{2}\sin\theta\delta\theta$$

Using these virtual changes and the expressions for the accelerations of the center of mass of the pendulum, the virtual work of the inertia forces of the pendulum can be written as

$$\delta W_i = m\ddot{R}_1\delta R_1 + m\ddot{R}_2\delta R_2 + I_c\ddot{\theta}\delta\theta$$

which reduces to

$$\delta W_i = I_o\ddot{\theta}\delta\theta$$

Equating this expression to the virtual work of the externally applied forces and keeping in mind that the virtual work of the constraint forces acting on the system is equal to zero (Shabana 1994a), one obtains the same equation of motion which was derived previously using Newton's second law and Lagrange's equation.

Elimination of the Constraint Forces In the pendulum example discussed in this section, the work and energy expressions were derived in terms of the system degree of freedom, which was selected to be the angular rotation of the pendulum θ. An alternate approach is to derive the kinetic energy and virtual work expressions in terms of the system coordinates R_1, R_2, and θ and use the variational form of Lagrange's equation of motion given by Eq. 59. One can then use the generalized coordinate partitioning to identify the independent coordinates and accordingly the matrix $\mathbf{B}_{di}$ of Eq. 60. This will eventually lead to a single second-order differential equation associated with the system-independent coordinate θ. For instance, since the kinetic energy of the pendulum is a quadratic form in the velocities, that is

$$T = \frac{1}{2}m(\dot{R}_1)^2 + \frac{1}{2}m(\dot{R}_2)^2 + \frac{1}{2}I_c(\dot{\theta})^2$$

using matrix notation, one can write the kinetic energy as

$$T = \frac{1}{2}[\dot{R}_1 \quad \dot{R}_2 \quad \dot{\theta}] \begin{bmatrix} m & 0 & 0 \\ 0 & m & 0 \\ 0 & 0 & I_c \end{bmatrix} \begin{bmatrix} \dot{R}_1 \\ \dot{R}_2 \\ \dot{\theta} \end{bmatrix}$$

and the virtual work as

$$\delta W = [F_1 \quad F_2 \quad M_c] \begin{bmatrix} \delta R_1 \\ \delta R_2 \\ \delta\theta \end{bmatrix}$$

which implies that the vector $\mathbf{Q}$ of the system generalized forces of Eq. 59 is given by

$$\mathbf{Q}^{\mathrm{T}} = [F_1 \quad F_2 \quad M_c]$$

We also have

$$(T_{\dot{\mathbf{q}}})^{\mathrm{T}} = \begin{bmatrix} m & 0 & 0 \\ 0 & m & 0 \\ 0 & 0 & I_c \end{bmatrix} \begin{bmatrix} \dot{R}_1 \\ \dot{R}_2 \\ \dot{\theta} \end{bmatrix}$$

where $\mathbf{q} = [R_1 \; R_2 \; \theta]^{\mathrm{T}}$ is the total vector of system coordinates; thus

$$\frac{d}{dt}(T_{\dot{\mathbf{q}}})^{\mathrm{T}} = \begin{bmatrix} m & 0 & 0 \\ 0 & m & 0 \\ 0 & 0 & I_c \end{bmatrix} \begin{bmatrix} \ddot{R}_1 \\ \ddot{R}_2 \\ \ddot{\theta} \end{bmatrix}$$

Since the kinetic energy does not, in this example, depend on the system coordinates, one can verify that

$$T_{\mathbf{q}} = \frac{\partial T}{\partial \mathbf{q}} = \mathbf{0}^{\mathrm{T}}$$

Therefore, Eq. 59 can be written as

$$\left\{ [\ddot{R}_1 \quad \ddot{R}_2 \quad \ddot{\theta}] \begin{bmatrix} m & 0 & 0 \\ 0 & m & 0 \\ 0 & 0 & I_c \end{bmatrix} - [F_1 \quad F_2 \quad M_c] \right\} \begin{bmatrix} \delta R_1 \\ \delta R_2 \\ \delta\theta \end{bmatrix} = 0$$

which can also be written as

$$[\ddot{\mathbf{q}}^{\mathrm{T}}\mathbf{M} - \mathbf{Q}^{\mathrm{T}}]\delta\mathbf{q} = 0$$

where $\mathbf{M}$ is the system mass matrix defined as

$$\mathbf{M} = \begin{bmatrix} m & 0 & 0 \\ 0 & m & 0 \\ 0 & 0 & I_c \end{bmatrix}$$

The terms between brackets in the system variational equations cannot be set equal to zero because δR_1, δR_2, and $\delta\theta$ are not independent. They are related by the constraint equations that describe the pin joint at O and are given by

$$R_1 - \frac{l}{2}\sin\theta = 0, \qquad R_2 + \frac{l}{2}\cos\theta = 0$$

For a virtual change in the system coordinates, those equations lead to

$$\delta R_1 - \frac{l}{2}\cos\theta\,\delta\theta = 0, \qquad \delta R_2 - \frac{l}{2}\sin\theta\,\delta\theta = 0$$

which can be written in a matrix form as

$$\begin{bmatrix} 1 & 0 & -\frac{l}{2}\cos\theta \\ 0 & 1 & -\frac{l}{2}\sin\theta \end{bmatrix} \begin{bmatrix} \delta R_1 \\ \delta R_2 \\ \delta\theta \end{bmatrix} = \begin{bmatrix} 0 \\ 0 \end{bmatrix}$$

This equation can also be written as

$$\mathbf{C_q}\delta\mathbf{q} = \mathbf{0}$$

where $\mathbf{C_q}$ is the system Jacobian matrix defined as

$$\mathbf{C_q} = \begin{bmatrix} 1 & 0 & -\frac{l}{2}\cos\theta \\ 0 & 1 & -\frac{l}{2}\sin\theta \end{bmatrix}$$

Since the constraint equations do not explicitly depend on time, one can also verify that

$$\mathbf{C_q}\dot{\mathbf{q}} = \mathbf{0}$$

that is,

$$\begin{bmatrix} 1 & 0 & -\frac{l}{2}\cos\theta \\ 0 & 1 & -\frac{l}{2}\sin\theta \end{bmatrix} \begin{bmatrix} \dot{R}_1 \\ \dot{R}_2 \\ \dot{\theta} \end{bmatrix} = \begin{bmatrix} 0 \\ 0 \end{bmatrix}$$

For this holonomic system, we may identify the independent and dependent coordinates as

$$\mathbf{q}_i = \theta, \quad \mathbf{q}_d = [R_1 \quad R_2]^{\mathrm{T}}$$

According to this partitioning, we can write

$$\begin{bmatrix} \delta R_1 \\ \delta R_2 \end{bmatrix} + \begin{bmatrix} -\frac{l}{2}\cos\theta \\ -\frac{l}{2}\sin\theta \end{bmatrix} \delta\theta = \begin{bmatrix} 0 \\ 0 \end{bmatrix}$$

where the matrix $\mathbf{C}_{\mathbf{q}_d}$ of Eq. 57 can be recognized as the identity matrix and the matrix $\mathbf{C}_{\mathbf{q}_i}$ as

$$\mathbf{C}_{\mathbf{q}_i} = \begin{bmatrix} -\frac{l}{2}\cos\theta \\ -\frac{l}{2}\sin\theta \end{bmatrix}$$

Therefore, the matrix $\mathbf{C}_{di}$ is the column vector defined as

$$\mathbf{C}_{di} = -\mathbf{C}_{\mathbf{q}_d}^{-1}\mathbf{C}_{\mathbf{q}_i} = \frac{l}{2}\begin{bmatrix} \cos\theta \\ \sin\theta \end{bmatrix}$$

that is,

$$\begin{bmatrix} \delta R_1 \\ \delta R_2 \\ \delta\theta \end{bmatrix} = \begin{bmatrix} \frac{l}{2}\cos\theta \\ \frac{l}{2}\sin\theta \\ 1 \end{bmatrix} \delta\theta$$

Substituting this in D'Alembert–Lagrange's equation, one obtains

$$[\ddot{R}_1 \quad \ddot{R}_2 \quad \ddot{\theta}] \begin{bmatrix} m & 0 & 0 \\ 0 & m & 0 \\ 0 & 0 & I_c \end{bmatrix} \begin{bmatrix} \frac{l}{2}\cos\theta \\ \frac{l}{2}\sin\theta \\ 1 \end{bmatrix} \delta\theta - [F_1 \quad F_2 \quad M_c] \begin{bmatrix} \frac{l}{2}\cos\theta \\ \frac{l}{2}\sin\theta \\ 1 \end{bmatrix} \delta\theta = 0$$

By differentiating the constraint equations twice with respect to time, we obtain

$$\begin{bmatrix} \ddot{R}_1 \\ \ddot{R}_2 \\ \ddot{\theta} \end{bmatrix} = \begin{bmatrix} \frac{l}{2}\cos\theta \\ \frac{l}{2}\sin\theta \\ 1 \end{bmatrix} \ddot{\theta} + \begin{bmatrix} -\sin\theta \\ \cos\theta \\ 0 \end{bmatrix} \frac{l}{2}(\dot{\theta})^2$$

Substituting this into the equation of motion leads to

$$\left[\left(m\frac{(l)^2}{4} + I_c\right)\ddot{\theta} - \left(F_1\frac{l}{2}\cos\theta + F_2\frac{l}{2}\sin\theta + M_c\right)\right]\delta\theta = 0$$

that is,

$$I_o\ddot{\theta} = M_o$$

which is the same as the equation obtained previously.

Use of Redundant Coordinates It is clear from the above discussion that when the dynamic equations are developed in terms of the system degrees of freedom using D'Alembert's principle or Lagrange's equation, the force of constraints is automatically eliminated. Another approach to solve the same problem is to keep both the dependent and independent coordinates in the final form of the dynamic equation. This can be achieved by using Lagrange's equation with the multipliers, which can be written as

$$\frac{d}{dt}\left(\frac{\partial T}{\partial \dot{\mathbf{q}}}\right)^{\mathrm{T}} - \left(\frac{\partial T}{\partial \mathbf{q}}\right)^{\mathrm{T}} + \mathbf{C}_{\mathbf{q}}^{\mathrm{T}}\boldsymbol{\lambda} = \mathbf{Q}$$

Using the kinetic energy, the constraint Jacobian, and the generalized forces previously developed, one can verify that the equations of motion of the pendulum shown in Fig. 15 can be written as

$$\begin{bmatrix} m & 0 & 0 \\ 0 & m & 0 \\ 0 & 0 & I_c \end{bmatrix} \begin{bmatrix} \ddot{R}_1 \\ \ddot{R}_2 \\ \ddot{\theta} \end{bmatrix} + \begin{bmatrix} 1 & 0 \\ 0 & 1 \\ -\frac{l}{2}\cos\theta & -\frac{l}{2}\sin\theta \end{bmatrix} \begin{bmatrix} \lambda_1 \\ \lambda_2 \end{bmatrix} = \begin{bmatrix} F_1 \\ F_2 \\ M_c \end{bmatrix}$$

which provides the following scalar equations:

$$m\ddot{R}_1 + \lambda_1 = F_1$$

$$m\ddot{R}_2 + \lambda_2 = F_2$$

$$I_c\ddot{\theta} - \lambda_1 \frac{l}{2}\cos\theta - \lambda_2 \frac{l}{2}\sin\theta = M_c$$

These are three differential equations in five unknowns, R_1, R_2, θ, λ_1, and λ_2. Two additional equations are needed in order to solve for these five unknowns. These equations can be obtained by using the kinematic equations that describe the revolute joint at O, that is

$$R_1 - \frac{l}{2}\sin\theta = 0, \qquad R_2 + \frac{l}{2}\cos\theta = 0$$

These are two nonlinear algebraic equations that can be solved simultaneously with the differential equations in order to determine the unknowns R_1, R_2, θ, λ_1, and λ_2. Methods for solving mixed systems of algebraic and differential equations are discussed in Chapter 5. It is important, however, to point out that the vector $\mathbf{C}_{\mathbf{q}}^{\mathrm{T}}\boldsymbol{\lambda}$ represents the generalized reaction forces associated with the system generalized coordinates. This vector may not be the vector of actual reaction forces at the joints. Let us write the differential equations of motion in the following form:

$$m\ddot{R}_1 = F_1 - \lambda_1$$

$$m\ddot{R}_2 = F_2 - \lambda_2$$

$$I_c\ddot{\theta} = M_c + \lambda_1 \frac{l}{2}\cos\theta + \lambda_2 \frac{l}{2}\sin\theta$$

Clearly, in this example, the generalized reaction forces associated with the coordinates R_1 and R_2 are, respectively, the Lagrange multipliers λ_1 and λ_2, while

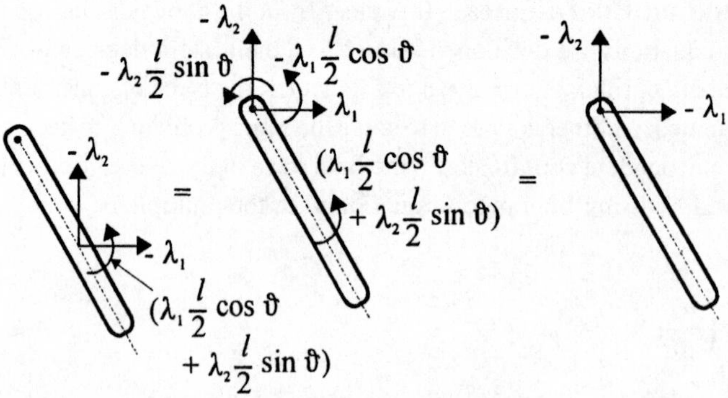

Figure 3.16 Generalized reaction forces.

the generalized moment associated with the angular rotation θ is $\lambda_1(l/2)\cos\theta + \lambda_2(l/2)\sin\theta$. This system of generalized reactions, however, must be equivalent (*equipollent*) to a force and zero moment at point O since the friction-free revolute joint is a workless constraint. This is, indeed, the case as shown in Fig. 16, where it is clear in this simple example that the actual reactions F_{c1} and F_{c2} are given by

$$F_{c1} = -\lambda_1, \qquad F_{c2} = -\lambda_2$$

That is, the actual reactions can be written as functions of the vector of Lagrange multipliers. In this simple example, it was found that the actual reactions are the negative of the Lagrange multipliers. In other applications, however, the actual reaction forces may be a nonlinear function of the system generalized coordinates as well.

3.5 CALCULUS OF VARIATIONS

In this section some techniques of the calculus of variations are presented. These techniques represent an alternative for deriving Lagrange's equation of motion from integral principles. One of the main problems of the calculus of variations is to find the curve for which some given integral is an extremum. First we will consider the one-dimensional form where the interest will be focused on finding a path $y = y(x)$ between two points such that the integral of some function $f(y, y', x)$, where $y' = dy/dx$, is an extremum. The integral is in the following form:

$$J = \int_{x_1}^{x_2} f(y, y', x)\,dx \tag{3.67}$$

The integral form of Eq. 67 is called a *functional*. Therefore, we can state the problem as follows. Find a path $y(x)$ between the two points (Fig. 17) such that the functional of Eq. 67 must be maximum or minimum. The function $f(y, y', x)$ is assumed to have continuous first and second (partial) derivatives with respect to all its arguments. The function $y(x)$, the solution of the problem, is assumed to be continuously differentiable for $x_1 \le x \le x_2$ and to satisfy the boundary conditions

$$y(x_1) = y_1, \qquad y(x_2) = y_2 \tag{3.68}$$

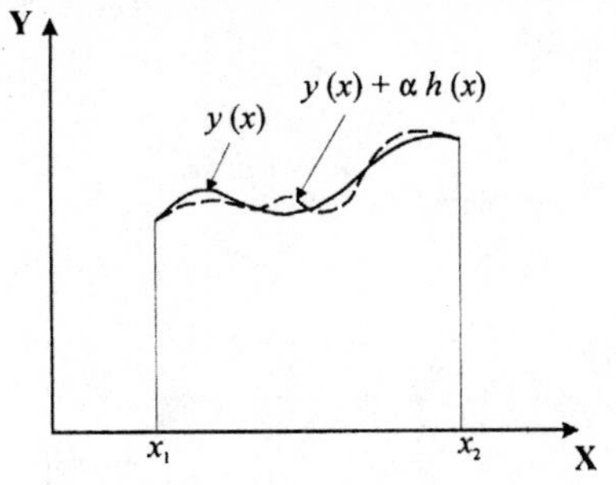

Figure 3.17 Calculus of variations.

Let $y(x)$ be the required curve, and suppose that we give $y(x)$ an increment $\alpha h(x)$ such that

$$y(x, \alpha) = y(x, 0) + \alpha h(x) \tag{3.69}$$

where α is a parameter that takes different values and $h(x)$ is a function that satisfies the following conditions:

$$h(x_1) = h(x_2) = 0 \tag{3.70}$$

These conditions assure us that $y(x, \alpha)$ is an admissible function. It is clear that when $\alpha = 0$, the curve of Eq. 69 coincides with the path that gives an extremum for the functional J of Eq. 67. In terms of the parameter α, Eq. 67 can be written as

$$J(\alpha) = \int_{x_1}^{x_2} f[y(x, \alpha), y'(x, \alpha), x]\, dx \tag{3.71}$$

The condition for obtaining an extremum is

$$\delta J = \left(\frac{\partial J}{\partial \alpha}\right)_{\alpha=0} \delta\alpha = 0 \tag{3.72}$$

When the chain rule of differentiation is used, Eq. 71 yields

$$\delta J = \int_{x_1}^{x_2} \left\{\frac{\partial f}{\partial y}\frac{\partial y}{\partial \alpha} + \frac{\partial f}{\partial y'}\frac{\partial y'}{\partial \alpha}\right\} \delta\alpha\, dx \tag{3.73}$$

in which

$$\int_{x_1}^{x_2} \frac{\partial f}{\partial y'}\frac{\partial y'}{\partial \alpha} dx = \int_{x_1}^{x_2} \frac{\partial f}{\partial y'}\frac{\partial^2 y}{\partial \alpha\, \partial x} dx$$

which on integrating by parts yields

$$\int_{x_1}^{x_2} \frac{\partial f}{\partial y'}\frac{\partial^2 y}{\partial x\, \partial \alpha} dx = \left.\frac{\partial f}{\partial y'}\frac{\partial y}{\partial \alpha}\right|_{x_1}^{x_2} - \int_{x_1}^{x_2} \frac{d}{dx}\left(\frac{\partial f}{\partial y'}\right)\frac{\partial y}{\partial \alpha} dx \tag{3.74}$$

Equation 69 implies

$$\frac{\partial y}{\partial \alpha} = h(x) \tag{3.75}$$

and accordingly, Eq. 70 gives

$$\left(\frac{\partial y}{\partial \alpha}\right)_{x=x_1} = \left(\frac{\partial y}{\partial \alpha}\right)_{x=x_2} = 0$$

Therefore, Eq. 74 can be written as

$$\int_{x_1}^{x_2} \frac{\partial f}{\partial y'} \frac{\partial^2 y}{\partial x\, \partial \alpha} dx = -\int_{x_1}^{x_2} \frac{d}{dx}\left(\frac{\partial f}{\partial y'}\right) \frac{\partial y}{\partial \alpha} dx$$

which on substitution in Eq. 73 yields

$$\delta J = \int_{x_1}^{x_2} \left\{\frac{\partial f}{\partial y} - \frac{d}{dx}\left(\frac{\partial f}{\partial y'}\right)\right\}\left(\frac{\partial y}{\partial \alpha}\right) \delta\alpha\, dx \tag{3.76}$$

To obtain the extremum, we evaluate the derivative at $\alpha = 0$, resulting in

$$\delta J = \left(\frac{\partial J}{\partial \alpha}\right)_{\alpha=0} \delta\alpha = \int_{x_1}^{x_2} \left\{\frac{\partial f}{\partial y} - \frac{d}{dx}\left(\frac{\partial f}{\partial y'}\right)\right\}\left(\frac{\partial y}{\partial \alpha}\right)_{\alpha=0} \delta\alpha\, dx \tag{3.77}$$

where

$$\left(\frac{\partial y}{\partial \alpha}\right)_{\alpha=0} \delta\alpha = \delta y$$

Therefore, Eq. 77 can be written as

$$\delta J = \int_{x_1}^{x_2} \left[\frac{\partial f}{\partial y} - \frac{d}{dx}\left(\frac{\partial f}{\partial y'}\right)\right] \delta y\, dx \tag{3.78}$$

Since δy is arbitrary, it follows from condition 72 that

$$\frac{\partial f}{\partial y} - \frac{d}{dx}\left(\frac{\partial f}{\partial y'}\right) = 0 \tag{3.79}$$

Therefore, the functional J is an extremum only for curves $y(x)$ that satisfy Eq. 79. Equation 79 is sometimes called *Euler's equation.* The curves that satisfy Euler's equation are called *extremals.*

Euler's equation is a second-order ordinary differential equation. The solution of Eq. 79 will, in general, depend on the boundary conditions of Eq. 68. Since Euler's equation plays a fundamental role in the calculus of variations, we discuss below some special cases.

Case 1 Suppose that the function does not depend on y; then the functional J can be written as

$$J = \int_{x_1}^{x_2} f(x, y')\, dx$$

In this case, Eq. 79 reduces to

$$\frac{d}{dx}(f_{y'}) = 0$$

which implies that

$$f_{y'} = C$$

where C is a constant.

Case 2 Suppose that the integrand f does not depend on x, that is

$$J = \int_{x_1}^{x_2} f(y, y')\,dx$$

then

$$f_y - \frac{d}{dx}(f_{y'}) = f_y - f_{y'y}y' - f_{y'y'}y''$$

If we multiply by y', we obtain

$$f_y y' - f_{y'y}y'^2 - f_{y'y'}y'y'' = \frac{d}{dx}(f - y' f_{y'})$$

Therefore, in this special case Euler's equation reduces to

$$f - y' f_{y'} = C$$

where C is a constant.

Case 3 Suppose that f does not depend on y'; then Euler's equation yields

$$f_y = 0$$

which is not a differential equation, but an algebraic equation that involves y and the parameter x.

Example 3.7 (*The brachistochrone problem*) This problem was first posed by John Bernoulli in 1696 and can be stated as follows. Find the curve joining two given points A and B such that the time taken by a particle to slide on this curve under the influence of gravity is minimum.

Let ds be an infinitesimal arc length along the required curve that joins points A and B. Let v be the speed along the curve. Then the time required for the particle to travel an arc length ds is

$$\Delta t = \frac{ds}{v}$$

The functional to be minimized can then be written as

$$t = \int \Delta t = \int_A^B \frac{ds}{v}$$

From Fig. 18 it is clear that

$$ds = [(dx)^2 + (dy)^2]^{1/2} = [1 + (y')^2]^{1/2} dx$$

where$y' = dy/dx$. Therefore, one may write t in a more explicit form as

$$t = \int_A^B \frac{[1 + (y')^2]^{1/2}}{v} dx$$

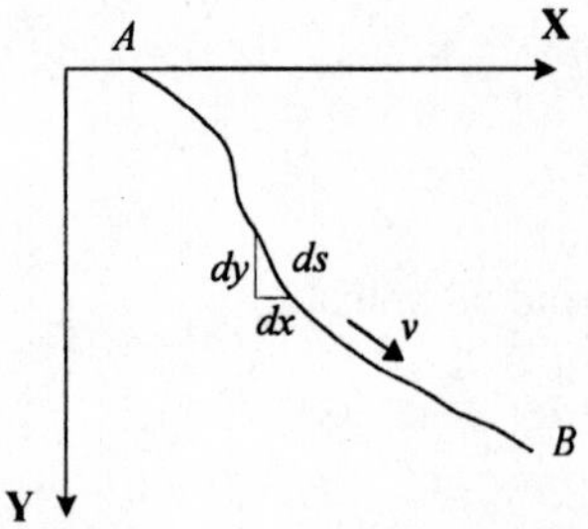

Figure 3.18 Brachistochrone problem.

Using the conservation of energy equation, we have

$$\frac{1}{2}m(v)^2 = mgy$$

where m is the mass of the particle, g is the gravitational constant, and y is measured down from the initial point of release. The conservation of energy equation yields

$$v = \sqrt{2gy}$$

and the time t can then be written as

$$t = \int_A^B \sqrt{\frac{1+(y')^2}{2gy}}dx$$

where the function f can be identified as

$$f = \sqrt{\frac{1+(y')^2}{2gy}}$$

It is left as an exercise to show that the general solution of the corresponding Euler equation consists of a family of *cycloids*.

Example 3.8 The Euler equation can also be used to find the shortest distance between two points A and B in a plane. As in the preceding example, the element arc length in a plane is

$$ds = \sqrt{(dx)^2 + (dy)^2}$$

and the functional to be minimized can be written as

$$s = \int_A^B ds = \int_{x_A}^{x_B} \sqrt{1+(y')^2}dx$$

where $y' = dy/dx$. Therefore, the function f in the Euler equation can be identified as

$$f = \sqrt{1+(y')^2}$$

Substituting the following in the Euler equation

$$\frac{\partial f}{\partial y} = 0, \qquad \frac{\partial f}{\partial y'} = \frac{y'}{\sqrt{1+(y')^2}}$$

we have

$$\frac{d}{dx}\left(\frac{\partial f}{\partial y'}\right) = \frac{d}{dx}\left(\frac{y'}{\sqrt{1+(y')^2}}\right) = 0$$

or

$$\frac{y'}{(1+(y')^2)^{1/2}} = C$$

where C is the constant of integration. One may verify that the solution of the preceding equation yields

$$y = C_1 x + C_2$$

where C_1 and C_2 are constants. The result of this example is familiar and shows that the straight line is the shortest distance between the two points.

Example 3.9 Among all the curves joining two given points (x_1, y_1) and (x_2, y_2), find the one that generates the surface of minimum area when rotated about the **X** axis.

As shown in Fig. 19, the area of the surface of revolution generated by rotating the curve $y = y(x)$ about the **X** axis is

$$I = 2\pi \int_1^2 y\, ds = 2\pi \int_{x_1}^{x_2} y\sqrt{1+(y')^2}\, dx$$

in which the function f in Euler's equation can be recognized as

$$f = y\sqrt{1+(y')^2}$$

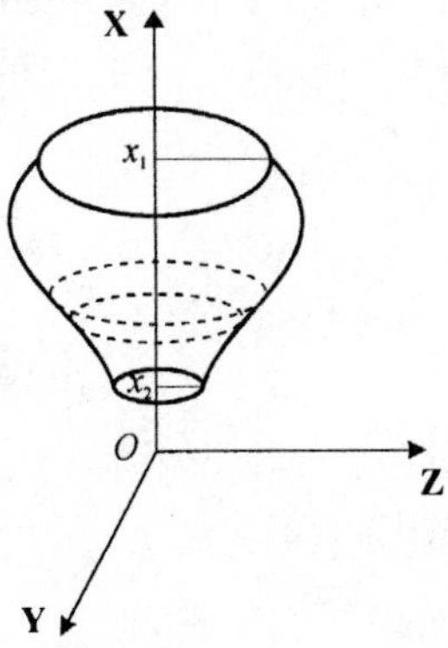

Figure 3.19 Surface of minimum area.

Since f does not depend explicitly on the parameter x (case 2), we have

$$f - y' f_{y'} = a$$

where a is a constant. Substituting the value of f in the preceding equation yields

$$y\sqrt{1+(y')^2} - \frac{y(y')^2}{\sqrt{1+(y')^2}} = a$$

or

$$y = a\sqrt{1+(y')^2}$$

so that

$$y' = \sqrt{\frac{(y)^2-(a)^2}{(a)^2}}$$

which by the separation of variables yields

$$dx = \frac{a\,dy}{\sqrt{(y)^2-(a)^2}}$$

or

$$x + a_1 = a \ln\left(\frac{y+\sqrt{(y)^2-(a)^2}}{a}\right)$$

where a_1 is a constant. The above equation can be written in another form as

$$y = a \cosh \frac{x+a_1}{a}$$

This is the equation of a *catenary* passing through the two given points.

3.6 EULER'S EQUATION IN THE CASE OF SEVERAL VARIABLES

In the previous section, only one independent variable $y = y(x)$ was considered. In this section, the preceding development is generalized to the case of several variables $y_1(x), y_2(x), \ldots, y_n(x)$, where x is the parametric variable. In this case the interest will be focused on finding the vector function $\mathbf{y}(x) = [y_1\ y_2 \cdots y_n]^{\mathrm{T}}$, which minimizes the following functional

$$J = \int_{x_1}^{x_2} f(y_1, y_2, \ldots, y_n, y_1', y_2', \ldots, y_n', x)\,dx \tag{3.80}$$

We will use a vector notation and write Eq. 80 as

$$J = \int_{x_1}^{x_2} f(\mathbf{y}, \mathbf{y}', x)\,dx \tag{3.81}$$

In a similar manner to the preceding development, we write

$$\mathbf{y}(x, \alpha) = \mathbf{y}(x, 0) + \alpha\mathbf{h}(x) \tag{3.82}$$

where α is the parameter described in the preceding section and $\mathbf{h}(x) = [h_1\, h_2 \cdots h_n]^{\mathrm{T}}$ is a vector function satisfying the following conditions at the endpoints (x_1, y_1) and (x_2, y_2):

$$\mathbf{h}(x_1) = \mathbf{h}(x_2) = 0 \tag{3.83}$$

The variation of Eq. 80 can then be written as

$$\delta J = \frac{\partial J}{\partial \alpha}\delta\alpha = \int_{x_1}^{x_2} \left(f_{\mathbf{y}} \frac{\partial \mathbf{y}}{\partial \alpha}\delta\alpha + f_{\mathbf{y}'} \frac{\partial \mathbf{y}'}{\partial \alpha}\delta\alpha \right) dx \tag{3.84}$$

where the vector subscript implies differentiation with respect to this vector, that is,

$$\begin{aligned} f_{\mathbf{y}} &= \left[f_{y_1} \quad f_{y_2} \quad \cdots \quad f_{y_n} \right] \\ f_{\mathbf{y}'} &= \left[f_{y_1'} \quad f_{y_2'} \quad \cdots \quad f_{y_n'} \right] \end{aligned}$$

By using the integration by parts, one can write the second term in the right-hand side of Eq. 84 as

$$\int_{x_1}^{x_2} f_{\mathbf{y}'} \frac{\partial^2 \mathbf{y}}{\partial \alpha\, \partial x} dx = f_{\mathbf{y}'} \frac{\partial \mathbf{y}}{\partial \alpha}\bigg|_{x_1}^{x_2} - \int_{x_1}^{x_2} \frac{d}{dx}(f_{\mathbf{y}'}) \frac{\partial \mathbf{y}}{\partial \alpha} dx$$

which, with the use of Eq. 83, yields

$$\int_{x_1}^{x_2} f_{\mathbf{y}'} \frac{\partial^2 \mathbf{y}}{\partial \alpha\, \partial x} dx = -\int_{x_1}^{x_2} \frac{d}{dx}(f_{\mathbf{y}'}) \frac{\partial \mathbf{y}}{\partial \alpha} dx$$

Substituting this into Eq. 84 gives

$$\delta J = \int_{x_1}^{x_2} \left(f_{\mathbf{y}} - \frac{d}{dx} f_{\mathbf{y}'} \right) \frac{\partial \mathbf{y}}{\partial \alpha} \delta\alpha\, dx \tag{3.85}$$

Since $\delta \mathbf{y} = (\partial \mathbf{y}/\partial \alpha)\, \delta\alpha$, Eq. 85 can be written as

$$\delta J = \int_{x_1}^{x_2} \left(f_{\mathbf{y}} - \frac{d}{dx} f_{\mathbf{y}'} \right) \delta \mathbf{y}\, dx \tag{3.86}$$

If the $\mathbf{y}$ variables are linearly independent, Eq. 86 yields

$$f_{\mathbf{y}} - \frac{d}{dx}(f_{\mathbf{y}'}) = \mathbf{0}^{\mathrm{T}} \tag{3.87}$$

Equation 87 is a set of differential equations called *Euler–Lagrange equations*. The solution of these differential equations defines the vector function $\mathbf{y}$, which minimizes the functional J of Eq. 80.

Hamilton's Principle The technique of the calculus of variations presented in this section provides other means of establishing the equations of motion by using scalar energy quantities. This can be done by using *Hamilton's principle*, which may be stated mathematically as

$$\delta \int_{t_1}^{t_2} (T - V)\, dt + \int_{t_1}^{t_2} \delta W_{nc}\, dt = 0 \tag{3.88}$$

where T is the *kinetic energy* of the system; V is the *potential energy*, which includes both strain energy and the potential of any conservative external forces; and δW_{nc} is the virtual work done by nonconservative forces acting on the system.

Hamilton's principle states that the variation of the kinetic and potential energy plus the line integral of the virtual work done by the nonconservative forces during any time interval t_1 to t_2 must be equal to zero.

One may define the following quantity

$$L = T - V \tag{3.89}$$

which is called the *Lagrangian*, and write Eq. 88 as

$$\delta \int_{t_1}^{t_2} L\, dt + \int_{t_1}^{t_2} \delta W_{nc}\, dt = 0 \tag{3.90}$$

In previous developments, it has been shown that the virtual work δW_{nc} can be written as the dot product of the vector of generalized forces and the vector of system virtual displacements, that is,

$$\delta W_{nc} = \mathbf{Q}_{nc}^{\mathrm{T}} \delta \mathbf{q} \tag{3.91}$$

where $\mathbf{Q}_{nc}$ is the vector of *nonconservative* generalized forces. Using the techniques of calculus of variations, one can show that

$$\delta \int_{t_1}^{t_2} L\, dt = \int_{t_1}^{t_2} \left[-\frac{d}{dt}\left(\frac{\partial L}{\partial \dot{\mathbf{q}}}\right) + \frac{\partial L}{\partial \mathbf{q}} \right] \delta \mathbf{q}\, dt \tag{3.92}$$

From Eqs. 91 and 92 it follows that

$$\int_{t_1}^{t_2} \left[\frac{d}{dt}\left(\frac{\partial L}{\partial \dot{\mathbf{q}}}\right) - \frac{\partial L}{\partial \mathbf{q}} \right] \delta \mathbf{q}\, dt - \int_{t_1}^{t_2} \mathbf{Q}_{nc}^{\mathrm{T}}\, \delta \mathbf{q}\, dt = 0 \tag{3.93}$$

or

$$\int_{t_1}^{t_2} \left[\frac{d}{dt}\left(\frac{\partial L}{\partial \dot{\mathbf{q}}}\right) - \frac{\partial L}{\partial \mathbf{q}} - \mathbf{Q}_{nc}^{\mathrm{T}} \right] \delta \mathbf{q}\, dt = 0 \tag{3.94}$$

If δq_j, $j = 1, 2, \ldots, n$, are linearly independent, Eq. 94 yields the system matrix equations of motion as

$$\frac{d}{dt}\left(\frac{\partial L}{\partial \dot{\mathbf{q}}}\right) - \frac{\partial L}{\partial \mathbf{q}} - \mathbf{Q}_{nc}^{\mathrm{T}} = \mathbf{0}^{\mathrm{T}} \tag{3.95}$$

In case δq_j, $j = 1, 2, \ldots, n$, are not linearly independent, one may use Lagrange multipliers and write the following mixed sets of differential and algebraic equations

$$\frac{d}{dt}\left(\frac{\partial L}{\partial \dot{\mathbf{q}}}\right) - \frac{\partial L}{\partial \mathbf{q}} + \boldsymbol{\lambda}^{\mathrm{T}} \mathbf{C}_{\mathbf{q}} = \mathbf{Q}_{nc}^{\mathrm{T}} \tag{3.96}$$

$$\mathbf{C}(\mathbf{q}, t) = \mathbf{0} \tag{3.97}$$

where $\boldsymbol{\lambda}$ is the vector of Lagrange multipliers, $\mathbf{C}$ is the vector of constraint functions, and $\mathbf{C}_{\mathbf{q}}$ is the constraint Jacobian matrix.

Hamilton's principle (Eq. 90) can be also stated differently as

$$\delta \int_{t_1}^{t_2} T dt + \int_{t_1}^{t_2} \delta W \, dt = 0 \tag{3.98}$$

where in this case

$$\delta W = \delta W_c + \delta W_{nc}$$

is the virtual work done by all forces acting on the system and δW_c is the virtual work done by the *conservative* forces. In this case one may write δW as

$$\delta W = \mathbf{Q}^{\mathrm{T}} \delta \mathbf{q} \tag{3.99}$$

where $\mathbf{Q}$ is the vector of generalized forces of both conservative and nonconservative forces. Equations 98 and 99 provide the following equivalent form to Eq. 96:

$$\frac{d}{dt}\left(\frac{\partial T}{\partial \dot{\mathbf{q}}}\right) - \frac{\partial T}{\partial \mathbf{q}} + \boldsymbol{\lambda}^{\mathrm{T}} \mathbf{C}_{\mathbf{q}} = \mathbf{Q}^{\mathrm{T}} \tag{3.100}$$

Conservative Forces In Eq. 98 it is important to note that

$$\int_{t_1}^{t_2} \delta W dt = \delta \int_{t_1}^{t_2} W dt \tag{3.101}$$

holds only when all the system forces are conservative, that is, when there exists a potential function V such that all the forces can be derived from this function. In this special case, the generalized force Q_j associated with the jth coordinate is determined by

$$Q_j = -\frac{\partial V}{\partial q_j} \tag{3.102}$$

Since

$$\frac{\partial V}{\partial q_i \partial q_j} = \frac{\partial V}{\partial q_j \partial q_i}$$

one can then write

$$\frac{\partial Q_j}{\partial q_i} = \frac{\partial Q_i}{\partial q_j}$$

which is equivalent to saying that V is an *exact differential*. In this case, one can write the virtual work δW as

$$\delta W = Q_1 \delta q_1 + Q_2 \delta q_2 + \cdots + Q_n \delta q_n = \sum_j Q_j \delta q_j$$

Using Eq. 102, δW can be written as

$$\delta W = -\sum_j \frac{\partial V}{\partial q_j} \delta q_j = -\delta V$$

where in this special case of conservative forces, the virtual work is equal to the negative of the variation of the potential energy. Nonconservative forces, however, cannot be derived from a potential function, and hence the virtual work is not equal to the variation of a certain function.

Example 3.10 Figure 20 shows a mass–spring–damper system. The stiffness coefficient is k and the damping coefficient is c. The kinetic energy of the system is

$$T = \frac{1}{2}m(\dot{x})^2$$

The potential energy V is

$$V = \frac{1}{2}k(x)^2$$

The virtual work of the nonconservative damping and external forces is

$$\delta W_{nc} = -c\dot{x}\delta x + F(t)\,\delta x$$

To use Eq. 90, we define the Lagrangian L as

$$L = T - V = \frac{1}{2}m(\dot{x})^2 - \frac{1}{2}k(x)^2$$

Substituting into Eq. 90 yields

$$\delta \int_{t_1}^{t_2} \left(\frac{1}{2}m(\dot{x})^2 - \frac{1}{2}k(x)^2\right) dt + \int_{t_1}^{t_2} [-c\dot{x}\delta x + F(t)\delta x]\, dt = 0$$

We first follow the formal procedure of the calculus of variations and later use Eq. 100 to confirm our results. By taking the variation of the preceding equation, one gets

$$\int_{t_1}^{t_2} (m\dot{x}\,\delta\dot{x} - kx\delta x)\, dt + \int_{t_1}^{t_2} [-c\dot{x} + F(t)]\,\delta x\, dt = 0$$

By integrating by parts the first integral, one obtains

$$\int_{t_1}^{t_2} m\dot{x}\,\delta\dot{x}\, dt = m\dot{x}\,\delta x\big|_{t_1}^{t_2} - \int_{t_1}^{t_2} m\ddot{x}\,\delta x\, dt$$

where

$$m\dot{x}\,\delta x\big|_{t_1}^{t_2} = 0$$

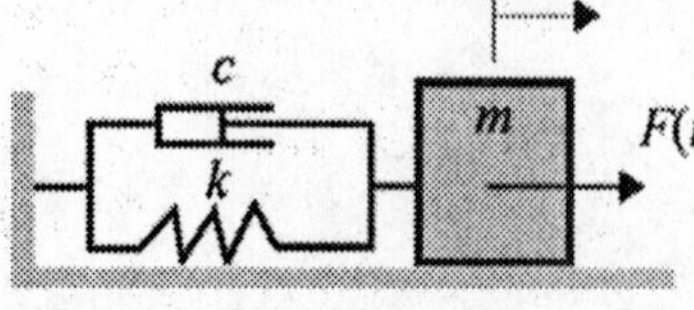

Figure 3.20 Mass–spring–damper system.

since the displacement is specified at the endpoints. Therefore, the equation of motion can be written as

$$\int_{t_1}^{t_2} (-m\ddot{x} - kx)\,\delta x\,dt + \int_{t_1}^{t_2} [-c\dot{x} + F(t)]\,\delta x\,dt = 0$$

or

$$\int_{t_1}^{t_2} [-m\ddot{x} - kx - c\dot{x} + F(t)]\,\delta x\,dt = 0$$

Since δx is an independent coordinate, we get the familiar equation of motion of this simple oscillatory system as

$$m\ddot{x} + c\dot{x} + kx = F(t)$$

This equation could have been derived directly by using Lagrange's equation. Note that

$$\frac{d}{dt}\left(\frac{\partial T}{\partial \dot{x}}\right) = m\ddot{x}, \qquad \frac{\partial T}{\partial x} = 0$$

The virtual work of all forces acting on the system can be written as

$$\delta W = -kx\,\delta x - c\dot{x}\,\delta x + F(t)\delta x = Q\,\delta x$$

where Q is given by

$$Q = -kx - c\dot{x} + F(t)$$

By using Eq. 100 for this system, one gets

$$m\ddot{x} = -kx - c\dot{x} + F(t)$$

which is the same equation of motion derived earlier.

Example 3.11 Figure 21 shows a hoop rolling without slipping down an inclined plane that makes an angle ϕ with the horizontal. This is a simple holonomic system (Goldstein 1950). For demonstration purposes, however, we treat it using the Lagrange multiplier technique. We select x and θ to be our generalized coordinates. These coordinates are related by the constraint equation

$$dx = r\,d\theta, \qquad \text{that is,}\ dx - r\,d\theta = 0$$

where r is the radius of the hoop. From the elementary rigid body analysis, the

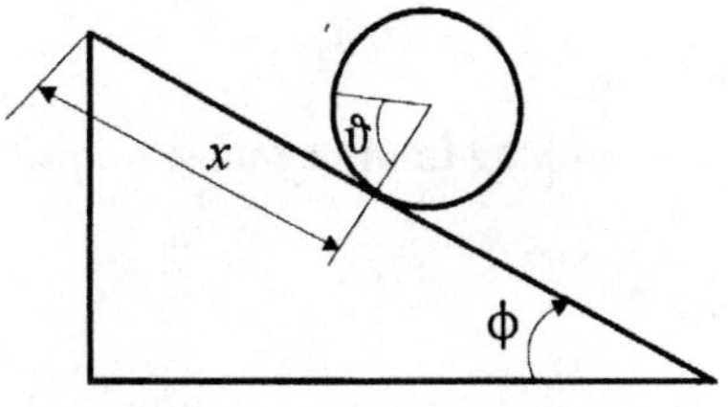

Figure 3.21 Rolling hoop.

kinetic energy is given by

$$T = \frac{1}{2}m(\dot{x})^2 + \frac{1}{2}I(\dot{\theta})^2$$

where m and I are, respectively, the mass and the mass moment of inertia of the hoop. Since $I = m(r)^2$, the kinetic energy T can be written as

$$T = \frac{1}{2}m(\dot{x})^2 + \frac{1}{2}m(r)^2(\dot{\theta})^2$$

The potential energy of the hoop is

$$V = mg(d - x)\sin\phi$$

where d is the length of the inclined plane and g is the gravitational constant. One can now write the Lagrangian of the system as

$$L = T - V = \frac{1}{2}(m(\dot{x})^2 + m(r)^2(\dot{\theta})^2) - mg(d - x)\sin\phi$$

Since we have two generalized coordinates, mainly x and θ, we can write the following two equations:

$$\frac{d}{dt}\left(\frac{\partial L}{\partial \dot{x}}\right) - \frac{\partial L}{\partial x} + C_x\lambda = 0$$

$$\frac{d}{dt}\left(\frac{\partial L}{\partial \dot{\theta}}\right) - \frac{\partial L}{\partial \theta} + C_\theta\lambda = 0$$

where C_x and C_θ are the elements of the constraint Jacobian matrix

$$\mathbf{C_q} = [C_x \quad C_\theta]$$

in which

$$C_x = 1, \qquad C_\theta = -r$$

One can then verify that the two Lagrange's equations yield

$$m\ddot{x} - mg\sin\phi + \lambda = 0$$

$$m(r)^2\ddot{\theta} - \lambda r = 0$$

These two equations, along with the constraint equation

$$\dot{x} - r\dot{\theta} = 0$$

represent the constrained system equations of motion that can be solved for the three unknowns, x, θ, and λ. In order to solve for the accelerations and the Lagrange multiplier λ, we differentiate the constraint equation with respect to time to get

$$\ddot{x} - r\ddot{\theta} = 0$$

This equation, along with the two equations resulting from Lagrange's equation, can be written in a matrix form as

$$\begin{bmatrix} m & 0 & 1 \\ 0 & m(r)^2 & -r \\ 1 & -r & 0 \end{bmatrix} \begin{bmatrix} \ddot{x} \\ \ddot{\theta} \\ \lambda \end{bmatrix} = \begin{bmatrix} mg\sin\phi \\ 0 \\ 0 \end{bmatrix}$$

which can be solved for $\ddot{x}$, $\ddot{\theta}$, and λ. One can then verify that

$$\ddot{x} = \frac{1}{2} g \sin\phi$$

$$\ddot{\theta} = \frac{g}{2r} \sin\phi$$

$$\lambda = \frac{1}{2} mg \sin\phi$$

The forces of constraint can then be evaluated as

$$R_x = C_x \lambda = \frac{1}{2} mg \sin\phi$$

$$R_\theta = C_\theta \lambda = -\frac{1}{2} mgr \sin\phi$$

The acceleration $\ddot{x}$ and $\ddot{\theta}$ can be integrated to evaluate the velocities at any distance b along the inclined plane. Since

$$\ddot{x} = \dot{x} \frac{d\dot{x}}{ds}$$

one may verify that

$$\dot{x} = \sqrt{gb \sin\phi}$$

and

$$\dot{\theta} = \frac{1}{r} \sqrt{gb \sin\phi}$$

Thus far we have been concerned with the formulation of the system equations of motion of mechanical systems. No mention has been made of the solution of these equations except for some simple examples. In many applications, closed-form solution is impossible. The resulting system of n second-order differential equations associated with n generalized coordinates is, in general, highly nonlinear. This system of equations can be solved numerically by using direct numerical integration methods, provided the initial values for the generalized coordinates and velocities are defined. When multibody systems are considered, the Lagrangian formulation presented in this chapter leads to a mixed set of nonlinear differential and algebraic constraint equations that have to be solved simultaneously for the state of the system. There are some numerical procedures for solving such systems; one of these is the procedure originated by Wehage (1980). In this solution procedure, a set of independent coordinates are identified and integrated forward in time using a direct numerical integration routine. Dependent coordinates are then determined by using the nonlinear kinematic constraint equations. This solution procedure can be used to solve for both holonomic and nonholonomic systems. We shall defer a detailed discussion of *Wehage's algorithm* until the dynamic formulation of flexible multibody systems is presented in Chapter 5.

3.7 EQUATIONS OF MOTION OF RIGID BODY SYSTEMS

We conclude this chapter by deriving, in this and the following sections, the dynamic equations of motion of multibody systems consisting of interconnected rigid bodies. To determine the configuration or state of the multibody system, it is first necessary to define generalized coordinates that specify the location of each point of any body in the multibody system.

Kinematic Equations For the general body i, we have previously shown that the set of coordinates $\mathbf{R}^i$ and $\boldsymbol{\theta}^i$ that, respectively, represent the translation of the origin of the body reference and the orientation of this reference with respect to the inertial frame can be selected as the generalized coordinates of the body. The global position of an arbitrary point P^i on the body i can be defined in terms of these generalized coordinates as

$$\mathbf{r}^i = \mathbf{R}^i + \mathbf{A}^i \bar{\mathbf{u}}^i \tag{3.103}$$

where $\mathbf{R}^i$ is the location of the origin of the body axes relative to the inertial frame, $\mathbf{A}^i$ is the transformation matrix from the ith body coordinates to the inertial frame, and $\bar{\mathbf{u}}^i$ is the location of point P^i with respect to the body coordinate system. In the planar analysis the vectors $\mathbf{R}^i$ and $\bar{\mathbf{u}}^i$ are the following vectors:

$$\mathbf{R}^i = \begin{bmatrix} R_1^i & R_2^i & 0 \end{bmatrix}^{\mathrm{T}}$$
$$\bar{\mathbf{u}}^i = \begin{bmatrix} \bar{u}_1^i & \bar{u}_2^i & 0 \end{bmatrix}^{\mathrm{T}} = \begin{bmatrix} x_1^i & x_2^i & 0 \end{bmatrix}^{\mathrm{T}}$$

The transformation matrix $\mathbf{A}^i$ in terms of the rotation angle θ^i about the $\mathbf{X}_3$ axis is given by

$$\mathbf{A}^i = \begin{bmatrix} \cos\theta^i & -\sin\theta^i & 0 \\ \sin\theta^i & \cos\theta^i & 0 \\ 0 & 0 & 1 \end{bmatrix}$$

In the spatial analysis $\mathbf{R}^i$ and $\bar{\mathbf{u}}^i$ are the three-dimensional vectors given by

$$\mathbf{R}^i = \begin{bmatrix} R_1^i & R_2^i & R_3^i \end{bmatrix}^{\mathrm{T}}$$
$$\bar{\mathbf{u}}^i = \begin{bmatrix} \bar{u}_1^i & \bar{u}_2^i & \bar{u}_3^i \end{bmatrix}^{\mathrm{T}} = \begin{bmatrix} x_1^i & x_2^i & x_3^i \end{bmatrix}^{\mathrm{T}}$$

In this case, the transformation matrix $\mathbf{A}^i$ can be formulated in terms of *Euler parameters, Rodriguez parameters*, or *Euler angles*. These different forms of the transformation matrix are derived in the preceding chapter, and we reproduce them in this section for convenience. In terms of the four Euler parameters θ_0^i, θ_1^i, θ_2^i, and θ_3^i, the transformation matrix is given by

$$\mathbf{A}^i = \begin{bmatrix} 1-2(\theta_2)^2-2(\theta_3)^2 & 2(\theta_1\theta_2-\theta_0\theta_3) & 2(\theta_1\theta_3+\theta_0\theta_2) \\ 2(\theta_1\theta_2+\theta_0\theta_3) & 1-2(\theta_1)^2-2(\theta_3)^2 & 2(\theta_2\theta_3-\theta_0\theta_1) \\ 2(\theta_1\theta_3-\theta_0\theta_2) & 2(\theta_2\theta_3+\theta_0\theta_1) & 1-2(\theta_1)^2-2(\theta_2)^2 \end{bmatrix}^i$$

where the four Euler parameters are related by

$$\sum_{k=0}^{3} (\theta_k^i)^2 = 1$$

In terms of the three Rodriguez parameters γ_1^i, γ_2^i, and γ_3^i, the transformation matrix $\mathbf{A}^i$ is given by

$$\mathbf{A}^i = \frac{1}{1+(\gamma)^2} \times \begin{bmatrix} 1+(\gamma_1)^2-(\gamma_2)^2-(\gamma_3)^2 & 2(\gamma_1\gamma_2-\gamma_3) & 2(\gamma_1\gamma_3+\gamma_2) \\ 2(\gamma_1\gamma_2+\gamma_3) & 1-(\gamma_1)^2+(\gamma_2)^2-(\gamma_3)^2 & 2(\gamma_2\gamma_3-\gamma_1) \\ 2(\gamma_1\gamma_3-\gamma_2) & 2(\gamma_2\gamma_3+\gamma_1) & 1-(\gamma_1)^2-(\gamma_2)^2+(\gamma_3)^2 \end{bmatrix}^i$$

where

$$(\gamma)^2 = \sum_{k=1}^{3} (\gamma_k^i)^2$$

In terms of the three Euler angles ϕ^i, θ^i, and ψ^i, the matrix $\mathbf{A}^i$ is given by

$$\mathbf{A}^i = \begin{bmatrix} \cos\psi\cos\phi-\cos\theta\sin\phi\sin\psi & -\sin\psi\cos\phi-\cos\theta\sin\phi\cos\psi & \sin\theta\sin\phi \\ \cos\psi\sin\phi+\cos\theta\cos\phi\sin\psi & -\sin\psi\sin\phi+\cos\theta\cos\phi\cos\psi & -\sin\theta\cos\phi \\ \sin\theta\sin\psi & \sin\theta\cos\psi & \cos\theta \end{bmatrix}^i$$

In order to have a unified development in this section and the chapters that follow, we henceforth denote the set of rotational coordinates of the ith body reference as $\boldsymbol{\theta}^i$, that is, in the case of Euler parameters

$$\boldsymbol{\theta}^i = (\theta_0, \theta_1, \theta_2, \theta_3)^i$$

In case of Rodriguez parameters, the set $\boldsymbol{\theta}^i$ is given by

$$\boldsymbol{\theta}^i = (\gamma_1, \gamma_2, \gamma_3)^i$$

Similarly, when Euler angles are used

$$\boldsymbol{\theta}^i = (\phi, \theta, \psi)^i$$

Differentiating Eq. 103 with respect to time yields the velocity vector

$$\dot{\mathbf{r}}^i = \dot{\mathbf{R}}^i + \dot{\mathbf{A}}^i\bar{\mathbf{u}}^i \tag{3.104}$$

where ($\dot{}$) denotes differentiation with respect to time. We have previously shown that the vector $\dot{\mathbf{A}}^i\bar{\mathbf{u}}^i$ can be written as

$$\dot{\mathbf{A}}^i\bar{\mathbf{u}}^i = \mathbf{A}^i(\bar{\omega}^i \times \bar{\mathbf{u}}^i) \tag{3.105}$$

where ω^i is the angular velocity vector defined with respect to the ith body coordinate system. Recall that

$$\bar{\omega}^i \times \bar{\mathbf{u}}^i = \tilde{\bar{\omega}}^i\bar{\mathbf{u}}^i = -\tilde{\bar{\mathbf{u}}}^i\bar{\omega}^i$$

where $\tilde{\bar{\omega}}^i$ and $\tilde{\bar{\mathbf{u}}}^i$ are the skew symmetric matrices defined as

$$\tilde{\bar{\omega}}^i = \begin{bmatrix} 0 & -\bar{\omega}_3^i & \bar{\omega}_2^i \\ \bar{\omega}_3^i & 0 & -\bar{\omega}_1^i \\ -\bar{\omega}_2^i & \bar{\omega}_1^i & 0 \end{bmatrix}, \qquad \tilde{\bar{\mathbf{u}}}^i = \begin{bmatrix} 0 & -x_3^i & x_2^i \\ x_3^i & 0 & -x_1^i \\ -x_2^i & x_1^i & 0 \end{bmatrix} \tag{3.106}$$

in which $\bar{\omega}_1^i, \bar{\omega}_2^i, \bar{\omega}_3^i$ and x_1^i, x_2^i, x_3^i are, respectively, the components of the vectors $\bar{\omega}^i$ and $\bar{\mathbf{u}}^i$. One can then write Eq. 105 as

$$\dot{\mathbf{A}}^i\bar{\mathbf{u}}^i = -\mathbf{A}^i\tilde{\bar{\mathbf{u}}}^i\bar{\omega}^i$$

Substituting this equation into Eq. 104 yields

$$\dot{\mathbf{r}}^i = \dot{\mathbf{R}}^i - \mathbf{A}^i\tilde{\bar{\mathbf{u}}}^i\bar{\omega}^i \tag{3.107}$$

In the preceding chapter, it was shown that, in general, the angular velocity vector $\bar{\omega}^i$ can be written in terms of the time derivative of the rotational coordinates of the body reference as

$$\bar{\omega}^i = \bar{\mathbf{G}}^i\dot{\theta}^i \tag{3.108}$$

where $\bar{\mathbf{G}}^i$ is a matrix that depends on the selected rotational coordinates of body i. The dimension of the matrix $\bar{\mathbf{G}}^i$ depends on whether two-dimensional or three-dimensional analysis is considered. It also depends on the selected rotational coordinates in the case of spatial analysis. For instance, in the planar analysis, the matrix $\bar{\mathbf{G}}^i$ reduces to a unit vector, that is

$$\bar{\omega}^i = \dot{\theta}^i[0 \quad 0 \quad 1]^{\mathrm{T}}$$

When Euler parameters are used to describe the orientation of the body in space, the matrix $\bar{\mathbf{G}}^i$ is a 3×4 matrix given by (see Chapter 2)

$$\bar{\mathbf{G}}^i = 2\begin{bmatrix} -\theta_1^i & \theta_0^i & \theta_3^i & -\theta_2^i \\ -\theta_2^i & -\theta_3^i & \theta_0^i & \theta_1^i \\ -\theta_3^i & \theta_2^i & -\theta_1^i & \theta_0^i \end{bmatrix} \tag{3.109}$$

When the three Rodriguez parameters γ_1^i, γ_2^i, and γ_3^i are used, $\bar{\mathbf{G}}^i$ is a 3×3 matrix given by

$$\bar{\mathbf{G}}^i = \frac{1}{1+(\gamma)^2}\begin{bmatrix} 1 & \gamma_3^i & -\gamma_2^i \\ -\gamma_3^i & 1 & \gamma_1^i \\ \gamma_2^i & -\gamma_1^i & 1 \end{bmatrix} \tag{3.110}$$

Similarly, when the three Euler angles ϕ^i, θ^i, and ψ^i are used, the matrix $\bar{\mathbf{G}}^i$ is a 3×3 matrix defined as

$$\bar{\mathbf{G}}^i = \begin{bmatrix} \sin\theta^i \sin\psi^i & \cos\psi^i & 0 \\ \sin\theta^i \cos\psi^i & -\sin\psi^i & 0 \\ \cos\theta^i & 0 & 1 \end{bmatrix} \tag{3.111}$$

Therefore, there is no loss of generality, by using Eq. 108.

Mass Matrix of the Rigid Bodies Substituting Eq. 108 into Eq. 107, one obtains

$$\dot{\mathbf{r}}^i = \dot{\mathbf{R}}^i - \mathbf{A}^i\tilde{\bar{\mathbf{u}}}^i\bar{\mathbf{G}}^i\dot{\theta}^i$$

which can be written in a partitioned form as

$$\dot{\mathbf{r}}^i = [\mathbf{I} \quad -\mathbf{A}^i\tilde{\bar{\mathbf{u}}}^i\bar{\mathbf{G}}^i]\begin{bmatrix}\dot{\mathbf{R}}^i \\ \dot{\theta}^i\end{bmatrix} \tag{3.112}$$

where $\mathbf{I}$ is the 3×3 identity matrix.

The kinetic energy of the rigid body i can be written as

$$T^i = \frac{1}{2}\int_{V^i} \rho^i \dot{\mathbf{r}}^{i\mathrm{T}}\dot{\mathbf{r}}^i \, dV^i \tag{3.113}$$

where ρ^i and V^i are, respectively, the mass density and the volume of body i. Substituting Eq. 112 into Eq. 113 yields

$$T^i = \frac{1}{2}\int_{V^i} \rho^i [\dot{\mathbf{R}}^{i\mathrm{T}} \quad \dot{\theta}^{i\mathrm{T}}]\begin{bmatrix}\mathbf{I} \\ -\bar{\mathbf{G}}^{i\mathrm{T}}\tilde{\bar{\mathbf{u}}}^{i\mathrm{T}}\mathbf{A}^{i\mathrm{T}}\end{bmatrix}[\mathbf{I} \quad -\mathbf{A}^i\tilde{\bar{\mathbf{u}}}^i\bar{\mathbf{G}}^i]\begin{bmatrix}\dot{\mathbf{R}}^i \\ \dot{\theta}^i\end{bmatrix} dV^i$$

Carrying out the matrix multiplication and utilizing the orthogonality of the transformation matrix, we can write the kinetic energy T^i of body i as

$$T^i = \frac{1}{2}[\dot{\mathbf{R}}^{i\mathrm{T}} \quad \dot{\theta}^{i\mathrm{T}}]\left\{\int_{V^i} \rho^i \begin{bmatrix}\mathbf{I} & -\mathbf{A}^i\tilde{\bar{\mathbf{u}}}^i\bar{\mathbf{G}}^i \\ \text{symmetric} & \bar{\mathbf{G}}^{i\mathrm{T}}\tilde{\bar{\mathbf{u}}}^{i\mathrm{T}}\tilde{\bar{\mathbf{u}}}^i\bar{\mathbf{G}}^i\end{bmatrix} dV^i\right\}\begin{bmatrix}\dot{\mathbf{R}}^i \\ \dot{\theta}^i\end{bmatrix}$$

which can be written as

$$T^i = \frac{1}{2}\dot{\mathbf{q}}_r^{i\mathrm{T}}\mathbf{M}^i\dot{\mathbf{q}}_r^i \tag{3.114}$$

where $\mathbf{q}_r^i = [\mathbf{R}^{i\mathrm{T}} \; \theta^{i\mathrm{T}}]^\mathrm{T}$ is the vector of generalized coordinates of body i and $\mathbf{M}^i$ is the mass matrix of the rigid body defined as

$$\mathbf{M}^i = \int_{V^i} \rho^i \begin{bmatrix}\mathbf{I} & -\mathbf{A}^i\tilde{\bar{\mathbf{u}}}^i\bar{\mathbf{G}}^i \\ \text{symmetric} & \bar{\mathbf{G}}^{i\mathrm{T}}\tilde{\bar{\mathbf{u}}}^{i\mathrm{T}}\tilde{\bar{\mathbf{u}}}^i\bar{\mathbf{G}}^i\end{bmatrix} dV^i \tag{3.115}$$

which can be written in a simplified form as

$$\mathbf{M}^i = \begin{bmatrix}\mathbf{m}_{RR}^i & \mathbf{m}_{R\theta}^i \\ \text{symmetric} & \mathbf{m}_{\theta\theta}^i\end{bmatrix} \tag{3.116}$$

where

$$\mathbf{m}_{RR}^i = \int_{V^i} \rho^i \mathbf{I} \, dV^i \tag{3.117}$$

$$\mathbf{m}_{R\theta}^i = -\int_{V^i} \rho^i \mathbf{A}^i\tilde{\bar{\mathbf{u}}}^i\bar{\mathbf{G}}^i \, dV^i \tag{3.118}$$

$$\mathbf{m}_{\theta\theta}^i = \int_{V^i} \rho^i \bar{\mathbf{G}}^{i\mathrm{T}}\tilde{\bar{\mathbf{u}}}^{i\mathrm{T}}\tilde{\bar{\mathbf{u}}}^i\bar{\mathbf{G}}^i \, dV^i \tag{3.119}$$

One can verify that the matrix $\mathbf{m}^i_{RR}$ of Eq. 117 can be written as

$$\mathbf{m}^i_{RR} = \int_{V^i} \rho^i \mathbf{I}\, dV^i = \begin{bmatrix} m^i & 0 & 0 \\ 0 & m^i & 0 \\ 0 & 0 & m^i \end{bmatrix} \tag{3.120}$$

where m^i is the total mass of the body. Thus, the matrix $\mathbf{m}^i_{RR}$ associated with the translation of the body reference is a constant diagonal matrix.

The matrix $\mathbf{m}^i_{R\theta}$, which represents the inertia coupling between the translation and rotation of the body reference, can be written as

$$\mathbf{m}^i_{R\theta} = -\int_{V^i} \rho^i \mathbf{A}^i \tilde{\bar{\mathbf{u}}}^i \bar{\mathbf{G}}^i\, dV^i = -\mathbf{A}^i \left[\int_{V^i} \rho^i \tilde{\bar{\mathbf{u}}}^i\, dV^i\right] \bar{\mathbf{G}}^i$$

since $\mathbf{A}^i$ and $\bar{\mathbf{G}}^i$ are not space-dependent. One may write the matrix $\mathbf{m}^i_{R\theta}$ in an abbreviated form as

$$\mathbf{m}^i_{R\theta} = -\mathbf{A}^i \tilde{\bar{\mathbf{U}}}^i \bar{\mathbf{G}}^i \tag{3.121}$$

where the skew symmetric matrix $\tilde{\bar{\mathbf{U}}}^i$ is given by

$$\tilde{\bar{\mathbf{U}}}^i = \int_{V^i} \rho^i \tilde{\bar{\mathbf{u}}}^i\, dV^i$$

From the definition of the skew symmetric matrix $\tilde{\bar{\mathbf{u}}}^i$ of Eq. 106, one concludes that if the origin of the body axes is attached to the center of mass of the body, then the matrix $\tilde{\bar{\mathbf{U}}}^i$ is the null matrix; this is because

$$\int_{V^i} \rho^i \bar{u}^i_k\, dV^i = \int_{V^i} \rho^i x^i_k\, dV^i = 0, \quad k = 1, 2, 3$$

in this special case. In this case, the translation and rotation of the body reference are decoupled. This is not, however, the case when the origin of the body reference is attached to a point different from the body center of mass.

We may also write the matrix $\mathbf{m}^i_{\theta\theta}$ associated with the rotational coordinates of the body reference as

$$\begin{aligned} \mathbf{m}^i_{\theta\theta} &= \int_{V^i} \rho^i \bar{\mathbf{G}}^{i\mathrm{T}} \tilde{\bar{\mathbf{u}}}^{i\mathrm{T}} \tilde{\bar{\mathbf{u}}}^i \bar{\mathbf{G}}^i\, dV^i \\ &= \bar{\mathbf{G}}^{i\mathrm{T}} \int_{V^i} \rho^i \tilde{\bar{\mathbf{u}}}^{i\mathrm{T}} \tilde{\bar{\mathbf{u}}}^i\, dV^i \bar{\mathbf{G}}^i \\ &= \bar{\mathbf{G}}^{i\mathrm{T}} \bar{\mathbf{I}}^i_{\theta\theta} \bar{\mathbf{G}}^i \end{aligned} \tag{3.122}$$

where $\bar{\mathbf{I}}^i_{\theta\theta}$, called the *inertia tensor* of the rigid body i, is defined as

$$\bar{\mathbf{I}}^i_{\theta\theta} = \int_{V^i} \rho^i \tilde{\bar{\mathbf{u}}}^{i\mathrm{T}} \tilde{\bar{\mathbf{u}}}^i\, dV^i \tag{3.123}$$

Substituting the matrix $\tilde{\bar{\mathbf{u}}}^i$ from Eq. 106 into Eq. 123, we obtain

$$\bar{\mathbf{I}}^i_{\theta\theta} = \begin{bmatrix} i_{11} & i_{12} & i_{13} \\ & i_{22} & i_{23} \\ \text{symmetric} & & i_{33} \end{bmatrix}^i \tag{3.124}$$

where

$$i_{11} = \int_{V^i} \rho^i \left[(x_2^i)^2 + (x_3^i)^2 \right] dV^i$$

$$i_{12} = -\int_{V^i} \rho^i x_1^i x_2^i \, dV^i$$

$$i_{13} = -\int_{V^i} \rho^i x_1^i x_3^i \, dV^i$$

$$i_{22} = \int_{V^i} \rho^i \left[(x_1^i)^2 + (x_3^i)^2 \right] dV^i$$

$$i_{23} = -\int_{V^i} \rho^i x_2^i x_3^i \, dV^i$$

$$i_{33} = \int_{V^i} \rho^i \left[(x_1^i)^2 + (x_2^i)^2 \right] dV^i$$

The elements i_{ij} are the *moments of inertia*. In particular, the elements i_{ij}, for $i \neq j$, are called the *products of inertia*. One may observe that in the case of rigid bodies, these elements are constant. In deformable body systems, however, these elements are time-dependent since they are explicit functions of the elastic generalized coordinates of the body.

According to the partitioning of generalized coordinates of the rigid body i, one may observe that the kinetic energy of the body i in the multibody system can be written as

$$T^i = T^i_{RR} + T^i_{R\theta} + T^i_{\theta\theta} \tag{3.125}$$

where

$$T^i_{RR} = \frac{1}{2} \dot{\mathbf{R}}^{i\mathrm{T}} \mathbf{m}^i_{RR} \dot{\mathbf{R}}^i$$

$$T^i_{R\theta} = \dot{\mathbf{R}}^{i\mathrm{T}} \mathbf{m}^i_{R\theta} \dot{\boldsymbol{\theta}}^i$$

$$T^i_{\theta\theta} = \frac{1}{2} \dot{\boldsymbol{\theta}}^{i\mathrm{T}} \mathbf{m}^i_{\theta\theta} \dot{\boldsymbol{\theta}}^i$$

with the understanding that $T^i_{R\theta} = 0$ if the origin of the body reference is attached to the center of mass of the body. While T^i_{RR} is called the *translational kinetic energy*, $T^i_{\theta\theta}$ is called the *rotational kinetic energy*. Using Eq. 108, we may write the rotational kinetic energy $T^i_{\theta\theta}$ in terms of the angular velocity vector and the inertia tensor as

$$T^i_{\theta\theta} = \frac{1}{2} \bar{\boldsymbol{\omega}}^{i\mathrm{T}} \bar{\mathbf{I}}^i_{\theta\theta} \bar{\boldsymbol{\omega}}^i \tag{3.126}$$

which is the familiar form given in elementary texts on the subject of dynamics of rigid bodies. Furthermore, since $\boldsymbol{\omega}^i = \mathbf{A}^i\bar{\boldsymbol{\omega}}^i$, that is $\bar{\boldsymbol{\omega}}^i = \mathbf{A}^{i\mathrm{T}}\boldsymbol{\omega}^i$, Eq. 126 yields

$$T^i_{\theta\theta} = \frac{1}{2}\boldsymbol{\omega}^{i\mathrm{T}}\mathbf{A}^i\bar{\mathbf{I}}^i_{\theta\theta}\mathbf{A}^{i\mathrm{T}}\boldsymbol{\omega}^i = \frac{1}{2}\boldsymbol{\omega}^{i\mathrm{T}}\mathbf{I}^i_{\theta\theta}\boldsymbol{\omega}^i$$

where

$$\mathbf{I}^i_{\theta\theta} = \mathbf{A}^i\bar{\mathbf{I}}^i_{\theta\theta}\mathbf{A}^{i\mathrm{T}}$$

Dynamic Equations We have observed that a significant simplification in the form of the mass matrix of the rigid body can be achieved if the origin of the body reference is attached to the mass center of the body. Therefore, for simplicity and to eliminate the inertia coupling between the translation and the rotation of the body reference, the origin of the rigid body reference is often attached to the center of mass of the body. In this case, the mass matrix of the rigid body i can be written as

$$\mathbf{M}^i = \begin{bmatrix} \mathbf{m}^i_{RR} & \mathbf{0} \\ \mathbf{0} & \mathbf{m}^i_{\theta\theta} \end{bmatrix} \tag{3.127}$$

and the kinetic energy T^i is

$$\begin{aligned} T^i &= \frac{1}{2}\dot{\mathbf{R}}^{i\mathrm{T}}\mathbf{m}^i_{RR}\dot{\mathbf{R}}^i + \frac{1}{2}\dot{\boldsymbol{\theta}}^{i\mathrm{T}}\mathbf{m}^i_{\theta\theta}\dot{\boldsymbol{\theta}}^i \\ &= \frac{1}{2}[\dot{\mathbf{R}}^{i\mathrm{T}} \quad \dot{\boldsymbol{\theta}}^{i\mathrm{T}}]\begin{bmatrix} \mathbf{m}^i_{RR} & \mathbf{0} \\ \mathbf{0} & \mathbf{m}^i_{\theta\theta} \end{bmatrix}\begin{bmatrix} \dot{\mathbf{R}}^i \\ \dot{\boldsymbol{\theta}}^i \end{bmatrix} \end{aligned} \tag{3.128}$$

Having determined the kinetic energy of the rigid body i, we proceed to write an expression for the virtual work of all externally applied forces acting on the body. This virtual work expression can be written as

$$\delta W^i = \mathbf{Q}^{i\mathrm{T}}_e\,\delta\mathbf{q}^i \tag{3.129}$$

where $\mathbf{Q}^i_e$ is the vector of generalized forces and $\delta\mathbf{q}^i$ is the virtual change in the vector of generalized coordinates. The virtual work of Eq. 129 can be written in a partitioned form as

$$\delta W^i = \left[(\mathbf{Q}^i_R)^{\mathrm{T}}_e \quad (\mathbf{Q}^i_\theta)^{\mathrm{T}}_e\right]\begin{bmatrix} \delta\mathbf{R}^i \\ \delta\boldsymbol{\theta}^i \end{bmatrix} \tag{3.130}$$

where $(\mathbf{Q}^i_R)_e$ and $(\mathbf{Q}^i_\theta)_e$ are the vectors of generalized forces associated, respectively, with the translation and rotation of the body reference.

Kinematic constraints between different components in the multibody system can be written in a vector form as

$$\mathbf{C}(\mathbf{q}, t) = \mathbf{0} \tag{3.131}$$

where $\mathbf{C}$ is the vector of linearly independent constraint equations, t is time, and $\mathbf{q}$ is the total vector of the multibody system generalized coordinates given by

$$\mathbf{q} = [\mathbf{q}^{1\mathrm{T}}_r \quad \mathbf{q}^{2\mathrm{T}}_r \quad \cdots \quad \mathbf{q}^{n_b\mathrm{T}}_r]^{\mathrm{T}} \tag{3.132}$$

in which n_b is the total number of bodies in the multibody system. Having defined the kinetic energy, the virtual work, and the vector of nonlinear algebraic constraint equations that describe mechanical joints in the system as well as specified motion trajectories, we can write the system equations of motion of the rigid body i in the multibody system using Lagrange's equation or Hamilton's principle as

$$\mathbf{M}^i \ddot{\mathbf{q}}_r^i + \mathbf{C}_{\mathbf{q}_r^i}^{\mathrm{T}} \boldsymbol{\lambda} = \mathbf{Q}_e^i + \mathbf{Q}_v^i \tag{3.133}$$

where $\mathbf{M}^i$ is the mass matrix, $\mathbf{C}_{\mathbf{q}_r^i}$ is the constraint Jacobian matrix, $\boldsymbol{\lambda}$ is the vector of Lagrange multipliers, $\mathbf{Q}_e^i$ is the vector of externally applied forces, and $\mathbf{Q}_v^i$ is a quadratic velocity vector that arises from differentiating the kinetic energy with respect to time and with respect to the generalized coordinates of body i. This quadratic velocity vector, as shown in Section 8, is given by

$$\mathbf{Q}_v^i = -\dot{\mathbf{M}}^i \dot{\mathbf{q}}_r^i + \left(\frac{\partial T^i}{\partial \mathbf{q}_r^i}\right)^{\mathrm{T}} = \left[(\mathbf{Q}_R^i)_v^{\mathrm{T}} \quad (\mathbf{Q}_\theta^i)_v^{\mathrm{T}}\right]^{\mathrm{T}} = \left[\mathbf{0}^{\mathrm{T}} \quad -2\bar{\boldsymbol{\omega}}^{i\mathrm{T}} \bar{\mathbf{I}}_{\theta\theta}^i \dot{\bar{\mathbf{G}}}^i\right]^{\mathrm{T}} \tag{3.134}$$

The differential equations of motion of the multibody system can then be written as

$$\mathbf{M}^i \ddot{\mathbf{q}}_r^i + \mathbf{C}_{\mathbf{q}_r^i}^{\mathrm{T}} \boldsymbol{\lambda} = \mathbf{Q}_e^i + \mathbf{Q}_v^i, \quad i = 1, 2, \ldots, n_b \tag{3.135}$$

Equation 135 can be written in a matrix form as

$$\mathbf{M}\ddot{\mathbf{q}} + \mathbf{C}_{\mathbf{q}}^{\mathrm{T}} \boldsymbol{\lambda} = \mathbf{Q}_e + \mathbf{Q}_v \tag{3.136}$$

where $\mathbf{q}$ is the total vector of the multibody system generalized coordinates defined by Eq. 132 and

$$\mathbf{M} = \begin{bmatrix} \mathbf{M}^1 & & & \\ & \mathbf{M}^2 & & \mathbf{0} \\ & & \ddots & \\ \mathbf{0} & & & \mathbf{M}^{n_b} \end{bmatrix}$$

$$\mathbf{C}_{\mathbf{q}}^{\mathrm{T}} = \begin{bmatrix} \mathbf{C}_{\mathbf{q}_r^1}^{\mathrm{T}} \\ \mathbf{C}_{\mathbf{q}_r^2}^{\mathrm{T}} \\ \vdots \\ \mathbf{C}_{\mathbf{q}_r^{n_b}}^{\mathrm{T}} \end{bmatrix}, \quad \mathbf{Q}_e = \begin{bmatrix} \mathbf{Q}_e^1 \\ \mathbf{Q}_e^2 \\ \vdots \\ \mathbf{Q}_e^{n_b} \end{bmatrix}, \quad \mathbf{Q}_v = \begin{bmatrix} \mathbf{Q}_v^1 \\ \mathbf{Q}_v^2 \\ \vdots \\ \mathbf{Q}_v^{n_b} \end{bmatrix}$$

The differential equations of Eq. 136 and the vector of kinematic constraints of Eq. 131 represent the dynamic equations of the constrained multibody system. These dynamic equations are, in general, nonlinear, and a closed-form solution of these equations is often difficult to obtain. A solution procedure for these dynamic equations is as follows. First, we differentiate Eq. 131 twice with respect to time to get

$$\mathbf{C}_{\mathbf{q}}\dot{\mathbf{q}} = -\mathbf{C}_t \tag{3.137}$$

$$\mathbf{C}_{\mathbf{q}}\ddot{\mathbf{q}} = -\mathbf{C}_{tt} - (\mathbf{C}_{\mathbf{q}}\dot{\mathbf{q}})_{\mathbf{q}}\dot{\mathbf{q}} - 2\mathbf{C}_{\mathbf{q}t}\dot{\mathbf{q}} \tag{3.138}$$

where $\mathbf{C}_t$ is the partial derivative of the constraint vector with respect to time. Let

$$\mathbf{Q}_c = -\mathbf{C}_{tt} - (\mathbf{C}_{\mathbf{q}}\dot{\mathbf{q}})_{\mathbf{q}}\dot{\mathbf{q}} - 2\mathbf{C}_{\mathbf{q}t}\dot{\mathbf{q}}$$

that is,

$$\mathbf{C}_{\mathbf{q}}\ddot{\mathbf{q}} = \mathbf{Q}_c \tag{3.139}$$

Equations 136 and 139 can then be combined in one matrix equation as

$$\begin{bmatrix} \mathbf{M} & \mathbf{C}_{\mathbf{q}}^{\mathrm{T}} \\ \mathbf{C}_{\mathbf{q}} & \mathbf{0} \end{bmatrix} \begin{bmatrix} \ddot{\mathbf{q}} \\ \boldsymbol{\lambda} \end{bmatrix} = \begin{bmatrix} \mathbf{Q}_e + \mathbf{Q}_v \\ \mathbf{Q}_c \end{bmatrix} \tag{3.140}$$

Equation 140 is a system of algebraic equations that can be solved for the acceleration vector $\ddot{\mathbf{q}}$ and the vector of Lagrange multipliers. Given a set of initial conditions, the acceleration vector can be integrated to obtain the velocities and the generalized coordinates (Shabana 1994a).

3.8 NEWTON–EULER EQUATIONS

In this section, the development of the preceding sections will be used to develop *Newton–Euler equations* of motion for a rigid body in the multibody system. To this end, many of the identities developed in Chapter 2 will be used, in particular, the relationships between the angular velocity vector and the time derivative of the orientational coordinates. Since many of these identities were developed in the case of Euler parameters, for convenience and without any loss of generality, we will refer to Euler parameters when we speak about the orientational coordinates. Newton–Euler equations, however, will be presented in a general form in terms of the angular velocity vector.

Summary of the Dynamic Equations In the preceding section it was shown that the kinetic energy of the rigid body i in the multibody system can be written as

$$T^i = \frac{1}{2}\dot{\mathbf{q}}_r^{i\mathrm{T}}\mathbf{M}^i\dot{\mathbf{q}}_r^i \tag{3.141}$$

where the mass matrix $\mathbf{M}^i$ is given by

$$\mathbf{M}^i = \begin{bmatrix} \mathbf{m}_{RR}^i & \mathbf{m}_{R\theta}^i \\ \text{symmetric} & \mathbf{m}_{\theta\theta}^i \end{bmatrix} \tag{3.142}$$

The submatrices $\mathbf{m}_{RR}^i$, $\mathbf{m}_{R\theta}^i$, and $\mathbf{m}_{\theta\theta}^i$ were defined in the preceding section.

It was also pointed out that in the special case in which the origin of the coordinate system of the rigid body is rigidly attached to the center of mass, the submatrix $\mathbf{m}_{R\theta}^i$ is the null matrix and the body i mass matrix reduces to

$$\mathbf{M}^i = \begin{bmatrix} \mathbf{m}_{RR}^i & \mathbf{0} \\ \mathbf{0} & \mathbf{m}_{\theta\theta}^i \end{bmatrix} \tag{3.143}$$

where $\mathbf{m}^i_{RR}$ is defined by Eq. 120 and $\mathbf{m}^i_{\theta\theta}$ is defined by Eq. 122. The kinetic energy T^i of the rigid body can be written in this case as

$$T^i = \frac{1}{2}\dot{\mathbf{q}}_r^{i\mathrm{T}}\mathbf{M}^i\dot{\mathbf{q}}_r^i = \frac{1}{2}\dot{\mathbf{R}}^{i\mathrm{T}}\mathbf{m}^i_{RR}\dot{\mathbf{R}}^i + \frac{1}{2}\dot{\boldsymbol{\theta}}^{i\mathrm{T}}\mathbf{m}^i_{\theta\theta}\dot{\boldsymbol{\theta}}^i \tag{3.144}$$

If the joint reaction forces are treated as externally applied forces, Lagrange's equation of motion can be written as

$$\frac{d}{dt}\left(\frac{\partial T^i}{\partial \dot{\mathbf{q}}_r^i}\right) - \frac{\partial T^i}{\partial \mathbf{q}_r^i} = \bar{\mathbf{Q}}^{i\mathrm{T}} \tag{3.145}$$

where $\bar{\mathbf{Q}}^i$ is defined as

$$\bar{\mathbf{Q}}^i = \mathbf{Q}_e^i + \mathbf{F}_c^i \tag{3.146}$$

in which $\mathbf{Q}_e^i$ is the generalized external force vector and $\mathbf{F}_c^i$ is the vector of generalized joint reaction forces.

Quadratic Velocity Vector Equation 144 yields

$$\frac{\partial T^i}{\partial \dot{\mathbf{q}}_r^i} = \left[\dot{\mathbf{R}}^{i\mathrm{T}}\mathbf{m}^i_{RR} \quad \dot{\boldsymbol{\theta}}^{i\mathrm{T}}\mathbf{m}^i_{\theta\theta}\right] \tag{3.147}$$

$$\frac{d}{dt}\left(\frac{\partial T^i}{\partial \dot{\mathbf{q}}_r^i}\right) = \left[\ddot{\mathbf{R}}^{i\mathrm{T}}\mathbf{m}^i_{RR} \quad (\ddot{\boldsymbol{\theta}}^{i\mathrm{T}}\mathbf{m}^i_{\theta\theta} + \dot{\boldsymbol{\theta}}^{i\mathrm{T}}\dot{\mathbf{m}}^i_{\theta\theta})\right] \tag{3.148}$$

where

$$\dot{\boldsymbol{\theta}}^{i\mathrm{T}}\dot{\mathbf{m}}^i_{\theta\theta} = \dot{\boldsymbol{\theta}}^{i\mathrm{T}}\dot{\bar{\mathbf{G}}}^{i\mathrm{T}}\bar{\mathbf{I}}^i_{\theta\theta}\bar{\mathbf{G}}^i + \dot{\boldsymbol{\theta}}^{i\mathrm{T}}\bar{\mathbf{G}}^{i\mathrm{T}}\bar{\mathbf{I}}^i_{\theta\theta}\dot{\bar{\mathbf{G}}}^i \tag{3.149}$$

It was shown in Chapter 2 that in the case of Euler parameters, one has

$$\dot{\bar{\mathbf{G}}}^i\dot{\boldsymbol{\theta}}^i = \mathbf{0} \quad \text{and} \quad \bar{\boldsymbol{\omega}}^i = \bar{\mathbf{G}}^i\dot{\boldsymbol{\theta}}^i \tag{3.150}$$

where $\bar{\boldsymbol{\omega}}^i$ is the angular velocity vector defined in the body coordinate system and $\bar{\mathbf{G}}^i$ is the matrix that relates the angular velocity vector to the time derivatives of the orientation coordinates. Equation 149 can then be written as

$$\dot{\boldsymbol{\theta}}^{i\mathrm{T}}\dot{\mathbf{m}}^i_{\theta\theta} = \bar{\boldsymbol{\omega}}^{i\mathrm{T}}\bar{\mathbf{I}}^i_{\theta\theta}\dot{\bar{\mathbf{G}}}^i \tag{3.151}$$

Substituting the above equation into Eq. 148, we get

$$\frac{d}{dt}\left(\frac{\partial T^i}{\partial \dot{\mathbf{q}}_r^i}\right) = \left[\ddot{\mathbf{R}}^{i\mathrm{T}}\mathbf{m}^i_{RR} \quad (\ddot{\boldsymbol{\theta}}^{i\mathrm{T}}\mathbf{m}^i_{\theta\theta} + \bar{\boldsymbol{\omega}}^{i\mathrm{T}}\bar{\mathbf{I}}^i_{\theta\theta}\dot{\bar{\mathbf{G}}}^i)\right] \tag{3.152}$$

The derivative of the kinetic energy with respect to the generalized coordinates $\mathbf{q}_r^i$ is

$$\frac{\partial T^i}{\partial \mathbf{q}_r^i} = \frac{1}{2}\frac{\partial}{\partial \mathbf{q}_r^i}\left[\dot{\boldsymbol{\theta}}^{i\mathrm{T}}\mathbf{m}^i_{\theta\theta}\dot{\boldsymbol{\theta}}^i\right] = \left[\mathbf{0}_3^{\mathrm{T}} \quad \frac{1}{2}\frac{\partial}{\partial \boldsymbol{\theta}^i}(\dot{\boldsymbol{\theta}}^{i\mathrm{T}}\mathbf{m}^i_{\theta\theta}\dot{\boldsymbol{\theta}}^i)\right] \tag{3.153}$$

where $\mathbf{0}_3$ is the three-dimensional null vector.

Using Eqs. 122 and 108, one has

$$\frac{\partial T^i}{\partial \mathbf{q}_r^i} = \left[\mathbf{0}_3^{\mathrm{T}} \quad \frac{1}{2}\frac{\partial}{\partial \theta^i}(\dot{\theta}^{i\mathrm{T}}\bar{\mathbf{G}}^{i\mathrm{T}}\bar{\mathbf{I}}_{\theta\theta}^i\bar{\mathbf{G}}^i\dot{\theta}^i)\right] = \left[\mathbf{0}_3^{\mathrm{T}} \quad \frac{1}{2}\frac{\partial}{\partial \theta^i}(\theta^{i\mathrm{T}}\dot{\bar{\mathbf{G}}}^{i\mathrm{T}}\bar{\mathbf{I}}_{\theta\theta}^i\dot{\bar{\mathbf{G}}}^i\theta^i)\right]$$
$$= [\mathbf{0}_3^{\mathrm{T}} \quad \theta^{i\mathrm{T}}\dot{\bar{\mathbf{G}}}^{i\mathrm{T}}\bar{\mathbf{I}}_{\theta\theta}^i\dot{\bar{\mathbf{G}}}^i] = [\mathbf{0}_3^{\mathrm{T}} \quad -\bar{\omega}^{i\mathrm{T}}\bar{\mathbf{I}}_{\theta\theta}^i\dot{\bar{\mathbf{G}}}^i] \tag{3.154}$$

Substituting Eqs. 152 and 154 into Lagrange's equation of Eq. 145, one obtains

$$\left[\ddot{\mathbf{R}}^{i\mathrm{T}}\mathbf{m}_{RR}^i \quad (\ddot{\theta}^{i\mathrm{T}}\mathbf{m}_{\theta\theta}^i + 2\bar{\omega}^{i\mathrm{T}}\bar{\mathbf{I}}_{\theta\theta}^i\dot{\bar{\mathbf{G}}}^i)\right] = [\bar{\mathbf{Q}}_R^{i\mathrm{T}} \quad \bar{\mathbf{Q}}_\theta^{i\mathrm{T}}] \tag{3.155}$$

where subscripts R and θ refer, respectively, to the body translation and rotation and

$$\bar{\mathbf{Q}}^i = [\bar{\mathbf{Q}}_R^{i\mathrm{T}} \quad \bar{\mathbf{Q}}_\theta^{i\mathrm{T}}]^{\mathrm{T}}$$

Equation 155 can be written as two uncoupled matrix equations. The first matrix equation is associated with the translation of the center of mass of the rigid body i, while the second equation is associated with the rotation of the body. These two matrix equations are

$$\mathbf{m}_{RR}^i\ddot{\mathbf{R}}^i = \bar{\mathbf{Q}}_R^i \tag{3.156}$$
$$\mathbf{m}_{\theta\theta}^i\ddot{\theta}^i = \bar{\mathbf{Q}}_\theta^i - 2\dot{\bar{\mathbf{G}}}^{i\mathrm{T}}\bar{\mathbf{I}}_{\theta\theta}^i\bar{\omega}^i \tag{3.157}$$

Generalized and Actual Forces Equations 122 and 157 yield

$$\bar{\mathbf{G}}^{i\mathrm{T}}\bar{\mathbf{I}}_{\theta\theta}^i\bar{\mathbf{G}}^i\ddot{\theta}^i = \bar{\mathbf{Q}}_\theta^i - 2\dot{\bar{\mathbf{G}}}^{i\mathrm{T}}\bar{\mathbf{I}}_{\theta\theta}^i\bar{\omega}^i$$

Multiplying both sides of the above equation by $\bar{\mathbf{G}}^i$ and using the identity of Eq. 23 in Chapter 2 with the understanding that $\bar{\mathbf{G}}^i = 2\bar{\mathbf{E}}^i$ in the case of Euler parameters, one gets

$$4\bar{\mathbf{I}}_{\theta\theta}^i\bar{\mathbf{G}}^i\ddot{\theta}^i = \bar{\mathbf{G}}^i\bar{\mathbf{Q}}_\theta^i - 2\bar{\mathbf{G}}^i\dot{\bar{\mathbf{G}}}^{i\mathrm{T}}\bar{\mathbf{I}}_{\theta\theta}^i\bar{\omega}^i \tag{3.158}$$

By differentiating Eq. 108 with respect to time, it can be shown that the angular acceleration vector $\bar{\alpha}^i$ defined in the coordinate system of the rigid body i is given by

$$\bar{\alpha}^i = \bar{\mathbf{G}}^i\ddot{\theta}^i \tag{3.159}$$

Furthermore, by using Eqs. 65 and 68 of Chapter 2, we obtain

$$2\bar{\mathbf{G}}^i\dot{\bar{\mathbf{G}}}^{i\mathrm{T}}\bar{\mathbf{I}}_{\theta\theta}^i\bar{\omega}^i = 4\tilde{\bar{\omega}}^i\bar{\mathbf{I}}_{\theta\theta}^i\bar{\omega}^i = 4\bar{\omega}^i \times (\bar{\mathbf{I}}_{\theta\theta}^i\bar{\omega}^i) \tag{3.160}$$

By substituting Eqs. 159 and 160 into Eq. 158, one can obtain the following equation:

$$\bar{\mathbf{I}}_{\theta\theta}^i\bar{\alpha}^i = \bar{\mathbf{F}}_\theta^i - \bar{\omega}^i \times (\bar{\mathbf{I}}_{\theta\theta}^i\bar{\omega}^i) \tag{3.161}$$

Clearly, the vector $\bar{\mathbf{F}}_\theta^i$ is the vector of the sum of the moments that act on the rigid body i. This vector is defined in the coordinate system of the rigid body i and given by

$$\bar{\mathbf{F}}_\theta^i = \frac{1}{4}\bar{\mathbf{G}}^i\bar{\mathbf{Q}}_\theta^i$$

This is the relationship between the vector of moments defined in the body Cartesian coordinates and the generalized forces $\bar{\mathbf{Q}}^i_\theta$ associated with the generalized orientational coordinates of the rigid body i.

Newton–Euler Matrix Equation In summary, the motion of the rigid body i in the multibody system is governed by six differential equations, Eqs. 156 and 161, which can be written using the following two matrix equations:

$$\mathbf{m}^i_{RR}\ddot{\mathbf{R}}^i = \bar{\mathbf{Q}}^i_R \tag{3.162}$$

$$\bar{\mathbf{I}}^i_{\theta\theta}\bar{\alpha}^i = \bar{\mathbf{F}}^i_\theta - \bar{\omega}^i \times (\bar{\mathbf{I}}^i_{\theta\theta}\bar{\omega}^i) \tag{3.163}$$

Equation 162 is a matrix equation consisting of three scalar equations that relate the forces and the accelerations of the center of mass of the rigid body. Equation 162 is called *Newton's equation*. Equation 163, on the other hand, defines the body orientation for a given set of moments $\bar{\mathbf{F}}^i_\theta$. This matrix equation consists also of three scalar equations and is called *Euler's equation*. Equations 162 and 163 together are called the *Newton–Euler equations* and can be combined in one matrix equation as

$$\begin{bmatrix} \mathbf{m}^i_{RR} & \mathbf{0} \\ \mathbf{0} & \bar{\mathbf{I}}^i_{\theta\theta} \end{bmatrix} \begin{bmatrix} \ddot{\mathbf{R}}^i \\ \bar{\alpha}^i \end{bmatrix} = \begin{bmatrix} \bar{\mathbf{Q}}^i_R \\ \bar{\mathbf{F}}^i_\theta - \bar{\omega}^i \times (\bar{\mathbf{I}}^i_{\theta\theta}\bar{\omega}^i) \end{bmatrix} \tag{3.164}$$

where $\bar{\alpha}^i$ is the angular acceleration vector of the rigid body i defined by Eq. 159.

Newton–Euler equations can be used to systematically develop a *recursive formulation* for constrained multibody systems (Shabana 1994a). In this case, the recursive kinematic equations can be first developed to express the absolute coordinates of the bodies in terms of the independent joint coordinates. Using these recursive kinematic equations, one can systematically obtain a minimum set of differential equations of the constrained multibody system expressed in terms of the joint variables, and as a consequence, the constraint forces are automatically eliminated from these dynamic equations. The recursive formulations can be developed for both open and closed kinematic chains (Shabana 1994a).

3.9 CONCLUDING REMARKS

In this chapter, methods were presented for developing the dynamic differential equations of motion of multibody systems consisting of a set of interconnected rigid bodies. The concepts of generalized coordinates, degrees of freedom, virtual work, and generalized forces were first introduced and later used to derive Lagrange's equation from D'Alembert's principle. The two cases of holonomic and nonholonomic systems were considered. Hamilton's principle was also discussed and the equivalence between Hamilton's principle and Lagrange's equation was demonstrated. Lagrange's equation was then used to derive the system differential equations of motion of a rigid body in the multibody system. These equations were presented in a matrix form, and the special case of planar motion was discussed. In the Lagrangian formulation presented in this chapter, the concept of the generalized coordinates, velocities, and forces was used with scalar quantities such as the kinetic energy, potential energy, or the Lagrangian to formulate the dynamic equations of multibody systems consisting of rigid bodies. It was, however, shown that this approach is equivalent to the Newtonian approach wherein vector quantities such as the angular velocity and angular acceleration are usually used. This equivalence was demonstrated by deriving the Newton–Euler equations from Lagrange's equations.

Because of space limitations, some of the important techniques for developing the dynamic equations of rigid body systems were not discussed – for instance, *Appell's equations* (Neimark and Fufaev 1972; Shabana 1994a), which can be used for deriving the equations of motion of holonomic and nonholonomic systems. In Appell's equations, a function S, called the *acceleration function* or *acceleration energy*, analogous to the kinetic energy in Lagrange's equation is introduced. This function by itself completely characterizes the dynamics of holonomic and nonholonomic systems in the same way as the kinetic energy in Lagrange's equation. It is, however, important to emphasize that, even though the form of Appell's equations is very simple, it is much more difficult to evaluate the acceleration function than the expression for the kinetic energy.

Absolute Coordinates and Recursive Methods The computer methods used in the automated dynamic analysis of multibody systems consisting of rigid bodies can, in general, be divided into two main approaches. In the first approach, the configuration of the system is identified by using a set of Cartesian coordinates that describe the locations and orientations of the bodies in the multibody systems. This approach has the advantage that the dynamic formulation of the equations of motion is straightforward. Moreover, this approach in general allows easy additions of complex force functions and constraint equations. For each spatial rigid body in the system, six coordinates are used to describe the body configuration. The connectivity between different bodies in the system are introduced to the dynamic formulation by applying a set of nonlinear algebraic constraint equations. This set of constraint equations can be adjoined to the system equations of motion by using Lagrange multipliers or can be used to identify a set of independent coordinates by use of the generalized coordinate partitioning of the constraint Jacobian matrix. In this approach, however, the relative joint coordinates and their time derivatives are not explicitly available.

In the second approach, relative joint variables are used to formulate a minimum set of differential equations of motion. This approach, in which the dynamic differential equations are written in terms of the system degrees of freedom, leads to a recursive formulation. Unlike the formulation based on the Cartesian coordinates, the incorporation of general forcing functions and constraint equations in the recursive formulation is difficult. Newton–Euler equations are often used to develop a recursive formulation for mechanical manipulators. The link velocities and accelerations are transformed from the base link to the end link, and the joint torques can thus be solved recursively from the end effector to the base. A similar recursive transformation concept can be applied by using the Lagrangian formulation (Hollerbach 1980).

In a hybrid formulation called the *velocity transformation* (Jerkovsky 1978; Kim and Vanderploeg 1986), the momentum (Newton–Euler) and velocity (Lagrangian) formulations were used to formulate the differential equations of motion for treelike multibody systems consisting of rigid bodies. The equations of motion of the rigid body in the multibody system were first formulated in terms of the Cartesian coordinates. A velocity transformation matrix is then developed in order to relate joint coordinates to the Cartesian coordinates. This velocity transformation is then used to reduce the number of differential equations and write these equations in terms of the relative joint variables.

Inertia Shape Integrals We have seen from the development presented in this chapter that the dynamic equations that govern the motion of multibody systems consisting of interconnected rigid bodies are highly nonlinear second-order ordinary differential equations. This is mainly because of the finite rotations of the rigid bodies. The dynamic formulation of large-scale multi-rigid-body systems, however, can be carried out to the stage of numerical calculations and can be automated in a fairly general way. In fact, there are in existence today many general-purpose computer programs that can be used for the automatic generation and the numerical solution of the differential equations of motion of large-scale multi-rigid-body systems. Even though these equations are highly nonlinear, the structure of these equations is well defined. In fact, as seen from the development presented in this chapter, these equations can be developed in a fairly systematic manner once the mass of the rigid body i and the following set of *inertia shape integrals* are defined:

$$\mathbf{I}_1^i = \int_{V^i} \rho^i \bar{\mathbf{u}}^i dV^i = \int_{V^i} \rho^i \left[x_1^i \quad x_2^i \quad x_3^i\right]^{\mathrm{T}} dV^i \tag{3.165}$$

$$I_{kl}^i = \int_{V^i} \rho^i x_k^i x_l^i dV^i, \quad k, l = 1, 2, 3 \tag{3.166}$$

where $\bar{\mathbf{u}}^i = [x_1^i \; x_2^i \; x_3^i]^{\mathrm{T}}$, ρ^i is the mass density and V^i is the volume of the rigid body i. The integral $\mathbf{I}_1^i$ represents the moment of mass of the rigid body. If the origin of the body reference is rigidly attached to the body center of mass, $\mathbf{I}_1^i$ is identically the null vector. The integrals I_{kl}^i are required in order to evaluate the inertia tensor $\bar{\mathbf{I}}_{\theta\theta}^i$ of Eq. 122. If the rigid body i has a complex geometric shape, the integrals of Eqs. 165 and 166 can be evaluated by computer or by using hand calculations at a preprocessing stage in advance for the dynamic analysis. They can also be evaluated using a *lumped mass technique* by assuming that the rigid body i consists of n_p particles. In this case the integrals of Eqs. 165 and 166 are given by

$$\mathbf{I}_1^i = \sum_{j=1}^{n_p} m^{ij} \bar{\mathbf{u}}^{ij} = \sum_{j=1}^{n_p} m^{ij} \left[x_1^{ij} \quad x_2^{ij} \quad x_3^{ij}\right]^{\mathrm{T}} \tag{3.167}$$

$$I_{kl}^i = \sum_{j=1}^{n_p} m^{ij} x_k^{ij} x_l^{ij}, \quad k, l = 1, 2, 3 \tag{3.168}$$

where m^{ij} is the mass of the jth particle on the rigid body i and $\bar{\mathbf{u}}^{ij} = [x_1^{ij} \; x_2^{ij} \; x_3^{ij}]^{\mathrm{T}}$ is the position vector of this particle defined in the body coordinate system.

It will be shown in Chapter 5 that the nonlinear dynamic equations of motion of deformable bodies in the multibody systems can also be written in terms of a set of inertia shape integrals that depend on the assumed displacement field. These integrals can also be evaluated by using consistent and lumped masses. Once these integrals are defined, the dynamic equations of deformable bodies can be developed in a systematic manner and can be carried out to the stage of numerical calculations and automated in a fairly general way. Before we provide a more detailed discussion on this subject, however, we first introduce some basic concepts and definitions related to the mechanics of deformable bodies. These important concepts and definitions discussed in the following chapter are basic to the understanding of the development presented in the following chapters in which the dynamic equations of multi-deformable-body systems are developed.

Problems

1. A rigid body rotates with an angular velocity 50 rad/sec about a fixed axis defined by the vector $\mathbf{a} = [0.0\ 2.0\ -1.0]^{\mathrm{T}}$. Assuming that the vector $\mathbf{a}$ is defined in the global coordinate system, derive the velocity constraints in terms of Euler parameters.

2. Show that the constraint equations of the system described in Problem 1 are of the holonomic type. Obtain an expression for the algebraic holonomic constraint equations expressed in terms of Euler parameters.

3. A rigid body rotates with an angular velocity 15 rad/sec about a fixed axis defined by the vector $\mathbf{a} = [1.0\ 0.0\ -1.0]^{\mathrm{T}}$. Assuming that the vector $\mathbf{a}$ is defined in the global coordinate system, derive the velocity constraints in terms of Euler angles. Show that these constraints are of the holonomic type, and obtain the algebraic constraint equations that relate the coordinates.

4. A rigid body rotates with an angular velocity 50 rad/sec about a follower axis defined by the vector $\mathbf{a} = [0.0\ 2.0\ -1.0]^{\mathrm{T}}$. Assuming that the vector $\mathbf{a}$ is defined in the body coordinate system, derive the velocity constraints in terms of Euler parameters. Are the resulting constraints of the holonomic or nonholonomic type?

5. Repeat Problem 4 using Euler angles.

6. For the slider crank mechanism discussed in Example 1, express the velocities of the links of the mechanism in terms of the independent velocity. Assume that the independent velocity is the angular velocity of the crankshaft.

7. Derive expressions for the velocities of the links of the slider crank mechanism of Example 1 in terms of the velocity of the slider block.

8. Using the absolute coordinates of the slider crank mechanism of Example 1, obtain the constraint Jacobian matrices associated with the dependent and independent coordinates when the independent coordinate is assumed to be the angle that defines the orientation of the connecting rod.

9. Derive the constraint equations of the spherical joint that connects two arbitrary rigid bodies in a multibody system. Express these constraints in terms of Euler angles, and determine the constraint Jacobian matrix. Formulate also the spherical joint constraints and the Jacobian matrix in terms of Euler parameters.

10. Using the absolute Cartesian coordinates, derive the constraint equations of the revolute joint that connects two rigid bodies in the three-dimensional analysis. Define also the constraint Jacobian matrix of this joint. Discuss the use of Euler angles and Euler parameters in formulating the revolute joint constraint equations.

11. Using the absolute Cartesian coordinates, derive the constraint equations of the prismatic joint that connects two rigid bodies in the three-dimensional analysis. Define also the constraint Jacobian matrix of this joint. Discuss the use of Euler angles and Euler parameters in formulating the prismatic joint constraint equations.

12. Formulate the constraint equations of the cylindrical joint that connects two rigid bodies in a multibody system. Obtain also the constraint Jacobian matrix of this joint using the absolute Cartesian coordinates.

13. A force vector $\mathbf{F} = [1.0\ 2.0\ -15.0]^{\mathrm{T}}$ N acts at a point whose coordinates are defined in the body reference by the vector $[0.0\ 0.1\ -0.2]^{\mathrm{T}}$ m. Define the generalized forces associated with the generalized coordinates of this body using Euler parameters as the orientation coordinates of the body.

14. Using Euler angles as the orientation coordinates of a rigid body, determine the generalized forces due to the application of the force $\mathbf{F} = [0.0\ 12.0\ -8.0]^{\mathrm{T}}$ N that acts at a point whose coordinates in the body coordinate system are defined by the vector $[0.1\ 0.0\ -0.1]^{\mathrm{T}}$ m.

15. Formulate the generalized forces of the spring–damper–actuator element in the case of planar motion.

16. Formulate the equations of motion of the slider crank mechanism of Example 1 using the embedding technique assuming that the independent coordinate is the angle that defines the orientation of the crankshaft.

17. Repeat Problem 16 assuming that the degree of freedom of the mechanism is the displacement of the slider block.

18. Formulate the equations of motion of the slider crank mechanism of Example 1 using the augmented formulation.

19. Show that the solution of the brachistochrone problem is a cycloid.

20. Find the curve $y = y(x)$ that minimizes the functional

$$I = \int_1^2 \frac{\sqrt{1 + (y')^2}}{x} dx, \quad y(1) = 0, \quad y(2) = 1$$

where $y' = dy/dx$.

21. Show that the solution of Euler's equation to the functional

$$I = \int_a^b (x - y)^2 dx$$

is a straight line.

22. Find the extremals of the following functional:

$$I_1 = \int_a^b \frac{(y')^2}{(x)^3} dx$$

$$I_2 = \int_a^b \{(y)^2 + (y')^2 + 2ye^x\}dx$$

23. Use the principle of virtual work in dynamics to formulate the mass matrix of the rigid body and the vector of centrifugal forces in terms of Euler parameters.

24. Formulate the mass matrix and the centrifugal forces of the rigid body in terms of Euler angles using the principle of virtual work in dynamics.

25. Discuss the use of Newton–Euler equations in developing recursive formulations for multibody systems in terms of the joint variables.

26. Discuss the computational advantages and disadvantages of using the augmented formulation and the recursive method in multibody simulations.

4 MECHANICS OF DEFORMABLE BODIES

Thus far, only the dynamics of multibody systems consisting of interconnected rigid bodies has been discussed. In Chapter 2, methods for the kinematic analysis of the rigid frames of reference were presented and many useful kinematic relationships and identities were developed. These kinematic equations were used in Chapter 3 to develop general formulations for the dynamic differential equations of motion of multi-rigid-body systems. In rigid body dynamics, it is assumed that the distance between two arbitrary points on the body remains constant. This implies that when a force is applied to any point on the rigid body, the resultant stresses set every other point in motion instantaneously, and as shown in the preceding chapter, the force can be considered as producing a linear acceleration for the whole body together with an angular acceleration about its center of mass. The dynamic motion of the body, in this case, can be described using *Newton–Euler equations*, developed in the preceding chapter.

In recent years, greater emphasis has been placed on the design of high-speed, lightweight, precision mechanical systems. These systems, in general, incorporate various types of driving, sensing, and controlling devices working together to achieve specified performance requirements under different loading conditions. In many of these industrial and technological applications, systems cannot be treated as collections of rigid bodies and the rigid body assumption is no longer valid. In such cases, a mechanical system can be modeled as a multibody system that consists of two collections of bodies. One collection consists of bulky compact solids that can be modeled as rigid bodies, while the second collection consists of relatively elastic bodies, such as rods, beams, plates, and shells, that may deform. Many of these structural components are used constantly in industrial and technological applications, such as high-speed robotic manipulators, vehicle systems, airplanes, and space structures.

Continuum mechanics is concerned with the mechanical behavior of solids on the macroscopic scale and treats material as uniformly distributed throughout regions of space. It is then possible to define quantities such as density, displacement, and velocity as continuous (or at least piecewise continuous) functions of position. The study of continuum mechanics is focused on the motion of deformable bodies, which can change their shape. For such bodies the relative motion of the particles is important, and this introduces as significant kinematic variables the spatial derivatives of displacement, velocity, and so on. For deformable bodies, the relative motion between particles that form the body is important and has a significant effect on the body dynamics. When a force is applied to a point on a body, other points are not set in motion instantaneously. The effect of the force must be considered in terms of the propagation of waves.

In this chapter, we briefly discuss the subject of continuum mechanics and introduce many concepts and definitions that are important in the development of computational methods for the dynamic analysis of multi-deformable body systems presented in subsequent chapters. First the kinematics of deformable bodies is discussed and important definitions such as the *Jacobian matrix*, the *gradient of the displacement vector*, the *strain tensor*, and the *rotation tensor* are introduced. These definitions are then used to express the strain vector in terms of the time derivatives of the displacements. The assumptions of the linear theory of elasticity are then stated and used to obtain linear strain displacement relationships. In Section 3, a brief discussion of the physical meaning of the strain components is provided, and in Section 4, other deformation measures are introduced. In Section 5, the *stress components* are defined and the important *Cauchy stress formula* is developed. The general form of the *partial differential equations of equilibrium* is derived and used to prove the symmetry of the *stress tensor* in Section 6. The kinematic and force relationships developed in the first six sections do not depend on the material of the body and, accordingly, apply equally to all materials. In Section 7, the *constitutive relationships* that serve to distinguish one material from another are discussed. Finally, an expression for the *virtual work* of the elastic forces in terms of the stress and strain components is developed in Section 8. Since in this chapter we will be concerned with deformation analysis of one body in the system, the superscript i which denotes the body number in the multibody system will be omitted for simplicity.

4.1 KINEMATICS OF DEFORMABLE BODIES

The deformation, or change of shape, of a body depends on the motion of each particle relative to its neighbors. Therefore, basic to any presentation of deformable body kinematics is the understanding of particle kinematics. We introduce a fixed rectangular Cartesian coordinate system $\mathbf{X}_1\mathbf{X}_2\mathbf{X}_3$ with origin O. Throughout this chapter and the chapters that follow, the global motion will be motion relative to this fixed frame of reference and, unless otherwise stated, all vector and tensor components defined globally are components in the $\mathbf{X}_1\mathbf{X}_2\mathbf{X}_3$ coordinate system. Suppose that at time $t = 0$ a deformable body occupies a fixed region of space B_o, which may be finite or infinite in extent. Suppose that the body moves so that at a subsequent time t it occupies a new continuous region of space B. An assumption (which is an essential feature of continuum mechanics) will be made that we can identify individual particles

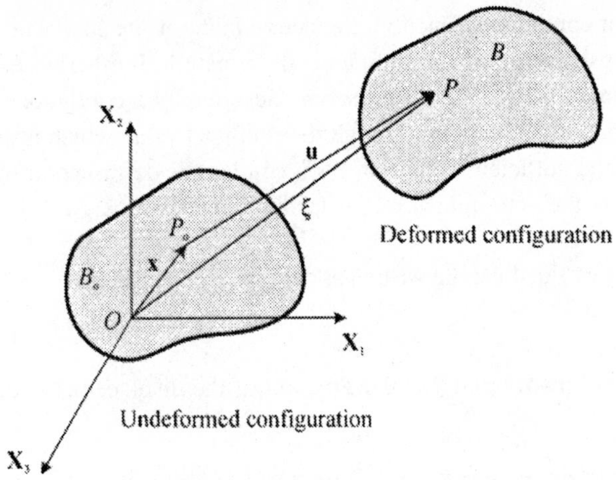

Figure 4.1 Deformed and undeformed configurations.

of the body; that is, we assume that we can identify a point P with position vector ξ, which is occupied at t by the particle that was at P_o at time $t = 0$. Then the final displacement of P can be written as

$$\mathbf{u} = \boldsymbol{\xi} - \mathbf{x} \tag{4.1}$$

where $\mathbf{x}$ is the position vector of P_o as shown in Fig. 1. If we define the vectors $\mathbf{u}$, $\mathbf{x}$, and $\boldsymbol{\xi}$ by their components as

$$\mathbf{u} = [u_1 \quad u_2 \quad u_3]^{\mathrm{T}}, \qquad \mathbf{x} = [x_1 \quad x_2 \quad x_3]^{\mathrm{T}}$$
$$\boldsymbol{\xi} = [\xi_1 \quad \xi_2 \quad \xi_3]^{\mathrm{T}}$$

we can write Eq. 1 in component form as

$$u_1 = \xi_1 - x_1, \qquad u_2 = \xi_2 - x_2, \qquad u_3 = \xi_3 - x_3 \tag{4.2}$$

In the theory of functions, it is shown that Eq. 1 has a single-valued continuous solution if and only if the following determinant does not vanish:

$$|\mathbf{J}| = \begin{vmatrix} \xi_{1,1} & \xi_{1,2} & \xi_{1,3} \\ \xi_{2,1} & \xi_{2,2} & \xi_{2,3} \\ \xi_{3,1} & \xi_{3,2} & \xi_{3,3} \end{vmatrix} \tag{4.3}$$

where $\xi_{i,j} = (\partial \xi_i / \partial x_j)$. Using Eq. 2, we can write the *Jacobian matrix* $\mathbf{J}$ as

$$\mathbf{J} = \begin{bmatrix} 1 + u_{1,1} & u_{1,2} & u_{1,3} \\ u_{2,1} & 1 + u_{2,2} & u_{2,3} \\ u_{3,1} & u_{3,2} & 1 + u_{3,3} \end{bmatrix} \tag{4.4}$$

where u_1, u_2, and u_3 are the components of the displacement vector and $u_{i,j} = \partial u_i / \partial x_j$. If the particles of the body are not displaced at all, the vector $\boldsymbol{\xi}$ is equal to the vector $\mathbf{x}$, and accordingly, the displacement vector is the zero vector. In this case, the Jacobian matrix $\mathbf{J}$ is the identity matrix. Since an assumption is made that the deformation is a continuous function, the determinant of the Jacobian matrix is expected to be positive for small continuous deformation. Furthermore, the determinant of the Jacobian matrix **J** cannot become negative by a continuous deformation of the medium without passing through the excluded value which is zero. Therefore, a necessary and sufficient condition for a continuous deformation to be physically possible is that the determinant of the Jacobian matrix **J** be greater than zero.

The Jacobian matrix of Eq. 4 can be written as

$$\mathbf{J} = \mathbf{I} + \bar{\mathbf{J}} \tag{4.5}$$

where **I** is a 3 × 3 identity matrix and $\bar{\mathbf{J}}$ is the *gradient* of the displacement vector defined as

$$\bar{\mathbf{J}} = \begin{bmatrix} \frac{\partial u_1}{\partial x_1} & \frac{\partial u_1}{\partial x_2} & \frac{\partial u_1}{\partial x_3} \\ \frac{\partial u_2}{\partial x_1} & \frac{\partial u_2}{\partial x_2} & \frac{\partial u_2}{\partial x_3} \\ \frac{\partial u_3}{\partial x_1} & \frac{\partial u_3}{\partial x_2} & \frac{\partial u_3}{\partial x_3} \end{bmatrix} = \begin{bmatrix} u_{1,1} & u_{1,2} & u_{1,3} \\ u_{2,1} & u_{2,2} & u_{2,3} \\ u_{3,1} & u_{3,2} & u_{3,3} \end{bmatrix} \tag{4.6}$$

The gradient of the displacement vector is a second-order tensor and can be represented as the sum of a symmetric tensor and antisymmetric tensor, that is,

$$\bar{\mathbf{J}} = \bar{\mathbf{J}}_s + \bar{\mathbf{J}}_r \tag{4.7}$$

where

$$\bar{\mathbf{J}}_s = \frac{1}{2}[\bar{\mathbf{J}} + \bar{\mathbf{J}}^{\mathrm{T}}] = \begin{bmatrix} e_{11} & e_{12} & e_{13} \\ e_{21} & e_{22} & e_{23} \\ e_{31} & e_{32} & e_{33} \end{bmatrix} \tag{4.8}$$

$$\bar{\mathbf{J}}_r = \frac{1}{2}[\bar{\mathbf{J}} - \bar{\mathbf{J}}^{\mathrm{T}}] = \begin{bmatrix} 0 & \omega_{12} & \omega_{13} \\ \omega_{21} & 0 & \omega_{23} \\ \omega_{31} & \omega_{32} & 0 \end{bmatrix} \tag{4.9}$$

in which

$$2e_{ij} = 2e_{ji} = u_{i,j} + u_{j,i}, \qquad 2\omega_{ij} = u_{i,j} - u_{j,i} = -2\omega_{ji}$$

and the subscript (, i) denotes the differentiation with respect to x_i. For small deformation, it will be shown later that $\bar{\mathbf{J}}_s$ describes the *strain* components at a point in the deformable body, whereas $\bar{\mathbf{J}}_r$ characterizes the *mean rotation* of a volume element.

Example 4.1 The displacement of a body is described in terms of the undeformed rectangular coordinates (x_1, x_2, x_3) as

$$u_1 = k_1 + k_2 x_1$$
$$u_2 = k_3 + k_4 x_1 + k_5 (x_1)^2 + k_6 (x_1)^3$$
$$u_3 = 0$$

where k_i, $(i = 1, \ldots, 6)$ are constants.

In this case, the spatial derivatives of the vector **u** are defined as

$$u_{1,1} = \frac{\partial u_1}{\partial x_1} = k_2, \quad u_{1,2} = \frac{\partial u_1}{\partial x_2} = 0, \quad u_{1,3} = \frac{\partial u_1}{\partial x_3} = 0$$

$$u_{2,1} = \frac{\partial u_2}{\partial x_1} = k_4 + 2k_5 x_1 + 3k_6 (x_1)^2$$

$$u_{2,2} = \frac{\partial u_2}{\partial x_2} = 0, \quad u_{2,3} = \frac{\partial u_2}{\partial x_3} = 0$$

$$u_{3,1} = \frac{\partial u_3}{\partial x_1} = 0, \quad u_{3,2} = \frac{\partial u_3}{\partial x_2} = 0, \quad u_{3,3} = \frac{\partial u_3}{\partial x_3} = 0$$

Therefore, the Jacobian matrix **J** is given by

$$\mathbf{J} = \begin{bmatrix} 1 + k_2 & 0 & 0 \\ [k_4 + 2k_5 x_1 + 3k_6 (x_1)^2] & 1 & 0 \\ 0 & 0 & 1 \end{bmatrix}$$

The Jacobian matrix can also be written as

$$\mathbf{J} = \mathbf{I} + \bar{\mathbf{J}}$$

where **I** is the identity matrix and $\bar{\mathbf{J}}$ is the gradient of the displacement given by

$$\bar{\mathbf{J}} = \begin{bmatrix} k_2 & 0 & 0 \\ [k_4 + 2k_5 x_1 + 3k_6 (x_1)^2] & 0 & 0 \\ 0 & 0 & 0 \end{bmatrix}$$

The gradient of the displacement vector $\bar{\mathbf{J}}$ can be written as the sum of the following symmetric tensor $\bar{\mathbf{J}}_s$ and the antisymmetric tensor $\bar{\mathbf{J}}_r$

$$\bar{\mathbf{J}}_s = \frac{1}{2}[\bar{\mathbf{J}} + \bar{\mathbf{J}}^{\mathrm{T}}] = \begin{bmatrix} e_{11} & e_{12} & e_{13} \\ e_{21} & e_{22} & e_{23} \\ e_{31} & e_{32} & e_{33} \end{bmatrix}$$

$$= \begin{bmatrix} k_2 & \frac{1}{2}[k_4 + 2k_5 x_1 + 3k_6 (x_1)^2] & 0 \\ \frac{1}{2}[k_4 + 2k_5 x_1 + 3k_6 (x_1)^2] & 0 & 0 \\ 0 & 0 & 0 \end{bmatrix}$$

and

$$\bar{\mathbf{J}}_r = \frac{1}{2}[\bar{\mathbf{J}} - \bar{\mathbf{J}}^{\mathrm{T}}] = \begin{bmatrix} 0 & \omega_{12} & \omega_{13} \\ \omega_{21} & 0 & \omega_{23} \\ \omega_{31} & \omega_{32} & 0 \end{bmatrix}$$

$$= \begin{bmatrix} 0 & -\frac{1}{2}[k_4 + 2k_5 x_1 + 3k_6 (x_1)^2] & 0 \\ \frac{1}{2}[k_4 + 2k_5 x_1 + 3k_6 (x_1)^2] & 0 & 0 \\ 0 & 0 & 0 \end{bmatrix}$$

4.2 STRAIN COMPONENTS

In this section, we introduce the *strain components* that arise naturally in the kinematic analysis of deformable bodies. We define δl_o to be the distance between two points P_o and Q_o in the undeformed state as shown in Fig. 2 and δl to be the distance between these two points in the deformed state. Since the coordinates of P_o

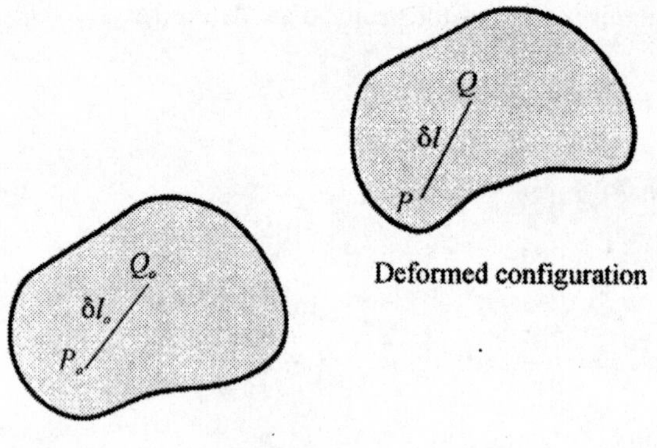

Figure 4.2 Strain components.

in the undeformed state are (x_1, x_2, x_3), we denote the coordinates of Q_o as $(x_1 + dx_1, x_2 + dx_2, x_3 + dx_3)$. Similarly, in the deformed state we denote the coordinates of P and Q as (ξ_1, ξ_2, ξ_3) and $(\xi_1 + d\xi_1, \xi_2 + d\xi_2, \xi_3 + d\xi_3)$, respectively. Therefore, the distances δl_o and δl can be determined according to

$$(\delta l_o)^2 = (d\mathbf{x})^{\mathrm{T}}(d\mathbf{x}) = (dx_1)^2 + (dx_2)^2 + (dx_3)^2 \tag{4.10}$$

$$(\delta l)^2 = (d\boldsymbol{\xi})^{\mathrm{T}}(d\boldsymbol{\xi}) = (d\xi_1)^2 + (d\xi_2)^2 + (d\xi_3)^2 \tag{4.11}$$

where $d\mathbf{x} = [dx_1\ dx_2\ dx_3]^{\mathrm{T}}$ and $d\boldsymbol{\xi} = [d\xi_1\ d\xi_2\ d\xi_3]^{\mathrm{T}}$.

One may write Eq. 1 as

$$\boldsymbol{\xi} = \mathbf{x} + \mathbf{u} \tag{4.12}$$

from which

$$\begin{aligned} d\boldsymbol{\xi} &= d\mathbf{x} + \frac{\partial \mathbf{u}}{\partial \mathbf{x}} d\mathbf{x} \\ &= \left(\mathbf{I} + \frac{\partial \mathbf{u}}{\partial \mathbf{x}}\right) d\mathbf{x} \end{aligned} \tag{4.13}$$

Since $(\partial \mathbf{u}/\partial \mathbf{x}) = \bar{\mathbf{J}}$ is the gradient of the displacement vector, Eq. 13 yields

$$d\boldsymbol{\xi} = (\mathbf{I} + \bar{\mathbf{J}})\, d\mathbf{x} = \mathbf{J}\, d\mathbf{x} \tag{4.14}$$

where $\mathbf{J}$ is the Jacobian matrix defined in Eq. 4. Substituting Eq. 14 into Eq. 11 yields

$$\begin{aligned} (\delta l)^2 &= (d\boldsymbol{\xi})^{\mathrm{T}}(d\boldsymbol{\xi}) = (d\mathbf{x})^{\mathrm{T}}\mathbf{J}^{\mathrm{T}}\mathbf{J}\, d\mathbf{x} \\ &= (d\mathbf{x})^{\mathrm{T}}[\mathbf{I} + \bar{\mathbf{J}}]^{\mathrm{T}}[\mathbf{I} + \bar{\mathbf{J}}]\, d\mathbf{x} \\ &= (d\mathbf{x})^{\mathrm{T}}[\mathbf{I} + (\bar{\mathbf{J}}^{\mathrm{T}} + \bar{\mathbf{J}}) + \bar{\mathbf{J}}^{\mathrm{T}}\bar{\mathbf{J}}]\, d\mathbf{x} \end{aligned} \tag{4.15}$$

Subtracting Eq. 10 from Eq. 15 yields

$$(\delta l)^2 - (\delta l_o)^2 = 2(d\mathbf{x})^{\mathrm{T}}\boldsymbol{\varepsilon}_m d\mathbf{x}$$

or

$$\frac{1}{2}[(\delta l)^2 - (\delta l_o)^2] = (d\mathbf{x})^{\mathrm{T}}\boldsymbol{\varepsilon}_m d\mathbf{x} \tag{4.16}$$

where ε_m is a 3×3 symmetric matrix called the *Lagrangian strain tensor*, and is defined as

$$\varepsilon_m = \frac{1}{2}\{[\bar{\mathbf{J}}^{\mathrm{T}} + \bar{\mathbf{J}}] + \bar{\mathbf{J}}^{\mathrm{T}}\bar{\mathbf{J}}\} \tag{4.17}$$

Using matrix multiplications, it can be verified that the components ε_{ij} of the matrix ε_m are given by

$$\varepsilon_{ij} = \frac{1}{2}\left(u_{i,j} + u_{j,i} + \sum_{k=1}^{3} u_{k,i}u_{k,j}\right), \quad i, j = 1, 2, 3 \tag{4.18}$$

The components ε_{ij} that arise naturally in the analysis of deformation are called the *strain components*. Therefore, the strain components are, in general, nonlinear functions of the spatial derivatives of the displacement. Because of the symmetry of the strain tensor, it is sufficient to identify only the following six components: $\varepsilon_{11}, \varepsilon_{22}, \varepsilon_{33}, \varepsilon_{12}, \varepsilon_{13}$, and ε_{23}, which form the *strain vector* $\boldsymbol{\varepsilon}$, that is,

$$\boldsymbol{\varepsilon} = [\varepsilon_{11} \quad \varepsilon_{22} \quad \varepsilon_{33} \quad \varepsilon_{12} \quad \varepsilon_{13} \quad \varepsilon_{23}]^{\mathrm{T}} \tag{4.19}$$

Thus, the strain vector $\boldsymbol{\varepsilon}$ can be written in a compact form as

$$\boldsymbol{\varepsilon} = \mathbf{D}\mathbf{u} \tag{4.20}$$

where $\mathbf{D}$ is a differential operator defined according to Eq. 18.

Small Strains In the preceding development, it has been shown that the gradient of the displacement vector $\bar{\mathbf{J}}$ can be written as

$$\bar{\mathbf{J}} = \bar{\mathbf{J}}_s + \bar{\mathbf{J}}_r \tag{4.21}$$

where $\bar{\mathbf{J}}_s$ is symmetric and $\bar{\mathbf{J}}_r$ is antisymmetric. In the case of small strains and rotations, the squares and products of $\bar{\mathbf{J}}_s$ and $\bar{\mathbf{J}}_r$ can be neglected, that is

$$\bar{\mathbf{J}}^{\mathrm{T}}\bar{\mathbf{J}} \approx \mathbf{0}$$

and to the same order of approximation the strain tensor reduces to

$$\varepsilon_m \approx \frac{1}{2}[\bar{\mathbf{J}}^{\mathrm{T}} + \bar{\mathbf{J}}] \tag{4.22}$$

which is the form of the strain tensor often used in engineering applications. In this special case, the differential operator of Eq. 20 reduces to

$$\mathbf{D} = \frac{1}{2}\begin{bmatrix} 2\frac{\partial}{\partial x_1} & 0 & 0 \\ 0 & 2\frac{\partial}{\partial x_2} & 0 \\ 0 & 0 & 2\frac{\partial}{\partial x_3} \\ \frac{\partial}{\partial x_2} & \frac{\partial}{\partial x_1} & 0 \\ \frac{\partial}{\partial x_3} & 0 & \frac{\partial}{\partial x_1} \\ 0 & \frac{\partial}{\partial x_3} & \frac{\partial}{\partial x_2} \end{bmatrix} \tag{4.23}$$

and ε_{11}, ε_{22}, and ε_{33} can be recognized as the *normal strains* while ε_{12}, ε_{13}, and ε_{23} are recognized as the *shear strains*. Note that $|\partial u_i/\partial x_j| \ll 1$ implies that the strains and rotations are small. There are, however, some applications in which the strains are small everywhere but the rotations are large. An example of these applications is the bending of a long thin flexible beam.

Example 4.2 For the displacement of the body given in Example 1, find the strain components.

Solution Using Eq. 18 and the spatial derivatives of the displacement given in Example 1, one has

$$\varepsilon_{11} = \frac{1}{2}[u_{1,1} + u_{1,1} + (u_{1,1})^2 + (u_{2,1})^2 + (u_{3,1})^2]$$

$$= \frac{1}{2}[2k_2 + (k_2)^2 + (k_4 + 2k_5x_1 + 3k_6(x_1)^2)^2]$$

$$\varepsilon_{22} = \frac{1}{2}[u_{2,2} + u_{2,2} + (u_{1,2})^2 + (u_{2,2})^2 + (u_{3,2})^2] = 0$$

$$\varepsilon_{33} = \frac{1}{2}[u_{3,3} + u_{3,3} + (u_{1,3})^2 + (u_{2,3})^2 + (u_{3,3})^2] = 0$$

$$\varepsilon_{12} = \frac{1}{2}[u_{1,2} + u_{2,1} + u_{1,1}u_{1,2} + u_{2,1}u_{2,2} + u_{3,1}u_{3,2}]$$

$$= \frac{1}{2}[k_4 + 2k_5x_1 + 3k_6(x_1)^2]$$

$$\varepsilon_{13} = \frac{1}{2}[u_{1,3} + u_{3,1} + u_{1,1}u_{1,3} + u_{2,1}u_{2,3} + u_{3,1}u_{3,3}] = 0$$

$$\varepsilon_{23} = \frac{1}{2}[u_{2,3} + u_{3,2} + u_{1,2}u_{1,3} + u_{2,2}u_{2,3} + u_{3,2}u_{3,3}] = 0$$

Therefore, the vector of strains $\boldsymbol{\varepsilon}$ of Eq. 19 is given by

$$\boldsymbol{\varepsilon} = [\varepsilon_{11} \quad \varepsilon_{22} \quad \varepsilon_{33} \quad \varepsilon_{12} \quad \varepsilon_{13} \quad \varepsilon_{23}]^{\mathrm{T}}$$

$$= \begin{bmatrix} \frac{1}{2}[2k_2 + (k_2)^2 + (k_4 + 2k_5x_1 + 3k_6(x_1)^2)^2] \\ 0 \\ 0 \\ \frac{1}{2}[k_4 + 2k_5x_1 + 3k_6(x_1)^2] \\ 0 \\ 0 \end{bmatrix}$$

If an assumption is made that the strains and rotations are small, one can use the differential operator of Eq. 23 to show that the strain components reduce to

$$\boldsymbol{\varepsilon} = \left[k_2 \quad 0 \quad 0 \quad \tfrac{1}{2}(k_4 + 2k_5x_1 + 3k_6(x_1)^2) \quad 0 \quad 0\right]^{\mathrm{T}}$$

4.3 PHYSICAL INTERPRETATION OF STRAINS

The physical interpretation of the strains can be provided in terms of the extension of the line element P_oQ_o (Fig. 2), defined as

$$e = \delta l - \delta l_o \tag{4.24}$$

The strain in this case is

$$\varepsilon = \frac{e}{\delta l_o} = \frac{\delta l}{\delta l_o} - 1 \tag{4.25}$$

Let $\mathbf{n}$ be the vector of direction cosines along the line $P_o Q_o$ in the undeformed state, that is

$$\mathbf{n} = \frac{d\mathbf{x}}{\delta l_o} = \frac{1}{\delta l_o}[dx_1 \quad dx_2 \quad dx_3]^{\mathrm{T}} \tag{4.26}$$

Then dividing Eq. 16 by $(\delta l_o)^2$ yields

$$\frac{1}{2}\left[\left(\frac{\delta l}{\delta l_o}\right)^2 - 1\right] = \frac{(d\mathbf{x})^{\mathrm{T}}}{\delta l_o}\boldsymbol{\varepsilon}_m \frac{d\mathbf{x}}{\delta l_o}$$

which on using Eq. 26 yields

$$\frac{1}{2}\left[\left(\frac{\delta l}{\delta l_o}\right)^2 - 1\right] = \mathbf{n}^{\mathrm{T}}\boldsymbol{\varepsilon}_m \mathbf{n} \tag{4.27}$$

Using Eq. 25, Eq. 27 reduces to

$$\varepsilon + \frac{1}{2}(\varepsilon)^2 = \mathbf{n}^{\mathrm{T}}\boldsymbol{\varepsilon}_m \mathbf{n}$$

which can be rearranged as

$$(\varepsilon)^2 + 2\varepsilon - \bar{\varepsilon}_m = 0 \tag{4.28}$$

where $\bar{\varepsilon}_m$ is given by

$$\bar{\varepsilon}_m = 2\mathbf{n}^{\mathrm{T}}\boldsymbol{\varepsilon}_m \mathbf{n} \tag{4.29}$$

Equation 28, which is quadratic in ε, has the solution

$$\varepsilon = -1 \pm (1 + \bar{\varepsilon}_m)^{1/2} \tag{4.30}$$

The second solution is physically impossible because it does not represent the rigid body motion. Hence

$$\begin{aligned} \varepsilon &= -1 + (1 + \bar{\varepsilon}_m)^{1/2} = -1 + 1 + \frac{1}{2}\bar{\varepsilon}_m - \frac{1}{8}(\bar{\varepsilon}_m)^2 + \cdots \\ &= \frac{1}{2}\bar{\varepsilon}_m - \frac{1}{8}(\bar{\varepsilon}_m)^2 + \cdots \end{aligned} \tag{4.31}$$

where the binomial theorem has been used. Equation 31 represents the strain in the general case of large deformation. If, however, the strain components are assumed to be small, that is,

$$(\bar{\varepsilon}_m)^2 \approx 0$$

Eq. 31 reduces to

$$\varepsilon \approx \frac{1}{2}\bar{\varepsilon}_m$$

which by using Eq. 29 yields

$$\varepsilon \approx \mathbf{n}^{\mathrm{T}} \varepsilon_m \mathbf{n} \tag{4.32}$$

One can also show by directly using Eq. 27 that the definition of strain in the case of large deformation theory (Eq. 31) does not differ greatly from the definition of Eq. 32 unless the relative elongation e of Eq. 24 is large. Equation 32 implies that the strain along a line element whose direction cosines in the undeformed state with respect to three orthogonal axes $\mathbf{X}_1, \mathbf{X}_2, \mathbf{X}_3$ are defined by the vector $\mathbf{n}$ can be determined if the strain components $\boldsymbol{\varepsilon} = [\varepsilon_{11}\ \varepsilon_{22}\ \varepsilon_{33}\ \varepsilon_{12}\ \varepsilon_{13}\ \varepsilon_{23}]^{\mathrm{T}}$ are known.

Simple Example The preceding development can be exemplified by considering the case in which the element and the extension are along the $\mathbf{X}_1$ direction. In this case, the vector $d\mathbf{x}$ has the components

$$d\mathbf{x} = [dx_1 \quad 0 \quad 0]^{\mathrm{T}}$$

The length of the line segment in the undeformed state can then be written as

$$\delta l_o = \sqrt{(d\mathbf{x})^{\mathrm{T}}(d\mathbf{x})} = dx_1$$

If higher-order terms are neglected in Eq. 31, the strain can be written as

$$\varepsilon = \frac{1}{2}\bar{\varepsilon}_m$$

or

$$\varepsilon = \mathbf{n}^{\mathrm{T}} \varepsilon_m \mathbf{n}$$

In this special case, one can verify that $\mathbf{n} = [1\ 0\ 0]^{\mathrm{T}}$ and

$$\varepsilon = \frac{1}{2}[2u_{1,1} + (u_{1,1})^2 + (u_{2,1})^2 + (u_{3,1})^2]$$

If the assumption that the displacement gradients are small is used, one may neglect second-order terms and write

$$\varepsilon = u_{1,1} = \frac{\partial u_1}{\partial x_1}$$

which is the same expression used in textbooks on the strength of materials.

4.4 RIGID BODY MOTION

In the case of a general rigid body displacement, the vector $\boldsymbol{\xi}$ can be written as

$$\boldsymbol{\xi} = \mathbf{R} + \mathbf{A}\mathbf{x}$$

where $\mathbf{R}$ is the translation of the reference point and $\mathbf{A}$ is the orthogonal transformation matrix that defines the body orientation. It follows, in the case of rigid body displacement, that

$$\mathbf{u} = \boldsymbol{\xi} - \mathbf{x} = \mathbf{R} + (\mathbf{A} - \mathbf{I})\mathbf{x}$$

Using the preceding two equations, it can be shown that

$$\mathbf{J} = \mathbf{A}, \quad \bar{\mathbf{J}} = \mathbf{A} - \mathbf{I}$$

which demonstrate that $\mathbf{J}$ and $\bar{\mathbf{J}}$ do not remain constant in the case of a rigid body motion, and therefore, they are not an appropriate measure of the deformation. Note that in this case, the Lagrangian strain tensor ε_m is given by

$$\varepsilon_m = \frac{1}{2}(\mathbf{J}^{\mathrm{T}}\mathbf{J} - \mathbf{I}) = \mathbf{0}$$

and, therefore, ε_m can be used as a deformation measure.

Other Deformation Measures In continuum mechanics, several other deformation measures are often used. To briefly introduce these measures, we use Eq. 15 to write

$$\left(\frac{\delta l}{\delta l_o}\right)^2 = \mathbf{n}^{\mathrm{T}}\mathbf{J}^{\mathrm{T}}\mathbf{J}\mathbf{n} = \mathbf{n}^{\mathrm{T}}\mathbf{C}_r\mathbf{n}$$

where

$$\mathbf{C}_r = \mathbf{J}^{\mathrm{T}}\mathbf{J}$$

is a symmetric tensor, called the *right Cauchy–Green deformation tensor*. The tensor $\mathbf{C}_r$ can be used as a measure of the deformation since in the case of a general rigid body displacement $\mathbf{C}_r = \mathbf{A}^{\mathrm{T}}\mathbf{A} = \mathbf{I}$, and as a consequence, $\mathbf{C}_r$ remains constant throughout a rigid body motion. The Lagrangian strain tensor can be expressed in terms of $\mathbf{C}_r$ as

$$\varepsilon_m = \frac{1}{2}(\mathbf{C}_r - \mathbf{I})$$

Another deformation measure is the *left Cauchy–Green deformation tensor* $\mathbf{C}_l$ defined as

$$\mathbf{C}_l = \mathbf{J}\mathbf{J}^{\mathrm{T}}$$

This tensor also remains constant and equal to the identity matrix in the case of rigid body motion. Another strain tensor ε_E, called the *Eulerian strain tensor*, is defined in terms of $\mathbf{C}_l$ as

$$\varepsilon_E = \frac{1}{2}(\mathbf{I} - \mathbf{C}_l^{-1})$$

In the case of rigid body motion, $\varepsilon_E = \varepsilon_m = \mathbf{0}$. Furthermore, in the case of infinitesimal strains (small displacement gradients),

$$\varepsilon_E = \varepsilon_m = \frac{1}{2}(\bar{\mathbf{J}} + \bar{\mathbf{J}}^{\mathrm{T}})$$

It is important, however, to point out that the infinitesimal strain tensor is not an exact measure of the deformation because it does not remain constant in the case of a rigid body motion. Recall that, in the case of rigid body motion, $\mathbf{J} = \mathbf{A}$, and

$$\varepsilon_m = \frac{1}{2}(\bar{\mathbf{J}}^{\mathrm{T}} + \bar{\mathbf{J}}) = \frac{1}{2}(\mathbf{A}^{\mathrm{T}} + \mathbf{A} - 2\mathbf{I})$$

It can be shown, however, that the elements of this tensor are of second order in the case of small rotations. For example, in the case of a simple rotation θ about the $\mathbf{X}_3$

axis, one has

$$\mathbf{A} = \begin{bmatrix} \cos\theta & -\sin\theta & 0 \\ \sin\theta & \cos\theta & 0 \\ 0 & 0 & 1 \end{bmatrix},$$

and the tensor ε_m in the case of small deformation is

$$\varepsilon_m = \frac{1}{2}(\mathbf{A}^{\mathrm{T}} + \mathbf{A} - 2\mathbf{I}) = \begin{bmatrix} \cos\theta - 1 & 0 & 0 \\ 0 & \cos\theta - 1 & 0 \\ 0 & 0 & 0 \end{bmatrix}$$

which is of second order in the rotation θ since

$$\cos\theta - 1 = -\frac{\theta^2}{2!} + \frac{\theta^4}{4!} + \cdots$$

Decomposition of Displacement Using the polar decomposition theorem (Spencer 1980), it can be shown that the Jacobian matrix $\mathbf{J}$ can be written as

$$\mathbf{J} = \mathbf{A}_J\mathbf{J}_r = \mathbf{J}_l\mathbf{A}_J$$

where $\mathbf{A}_J$ is an orthogonal *rotation matrix*, and $\mathbf{J}_r$ and $\mathbf{J}_l$ are symmetric positive definite matrices. The matrices $\mathbf{J}_r$ and $\mathbf{J}_l$ are called the *right stretch* and *left stretch tensors*, respectively. It follows from the preceding equation that

$$\mathbf{J}_r = \mathbf{A}_J^{\mathrm{T}}\mathbf{J}_l\mathbf{A}_J, \qquad \mathbf{J}_l = \mathbf{A}_J\mathbf{J}_r\mathbf{A}_J^{\mathrm{T}}$$

In the special case of *homogeneous motion*, the Jacobian matrix $\mathbf{J}$ is assumed to be constant and independent of the spatial coordinates. In this special case, one has

$$\boldsymbol{\xi} = \mathbf{J}\mathbf{x}$$

The motion of the body from the initial configuration $\mathbf{x}$ to the final configuration $\boldsymbol{\xi}$ can be considered as two successive homogeneous motions. In the first motion, the coordinate vector $\mathbf{x}$ changes to $\mathbf{x}_i$, and in the second motion, the coordinate vector $\mathbf{x}_i$ changes to $\boldsymbol{\xi}$, such that

$$\mathbf{x}_i = \mathbf{J}_r\mathbf{x}, \qquad \boldsymbol{\xi} = \mathbf{A}_J\mathbf{x}_i$$

It follows that

$$\boldsymbol{\xi} = \mathbf{A}_J\mathbf{x}_i = \mathbf{A}_J\mathbf{J}_r\mathbf{x} = \mathbf{J}\mathbf{x}$$

Therefore, any homogeneous displacement can be decomposed into a deformation described by the tensor $\mathbf{J}_r$ followed by a rotation described by the orthogonal tensor $\mathbf{A}_J$. Similarly, if $\mathbf{J}_l$ is used instead of $\mathbf{J}_r$, the displacement of the body can be considered as a rotation described by the orthogonal tensor $\mathbf{A}_J$ followed by a deformation defined by the tensor $\mathbf{J}_l$.

In the case of *nonhomogeneous deformation*, one can write the relationship between the change in coordinates as (Spencer 1980)

$$d\boldsymbol{\xi} = \mathbf{J}d\mathbf{x}$$

While $\mathbf{J}$, in this case, is a function of the spatial coordinates, the polar decomposition theorem can still be applied. In this case, the matrices $\mathbf{A}_J$, $\mathbf{J}_r$, and $\mathbf{J}_l$ are functions of the spatial coordinates, and the decomposition of the displacement can be regarded as decomposition of the displacements of infinitesimal volumes of the body.

Note that the deformation measures $\mathbf{C}_r$ and $\mathbf{C}_l$ can be written as

$$\mathbf{C}_r = \mathbf{J}^{\mathrm{T}}\mathbf{J} = \mathbf{J}_r\mathbf{A}_J^{\mathrm{T}}\mathbf{A}_J\mathbf{J}_r = \mathbf{J}_r^2$$
$$\mathbf{C}_l = \mathbf{J}\mathbf{J}^{\mathrm{T}} = \mathbf{J}_l\mathbf{A}_J\mathbf{A}_J^{\mathrm{T}}\mathbf{J}_l = \mathbf{J}_l^2$$

Therefore, $\mathbf{C}_r$ is equivalent to $\mathbf{J}_r$, while $\mathbf{C}_l$ is equivalent to $\mathbf{J}_l$. It is, however, easier and more efficient to calculate $\mathbf{C}_r$ and $\mathbf{C}_l$ for a given $\mathbf{J}$ than to evaluate $\mathbf{J}_r$ and $\mathbf{J}_l$ from the polar decomposition theorem. For this reason $\mathbf{C}_r$ and $\mathbf{C}_l$ are often used, instead of $\mathbf{J}_r$ and $\mathbf{J}_l$, as the deformation measures.

Small Strains and Rotations Using Eq. 7, the matrix $\mathbf{J}$ can be written as

$$\mathbf{J} = \mathbf{I} + \bar{\mathbf{J}} = \mathbf{I} + \bar{\mathbf{J}}_s + \bar{\mathbf{J}}_r$$

where $\bar{\mathbf{J}}_s$ and $\bar{\mathbf{J}}_r$ are defined by Eqs. 8 and 9. In the case of small strains and rotations, higher order terms can be neglected, and the matrix $\mathbf{C}_r$ can be defined as

$$\mathbf{C}_r = \mathbf{J}^{\mathrm{T}}\mathbf{J} = (\mathbf{I} + \bar{\mathbf{J}}_s - \bar{\mathbf{J}}_r)(\mathbf{I} + \bar{\mathbf{J}}_s + \bar{\mathbf{J}}_r) \approx \mathbf{I} + 2\bar{\mathbf{J}}_s,$$

which, upon using the same order of approximation, yields

$$\mathbf{J}_r \approx \mathbf{I} + \bar{\mathbf{J}}_s, \qquad \mathbf{J}_r^{-1} \approx \mathbf{I} - \bar{\mathbf{J}}_s$$

The first of these two equations implies that $\mathbf{J}_r - \mathbf{I}$ reduces to the infinitesimal strain tensor in the case of small deformations. Using the same assumption, it can be shown that $\mathbf{J}_l - \mathbf{I} = \mathbf{J}_r - \mathbf{I}$. Note also that

$$\mathbf{A}_J = \mathbf{J}\mathbf{J}_r^{-1} \approx (\mathbf{I} + \bar{\mathbf{J}}_s + \bar{\mathbf{J}}_r)(\mathbf{I} - \bar{\mathbf{J}}_s) \approx \mathbf{I} + \bar{\mathbf{J}}_r,$$

and, as a consequence,

$$\mathbf{A}_J - \mathbf{I} = \bar{\mathbf{J}}_r$$

in the case of small rotations.

4.5 STRESS COMPONENTS

In this section, we consider the forces acting in the interior of a continuous body. Let P be a point on the surface of the body, $\mathbf{n}$ be a unit vector directed along the outward normal to the surface at P, and δS be the area of an element of the surface that contains P. It is assumed that on the surface element with area δS, the material outside the region under consideration exerts a force (Fig. 3)

$$\mathbf{f} = \boldsymbol{\sigma}_n \, \delta S \tag{4.33}$$

on the material in the region under consideration. The force vector $\mathbf{f}$ is called the *surface force* and the vector $\boldsymbol{\sigma}_n$ is called the *mean surface traction* transmitted across the element of area δS from the outside to the inside of the region under consideration. A surface traction equal in magnitude and opposite in direction to $\boldsymbol{\sigma}_n$ is transmitted across the element with area δS from the inside to the outside of the part of the body under consideration. We make the assumption that as δS tends to zero, $\boldsymbol{\sigma}_n$ tends to a finite limit that is independent of the shape of the element with area δS. The elastic

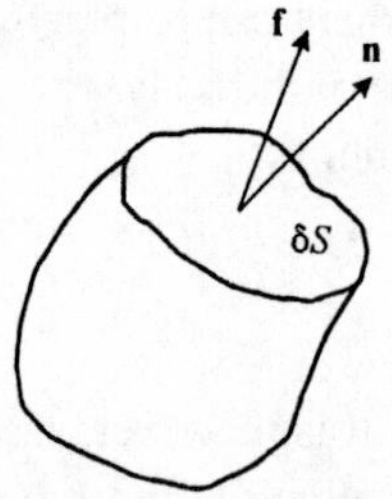

Figure 4.3 Surface force.

force on an arbitrary surface through point P can be written in terms of the elastic forces acting on three perpendicular surfaces of an infinitesimal volume containing point P. To do this, we examine the forces acting on the elementary tetrahedron shown in Fig. 4. Let $\mathbf{f}_1$, $\mathbf{f}_2$, and $\mathbf{f}_3$ be, respectively, the force vectors acting on the surfaces whose outward normal is parallel to $\mathbf{X}_1$, $\mathbf{X}_2$, and $\mathbf{X}_3$. Let $\mathbf{n}$ be the vector of direction cosines of the outward normal to the arbitrary surface δS. Then the areas of the other faces are

$$\delta S_1 = n_1 \delta S, \qquad \delta S_2 = n_2 \delta S, \qquad \delta S_3 = n_3 \delta S \tag{4.34}$$

where n_i, $i = 1, 2, 3$ are the components of $\mathbf{n}$. The elastic force vectors exerted on the tetrahedron across its four faces are

$$\left.\begin{array}{ll} \mathbf{f} = \boldsymbol{\sigma}_n \delta S, & \mathbf{f}_1 = -\boldsymbol{\sigma}_1 n_1 \delta S \\ \mathbf{f}_2 = -\boldsymbol{\sigma}_2 n_2 \delta S, & \mathbf{f}_3 = -\boldsymbol{\sigma}_3 n_3 \delta S \end{array}\right\} \tag{4.35}$$

where $\boldsymbol{\sigma}_1$, $\boldsymbol{\sigma}_2$, and $\boldsymbol{\sigma}_3$ are, respectively, the vectors of mean surface traction acting on the surfaces whose normals are in the directions $\mathbf{X}_1$, $\mathbf{X}_2$, and $\mathbf{X}_3$. The components

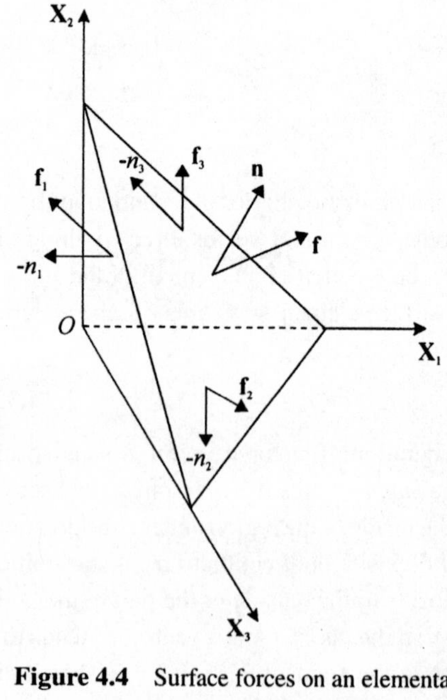

Figure 4.4 Surface forces on an elementary tetrahedron.

of each surface traction $\boldsymbol{\sigma}_i$ will be denoted as σ_{ij}, $j = 1, 2, 3$, that is,

$$\boldsymbol{\sigma}_i = [\sigma_{i1} \quad \sigma_{i2} \quad \sigma_{i3}]^{\mathrm{T}}, \quad i = 1, 2, 3 \tag{4.36}$$

Equation 36 can be written in a more explicit form as

$$\left.\begin{aligned} \boldsymbol{\sigma}_1 &= \sigma_{11}\mathbf{i}_1 + \sigma_{12}\mathbf{i}_2 + \sigma_{13}\mathbf{i}_3 \\ \boldsymbol{\sigma}_2 &= \sigma_{21}\mathbf{i}_1 + \sigma_{22}\mathbf{i}_2 + \sigma_{23}\mathbf{i}_3 \\ \boldsymbol{\sigma}_3 &= \sigma_{31}\mathbf{i}_1 + \sigma_{32}\mathbf{i}_2 + \sigma_{33}\mathbf{i}_3 \end{aligned}\right\} \tag{4.37}$$

where $\mathbf{i}_1$, $\mathbf{i}_2$, and $\mathbf{i}_3$ are unit vectors in the $\mathbf{X}_1$, $\mathbf{X}_2$, and $\mathbf{X}_3$ directions. It is also recognized that there is a *body force* whose mean value over the tetrahedron is $\mathbf{f}_b$ per unit volume. Examples of this kind of force are the gravitational and the magnetic forces. According to Newton's second law, which states that the rate of change of momentum is proportional to the resultant force acting on the system, the equation of equilibrium of the tetrahedron can be written as

$$\mathbf{f}_1 + \mathbf{f}_2 + \mathbf{f}_3 + \mathbf{f} + \mathbf{f}_b \delta V = \rho \mathbf{a}\, \delta V \tag{4.38}$$

where $\mathbf{a}$, ρ, and δV are, respectively, the acceleration, mass density, and volume of the tetrahedron. Substituting Eq. 35 into Eq. 38 yields

$$\boldsymbol{\sigma}_n = \boldsymbol{\sigma}_1 n_1 + \boldsymbol{\sigma}_2 n_2 + \boldsymbol{\sigma}_3 n_3 + \frac{\delta V}{\delta S}(\rho \mathbf{a} - \mathbf{f}_b) \tag{4.39}$$

We assume $\mathbf{n}$ and the point P to be fixed and let δS and δV tend to zero. Since δV is proportional to the cube and δS is proportional to the square of the linear dimension of the tetrahedron, we conclude that $\delta V/\delta S$ tends to zero as δS approaches zero. Thus, in the limit one has

$$\boldsymbol{\sigma}_n = \boldsymbol{\sigma}_1 n_1 + \boldsymbol{\sigma}_2 n_2 + \boldsymbol{\sigma}_3 n_3 \tag{4.40}$$

where $\boldsymbol{\sigma}_1$, $\boldsymbol{\sigma}_2$, $\boldsymbol{\sigma}_3$, and $\boldsymbol{\sigma}_n$ are evaluated at P. Equation 40 can be written in matrix form as

$$\boldsymbol{\sigma}_n = \boldsymbol{\sigma}_m \mathbf{n} \tag{4.41}$$

where $\boldsymbol{\sigma}_m$ is a 3×3 matrix defined as

$$\boldsymbol{\sigma}_m = \begin{bmatrix} \sigma_{11} & \sigma_{21} & \sigma_{31} \\ \sigma_{12} & \sigma_{22} & \sigma_{32} \\ \sigma_{13} & \sigma_{23} & \sigma_{33} \end{bmatrix}$$

Therefore, one can write the elastic force vector of Eq. 33 as

$$\mathbf{f} = \boldsymbol{\sigma}_m \mathbf{n}\, \delta S \tag{4.42}$$

Therefore, the surface force $\mathbf{f}$ can be expressed in terms of the elements of the matrix $\boldsymbol{\sigma}_m$. These elements σ_{ij}, $(i, j = 1, 2, 3)$ are called the *stress components*. The components of the stress vectors $\boldsymbol{\sigma}_1$, $\boldsymbol{\sigma}_2$, and $\boldsymbol{\sigma}_3$ represent the stress on the planes that are perpendicular, respectively, to the $\mathbf{X}_1$, $\mathbf{X}_2$, and $\mathbf{X}_3$ axes (Eq. 37). Equation 40, which is called the *Cauchy stress formula*, gives the stress vector on an oblique plane with unit normal $\mathbf{n}$.

By using Eq. 40 or 41, it can be shown that $\boldsymbol{\sigma}_m$ is a tensor quantity. In the foregoing discussion, the stress components were defined with respect to the coordinate system

$\mathbf{X}_1$, $\mathbf{X}_2$, and $\mathbf{X}_3$. It is expected that the choice of the coordinate system will lead to a different set of stress components. Let $\bar{\mathbf{X}}_1\bar{\mathbf{X}}_2\bar{\mathbf{X}}_3$ be another coordinate system. We now examine the relationship between the stress components σ_{ij} associated with the coordinate system $\mathbf{X}_1\mathbf{X}_2\mathbf{X}_3$ and the stress components $\bar{\sigma}_{ij}$ at the same point defined with respect to the coordinate system $\bar{\mathbf{X}}_1\bar{\mathbf{X}}_2\bar{\mathbf{X}}_3$. Let $\mathbf{A}$ be an orthogonal transformation matrix that defines the orientation of the coordinate system $\bar{\mathbf{X}}_1\bar{\mathbf{X}}_2\bar{\mathbf{X}}_3$ with respect to the coordinate system $\mathbf{X}_1\mathbf{X}_2\mathbf{X}_3$. One can then write the following equation:

$$\bar{\boldsymbol{\sigma}}_n = \mathbf{A}^{\mathrm{T}}\boldsymbol{\sigma}_n = \mathbf{A}^{\mathrm{T}}\boldsymbol{\sigma}_m\mathbf{n} = \mathbf{A}^{\mathrm{T}}\boldsymbol{\sigma}_m\mathbf{A}\bar{\mathbf{n}} \tag{4.43}$$

where $\bar{\mathbf{n}} = [\bar{n}_1\ \bar{n}_2\ \bar{n}_3]^{\mathrm{T}}$ is the normal to the surface whose components are defined with respect to the $\bar{\mathbf{X}}_1\bar{\mathbf{X}}_2\bar{\mathbf{X}}_3$ coordinate system. Equation 43 can be written in a compact form as

$$\bar{\boldsymbol{\sigma}}_n = \bar{\boldsymbol{\sigma}}_m\bar{\mathbf{n}} \tag{4.44}$$

where $\bar{\boldsymbol{\sigma}}_m$ is given by

$$\bar{\boldsymbol{\sigma}}_m = \mathbf{A}^{\mathrm{T}}\boldsymbol{\sigma}_m\mathbf{A} \tag{4.45}$$

which demonstrates that $\boldsymbol{\sigma}_m$ is indeed a second-order tensor.

4.6 EQUATIONS OF EQUILIBRIUM

In studying the mechanics of deformable bodies, a distinction is made between two kinds of forces: *body forces* acting on the element of volume (or mass) of the body such as gravitational, magnetic, and inertia forces, and *surface forces* acting on surface elements inside or on the boundary of the body such as contact forces and hydrostatic pressure. The resultant of the first kind of force follows from integration over the volume, whereas the second kind is the result of a surface integral. Thus, the condition for the dynamic equilibrium can be mathematically stated as

$$\int_S \boldsymbol{\sigma}_n\, dS + \int_V \mathbf{f}_b\, dV = \int_V \rho\mathbf{a}\, dV \tag{4.46}$$

Substituting Eq. 41 into Eq. 46 yields

$$\int_S \boldsymbol{\sigma}_m\mathbf{n}\, dS + \int_V \mathbf{f}_b\, dV = \int_V \rho\mathbf{a}\, dV \tag{4.47}$$

The surface integral can be transformed into a volume integral by use of the *divergence theorem* (Greenberg 1978), that is

$$\int_S \boldsymbol{\sigma}_m\mathbf{n}\, dS = \int_V \boldsymbol{\sigma}_s\, dV \tag{4.48}$$

where $\boldsymbol{\sigma}_s = [\sigma_{s1}\ \ \sigma_{s2}\ \ \sigma_{s3}]^{\mathrm{T}}$ is a vector whose components are defined according to

$$\sigma_{si} = \sum_{j=1}^{3} \frac{\partial \sigma_{ji}}{\partial x_j} \tag{4.49}$$

Substituting Eq. 48 into Eq. 47 yields

$$\int_V [\boldsymbol{\sigma}_s + \mathbf{f}_b - \rho\mathbf{a}]\, dV = \mathbf{0} \tag{4.50}$$

This equation must hold in every region in the body, and hence the integrand must be zero throughout the body. This leads to

$$\boldsymbol{\sigma}_s + \mathbf{f}_b = \rho\mathbf{a} \tag{4.51}$$

which is known as the *equation of equilibrium*. By using Eq. 49, we can write the components of Eq. 51 as

$$\left.\begin{aligned} \sigma_{11,1} + \sigma_{21,2} + \sigma_{31,3} + f_{b1} &= \rho a_1 \\ \sigma_{12,1} + \sigma_{22,2} + \sigma_{32,3} + f_{b2} &= \rho a_2 \\ \sigma_{13,1} + \sigma_{23,2} + \sigma_{33,3} + f_{b3} &= \rho a_3 \end{aligned}\right\} \tag{4.52}$$

where $(, i)$ denotes differentiation with respect to the spatial coordinate x_i; a_1, a_2, and a_3 are the components of the acceleration vector; and f_{b1}, f_{b2}, and f_{b3} are the components of the vector of the body force. It is important to note that the equations of equilibrium contain both time and spatial derivatives.

Symmetry of the Stress Tensor In developing the differential equations of equilibrium we used the equilibrium of the forces. The condition that the resultant couple about the origin must be equal to zero can be used to prove the symmetry of the stress tensor. This condition can be expressed mathematically as

$$\int_S \mathbf{x} \times \boldsymbol{\sigma}_n dS + \int_V \mathbf{x} \times (\mathbf{f}_b - \rho\mathbf{a})\, dV = \mathbf{0} \tag{4.53}$$

Recall that

$$\mathbf{x} \times \boldsymbol{\sigma}_n = \tilde{\mathbf{x}}\boldsymbol{\sigma}_n \tag{4.54}$$

where $\tilde{\mathbf{x}}$ is a skew symmetric matrix defined as

$$\tilde{\mathbf{x}} = \begin{bmatrix} 0 & -x_3 & x_2 \\ x_3 & 0 & -x_1 \\ -x_2 & x_1 & 0 \end{bmatrix} \tag{4.55}$$

Using Eqs. 41 and 54, we can write the first integral of Eq. 53 as

$$\int_S \mathbf{x} \times \boldsymbol{\sigma}_n dS = \int_S \tilde{\mathbf{x}}\boldsymbol{\sigma}_m \mathbf{n}\, dS \tag{4.56}$$

If we define the matrix $\mathbf{B}$ and the vector $\mathbf{b}$ as

$$\mathbf{B} = \tilde{\mathbf{x}}\boldsymbol{\sigma}_m, \qquad \mathbf{b} = [b_1 \quad b_2 \quad b_3]^{\mathrm{T}} \tag{4.57}$$

where

$$b_i = \sum_{j=1}^{3} \frac{\partial B_{ji}}{\partial x_j}, \quad i = 1, 2, 3 \tag{4.58}$$

then on using the divergence theorem, Eq. 56 yields

$$\int_S \tilde{\mathbf{x}} \sigma_m \mathbf{n}\, dS = \int_S \mathbf{B}\mathbf{n}\, dS = \int_V \mathbf{b}\, dV \tag{4.59}$$

One can verify from the definitions of Eqs. 57 and 59 that

$$\mathbf{b} = \tilde{\mathbf{x}} \boldsymbol{\sigma}_s + \mathbf{b}_s = \mathbf{x} \times \boldsymbol{\sigma}_s + \mathbf{b}_s \tag{4.60}$$

where the components of the vector $\boldsymbol{\sigma}_s$ are defined by Eq. 49 and the vector $\mathbf{b}_s$ is given by

$$\mathbf{b}_s = \begin{bmatrix} \sigma_{32} - \sigma_{23} \\ \sigma_{13} - \sigma_{31} \\ \sigma_{21} - \sigma_{12} \end{bmatrix} \tag{4.61}$$

Substituting Eq. 60 into Eq. 59 yields

$$\int_S \tilde{\mathbf{x}} \sigma_m \mathbf{n}\, dS = \int_V (\mathbf{x} \times \boldsymbol{\sigma}_s + \mathbf{b}_s)\, dV$$

which on substituting into Eq. 53 and using Eq. 56 yields

$$\int_V \mathbf{b}_s\, dV + \int_V \mathbf{x} \times (\boldsymbol{\sigma}_s + \mathbf{f}_b - \rho \mathbf{a})\, dV = \mathbf{0} \tag{4.62}$$

By using Eq. 51, Eq. 62 becomes

$$\int_V \mathbf{b}_s\, dV = \mathbf{0} \tag{4.63}$$

This equation must hold in every region in the body, and hence the integrand must be zero throughout the body. This leads to

$$\mathbf{b}_s = \mathbf{0}$$

Using this equation and Eq. 61, one obtains

$$\sigma_{32} = \sigma_{23}, \qquad \sigma_{13} = \sigma_{31}, \qquad \sigma_{21} = \sigma_{12}$$

This result can be written in a compact form as

$$\sigma_{ij} = \sigma_{ji} \tag{4.64}$$

which implies that the stress tensor is symmetric.

4.7 CONSTITUTIVE EQUATIONS

The stress and strain tensors are insufficient for description of the mechanical behavior of deformable bodies. Body deformations depend on the applied forces, and the force-displacement relationship depends on the material of the body. To complete the specification of the mechanical properties of a material we require additional equations. These equations are called the *constitutive equations* and serve to distinguish one material from another. For convenience, we reproduce the stress and strain vectors, which are essential in the discussion that follows:

$$\boldsymbol{\sigma} = [\sigma_{11} \quad \sigma_{22} \quad \sigma_{33} \quad \sigma_{12} \quad \sigma_{13} \quad \sigma_{23}]^{\mathrm{T}} \tag{4.65}$$

$$\boldsymbol{\varepsilon} = [\varepsilon_{11} \quad \varepsilon_{22} \quad \varepsilon_{33} \quad \varepsilon_{12} \quad \varepsilon_{13} \quad \varepsilon_{23}]^{\mathrm{T}} \tag{4.66}$$

It has been found experimentally that for most solid materials, the measured strains are proportional to the applied forces, provided the load does not exceed a given value, known as the *elastic limit*. This experimental observation can be stated as follows: The stress components at any point in the body are a linear function of the strain components. This statement is a generalization of *Hooke's law* and does not apply to viscoelastic, plastic, or viscoplastic materials. The generalized form of Hooke's law may thus be written as

$$\left.\begin{aligned}
\sigma_{11} &= e_{11}\varepsilon_{11} + e_{12}\varepsilon_{22} + e_{13}\varepsilon_{33} + e_{14}\varepsilon_{12} + e_{15}\varepsilon_{13} + e_{16}\varepsilon_{23} \\
\sigma_{22} &= e_{21}\varepsilon_{11} + e_{22}\varepsilon_{22} + e_{23}\varepsilon_{33} + e_{24}\varepsilon_{12} + e_{25}\varepsilon_{13} + e_{26}\varepsilon_{23} \\
\sigma_{33} &= e_{31}\varepsilon_{11} + e_{32}\varepsilon_{22} + e_{33}\varepsilon_{33} + e_{34}\varepsilon_{12} + e_{35}\varepsilon_{13} + e_{36}\varepsilon_{23} \\
\sigma_{12} &= e_{41}\varepsilon_{11} + e_{42}\varepsilon_{22} + e_{43}\varepsilon_{33} + e_{44}\varepsilon_{12} + e_{45}\varepsilon_{13} + e_{46}\varepsilon_{23} \\
\sigma_{13} &= e_{51}\varepsilon_{11} + e_{52}\varepsilon_{22} + e_{53}\varepsilon_{33} + e_{54}\varepsilon_{12} + e_{55}\varepsilon_{13} + e_{56}\varepsilon_{23} \\
\sigma_{23} &= e_{61}\varepsilon_{11} + e_{62}\varepsilon_{22} + e_{63}\varepsilon_{33} + e_{64}\varepsilon_{12} + e_{65}\varepsilon_{13} + e_{66}\varepsilon_{23}
\end{aligned}\right\} \tag{4.67}$$

which can be written in a compact form as

$$\boldsymbol{\sigma} = \mathbf{E}\boldsymbol{\varepsilon} \tag{4.68}$$

where $\mathbf{E}$ is the matrix of the *elastic constants* of the material given by

$$\mathbf{E} = \begin{bmatrix}
e_{11} & e_{12} & e_{13} & e_{14} & e_{15} & e_{16} \\
e_{21} & e_{22} & e_{23} & e_{24} & e_{25} & e_{26} \\
e_{31} & e_{32} & e_{33} & e_{34} & e_{35} & e_{36} \\
e_{41} & e_{42} & e_{43} & e_{44} & e_{45} & e_{46} \\
e_{51} & e_{52} & e_{53} & e_{54} & e_{55} & e_{56} \\
e_{61} & e_{62} & e_{63} & e_{64} & e_{65} & e_{66}
\end{bmatrix} \tag{4.69}$$

Anisotropic Linearly Elastic Material Let U be the *strain energy* per unit volume that represents the work done by internal stresses. On a unit cube, stresses represent forces, whereas strains represent displacements. Therefore, the work done by a force $\boldsymbol{\sigma}$ during the motion $d\boldsymbol{\varepsilon}$ can be written as

$$dU = \boldsymbol{\sigma}^{\mathrm{T}} d\boldsymbol{\varepsilon}$$

which implies that

$$\boldsymbol{\sigma} = \left(\frac{\partial U}{\partial \boldsymbol{\varepsilon}}\right)^{\mathrm{T}} \tag{4.70}$$

or in a more explicit form as

$$\sigma_{ij} = \frac{\partial U}{\partial \varepsilon_{ij}}, \quad i, j = 1, 2, 3 \tag{4.71}$$

that is,

$$\sigma_{11} = \frac{\partial U}{\partial \varepsilon_{11}}, \qquad \sigma_{22} = \frac{\partial U}{\partial \varepsilon_{22}}, \qquad \sigma_{33} = \frac{\partial U}{\partial \varepsilon_{33}}$$

$$\sigma_{12} = \frac{\partial U}{\partial \varepsilon_{12}}, \qquad \sigma_{13} = \frac{\partial U}{\partial \varepsilon_{13}}, \qquad \sigma_{23} = \frac{\partial U}{\partial \varepsilon_{23}}$$

Equations 67 and 71 yield

$$\left.\begin{aligned} \frac{\partial U}{\partial \varepsilon_{11}} &= \sigma_{11} = e_{11}\varepsilon_{11} + \cdots + e_{16}\varepsilon_{23} \\ \frac{\partial U}{\partial \varepsilon_{22}} &= \sigma_{22} = e_{21}\varepsilon_{11} + \cdots + e_{26}\varepsilon_{23} \\ &\vdots \\ \frac{\partial U}{\partial \varepsilon_{23}} &= \sigma_{23} = e_{61}\varepsilon_{11} + \cdots + e_{66}\varepsilon_{23} \end{aligned}\right\} \tag{4.72}$$

Differentiation of Eq. 72 yields

$$\frac{\partial^2 U}{\partial \varepsilon_{11}\partial \varepsilon_{22}} = e_{12} = \frac{\partial^2 U}{\partial \varepsilon_{22}\partial \varepsilon_{11}} = e_{21}$$

$$\frac{\partial^2 U}{\partial \varepsilon_{11}\partial \varepsilon_{33}} = e_{13} = \frac{\partial^2 U}{\partial \varepsilon_{33}\partial \varepsilon_{11}} = e_{31}$$

$$\vdots$$

$$\frac{\partial^2 U}{\partial \varepsilon_{33}\partial \varepsilon_{23}} = e_{36} = \frac{\partial^2 U}{\partial \varepsilon_{23}\partial \varepsilon_{33}} = e_{63}$$

That is,

$$e_{ij} = e_{ji} \tag{4.73}$$

which shows that the matrix of the elastic coefficients is symmetric. Therefore, there are only 21 distinct elastic coefficients for a general *anisotropic* linearly elastic material. In terms of these coefficients, the matrix of elastic coefficients $\mathbf{E}$ can be written as

$$\mathbf{E} = \begin{bmatrix} e_{11} & & & & & \\ e_{21} & e_{22} & & \text{symmetric} & & \\ e_{31} & e_{32} & e_{33} & & & \\ e_{41} & e_{42} & e_{43} & e_{44} & & \\ e_{51} & e_{52} & e_{53} & e_{54} & e_{55} & \\ e_{61} & e_{62} & e_{63} & e_{64} & e_{65} & e_{66} \end{bmatrix} \tag{4.74}$$

Material Symmetry In some structural materials, special kinds of symmetry may exist. The elastic coefficients, for example, may remain invariant under a coordinate transformation. For instance, consider the reflection with respect to the $\mathbf{X}_1\mathbf{X}_2$ plane given by the following transformation:

$$\mathbf{A} = \begin{bmatrix} 1 & 0 & 0 \\ 0 & 1 & 0 \\ 0 & 0 & -1 \end{bmatrix}$$

The transformed stresses and strains $\boldsymbol{\sigma}'_m$ and $\boldsymbol{\varepsilon}'_m$ are given, respectively, by

$$\boldsymbol{\sigma}'_m = \mathbf{A}^{\mathrm{T}}\boldsymbol{\sigma}_m\mathbf{A} \tag{4.75}$$

$$\boldsymbol{\varepsilon}'_m = \mathbf{A}^{\mathrm{T}}\boldsymbol{\varepsilon}_m\mathbf{A} \tag{4.76}$$

where $\boldsymbol{\sigma}_m$ and $\boldsymbol{\varepsilon}_m$ are given by

$$\boldsymbol{\sigma}_m = \begin{bmatrix} \sigma_{11} & \sigma_{12} & \sigma_{13} \\ \sigma_{21} & \sigma_{22} & \sigma_{23} \\ \sigma_{31} & \sigma_{32} & \sigma_{33} \end{bmatrix} \tag{4.77}$$

$$\boldsymbol{\varepsilon}_m = \begin{bmatrix} \varepsilon_{11} & \varepsilon_{12} & \varepsilon_{13} \\ \varepsilon_{21} & \varepsilon_{22} & \varepsilon_{23} \\ \varepsilon_{31} & \varepsilon_{32} & \varepsilon_{33} \end{bmatrix} \tag{4.78}$$

Equation 75 yields

$$\left.\begin{aligned} \sigma'_{11} &= \sigma_{11}, \quad & \sigma'_{22} &= \sigma_{22}, \quad & \sigma'_{33} &= \sigma_{33} \\ \sigma'_{12} &= \sigma_{12}, \quad & \sigma'_{13} &= -\sigma_{13}, \quad & \sigma'_{23} &= -\sigma_{23} \end{aligned}\right\} \tag{4.79}$$

and Eq. 76 yields

$$\left.\begin{aligned} \varepsilon'_{11} &= \varepsilon_{11}, \quad & \varepsilon'_{22} &= \varepsilon_{22}, \quad & \varepsilon'_{33} &= \varepsilon_{33} \\ \varepsilon'_{12} &= \varepsilon_{12}, \quad & \varepsilon'_{13} &= -\varepsilon_{13}, \quad & \varepsilon'_{23} &= -\varepsilon_{23} \end{aligned}\right\} \tag{4.80}$$

Therefore, under the transformation of Eqs. 75 and 76, one can write, for example, σ'_{11} as

$$\sigma'_{11} = e_{11}\varepsilon'_{11} + e_{12}\varepsilon'_{22} + e_{13}\varepsilon'_{33} + e_{14}\varepsilon'_{12} + e_{15}\varepsilon'_{13} + e_{16}\varepsilon'_{23} \tag{4.81}$$

which on using Eqs. 79 and 80 yields

$$\sigma_{11} = \sigma'_{11} = e_{11}\varepsilon_{11} + e_{12}\varepsilon_{22} + e_{13}\varepsilon_{33} + e_{14}\varepsilon_{12} - e_{15}\varepsilon_{13} - e_{16}\varepsilon_{23} \tag{4.82}$$

By comparing Eqs. 81 and 82 and using Eqs. 79 and 80, one gets

$$e_{15} = -e_{15}, \qquad e_{16} = -e_{16}, \qquad \text{or} \qquad e_{15} = e_{16} = 0$$

In a similar manner by considering other stress components, we find

$$e_{25} = e_{26} = e_{35} = e_{36} = e_{45} = e_{46} = 0$$

Therefore, the elastic constants for a material that possesses a plane of elastic symmetry reduce to 13 elastic coefficients. If this plane of symmetry is the $\mathbf{X}_1\mathbf{X}_2$ plane, that is, the elastic properties are invariant under a reflection with respect to the $\mathbf{X}_1\mathbf{X}_2$ plane, the matrix $\mathbf{E}$ of elastic coefficients can be written as

$$\mathbf{E} = \begin{bmatrix} e_{11} & & & & & \text{symmetric} \\ e_{21} & e_{22} & & & & \\ e_{31} & e_{32} & e_{33} & & & \\ e_{41} & e_{42} & e_{43} & e_{44} & & \\ 0 & 0 & 0 & 0 & e_{55} & \\ 0 & 0 & 0 & 0 & e_{65} & e_{66} \end{bmatrix} \tag{4.83}$$

If the material has two mutually orthogonal planes of elastic symmetry, one can show that $e_{41} = e_{42} = e_{43} = e_{65} = 0$ and the matrix of elastic coefficients reduces to

$$\mathbf{E} = \begin{bmatrix} e_{11} & & & & & \text{symmetric} \\ e_{21} & e_{22} & & & & \\ e_{31} & e_{32} & e_{33} & & & \\ 0 & 0 & 0 & e_{44} & & \\ 0 & 0 & 0 & 0 & e_{55} & \\ 0 & 0 & 0 & 0 & 0 & e_{66} \end{bmatrix} \tag{4.84}$$

In some materials, the elastic coefficients e_{ij} remain invariant under a rotation through an angle α about one of the axes, that is, the values of these coefficients are independent of the set of rectangular axes chosen. The transformation matrix **A** in this case is given by

$$\mathbf{A} = \begin{bmatrix} \cos\alpha & -\sin\alpha & 0 \\ \sin\alpha & \cos\alpha & 0 \\ 0 & 0 & 1 \end{bmatrix}$$

One may then write two equations similar to Eqs. 75 and 76 and proceed as in the above case for different values of α to show that in the case of an *isotropic* solid there are only two independent constants, denoted as λ and μ. We then have

$$\left.\begin{aligned} & e_{12} = e_{13} = e_{21} = e_{23} = e_{31} = e_{32} = \lambda \\ & e_{44} = e_{55} = e_{66} = 2\mu \\ & e_{11} = e_{22} = e_{33} = \lambda + 2\mu \end{aligned}\right\} \tag{4.85}$$

The two elastic constants, λ and μ, are known as *Lame's constants.*

Homogeneous Isotropic Material If the material is *homogeneous*, λ and μ are constants at all points. The matrix **E** of elastic coefficients can be written in the case of an isotropic material in terms of Lame's constants as

$$\mathbf{E} = \begin{bmatrix} \lambda+2\mu & \lambda & \lambda & 0 & 0 & 0 \\ \lambda & \lambda+2\mu & \lambda & 0 & 0 & 0 \\ \lambda & \lambda & \lambda+2\mu & 0 & 0 & 0 \\ 0 & 0 & 0 & 2\mu & 0 & 0 \\ 0 & 0 & 0 & 0 & 2\mu & 0 \\ 0 & 0 & 0 & 0 & 0 & 2\mu \end{bmatrix} \tag{4.86}$$

Using Eq. 68, one can then write the stress-strain relations in the following explicit form:

$$\begin{aligned} & \sigma_{11} = \lambda\varepsilon_t + 2\mu\varepsilon_{11}, \quad \sigma_{22} = \lambda\varepsilon_t + 2\mu\varepsilon_{22}, \quad \sigma_{33} = \lambda\varepsilon_t + 2\mu\varepsilon_{33} \\ & \sigma_{12} = 2\mu\varepsilon_{12}, \quad \sigma_{13} = 2\mu\varepsilon_{13}, \quad \sigma_{23} = 2\mu\varepsilon_{23} \end{aligned} \tag{4.87}$$

where $\varepsilon_t = \varepsilon_{11} + \varepsilon_{22} + \varepsilon_{33}$, which represents the change in volume of a unit cube, is called the *dilation*. The inverse of Eq. 87 gives

$$\left.\begin{aligned} \varepsilon_{11} &= \frac{1}{E}[(1+\gamma)\sigma_{11} - \gamma\sigma_t], \quad & \varepsilon_{22} &= \frac{1}{E}[(1+\gamma)\sigma_{22} - \gamma\sigma_t] \\ \varepsilon_{33} &= \frac{1}{E}[(1+\gamma)\sigma_{33} - \gamma\sigma_t], \quad & \varepsilon_{12} &= \frac{1}{2\mu}\sigma_{12} = \frac{1+\gamma}{E}\sigma_{12} \\ \varepsilon_{13} &= \frac{1}{2\mu}\sigma_{13} = \frac{1+\gamma}{E}\sigma_{13}, \quad & \varepsilon_{23} &= \frac{1}{2\mu}\sigma_{23} = \frac{1+\gamma}{E}\sigma_{23} \end{aligned}\right\} \tag{4.88}$$

where

$$\begin{aligned} \sigma_t &= \sigma_{11} + \sigma_{22} + \sigma_{33} \\ E &= \frac{\mu(3\lambda + 2\mu)}{\lambda + \mu}, \quad \gamma = \frac{\lambda}{2(\lambda + \mu)} \end{aligned} \tag{4.89}$$

The constants μ, E, and γ are, respectively, called the *modulus of rigidity*, *Young's modulus*, and *Poisson's ratio*.

Example 4.3 In the case of two-dimensional analysis we have

$$\sigma_{23} = \sigma_{31} = 0, \qquad \varepsilon_{23} = \varepsilon_{31} = 0$$

If we further consider a plane stress problem, then

$$\sigma_{33} = 0$$

In this case, the strain components are related to the stress components by the relations

$$\varepsilon_{11} = \frac{1}{E}(\sigma_{11} - \gamma\sigma_{22})$$

$$\varepsilon_{22} = \frac{1}{E}(-\gamma\sigma_{11} + \sigma_{22})$$

$$\varepsilon_{12} = \frac{\sigma_{12}}{2\mu} = \frac{(1+\gamma)}{E}\sigma_{12}$$

In this case, the matrix of elastic coefficients can be recognized as

$$\mathbf{E} = \frac{E}{1-(\gamma)^2}\begin{bmatrix} 1 & \gamma & 0 \\ \gamma & 1 & 0 \\ 0 & 0 & 1-\gamma \end{bmatrix}$$

which relates the strain vector $\boldsymbol{\varepsilon} = [\varepsilon_{11}\ \varepsilon_{22}\ \varepsilon_{12}]^{\mathrm{T}}$ to the stress vector $\boldsymbol{\sigma} = [\sigma_{11}\ \sigma_{22}\ \sigma_{12}]^{\mathrm{T}}$. This relation can be written as

$$\boldsymbol{\sigma} = \mathbf{E}\boldsymbol{\varepsilon}$$

In case of plane strain problems, $\varepsilon_{33} = 0$, and for an isotropic material the matrix $\mathbf{E}$ can be written as

$$\mathbf{E} = \frac{E}{(1+\gamma)(1-2\gamma)}\begin{bmatrix} 1-\gamma & \gamma & 0 \\ \gamma & 1-\gamma & 0 \\ 0 & 0 & 1-2\gamma \end{bmatrix}$$

4.8 VIRTUAL WORK OF THE ELASTIC FORCES

In this section, the equation of equilibrium (Eq. 51 or 52) is employed to derive the *virtual work* of the elastic forces. To this end, we define the acceleration **a** of Eq. 51 as

$$\mathbf{a} = \ddot{\mathbf{u}}$$

where **u** is the displacement vector, and write Eq. 51 using the *summation convention* and the symmetry of the stress tensor as

$$\sigma_{jk,k} + f_{bj} = \rho \ddot{u}_j, \quad j = 1, 2, 3 \tag{4.90}$$

Multiplying Eq. 90 by the virtual displacement δu_j and integrating over the volume of the body leads to

$$\int_V \sigma_{jk,k} \delta u_j dV + \int_V f_{bj} \delta u_j dV = \int_V \rho \ddot{u}_j \delta u_j dV \tag{4.91}$$

where V is the volume of the deformable body.

By using *Cauchy's stress formula* (Eq. 41), one can write

$$\int_S \sigma_j \delta u_j dS = \int_S \sigma_{jk} n_k \delta u_j dS, \quad j = 1, 2, 3 \tag{4.92}$$

where $\mathbf{n} = [n_1 \; n_2 \; n_3]^{\mathrm{T}}$ is a unit vector directed along the outward normal to the surface whose area is S, and σ_j are the stress components defined by Eq. 37. Using the divergence theorem, one can write Eq. 92 as

$$\int_S \sigma_j \delta u_j dS = \int_V (\sigma_{jk} \delta u_j)_{,k} \, dV = \int_V \sigma_{jk,k} \delta u_j dV + \int_V \sigma_{jk} \delta u_{j,k} dV \tag{4.93}$$

in which

$$\delta u_{j,k} = \delta \varepsilon_{jk} + \delta \omega_{jk} \tag{4.94}$$

where ε_{jk} and ω_{jk} are the components of the strain and rotation tensor, respectively. Using the fact that ω_{jk} is a skew symmetric tensor and σ_{jk} is a symmetric tensor (Eqs. 8 and 9), one can write

$$\sigma_{jk} \delta \omega_{jk} = 0 \tag{4.95}$$

Equations 93–95 yield

$$\int_S \sigma_j \delta u_j dS = \int_V \sigma_{jk,k} \delta u_j dV + \int_V \sigma_{jk} \delta \varepsilon_{jk} dV \tag{4.96}$$

Substituting Eq. 96 into Eq. 91 yields

$$\int_V f_{bj} \delta u_j dV + \int_S \sigma_j \delta u_j dS = \int_V \rho \ddot{u}_j \delta u_j dV + \int_V \sigma_{jk} \delta \varepsilon_{jk} dV \tag{4.97}$$

Equation 97 can also be written as

$$\int_V f_{bj} \delta u_j dV + \int_S \sigma_j \delta u_j dS - \int_V \sigma_{jk} \delta \varepsilon_{jk} dV = \int_V \rho \ddot{u}_j \delta u_j dV \tag{4.98}$$

which can be written in the following form

$$\delta W_i = \delta W_e + \delta W_s \tag{4.99}$$

where δW_i is the virtual work of the inertia forces, δW_e is the virtual work of the applied forces, and δW_s is the virtual work of the elastic forces. These virtual work components are defined as

$$\delta W_i = \int_V \rho \ddot{u}_j \delta u_j dV \tag{4.100}$$

$$\delta W_e = \int_V f_{bj} \delta u_j dV + \int_S \sigma_j \delta u_j dS \tag{4.101}$$

$$\delta W_s = -\int_V \sigma_{jk} \delta \varepsilon_{jk} dV \tag{4.102}$$

The virtual work of the elastic forces can be written in vector form as

$$\delta W_s = -\int_V \boldsymbol{\sigma}^{\mathrm{T}} \delta \boldsymbol{\varepsilon} dV \tag{4.103}$$

Using the stress-strain relation of Eq. 68, we can write the virtual work of the elastic forces as

$$\delta W_s = -\int_V \boldsymbol{\varepsilon}^{\mathrm{T}} \mathbf{E} \, \delta \boldsymbol{\varepsilon} \, dV \tag{4.104}$$

Substituting the strain displacement relation of Eq. 20 into Eq. 104 yields

$$\delta W_s = -\int_V (\mathbf{Du})^{\mathrm{T}} \mathbf{E} (\mathbf{D} \delta \mathbf{u}) \, dV \tag{4.105}$$

Equations 103–105 are used in the following chapters to define the elastic forces in multibody systems consisting of interconnected deformable bodies.

Problems

1. Determine whether $\mathbf{u} = [k(x_2 - x_1), k(x_1 - x_2), kx_1x_2]$, where k is a constant, represents continuously possible displacement components for a continuous medium. Consider (x_1, x_2, x_3) to be rectangular Cartesian coordinates of a point in the body.

2. The deformation of a body is defined in terms of the undeformed rectangular coordinates (x_1, x_2, x_3) as

$$\mathbf{u} = [k(3(x_1)^2 + x_2), k(2(x_2)^2 + x_3), k(4(x_3)^2 + x_1)]$$

where k is a positive constant. Compute the strain of a line element that passes through the point (2, 2, 2) and has direction cosines $n_1 = n_2 = n_3 = \frac{1}{3}$.

3. The displacement components for a body are

$$u_1 = 2x_1 + x_2, \quad u_2 = x_3, \quad u_3 = x_3 - x_2$$

Verify that this displacement vector is physically possible for a continuous deformable body and determine the strain in the direction $n_1 = n_2 = n_3 = \frac{1}{3}$.

4. The stress tensor components at a point P are given by

$$\sigma_m = \begin{bmatrix} 2 & 4 & 6 \\ 4 & 8 & 12 \\ 6 & 12 & 5 \end{bmatrix}$$

Find the traction $\boldsymbol{\sigma}_n$ at point p on the plane whose outward normal has the vector of direction cosines $\mathbf{n} = [\frac{1}{3}, \frac{1}{3}, \frac{1}{3}]^{\mathrm{T}}$.

5. In the previous problem find the traction vector $\boldsymbol{\sigma}_n$ on the plane through P parallel to the plane $x_1 - 4x_2 - x_3 = 0$.

6. Show that $\sum_{i=1}^{3} \sigma_{ii} =$ constant; that is, the sum of the normal stress components is a constant in all rectangular coordinate systems.

7. Show that for a body subjected to hydrostatic pressure P

$$\sum_{i=1}^{3} \sigma_{ii} = -3P$$

8. The stress tensor $\boldsymbol{\sigma}_m$ at a point P is given by

$$\sigma_m = \begin{bmatrix} 3 & 2 & 2 \\ 2 & 4 & 0 \\ 2 & 0 & 2 \end{bmatrix}$$

Find the principal stresses and principal directions. Also find the stress vector at point P on a plane through P parallel to the plane $2x_1 - 2x_2 - x_3 = 0$.

9. The components of the stress tensor $\boldsymbol{\sigma}_m$ are given by

$$\sigma_{11} = \frac{12k_1x_1x_2}{(k_2)^3k_3}, \qquad \sigma_{12} = \frac{3k_1[(k_2)^2 - 4(x_2)^2]}{2(k_2)^3k_3}$$
$$\sigma_{22} = \sigma_{33} = \sigma_{13} = \sigma_{23} = 0$$

where k_1, k_2, and k_3 are constants. Determine whether or not these stress components satisfy the equations of equilibrium.

5 FLOATING FRAME OF REFERENCE FORMULATION

In this chapter, approximation methods are used to formulate a finite set of dynamic equations of motion of multibody systems that contain interconnected deformable bodies. As shown in Chapter 3, the dynamic equations of motion of the rigid bodies in the multibody system can be defined in terms of the mass of the body, the inertia tensor, and the generalized forces acting on the body. On the other hand, the dynamic formulation of the system equations of motion of linear structural systems requires the definition of the system mass and stiffness matrices as well as the vector of generalized forces. In this chapter, the dynamic formulation of the equations of motion of deformable bodies that undergo large translational and rotational displacements are developed using the floating frame of reference formulation. It will be shown that the equations of motion of such systems can be written in terms of a set of *inertia shape integrals* in addition to the mass of the body, the inertia tensor, and the generalized forces that appear in the dynamic formulation of rigid body system equations of motion and the mass and stiffness matrices and the vector of generalized forces that appear in the dynamic equations of linear structural systems. These inertia shape integrals that depend on the assumed displacement field appear in the nonlinear terms that represent the inertia coupling between the reference motion and the elastic deformation of the body. It will be also shown that the deformable body inertia tensor depends on the elastic deformation of the body, and accordingly it is an implicit function of time.

In the floating frame of reference formulation presented in this chapter, the configuration of each deformable body in the multibody system is identified by using two sets of coordinates: *reference* and *elastic* coordinates. Reference coordinates define the location and orientation of a selected body reference. Elastic coordinates, on the other hand, describe the body deformation with respect to the body reference. In order to avoid the computational difficulties associated with infinite-dimensional spaces, these coordinates are introduced by using classical approximation techniques such as Rayleigh–Ritz methods. The global position of an arbitrary point on the deformable body is thus defined by using a coupled set of reference and elastic coordinates. The kinetic energy of the deformable body is then developed and the inertia coupling between the reference motion and the elastic deformation is identified. The kinetic energy as well as the virtual work of the forces acting on the body are written in terms of the coupled sets of reference and elastic coordinates. Mechanical joints in the multibody system are formulated by using a set of nonlinear algebraic constraint equations that depend on the reference and elastic coordinates and possibly on time. These algebraic constraint equations can be used to identify a set of independent coordinates (system degrees of freedom) by using the generalized coordinate partitioning of the constraint Jacobian matrix, or can be adjoined to the system differential equations of motion by using the vector of Lagrange multipliers.

5.1 KINEMATIC DESCRIPTION

Multibody systems in general include two collections of bodies. One collection consists of bulky and compact solids that can be treated as rigid bodies, while the other collection includes typical structural components such as rods, beams, plates, and shells. As pointed out in previous chapters, rigid bodies have a finite number of degrees of freedom; for instance, a rigid body in space has six degrees of freedom that describe the location and orientation of the body with respect to the fixed frame of reference. On the other hand, structural components such as beams, plates, and shells have an infinite number of degrees of freedom that describe the displacement of each point on the component. As was shown in the preceding chapter, the behavior of such components is governed by a set of simultaneous partial differential equations. Using the separation of variables, the solution of these equations, if possible, leads to representation of the displacement field in terms of infinite series that can be written in the following form:

$$\left.\begin{aligned} \bar{u}_{f1} &= \sum_{k=1}^{\infty} a_k f_k \quad \text{where } f_k = f_k(x_1, x_2, x_3) \\ \bar{u}_{f2} &= \sum_{k=1}^{\infty} b_k g_k \quad \text{where } g_k = g_k(x_1, x_2, x_3) \\ \bar{u}_{f3} &= \sum_{k=1}^{\infty} c_k h_k \quad \text{where } h_k = h_k(x_1, x_2, x_3) \end{aligned}\right\} \tag{5.1}$$

where $\bar{u}_{f1}, \bar{u}_{f2}$, and $\bar{u}_{f3}$ are the components of the displacement of an arbitrary point that has coordinates (x_1, x_2, x_3) in the undeformed state. The vector of displacement $\bar{\mathbf{u}}_f = [\bar{u}_{f1}\ \bar{u}_{f2}\ \bar{u}_{f3}]^{\mathrm{T}}$ is space- and time-dependent. The coefficients a_k, b_k, and c_k are assumed to depend only on time. These coefficients are called the *coordinates*, and the functions f_k, g_k, and h_k are called the *base functions*. Each of the functions f_k, g_k, and h_k must be admissible; that is, the function has to satisfy the kinematic constraints imposed on the boundary of the deformable body. It is also required that the infinite series of Eq. 1 converge to the limit functions $\bar{u}_{f1}$, $\bar{u}_{f2}$, and $\bar{u}_{f3}$ and that these limit functions give an accurate representation to the deformed shape.

Rayleigh–Ritz Approximation A simple example of Eq. 1 is the displacement representation that arises when one writes the partial differential equation of a vibrating beam and uses the separation of variables technique to solve this equation. In this particular case, the base functions are the *eigenfunctions* and the coordinates that are infinite in dimension are the time-dependent *modal coordinates*. Because of the computational difficulties encountered in dealing with infinite-dimensional spaces, classical approximation methods such as the *Rayleigh–Ritz method* and the *Galerkin method* are employed wherein the displacement of each point is expressed in terms of a finite number of coordinates. In this case the series of Eq. 1 are truncated, and this leads to

$$\left.\begin{aligned} \bar{u}_{f1} &\approx \sum_{k=1}^{l} a_k f_k \\ \bar{u}_{f2} &\approx \sum_{k=1}^{m} b_k g_k \\ \bar{u}_{f3} &\approx \sum_{k=1}^{n} c_k h_k \end{aligned}\right\} \tag{5.2}$$

The functions $\bar{u}_{f1}$, $\bar{u}_{f2}$, and $\bar{u}_{f3}$ represent, in this case, partial sums of the series of Eq. 1. For the approximation of Eq. 2 to be valid, the sequences of partial sums of Eq. 2 must converge to the limit functions of Eq. 1. In other words, we require the sequences of partial sums to be *Cauchy sequences*. A sequence of functions $(s_1, s_2, \ldots)$ is said to be a Cauchy sequence if, given a small number $\varepsilon > 0$, there exists a natural number $M(\varepsilon)$ such that if n and m are two arbitrary natural numbers that are greater than or equal to $M(\varepsilon)$ and $m > n$, we have

$$|s_m - s_n| < \varepsilon$$

By assuming that the sequences of partial sums of the series in Eq. 1 are Cauchy sequences, and provided l, m, and n of Eq. 2 are relatively large, we are guaranteed that the approximation of Eq. 2 is acceptable.

Equation 2 implies also that approximations of the limit functions $\bar{u}_{f1}$, $\bar{u}_{f2}$, and $\bar{u}_{f3}$ can be obtained as linear combinations of the base functions f_k, g_k, and h_k, respectively. This property, in addition to the fact that the sequences of partial sums of the series of Eq. 1 are Cauchy sequences, is called *completeness*; that is, completeness is achieved if the exact displacements, and their derivatives, can be matched arbitrarily closely if enough coordinates appear in the assumed displacement field. The assumed displacement field is, in general, either exact or stiff. This is mainly because the structure is permitted to deform only into the shapes described by the assumed displacement field.

Floating Frame of Reference In the development presented in the subsequent sections, we assume that the displacement field of Eq. 2 describes the deformation of the body with respect to a selected body reference as shown in Fig. 1. The motion of the body is then defined as the motion of its reference plus the motion of the material points on the body with respect to its reference. If the assumed displacement field contains rigid body modes, a set of reference conditions has to be imposed to define a unique displacement field with respect to the selected body reference. This

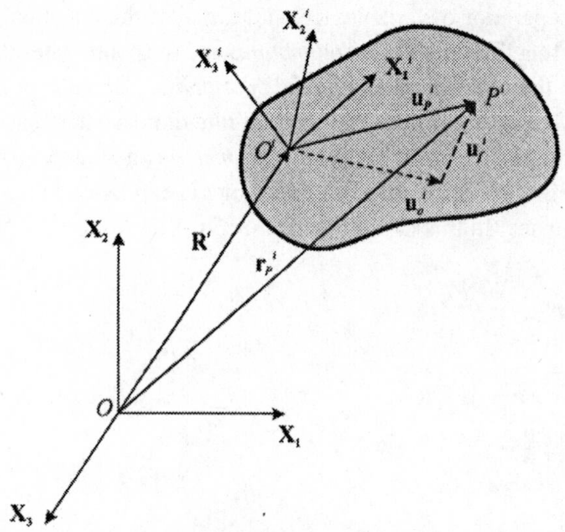

Figure 5.1 Deformable body coordinates.

subject is discussed in more detail in the following chapter where a finite element floating frame of reference formulation is presented.

One may write Eq. 2 in the following matrix form:

$$\bar{\mathbf{u}}_f = \mathbf{S}\mathbf{q}_f \tag{5.3}$$

where $\bar{\mathbf{u}}_f = [\bar{u}_{f1}\ \bar{u}_{f2}\ \bar{u}_{f3}]^{\mathrm{T}}$ is the deformation vector; $\mathbf{S}$ is the *shape matrix* whose elements are the base functions f_k, g_k, and h_k; and $\mathbf{q}_f$ is the vector of *elastic coordinates* that contains the time dependent coefficients a_k, b_k, and c_k.

To identify the configuration of deformable bodies, a set of generalized coordinates should be selected such that the location of an arbitrary point on the body can be described in terms of these generalized coordinates. To this end, we select a global coordinate system that is fixed in time and forms a single standard and as such serves to define the connectivity between different bodies in the multibody system. For an arbitrary body in the system, say, body i, we select a body reference $\mathbf{X}_1^i\mathbf{X}_2^i\mathbf{X}_3^i$ whose location and orientation with respect to the global coordinate system are defined by a set of coordinates called *reference coordinates* and denoted as $\mathbf{q}_r^i$. The vector $\mathbf{q}_r^i$ can be written in a partitioned form as

$$\mathbf{q}_r^i = [\mathbf{R}^{i\mathrm{T}} \quad \boldsymbol{\theta}^{i\mathrm{T}}]^{\mathrm{T}} \tag{5.4}$$

where $\mathbf{R}^i$ is a set of Cartesian coordinates that define the location of the origin of the body reference (Fig. 1) and $\boldsymbol{\theta}^i$ is a set of rotational coordinates that describe the orientation of the selected body reference. The body coordinate system $\mathbf{X}_1^i\mathbf{X}_2^i\mathbf{X}_3^i$ is the floating frame of reference. The origin of this reference frame does not have to be rigidly attached to a material point on the deformable body. It is required, however, that there be no rigid body motion between the body and its coordinate system. It is also important to point out that the reference motion should not be interpreted as the rigid body motion since different coordinate systems can be selected for the deformable body (Shabana 1996a).

Position Coordinates In this and the following chapter, the set of Cartesian reference coordinates is used to maintain the generality of the development. Other sets of coordinates such as joint variables can also be used with the formulation presented in this chapter by establishing the proper coordinate transformation. As pointed out in Chapter 2, three coordinates are required to define the location and orientation of the body reference in the two-dimensional analysis. These coordinates can be selected to be R_1^i, R_2^i, and θ^i, where R_1^i and R_2^i are the coordinates of the origin of the body reference and θ^i is the angular rotation of the body about the axis of rotation. In three-dimensional analysis, however, six independent coordinates are required. Three coordinates, R_1^i, R_2^i, and R_3^i, define the location of the origin of the body reference, and three independent rotational coordinates define the orientation of this reference. This subject has been thoroughly investigated in Chapter 2, where it is pointed out that the orientation of the body reference can be identified using the three independent *Euler angles*, *Rodriguez parameters*, or the four dependent *Euler parameters*. If the body is rigid, the reference coordinates are sufficient for definition of the location of an arbitrary point on the body, and accordingly these coordinates completely describe the body kinematics. For rigid bodies, therefore, the configuration space of the body and the configuration space of its reference are the same and no conceptual difficulties arise in selecting the local reference frame of rigid bodies. For example, in the case of a rigid body, the global position of an arbitrary point P on the rigid body can be written in the planar analysis as

$$\mathbf{r}_P^i = \mathbf{R}^i + \mathbf{A}^i \bar{\mathbf{u}}^i \tag{5.5}$$

where $\bar{\mathbf{u}}^i$ is the local position vector of point P and $\mathbf{A}^i$ is the transformation matrix defined as

$$\mathbf{A}^i = \begin{bmatrix} \cos\theta^i & -\sin\theta^i \\ \sin\theta^i & \cos\theta^i \end{bmatrix} \tag{5.5a}$$

Since the assumption of rigidity of the body i implies that the distance between two arbitrary points on the body remains constant, one may conclude that the length of the vector $\bar{\mathbf{u}}^i$ remains constant and, as such, the components of this vector relative to the body coordinate system remain unchanged. Similar comments apply for the spatial analysis.

When deformable bodies are considered, the distance between two arbitrary points on the deformable body does not, in general, remain constant because of the relative motion between the particles forming the body. In this case, the vector $\bar{\mathbf{u}}^i$ can be written as

$$\bar{\mathbf{u}}^i = \bar{\mathbf{u}}_o^i + \bar{\mathbf{u}}_f^i = \bar{\mathbf{u}}_o^i + \mathbf{S}^i \mathbf{q}_f^i \tag{5.6}$$

where $\bar{\mathbf{u}}_o^i$ is the position of point P in the undeformed state, $\mathbf{S}^i = \mathbf{S}^i(x_1^i, x_2^i, x_3^i)$ is a space-dependent shape matrix, and $\mathbf{q}_f^i$ is the vector of time-dependent *elastic generalized coordinates* of the deformable body i. One can then write the global position of an arbitrary point P on body i in the planar or the spatial case as

$$\begin{aligned} \mathbf{r}_P^i &= \mathbf{R}^i + \mathbf{A}^i \bar{\mathbf{u}}^i \\ &= \mathbf{R}^i + \mathbf{A}^i (\bar{\mathbf{u}}_o^i + \mathbf{S}^i \mathbf{q}_f^i) \end{aligned} \tag{5.7}$$

in which the global position of point P is written in terms of the generalized reference and elastic coordinates of body i. Therefore, we define the coordinates of body i as

$$\mathbf{q}^i = \begin{bmatrix} \mathbf{q}_r^i \\ \mathbf{q}_f^i \end{bmatrix} \tag{5.8}$$

or by using the partition of Eq. 4, we can write $\mathbf{q}^i$ in a more explicit form as

$$\mathbf{q}^i = \begin{bmatrix} \mathbf{R}^i \\ \theta^i \\ \mathbf{q}_f^i \end{bmatrix} \tag{5.9}$$

where $\mathbf{R}^i$ and $\boldsymbol{\theta}^i$ are the reference coordinates and $\mathbf{q}_f^i$ is the vector of elastic coordinates. Note that the vector $\bar{\mathbf{u}}_o^i$ of Eq. 6 can be written as

$$\bar{\mathbf{u}}_o^i = [x_1^i \quad x_2^i \quad x_3^i]^{\mathrm{T}} \tag{5.10}$$

where x_1^i, x_2^i, and x_3^i are the coordinates of point P, in the undeformed state, defined with respect to the body reference. Equation 6 can then be written as

$$\bar{\mathbf{u}}^i = \begin{bmatrix} x_1^i \\ x_2^i \\ x_3^i \end{bmatrix} + \begin{bmatrix} \mathbf{S}_1^i \\ \mathbf{S}_2^i \\ \mathbf{S}_3^i \end{bmatrix} \mathbf{q}_f^i \tag{5.11}$$

where $\mathbf{S}_k^i$ is the kth row of the body shape function.

Example 5.1 The beam shown in Fig. 2 has length $l = 0.5$ m. The beam is initially straight, and its axis is parallel to the global $\mathbf{X}_1$ axis. (Since we are considering only one beam in this example the superscript i is omitted for simplicity.) The origin of the beam reference is assumed to be rigidly attached to point O^i, while the displacement field defined in the body coordinate system is

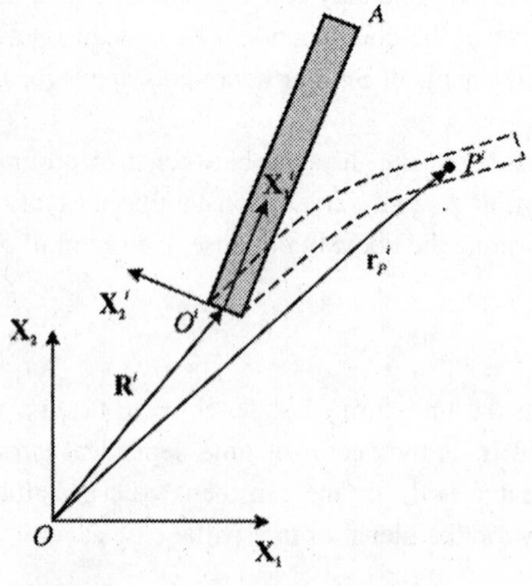

Figure 5.2 Two-dimensional beam.

assumed to be

$$\bar{\mathbf{u}}_f = \mathbf{S}\mathbf{q}_f = \begin{bmatrix} \bar{u}_{f1} \\ \bar{u}_{f2} \end{bmatrix} = \begin{bmatrix} \xi & 0 \\ 0 & 3(\xi)^2 - 2(\xi)^3 \end{bmatrix} \begin{bmatrix} q_{f1} \\ q_{f2} \end{bmatrix} \tag{5.12}$$

where $\bar{u}_{f1}$ and $\bar{u}_{f2}$ are the components of the displacement vector at any arbitrary point $x_1 = x$, $\mathbf{q}_f = [q_{f1}\ q_{f2}]^{\mathrm{T}}$ is the vector of elastic coordinates, ξ is a dimensionless quantity defined as $\xi = (x/l)$, and the body shape function $\mathbf{S}$ is defined as

$$\mathbf{S} = \begin{bmatrix} \xi & 0 \\ 0 & 3(\xi)^2 - 2(\xi)^3 \end{bmatrix}$$

The location and orientation of the beam reference is defined by using the Cartesian coordinates $\mathbf{q}_r = [R_1\ R_2\ \theta]^{\mathrm{T}}$. Therefore, the total vector of the beam coordinates $\mathbf{q} = [\mathbf{q}_r^{\mathrm{T}}\ \mathbf{q}_f^{\mathrm{T}}]^{\mathrm{T}}$ is defined as

$$\mathbf{q} = \left[\mathbf{q}_r^{\mathrm{T}} \quad \mathbf{q}_f^{\mathrm{T}}\right]^{\mathrm{T}} = [R_1 \quad R_2 \quad \theta \quad q_{f1} \quad q_{f2}]^{\mathrm{T}}$$

At a given instant of time t, let the components of the vector $\mathbf{q}$ have the following numerical values:

$$\mathbf{q} = [1.0 \quad 0.5 \quad 30^\circ \quad 0.001 \quad 0.01]^{\mathrm{T}}$$

Determine the global position of the tip point and the center of mass of the beam C.

Solution At this instant of time, the transformation matrix $\mathbf{A}$ of Eq. 5a is given by

$$\mathbf{A} = \begin{bmatrix} \cos\theta & -\sin\theta \\ \sin\theta & \cos\theta \end{bmatrix} = \begin{bmatrix} 0.8660 & -0.500 \\ 0.500 & 0.8660 \end{bmatrix}$$

The global position of point A can then be written as

$$\mathbf{r}_A = \mathbf{R} + \mathbf{A}\bar{\mathbf{u}}_A$$

where the vector $\bar{\mathbf{u}}_A$ is the local position of the tip point and can be written by using Eq. 6 as

$$\bar{\mathbf{u}}_A = \bar{\mathbf{u}}_o + \bar{\mathbf{u}}_f$$

where $\bar{\mathbf{u}}_o$ is the undeformed position of point A given by

$$\bar{\mathbf{u}}_o = \begin{bmatrix} l \\ 0 \end{bmatrix} = \begin{bmatrix} 0.5 \\ 0 \end{bmatrix}$$

The vector $\bar{\mathbf{u}}_f$ is the elastic deformation of point A and can be evaluated, since $\xi = 1$ at point A, as

$$\bar{\mathbf{u}}_f = \begin{bmatrix} \xi & 0 \\ 0 & 3(\xi)^2 - 2(\xi)^3 \end{bmatrix} \begin{bmatrix} q_{f1} \\ q_{f2} \end{bmatrix} = \begin{bmatrix} 1 & 0 \\ 0 & 1 \end{bmatrix} \begin{bmatrix} 0.001 \\ 0.01 \end{bmatrix} = \begin{bmatrix} 0.001 \\ 0.01 \end{bmatrix}$$

and accordingly

$$\bar{\mathbf{u}}_A = \begin{bmatrix} 0.501 \\ 0.01 \end{bmatrix}$$

The position vector $\mathbf{r}_A$ can be then written as

$$\mathbf{r}_A = \begin{bmatrix} 1.0 \\ 0.5 \end{bmatrix} + \begin{bmatrix} 0.8660 & -0.500 \\ 0.500 & 0.8660 \end{bmatrix} \begin{bmatrix} 0.501 \\ 0.01 \end{bmatrix} = \begin{bmatrix} 1.4289 \\ 0.75916 \end{bmatrix}$$

At point C, $\xi = 0.5$ and $\bar{\mathbf{u}}_o = [(l/2)\ 0]^{\mathrm{T}} = [0.25\ 0]^{\mathrm{T}}$. The deformation vector $\bar{\mathbf{u}}_f$ at C is given by

$$\bar{\mathbf{u}}_f = \begin{bmatrix} 0.5 & 0 \\ 0 & 3(0.5)^2 - 2(0.5)^3 \end{bmatrix} \begin{bmatrix} 0.001 \\ 0.01 \end{bmatrix} = \begin{bmatrix} 0.0005 \\ 0.005 \end{bmatrix}$$

and the local position of point C is

$$\bar{\mathbf{u}}_C = \bar{\mathbf{u}}_o + \bar{\mathbf{u}}_f = \begin{bmatrix} 0.25 \\ 0 \end{bmatrix} + \begin{bmatrix} 0.0005 \\ 0.005 \end{bmatrix} = \begin{bmatrix} 0.2505 \\ 0.005 \end{bmatrix}$$

The global position $\mathbf{r}_C$ can then be determined as

$$\mathbf{r}_C = \begin{bmatrix} 1.0 \\ 0.5 \end{bmatrix} + \begin{bmatrix} 0.8660 & -0.500 \\ 0.500 & 0.8660 \end{bmatrix} \begin{bmatrix} 0.2505 \\ 0.005 \end{bmatrix} = \begin{bmatrix} 1.2144 \\ 0.62958 \end{bmatrix}$$

Velocity Equations Differentiating Eq. 7 with respect to time yields

$$\dot{\mathbf{r}}_P^i = \dot{\mathbf{R}}^i + \dot{\mathbf{A}}^i \bar{\mathbf{u}}^i + \mathbf{A}^i \dot{\bar{\mathbf{u}}}^i \tag{5.13}$$

where ($\dot{}$) denotes differentiation with respect to time. Using Eq. 6, one can write $\dot{\bar{\mathbf{u}}}^i$ in terms of the time derivatives of the elastic coordinates of body i as

$$\dot{\bar{\mathbf{u}}}^i = \mathbf{S}^i \dot{\mathbf{q}}_f^i \tag{5.14}$$

where $\mathbf{S}^i = \mathbf{S}^i(x_1^i, x_2^i, x_3^i)$ is the body shape matrix and $\dot{\mathbf{q}}_f^i$ is the vector of elastic generalized velocities of body i. Substituting Eq. 14 into Eq. 13 yields

$$\dot{\mathbf{r}}_P^i = \dot{\mathbf{R}}^i + \dot{\mathbf{A}}^i \bar{\mathbf{u}}^i + \mathbf{A}^i \mathbf{S}^i \dot{\mathbf{q}}_f^i \tag{5.15}$$

where the equation $\dot{\bar{\mathbf{u}}}_o^i = \mathbf{0}$ is used. To isolate velocity terms, the central term on the right-hand side of Eq. 15 can, in general, be written as

$$\dot{\mathbf{A}}^i \bar{\mathbf{u}}^i = \mathbf{B}^i \dot{\boldsymbol{\theta}}^i \tag{5.16}$$

where $\dot{\boldsymbol{\theta}}^i$ is the vector whose elements $\dot{\theta}_k^i$ are the time derivatives of the rotational coordinates of the body reference and $\mathbf{B}^i = \mathbf{B}^i(\boldsymbol{\theta}^i, \mathbf{q}_f^i)$ is defined as

$$\mathbf{B}^i = \left[\frac{\partial}{\partial \theta_1^i}(\mathbf{A}^i \bar{\mathbf{u}}^i) \cdots \frac{\partial}{\partial \theta_{n_r}^i}(\mathbf{A}^i \bar{\mathbf{u}}^i) \right] \tag{5.17}$$

where n_r is the total number of rotational coordinates of the reference of body i. Equation 17 follows from using the chain rule of differentiation, which yields

$$\dot{\mathbf{A}}^i \bar{\mathbf{u}}^i = \sum_{k=1}^{n_r} \frac{\partial}{\partial \theta_k^i}(\mathbf{A}^i \bar{\mathbf{u}}^i) \dot{\theta}_k^i \tag{5.18}$$

Substituting Eq. 16 into Eq. 15, one gets

$$\dot{\mathbf{r}}^i_P = \dot{\mathbf{R}}^i + \mathbf{B}^i\dot{\theta}^i + \mathbf{A}^i\mathbf{S}^i\dot{\mathbf{q}}^i_f \tag{5.19}$$

In partitioned form, the absolute velocity vector of Eq. 19 can be written as

$$\dot{\mathbf{r}}^i_P = [\mathbf{I} \quad \mathbf{B}^i \quad \mathbf{A}^i\mathbf{S}^i] \begin{bmatrix} \dot{\mathbf{R}}^i \\ \dot{\theta}^i \\ \dot{\mathbf{q}}^i_f \end{bmatrix} \tag{5.20}$$

where $\mathbf{I}$ is a 3×3 identity matrix. Equation 20 can also be written as

$$\dot{\mathbf{r}}^i_P = \mathbf{L}^i\dot{\mathbf{q}}^i \tag{5.21}$$

where $\dot{\mathbf{q}}^i = [\dot{\mathbf{q}}^{i\mathrm{T}}_r \; \dot{\mathbf{q}}^{i\mathrm{T}}_f]^\mathrm{T} = [\dot{\mathbf{R}}^{i\mathrm{T}} \; \dot{\theta}^{i\mathrm{T}} \; \dot{\mathbf{q}}^{i\mathrm{T}}_f]^\mathrm{T}$ is the total vector of generalized velocities of body i, and $\mathbf{L}^i$ is the matrix

$$\mathbf{L}^i = [\mathbf{I} \quad \mathbf{B}^i \quad \mathbf{A}^i\mathbf{S}^i] \tag{5.22}$$

Before proceeding in our development, perhaps it is important to explain the nature of the terms appearing in the right-hand side of Eq. 19. The vector $\dot{\mathbf{R}}^i$ is the absolute velocity vector of the origin of the body reference, while the last term, $\mathbf{A}^i\mathbf{S}^i\dot{\mathbf{q}}^i_f$, is the velocity of point P due to the deformation of the body, defined with respect to an observer stationed on the body. If the body were rigid, the term $\mathbf{A}^i\mathbf{S}^i\dot{\mathbf{q}}^i_f$ would be identical to zero. The central term, $\mathbf{B}^i\dot{\theta}^i$, is the result of differentiation of the transformation matrix with respect to time. This term depends on the reference rotation as well as the elastic deformation of the body. In the case of rigid body translation this term vanishes, and accordingly the velocity of any point on the body is equal to the velocity $\dot{\mathbf{R}}^i$ of the origin of the body reference. In Chapter 2, it was shown that

$$\dot{\mathbf{A}}\bar{\mathbf{u}}^i = \mathbf{B}^i\dot{\theta}^i = \mathbf{A}^i(\bar{\omega}^i \times \bar{\mathbf{u}}^i) = -\mathbf{A}^i(\bar{\mathbf{u}}^i \times \bar{\omega}^i) \tag{5.23}$$

where $\bar{\omega}^i$ is the angular velocity vector defined in the body reference. Alternatively, if we define

$$\mathbf{u}^i = \mathbf{A}^i\bar{\mathbf{u}}^i \tag{5.24}$$

we may write Eq. 23 as

$$\dot{\mathbf{A}}^i\bar{\mathbf{u}}^i = \mathbf{B}^i\dot{\theta}^i = \omega^i \times \mathbf{u}^i = -\mathbf{u}^i \times \omega^i \tag{5.25}$$

where ω^i is the angular velocity vector defined in the global, fixed frame of reference. It is clear from Eq. 25 that the central term on the right-hand side of Eq. 19 is a vector that is perpendicular to both ω^i and the vector $\mathbf{u}^i$, which represents the position of P relative to the origin of the body reference.

Knowing that

$$-\mathbf{u}^i \times \omega^i = -\tilde{\mathbf{u}}^i\omega^i = \tilde{\mathbf{u}}^{i\mathrm{T}}\omega^i \tag{5.26}$$

where $\tilde{\mathbf{u}}^{i\mathrm{T}}$ is the skew symmetric matrix defined as

$$\tilde{\mathbf{u}}^{i\mathrm{T}} = \begin{bmatrix} 0 & u^i_3 & -u^i_2 \\ -u^i_3 & 0 & u^i_1 \\ u^i_2 & -u^i_1 & 0 \end{bmatrix} \tag{5.27}$$

and u_1^i, u_2^i, u_3^i are the components of the vector $\mathbf{u}^i$, one can write the velocity vector of Eq. 19 in the form

$$\dot{\mathbf{r}}_p^i = \dot{\mathbf{R}}^i + \tilde{\mathbf{u}}^{i\mathrm{T}}\boldsymbol{\omega}^i + \mathbf{A}^i\mathbf{S}^i\dot{\mathbf{q}}_f^i \tag{5.28}$$

or alternatively

$$\dot{\mathbf{r}}_P^i = [\mathbf{I} \quad \tilde{\mathbf{u}}^{i\mathrm{T}} \quad \mathbf{A}^i\mathbf{S}^i] \begin{bmatrix} \dot{\mathbf{R}}^i \\ \boldsymbol{\omega}^i \\ \dot{\mathbf{q}}_f^i \end{bmatrix} \tag{5.29}$$

Equation 28 (or Eq. 29) describes the velocity vector in terms of the angular velocity vector $\boldsymbol{\omega}^i$. We will, in general, use Eq. 19 or 20 instead of Eqs. 28 and 29 since we prefer to develop our equations in terms of the generalized coordinates of the body. Therefore, the definition of the matrix $\mathbf{B}^i$ is important in the development that follows.

It was shown in Chapters 2 and 3 that irrespective of the reference rotational coordinates used, the vector $\bar{\boldsymbol{\omega}}^i$ of Eq. 23 can be written in terms of the rotational coordinates and velocities of the body reference as

$$\bar{\boldsymbol{\omega}}^i = \bar{\mathbf{G}}^i\dot{\boldsymbol{\theta}}^i \tag{5.30}$$

where $\bar{\mathbf{G}}^i = \bar{\mathbf{G}}^i(\boldsymbol{\theta}^i)$ is a matrix given in Chapters 2 and 3. One can then write Eq. 23 as

$$\dot{\mathbf{A}}^i\bar{\mathbf{u}}^i = -\mathbf{A}^i(\bar{\mathbf{u}}^i \times \bar{\boldsymbol{\omega}}^i) = -\mathbf{A}^i\tilde{\bar{\mathbf{u}}}^i\bar{\boldsymbol{\omega}}^i \tag{5.31}$$

or

$$\dot{\mathbf{A}}^i\bar{\mathbf{u}}^i = -\mathbf{A}^i\tilde{\bar{\mathbf{u}}}^i\bar{\mathbf{G}}^i\dot{\boldsymbol{\theta}}^i \tag{5.32}$$

from which we identify the matrix $\mathbf{B}^i$ of Eqs. 20 and 22 as

$$\mathbf{B}^i = -\mathbf{A}^i\tilde{\bar{\mathbf{u}}}^i\bar{\mathbf{G}}^i \tag{5.33}$$

Since $\bar{\mathbf{u}}^i$ is the vector of the coordinates of an arbitrary point of body i that can be written as

$$\bar{\mathbf{u}}^i = [\bar{u}_1^i \quad \bar{u}_2^i \quad \bar{u}_3^i]^{\mathrm{T}}, \tag{5.34}$$

the skew symmetric matrix $\tilde{\bar{\mathbf{u}}}^i$ of Eq. 33 is defined as

$$\tilde{\bar{\mathbf{u}}}^i = \begin{bmatrix} 0 & -\bar{u}_3^i & \bar{u}_2^i \\ \bar{u}_3^i & 0 & -\bar{u}_1^i \\ -\bar{u}_2^i & \bar{u}_1^i & 0 \end{bmatrix} \tag{5.35}$$

Using Eq. 33, one can write the matrix $\mathbf{L}^i$ of Eq. 22 as

$$\mathbf{L}^i = [\mathbf{I} \quad -\mathbf{A}^i\tilde{\bar{\mathbf{u}}}^i\bar{\mathbf{G}}^i \quad \mathbf{A}^i\mathbf{S}^i] \tag{5.36}$$

The form of the velocity vector of Eq. 21 with $\mathbf{L}^i$ defined by Eq. 36 will be used in the development of the kinetic energy in the following section.

Acceleration Equations The acceleration of point P can be determined by direct differentiation of Eq. 21. This leads to

$$\ddot{\mathbf{r}}_P^i = \dot{\mathbf{L}}^i\dot{\mathbf{q}}^i + \mathbf{L}^i\ddot{\mathbf{q}}^i \tag{5.37}$$

where $\dot{\mathbf{L}}^i\dot{\mathbf{q}}^i$ is a quadratic velocity vector that contains the *Coriolis component*. Using the identities presented in Chapter 2, one can verify that the acceleration vector of Eq. 37 can be written as

$$\ddot{\mathbf{r}}_P^i = \ddot{\mathbf{R}}^i + \boldsymbol{\omega}^i \times (\boldsymbol{\omega}^i \times \mathbf{u}^i) + \boldsymbol{\alpha}^i \times \mathbf{u}^i + 2\boldsymbol{\omega}^i \times (\mathbf{A}^i\dot{\bar{\mathbf{u}}}^i) + \mathbf{A}^i\ddot{\bar{\mathbf{u}}}^i \tag{5.38}$$

In this equation, $\boldsymbol{\alpha}^i$ is the angular acceleration vector. The term $\ddot{\mathbf{R}}^i$ is the absolute acceleration of the origin of the body reference. The second term, $\boldsymbol{\omega}^i \times (\boldsymbol{\omega}^i \times \mathbf{u}^i)$, is the *normal component* of the acceleration of point P' that instantaneously coincides with P and does not undergo deformation. This component of the acceleration is directed along the straight line connecting the two points O and P. The third component, $\boldsymbol{\alpha}^i \times \mathbf{u}^i$, is the *tangential component* of the acceleration of P^i relative to O. The direction of this component is perpendicular to both the angular acceleration vector $\boldsymbol{\alpha}^i$ and the vector $\mathbf{u}^i$. The fourth term, $2\boldsymbol{\omega}^i \times (\mathbf{A}^i\dot{\bar{\mathbf{u}}}^i)$, is the Coriolis component of the acceleration, and the fifth term, $\mathbf{A}^i\ddot{\bar{\mathbf{u}}}^i$ is the acceleration of point P due to the deformation relative to the body reference. If the body is rigid, the fourth and fifth components vanish. Equation 38, however, is used in rigid body dynamics to describe the acceleration of particles that move relative to the body, provided the motion of these moving particles is appropriately described by the vector $\mathbf{u}^i$.

Example 5.2 The reference of the beam of Example 1 rotates with a constant angular velocity $\omega = \dot{\theta} = 5$ rad/sec. Determine the absolute velocity and acceleration of the tip point A at the instant of time t at which the beam coordinates, velocities, and accelerations are given by

$$\mathbf{q} = [\mathbf{q}_r^{\mathrm{T}} \quad \mathbf{q}_f^{\mathrm{T}}]^{\mathrm{T}} = [R_1 \quad R_2 \quad \theta \quad q_{f1} \quad q_{f2}]^{\mathrm{T}}$$
$$= [1.0 \quad 0.5 \quad 30^\circ \quad 0.001 \quad 0.01]^{\mathrm{T}}$$
$$\dot{\mathbf{q}} = [\dot{\mathbf{q}}_r^{\mathrm{T}} \quad \dot{\mathbf{q}}_f^{\mathrm{T}}]^{\mathrm{T}} = [\dot{R}_1 \quad \dot{R}_2 \quad \dot{\theta} \quad \dot{q}_{f1} \quad \dot{q}_{f2}]^{\mathrm{T}}$$
$$= [0.1 \quad 1.0 \quad 5 \quad 2 \quad 3]^{\mathrm{T}}$$
$$\ddot{\mathbf{q}} = [\ddot{\mathbf{q}}_r^{\mathrm{T}} \quad \ddot{\mathbf{q}}_f^{\mathrm{T}}]^{\mathrm{T}} = [\ddot{R}_1 \quad \ddot{R}_2 \quad \ddot{\theta} \quad \ddot{q}_{f1} \quad \ddot{q}_{f2}]^{\mathrm{T}}$$
$$= [2 \quad 0 \quad 0 \quad 10 \quad 20]^{\mathrm{T}}$$

Solution In this case the matrix $\mathbf{L}$ of Eq. 22 is the 2×5 matrix

$$\mathbf{L} = [\mathbf{I} \quad \mathbf{B} \quad \mathbf{AS}]$$

where $\mathbf{I}$ is the 2×2 identity matrix

$$\mathbf{I} = \begin{bmatrix} 1 & 0 \\ 0 & 1 \end{bmatrix}$$

One can verify that $\mathbf{B}$, in this case, is a two-dimensional vector defined as

$$\mathbf{B} = \mathbf{A}_\theta \bar{\mathbf{u}}_A$$

where $\mathbf{A}_\theta$ is the partial derivative of the transformation matrix $\mathbf{A}$ with respect to the rotational coordinate θ, that is,

$$\mathbf{A}_\theta = \begin{bmatrix} -\sin\theta & -\cos\theta \\ \cos\theta & -\sin\theta \end{bmatrix} = \begin{bmatrix} -0.5 & -0.866 \\ 0.866 & -0.5 \end{bmatrix}$$

From Example 1, the vector $\bar{\mathbf{u}}_A$, which is the position of the tip point A defined in the beam coordinate system, is given by

$$\bar{\mathbf{u}}_A = [0.501 \quad 0.01]^{\mathrm{T}}$$

Therefore, the vector $\mathbf{B}$ can be evaluated as

$$\mathbf{B} = \mathbf{A}_\theta \bar{\mathbf{u}}_A = \begin{bmatrix} -0.5 & -0.866 \\ 0.866 & -0.5 \end{bmatrix} \begin{bmatrix} 0.501 \\ 0.01 \end{bmatrix} = \begin{bmatrix} -0.25916 \\ 0.42886 \end{bmatrix}$$

Since at point A, $\xi = (x/l) = 1$, the shape matrix $\mathbf{S}$ evaluated at point A is given by

$$\mathbf{S} = \begin{bmatrix} \xi & 0 \\ 0 & 3(\xi)^2 - 2(\xi)^3 \end{bmatrix} = \begin{bmatrix} 1 & 0 \\ 0 & 1 \end{bmatrix}$$

and using the transformation matrix $\mathbf{A}$ evaluated in Example 1, one gets

$$\mathbf{AS} = \begin{bmatrix} 0.8660 & -0.500 \\ 0.5 & 0.8660 \end{bmatrix} \begin{bmatrix} 1 & 0 \\ 0 & 1 \end{bmatrix} = \begin{bmatrix} 0.8660 & -0.5000 \\ 0.5000 & 0.8660 \end{bmatrix}$$

The matrix $\mathbf{L}$ can then be defined as

$$\mathbf{L} = [\mathbf{I} \quad \mathbf{B} \quad \mathbf{AS}] = \begin{bmatrix} 1 & 0 & -0.25916 & 0.8660 & -0.5000 \\ 0 & 1 & 0.42886 & 0.5000 & 0.8660 \end{bmatrix}$$

and accordingly, the global velocity vector of point A is given by

$$\dot{\mathbf{r}}_A = \mathbf{L}\dot{\mathbf{q}} = \begin{bmatrix} 1 & 0 & -0.25916 & 0.8660 & -0.5000 \\ 0 & 1 & 0.42886 & 0.5000 & 0.8660 \end{bmatrix} \begin{bmatrix} 0.1 \\ 1.0 \\ 5 \\ 2 \\ 3 \end{bmatrix}$$

$$= \begin{bmatrix} -0.96380 \\ 6.74235 \end{bmatrix} \text{m/sec}$$

The acceleration of point A is given by

$$\ddot{\mathbf{r}}_A = \mathbf{L}\ddot{\mathbf{q}} + \dot{\mathbf{L}}\dot{\mathbf{q}}$$

where

$$\mathbf{L}\ddot{\mathbf{q}} = \begin{bmatrix} 1 & 0 & -0.25916 & 0.8660 & -0.5000 \\ 0 & 1 & 0.42886 & 0.500 & 0.8660 \end{bmatrix} \begin{bmatrix} 2 \\ 0 \\ 0 \\ 10 \\ 20 \end{bmatrix}$$

$$= \begin{bmatrix} 0.66 \\ 22.32 \end{bmatrix} \text{m/sec}^2$$

One can verify that the matrix $\dot{\mathbf{L}}$ is

$$\begin{aligned} \dot{\mathbf{L}} &= [\mathbf{0}_2 \quad \dot{\mathbf{B}} \quad \dot{\mathbf{A}}\mathbf{S}] \\ &= [\mathbf{0}_2 \quad (-\mathbf{A}\bar{\mathbf{u}}_A\dot{\theta} + \mathbf{A}_\theta \mathbf{S}\dot{\mathbf{q}}_f) \quad \mathbf{A}_\theta \mathbf{S}\dot{\theta}] \end{aligned}$$

where $\mathbf{0}_2$ is a 2×2 null matrix. Using the results obtained in Example 1, we can calculate the vector $\mathbf{A}\bar{\mathbf{u}}_A\dot{\theta}$ and $\mathbf{A}_\theta\mathbf{S}\dot{\mathbf{q}}_f$ as

$$\mathbf{A}\bar{\mathbf{u}}_A\dot{\theta} = \begin{bmatrix} 0.4289 \\ 0.25916 \end{bmatrix}(5) = \begin{bmatrix} 2.1445 \\ 1.2958 \end{bmatrix}$$

$$\mathbf{A}_\theta\mathbf{S}\dot{\mathbf{q}}_f = \begin{bmatrix} -0.5 & -0.866 \\ 0.866 & -0.5 \end{bmatrix}\begin{bmatrix} 1 & 0 \\ 0 & 1 \end{bmatrix}\begin{bmatrix} 2 \\ 3 \end{bmatrix} = \begin{bmatrix} -3.598 \\ 0.232 \end{bmatrix}$$

and

$$\mathbf{A}_\theta\mathbf{S}\dot{\theta} = \begin{bmatrix} -0.5 & -0.866 \\ 0.866 & -0.5 \end{bmatrix}\begin{bmatrix} 1 & 0 \\ 0 & 1 \end{bmatrix}(5) = \begin{bmatrix} -2.5 & -4.33 \\ 4.33 & -2.5 \end{bmatrix}$$

The vector $\dot{\mathbf{L}}\dot{\mathbf{q}}$ can then be evaluated as

$$\dot{\mathbf{L}}\dot{\mathbf{q}} = \begin{bmatrix} 0 & 0 & -5.7425 & -2.5 & -4.33 \\ 0 & 0 & -1.0638 & 4.33 & -2.5 \end{bmatrix}\begin{bmatrix} 0.1 \\ 1.0 \\ 5 \\ 2 \\ 3 \end{bmatrix}$$

$$= \begin{bmatrix} -46.702 \\ -4.159 \end{bmatrix} \text{m/sec}^2$$

The acceleration vector of point A is the sum of the two vectors

$$\ddot{\mathbf{r}}_A = \mathbf{L}\ddot{\mathbf{q}} + \dot{\mathbf{L}}\dot{\mathbf{q}} = \begin{bmatrix} 0.66 \\ 22.32 \end{bmatrix} + \begin{bmatrix} -46.702 \\ -4.159 \end{bmatrix} = \begin{bmatrix} -46.042 \\ 18.161 \end{bmatrix} \text{m/sec}^2$$

5.2 INERTIA OF DEFORMABLE BODIES

In this section, we develop the kinetic energy of deformable bodies and point out the differences between the inertia properties of deformable bodies that undergo finite rotations and the inertia properties of both rigid and structural systems. In addition to the inertia tensor and the conventional mass matrix that appear, respectively, in rigid body dynamics and the dynamics of linear structural systems, it will be shown that a set of inertia shape integrals that depend on the assumed displacement field has to be evaluated in order to completely define the mass matrix of deformable bodies that undergo large reference rotations. These inertia shape integrals appear in the components of the mass matrix that represent the inertia coupling between the reference motion and the elastic deformation of the deformable body. Moreover, the body inertia tensor depends on the elastic deformation and, as a consequence, is time-variant.

Mass Matrix The following definition of the kinetic energy is used:

$$T^i = \frac{1}{2}\int_{V^i} \rho^i \dot{\mathbf{r}}_p^{i\mathrm{T}} \dot{\mathbf{r}}_p^i \, dV^i \tag{5.39}$$

where T^i is the kinetic energy of body i in the system; ρ^i and V^i are, respectively, the mass density and volume of body i; and $\dot{\mathbf{r}}^i_P$ is the global velocity vector of an arbitrary point P on the body. Using the expression of the velocity vector of Eq. 21 given in the preceding section, one can write the kinetic energy of Eq. 39 as

$$T^i = \frac{1}{2}\int_{V^i} \rho^i \dot{\mathbf{q}}^{i\mathrm{T}} \mathbf{L}^{i\mathrm{T}} \mathbf{L}^i \dot{\mathbf{q}}^i \, dV^i \tag{5.40}$$

Since the total vector of generalized coordinates $\mathbf{q}^i$ is assumed to be time-dependent and the mass density ρ^i may depend on the location of point P, we can write Eq. 40 as

$$T^i = \frac{1}{2}\dot{\mathbf{q}}^{i\mathrm{T}} \left[\int_{V^i} \rho^i \mathbf{L}^{i\mathrm{T}} \mathbf{L}^i \, dV^i\right] \dot{\mathbf{q}}^i \tag{5.41}$$

or in a more compact form as

$$T^i = \frac{1}{2}\dot{\mathbf{q}}^{i\mathrm{T}} \mathbf{M}^i \dot{\mathbf{q}}^i \tag{5.42}$$

where $\mathbf{M}^i$ is recognized as the symmetric mass matrix of body i in the multibody system and is defined as

$$\mathbf{M}^i = \int_{V^i} \rho^i \mathbf{L}^{i\mathrm{T}} \mathbf{L}^i \, dV^i \tag{5.43}$$

Using the definition of $\mathbf{L}^i$ of Eq. 22, one can write the mass matrix of body i in a more explicit form as

$$\begin{aligned}
\mathbf{M}^i &= \int_{V^i} \rho^i \begin{bmatrix} \mathbf{I} \\ \mathbf{B}^{i\mathrm{T}} \\ (\mathbf{A}^i \bar{\mathbf{S}}^i)^{\mathrm{T}} \end{bmatrix} [\mathbf{I} \quad \mathbf{B}^i \quad \mathbf{A}^i \bar{\mathbf{S}}^i] \, dV^i \\
&= \int_{V^i} \rho^i \begin{bmatrix} \mathbf{I} & \mathbf{B}^i & \mathbf{A}^i \bar{\mathbf{S}}^i \\ & \mathbf{B}^{i\mathrm{T}}\mathbf{B}^i & \mathbf{B}^{i\mathrm{T}}\mathbf{A}^i \bar{\mathbf{S}}^i \\ \text{symmetric} & & \bar{\mathbf{S}}^{i\mathrm{T}} \bar{\mathbf{S}}^i \end{bmatrix} dV^i
\end{aligned} \tag{5.44}$$

where the orthogonality of the transformation matrix, that is, $\mathbf{A}^{i\mathrm{T}}\mathbf{A}^i = \mathbf{I}$, is used in order to simplify the submatrix in the lower right-hand corner of Eq. 44. The mass matrix of Eq. 44 can be written in a symbolic form as

$$\mathbf{M}^i = \begin{bmatrix} \mathbf{m}^i_{RR} & \mathbf{m}^i_{R\theta} & \mathbf{m}^i_{Rf} \\ & \mathbf{m}^i_{\theta\theta} & \mathbf{m}^i_{\theta f} \\ \text{symmetric} & & \mathbf{m}^i_{ff} \end{bmatrix} \tag{5.45}$$

where

$$\mathbf{m}^i_{RR} = \int_{V^i} \rho^i \mathbf{I} \, dV^i, \qquad \mathbf{m}^i_{R\theta} = \int_{V^i} \rho^i \mathbf{B}^i \, dV^i \tag{5.46a,b}$$

$$\mathbf{m}^i_{Rf} = \mathbf{A}^i \int_{V^i} \rho^i \bar{\mathbf{S}}^i \, dV^i, \qquad \mathbf{m}^i_{\theta\theta} = \int_{V^i} \rho^i \mathbf{B}^{i\mathrm{T}} \mathbf{B}^i \, dV^i \tag{5.46c,d}$$

$$\mathbf{m}^i_{\theta f} = \int_{V^i} \rho^i \mathbf{B}^{i\mathrm{T}} \mathbf{A}^i \bar{\mathbf{S}}^i \, dV^i, \qquad \mathbf{m}^i_{ff} = \int_{V^i} \rho^i \bar{\mathbf{S}}^{i\mathrm{T}} \bar{\mathbf{S}}^i \, dV^i \tag{5.46e,f}$$

Note that the two submatrices $\mathbf{m}^i_{RR}$ and $\mathbf{m}^i_{ff}$ associated, respectively, with the translational reference and elastic coordinates, are constant. Other matrices, however, depend on the system generalized coordinates, and as a result they are implicit functions of time. In terms of the submatrices defined in Eq. 46, one can write the kinetic energy of the deformable body i as

$$T^i = \frac{1}{2}(\dot{\mathbf{R}}^{i\mathrm{T}}\mathbf{m}^i_{RR}\dot{\mathbf{R}}^i + 2\dot{\mathbf{R}}^{i\mathrm{T}}\mathbf{m}^i_{R\theta}\dot{\boldsymbol{\theta}}^i + 2\dot{\mathbf{R}}^{i\mathrm{T}}\mathbf{m}^i_{Rf}\dot{\mathbf{q}}^i_f + \dot{\boldsymbol{\theta}}^{i\mathrm{T}}\mathbf{m}^i_{\theta\theta}\dot{\boldsymbol{\theta}}^i + 2\dot{\boldsymbol{\theta}}^{i\mathrm{T}}\mathbf{m}^i_{\theta f}\dot{\mathbf{q}}^i_f + \dot{\mathbf{q}}^{i\mathrm{T}}_f\mathbf{m}^i_{ff}\dot{\mathbf{q}}^i_f) \tag{5.47}$$

If the body is *rigid*, the vector $\dot{\mathbf{q}}^i_f$ of elastic coordinates of body i vanishes and the kinetic energy reduces to

$$T^i = \frac{1}{2}[\dot{\mathbf{R}}^{i\mathrm{T}}\mathbf{m}^i_{RR}\dot{\mathbf{R}}^i + 2\dot{\mathbf{R}}^{i\mathrm{T}}\mathbf{m}^i_{R\theta}\dot{\boldsymbol{\theta}}^i + \dot{\boldsymbol{\theta}}^{i\mathrm{T}}\mathbf{m}^i_{\theta\theta}\dot{\boldsymbol{\theta}}^i] \tag{5.48}$$

which can be written in a partitioned form as

$$T^i = \frac{1}{2}\begin{bmatrix}\dot{\mathbf{R}}^{i\mathrm{T}} & \dot{\boldsymbol{\theta}}^{i\mathrm{T}}\end{bmatrix}\begin{bmatrix}\mathbf{m}^i_{RR} & \mathbf{m}^i_{R\theta} \\ \mathbf{m}^i_{\theta R} & \mathbf{m}^i_{\theta\theta}\end{bmatrix}\begin{bmatrix}\dot{\mathbf{R}}^i \\ \dot{\boldsymbol{\theta}}^i\end{bmatrix} \tag{5.49}$$

where the mass matrix in the case of a rigid body motion can be recognized as

$$\mathbf{M}^i = \begin{bmatrix}\mathbf{m}^i_{RR} & \mathbf{m}^i_{R\theta} \\ \mathbf{m}^i_{\theta R} & \mathbf{m}^i_{\theta\theta}\end{bmatrix} \tag{5.50}$$

The matrix $\mathbf{m}^i_{R\theta}$ and its transpose $\mathbf{m}^i_{\theta R}$ represent the inertia coupling between the rigid body translation and the rigid body rotation. It is shown in Chapter 3 that this coupling disappears, in rigid body dynamics, if the origin of the body reference is rigidly attached to the mass center of the body. The term $\mathbf{m}^i_{RR}$ represents the mass matrix associated with the translational coordinates of the body reference. This matrix is diagonal, and the diagonal elements are equal to the total mass of the body. The matrix $\mathbf{m}^i_{\theta\theta}$ is associated with the rotational coordinates of the body reference.

In the case of a deformable body system wherein the reference motion is not allowed, the reference coordinates remain constant with respect to time, that is

$$\dot{\mathbf{R}}^i = \mathbf{0}, \quad \dot{\boldsymbol{\theta}}^i = \mathbf{0}$$

The kinetic energy of Eq. 47 reduces to

$$T^i = \frac{1}{2}\dot{\mathbf{q}}^{i\mathrm{T}}\mathbf{m}^i_{ff}\dot{\mathbf{q}}^i_f \tag{5.51}$$

and the mass matrix of the body can be recognized in this case as the constant matrix $\mathbf{m}^i_{ff}$, which appears in the dynamic formulation of *linear structural systems.*

When a deformable body undergoes rigid body motion, the mass matrix is defined by Eq. 45 and the submatrices $\mathbf{m}^i_{Rf}$ and $\mathbf{m}^i_{\theta f}$ and their transpose represent the coupling between the reference motion and elastic deformation. In this case these matrices as well as the matrices $\mathbf{m}^i_{R\theta}$ and $\mathbf{m}^i_{\theta\theta}$ depend on both the rotational reference coordinates and the elastic coordinates of the body i.

Spatial Motion To develop in detail the form of the components of the mass matrix in the general case of a spatial deformable body that undergoes rigid body motion, we use the general definition of the matrix $\mathbf{B}^i$ given by Eq. 33. We start with the submatrix $\mathbf{m}^i_{RR}$ associated with the translations of the origin of the body reference given by Eq. 46a. This matrix can be determined as

$$\mathbf{m}^i_{RR} = \int_{V^i} \rho^i \mathbf{I}\, dV^i = \int_{V^i} \rho^i \begin{bmatrix} 1 & 0 & 0 \\ 0 & 1 & 0 \\ 0 & 0 & 1 \end{bmatrix} dV^i$$

which by integration yields

$$\mathbf{m}^i_{RR} = \begin{bmatrix} m^i & 0 & 0 \\ 0 & m^i & 0 \\ 0 & 0 & m^i \end{bmatrix} \tag{5.52}$$

where m^i is the total mass of body i. Because of the conservation of mass, this matrix is the same for both cases of rigid and deformable bodies.

Using Eq. 33, one can determine the submatrix $\mathbf{m}^i_{R\theta}$ of Eq. 46b as

$$\mathbf{m}^i_{R\theta} = \mathbf{m}^{i\mathrm{T}}_{\theta R} = \int_{V^i} \rho^i \mathbf{B}^i dV^i = -\int_{V^i} \rho^i \mathbf{A}^i \tilde{\bar{\mathbf{u}}}^i \bar{\mathbf{G}}^i dV^i \tag{5.53}$$

Because the matrices $\mathbf{A}^i$ and $\bar{\mathbf{G}}^i$ are not space-dependent, one can write the integral of Eq. 53 in the form

$$\mathbf{m}^i_{R\theta} = \mathbf{m}^{i\mathrm{T}}_{\theta R} = -\mathbf{A}^i \left[\int_{V^i} \rho^i \tilde{\bar{\mathbf{u}}}^i dV^i \right] \bar{\mathbf{G}}^i \tag{5.54}$$

which can be written as

$$\mathbf{m}^i_{R\theta} = \mathbf{m}^{i\mathrm{T}}_{\theta R} = -\mathbf{A}^i \tilde{\bar{\mathbf{S}}}^i_t \bar{\mathbf{G}}^i \tag{5.55}$$

in which the skew symmetric matrix $\tilde{\bar{\mathbf{S}}}^i_t$ is defined by

$$\tilde{\bar{\mathbf{S}}}^i_t = \int_{V^i} \rho^i \tilde{\bar{\mathbf{u}}}^i dV^i \tag{5.56}$$

Using the definition of the skew symmetric matrix $\tilde{\bar{\mathbf{u}}}^i$ of Eq. 35, one can verify that the matrix $\tilde{\bar{\mathbf{S}}}^i_t$ is given by

$$\tilde{\bar{\mathbf{S}}}^i_t = \begin{bmatrix} 0 & -\bar{s}_3 & \bar{s}_2 \\ \bar{s}_3 & 0 & -\bar{s}_1 \\ -\bar{s}_2 & \bar{s}_1 & 0 \end{bmatrix}^i \tag{5.57}$$

where

$$\bar{s}^i_k = \int_{V^i} \rho^i \bar{u}^i_k dV^i, \quad k = 1, 2, 3 \tag{5.58}$$

or in a vector form as

$$\bar{\mathbf{S}}^i_t = \int_{V^i} \rho^i \bar{\mathbf{u}}^i dV^i \tag{5.59}$$

The vector $\bar{\mathbf{S}}_t^i$ represents the components of the moment of mass of the body i about the axes of the body coordinate system. In the general case of a deformable body this vector can be written as

$$\bar{\mathbf{S}}_t^i = \int_{V^i} \rho^i \left[\bar{\mathbf{u}}_o^i + \bar{\mathbf{u}}_f^i\right] dV^i \tag{5.60}$$

if the body is rigid $\bar{\mathbf{S}}_t^i$ is a constant vector. Furthermore, if the origin of the body reference is attached to the center of mass, one has

$$\mathbf{I}_1^i = \int_{V^i} \rho^i \bar{\mathbf{u}}_o^i \, dV^i = \int_{V^i} \rho^i \left[x_1^i \quad x_2^i \quad x_3^i\right]^{\mathrm{T}} dV^i = \mathbf{0} \tag{5.61}$$

which shows that the matrix $\mathbf{m}_{R\theta}^i$ in the case of rigid body dynamics is the null matrix. In this case the translation and rotation of the rigid body are dynamically decoupled. This, however, is not the case when deformable bodies are considered. This fact can be demonstrated by writing Eq. 60 in a more explicit form as

$$\bar{\mathbf{S}}_t^i = \int_{V^i} \rho^i \left[\bar{\mathbf{u}}_o^i + \mathbf{S}^i \mathbf{q}_f^i\right] dV^i \tag{5.62}$$

One can see that in the special case in which the origin of the body reference is rigidly attached to the center of mass in the undeformed state, which is the case in which Eq. 61 is satisfied, there is no guarantee that $\bar{\mathbf{S}}_t^i$ is the null matrix because of the deformation of the body. Therefore, $\bar{\mathbf{S}}_t^i$ and the submatrix $\mathbf{m}_{R\theta}^i$ must be iteratively updated. It is also clear from Eq. 62 that, in addition to evaluating the moment of mass in the undeformed state, one needs to evaluate the following inertia shape integrals:

$$\bar{\mathbf{S}}^i = \int_{V^i} \rho^i \mathbf{S}^i \, dV^i \tag{5.63}$$

This matrix is also required for the evaluation of the matrix $\mathbf{m}_{Rf}^i$ of Eq. 46c, since

$$\mathbf{m}_{Rf}^i = \mathbf{A}^i \int_{V^i} \rho^i \mathbf{S}^i \, dV^i = \mathbf{A}^i \bar{\mathbf{S}}^i \tag{5.64}$$

The submatrix $\mathbf{m}_{\theta\theta}^i$, of Eq. 46d, associated with the rotation of the body reference can be defined by using the matrix $\mathbf{B}^i$ of Eq. 33, as

$$\mathbf{m}_{\theta\theta}^i = \int_{V^i} \rho^i \mathbf{B}^{i\mathrm{T}} \mathbf{B}^i \, dV^i = \int_{V^i} \rho^i (\mathbf{A}^i \tilde{\bar{\mathbf{u}}}^i \bar{\mathbf{G}}^i)^{\mathrm{T}} (\mathbf{A}^i \tilde{\bar{\mathbf{u}}}^i \bar{\mathbf{G}}^i) dV^i \tag{5.65}$$

Using the orthogonality of the transformation matrix $\mathbf{A}^i$, that is, $\mathbf{A}^{i\mathrm{T}} \mathbf{A}^i = \mathbf{I}$, one can write $\mathbf{m}_{\theta\theta}^i$ as

$$\mathbf{m}_{\theta\theta}^i = \int_{V^i} \rho^i \bar{\mathbf{G}}^{i\mathrm{T}} \tilde{\bar{\mathbf{u}}}^{i\mathrm{T}} \tilde{\bar{\mathbf{u}}}^i \bar{\mathbf{G}}^i \, dV^i$$

By factoring out terms that are not space-dependent, one gets

$$\mathbf{m}_{\theta\theta}^i = \bar{\mathbf{G}}^{i\mathrm{T}} \left[\int_{V^i} \rho^i \tilde{\bar{\mathbf{u}}}^{i\mathrm{T}} \tilde{\bar{\mathbf{u}}}^i \, dV^i\right] \bar{\mathbf{G}}^i \tag{5.66}$$

Thus $\mathbf{m}^i_{\theta\theta}$ depends on both rotation of the body reference and the elastic deformation. The matrix $\mathbf{m}^i_{\theta\theta}$ can be written as

$$\mathbf{m}^i_{\theta\theta} = \bar{\mathbf{G}}^{i\mathrm{T}}\bar{\mathbf{I}}^i_{\theta\theta}\bar{\mathbf{G}}^i \tag{5.67}$$

where $\bar{\mathbf{I}}^i_{\theta\theta}$ is called the *inertia tensor* of the deformable body i and is defined as

$$\bar{\mathbf{I}}^i_{\theta\theta} = \int_{V^i} \rho^i \tilde{\bar{\mathbf{u}}}^{i\mathrm{T}}\tilde{\bar{\mathbf{u}}}^i dV^i \tag{5.68}$$

Using Eq. 35 and noting that $\tilde{\bar{\mathbf{u}}}^{i\mathrm{T}} = -\tilde{\bar{\mathbf{u}}}^i$, we note that Eq. 68 yields

$$\bar{\mathbf{I}}^i_{\theta\theta} = \int_{V^i} \rho^i \begin{bmatrix} (\bar{u}^i_2)^2 + (\bar{u}^i_3)^2 & -\bar{u}^i_2\bar{u}^i_1 & -\bar{u}^i_3\bar{u}^i_1 \\ & (\bar{u}^i_1)^2 + (\bar{u}^i_3)^2 & -\bar{u}^i_3\bar{u}^i_2 \\ \text{symmetric} & & (\bar{u}^i_1)^2 + (\bar{u}^i_2)^2 \end{bmatrix} dV^i \tag{5.69}$$

It can be shown that, in order to evaluate the inertia tensor, the following distinctive inertia shape integrals are required:

$$\begin{aligned} I^i_{kl} &= \int_{V^i} \rho^i x^i_k x^i_l \, dV^i, \quad \bar{\mathbf{I}}^i_{kl} = \int_{V^i} \rho^i x^i_k \mathbf{S}^i_l \, dV^i, \\ \bar{S}^i_{kl} &= \int_{V^i} \rho^i \mathbf{S}^{i\mathrm{T}}_k \mathbf{S}^i_l dV^i, \quad k, l = 1, 2, 3 \end{aligned} \tag{5.70}$$

where $\mathbf{S}^i_k$ is the kth row of the body shape function $\mathbf{S}^i$. In the case of rigid body analysis the inertia tensor $\bar{\mathbf{I}}^i_{\theta\theta}$ is a constant matrix. In deformable body dynamics, however, the inertia tensor depends on the elastic coordinates of the body. This is clear since the vector $\bar{\mathbf{u}}^i$ is the sum of two vectors; the first is the undeformed position vector of the arbitrary point P denoted as $\bar{\mathbf{u}}^i_o$, while the second is the deformation vector $\mathbf{S}^i\mathbf{q}^i_f$.

We proceed a step further to show that the inertia shape integrals of Eq. 70 are required for the evaluation of the matrix $\mathbf{m}^i_{\theta f}$ of Eq. 46e. Using Eq. 33 and the orthogonality of the transformation matrix, we may write $\mathbf{m}^i_{\theta f}$ as

$$\mathbf{m}^i_{\theta f} = -\int_{V^i} \rho^i \bar{\mathbf{G}}^{i\mathrm{T}}\tilde{\bar{\mathbf{u}}}^{i\mathrm{T}}\mathbf{A}^{i\mathrm{T}}\mathbf{A}^i\mathbf{S}^i dV^i = \bar{\mathbf{G}}^{i\mathrm{T}} \int_{V^i} \rho^i \tilde{\bar{\mathbf{u}}}^i \mathbf{S}^i dV^i \tag{5.71}$$

where the fact that $\tilde{\bar{\mathbf{u}}}^{i\mathrm{T}} = -\tilde{\bar{\mathbf{u}}}^i$ is used. Equation 71 can be written in an abbreviated form as

$$\mathbf{m}^i_{\theta f} = \bar{\mathbf{G}}^{i\mathrm{T}}\bar{\mathbf{I}}^i_{\theta f} \tag{5.72}$$

where $\bar{\mathbf{I}}^i_{\theta f}$ is defined as

$$\bar{\mathbf{I}}^i_{\theta f} = \int_{V^i} \rho^i \tilde{\bar{\mathbf{u}}}^i \mathbf{S}^i dV^i \tag{5.73}$$

Using Eq. 35, it can be shown that $\bar{\mathbf{I}}^i_{\theta f}$ is the matrix

$$\bar{\mathbf{I}}^i_{\theta f} = \int_{V^i} \rho^i \begin{bmatrix} \bar{u}^i_2\mathbf{S}^i_3 - \bar{u}^i_3\mathbf{S}^i_2 \\ \bar{u}^i_3\mathbf{S}^i_1 - \bar{u}^i_1\mathbf{S}^i_3 \\ \bar{u}^i_1\mathbf{S}^i_2 - \bar{u}^i_2\mathbf{S}^i_1 \end{bmatrix} dV^i$$

which on using Eq. 6 yields

$$\bar{\mathbf{I}}^i_{\theta f} = \int_{V^i} \rho^i \begin{bmatrix} \mathbf{q}_f^{iT}(\mathbf{S}_2^{iT}\mathbf{S}_3^i - \mathbf{S}_3^{iT}\mathbf{S}_2^i) \\ \mathbf{q}_f^{iT}(\mathbf{S}_3^{iT}\mathbf{S}_1^i - \mathbf{S}_1^{iT}\mathbf{S}_3^i) \\ \mathbf{q}_f^{iT}(\mathbf{S}_1^{iT}\mathbf{S}_2^i - \mathbf{S}_2^{iT}\mathbf{S}_1^i) \end{bmatrix} dV^i + \int_{V^i} \rho^i \begin{bmatrix} x_2^i\mathbf{S}_3^i - x_3^i\mathbf{S}_2^i \\ x_3^i\mathbf{S}_1^i - x_1^i\mathbf{S}_3^i \\ x_1^i\mathbf{S}_2^i - x_2^i\mathbf{S}_1^i \end{bmatrix} dV^i$$

where $\bar{\mathbf{u}}_o^i = [x_1^i \;\; x_2^i \;\; x_3^i]^{\mathrm{T}}$ is the undeformed position of the arbitrary point. Using Eq. 70, one can verify the following:

$$\begin{bmatrix} \mathbf{q}_f^{iT}(\bar{\mathbf{S}}_{23}^i - \bar{\mathbf{S}}_{23}^{iT}) \\ \mathbf{q}_f^{iT}(\bar{\mathbf{S}}_{31}^i - \bar{\mathbf{S}}_{31}^{iT}) \\ \mathbf{q}_f^{iT}(\bar{\mathbf{S}}_{12}^i - \bar{\mathbf{S}}_{12}^{iT}) \end{bmatrix} = \begin{bmatrix} \mathbf{q}_f^{iT}\tilde{\bar{\mathbf{S}}}_{23}^i \\ \mathbf{q}_f^{iT}\tilde{\bar{\mathbf{S}}}_{31}^i \\ \mathbf{q}_f^{iT}\tilde{\bar{\mathbf{S}}}_{12}^i \end{bmatrix} \tag{5.74}$$

where $\tilde{\bar{\mathbf{S}}}_{12}^i$, $\tilde{\bar{\mathbf{S}}}_{23}^i$, and $\tilde{\bar{\mathbf{S}}}_{31}^i$ are the skew symmetric matrices defined as

$$\left.\begin{aligned} \tilde{\bar{\mathbf{S}}}_{12}^i &= \bar{\mathbf{S}}_{12}^i - \bar{\mathbf{S}}_{12}^{iT} \\ \tilde{\bar{\mathbf{S}}}_{23}^i &= \bar{\mathbf{S}}_{23}^i - \bar{\mathbf{S}}_{23}^{iT} \\ \tilde{\bar{\mathbf{S}}}_{31}^i &= \bar{\mathbf{S}}_{31}^i - \bar{\mathbf{S}}_{31}^{iT} \end{aligned}\right\} \tag{5.75}$$

and the matrices $\bar{\mathbf{S}}_{kl}^i$ are given by Eq. 70.

Finally, the submatrix $\mathbf{m}_{ff}^i$ of Eq. 46f is independent of the generalized coordinates of the body and, therefore, is constant. This matrix can be written in terms of the inertia shape integrals of Eq. 70 as

$$\mathbf{m}_{ff}^i = \int_{V^i} \rho^i \mathbf{S}^{iT}\mathbf{S}^i \, dV^i = \bar{\mathbf{S}}_{11}^i + \bar{\mathbf{S}}_{22}^i + \bar{\mathbf{S}}_{33}^i \tag{5.76}$$

This completes the formulation of the components of the mass matrix of the deformable body in the spatial analysis.

Planar Motion A special case of three-dimensional motion is the *planar motion* of deformable bodies. In this special case, the reference coordinates are

$$\mathbf{q}_r^i = [\mathbf{R}^{iT} \;\; \theta^i]^{\mathrm{T}} \tag{5.77}$$

where $\mathbf{R}^i = [R_1^i \; R_2^i]^{\mathrm{T}}$ is the vector of Cartesian coordinates that define the location of the body reference. In the case of planar motion, the matrix $\mathbf{m}_{RR}^i$ can be defined as

$$\mathbf{m}_{RR}^i = \int_{V^i} \rho^i \mathbf{I} \, dV^i = \begin{bmatrix} m^i & 0 \\ 0 & m^i \end{bmatrix} \tag{5.78}$$

where $\mathbf{I}$ is a 2×2 identity matrix and m^i is the mass of the deformable body i in the multibody system. Using the planar transformation of Eq. 5a, one can define the matrix $\mathbf{B}^i$ as

$$\mathbf{B}^i = \mathbf{A}_\theta^i \bar{\mathbf{u}}^i \tag{5.79}$$

where $\mathbf{A}_\theta^i$ is the partial derivative of the transformation matrix $\mathbf{A}^i$ with respect to the rotational coordinate θ^i, that is,

$$\mathbf{A}_\theta^i = \begin{bmatrix} -\sin\theta^i & -\cos\theta^i \\ \cos\theta^i & -\sin\theta^i \end{bmatrix} \tag{5.80}$$

Using Eq. 79, we can write the submatrix $\mathbf{m}_{R\theta}^i$ of Eq. 46b as

$$\begin{aligned}\mathbf{m}_{R\theta}^i &= \int_{V^i} \rho^i \mathbf{B}^i dV^i = \mathbf{A}_\theta^i \int_{V^i} \rho^i \bar{\mathbf{u}}^i dV^i = \mathbf{A}_\theta^i \int_{V^i} \rho^i \left[\bar{\mathbf{u}}_o^i + \bar{\mathbf{u}}_f^i\right] dV^i \\ &= \mathbf{A}_\theta^i \left[\mathbf{I}_1^i + \bar{\mathbf{S}}^i \mathbf{q}_f^i\right]\end{aligned} \tag{5.81}$$

where the matrices $\mathbf{I}_1^i$ and $\bar{\mathbf{S}}^i$ are defined as

$$\mathbf{I}_1^i = \int_{V^i} \rho^i \bar{\mathbf{u}}_o^i dV^i, \qquad \bar{\mathbf{S}}^i = \int_{V^i} \rho^i \mathbf{S}^i dV^i \tag{5.82}$$

The vector $\mathbf{I}_1^i$ is the moment of mass of the body about the axes of the body reference in the undeformed state. Therefore, if the origin of the body reference is initially attached to the body center of mass, the vector $\mathbf{I}_1^i$ vanishes. The vector $\bar{\mathbf{S}}^i \mathbf{q}_f^i$ represents the change in the moment of mass due to the deformation.

Using Eq. 46c, one can verify that

$$\mathbf{m}_{Rf}^i = \mathbf{A}^i \bar{\mathbf{S}}^i \tag{5.83}$$

Because $\mathbf{A}_\theta^i$ of Eq. 80 is an orthogonal matrix, that is, $\mathbf{A}_\theta^{i\mathrm{T}} \mathbf{A}_\theta^i = \mathbf{I}$, Eq. 46d yields

$$\mathbf{m}_{\theta\theta}^i = \int_{V^i} \rho^i \mathbf{B}^{i\mathrm{T}} \mathbf{B}^i dV^i = \int_{V^i} \rho^i \bar{\mathbf{u}}^{i\mathrm{T}} \mathbf{A}_\theta^{i\mathrm{T}} \mathbf{A}_\theta^i \bar{\mathbf{u}}^i dV^i = \int_{V^i} \rho^i \bar{\mathbf{u}}^{i\mathrm{T}} \bar{\mathbf{u}}^i dV^i \tag{5.84}$$

Since $\bar{\mathbf{u}}^i = \bar{\mathbf{u}}_o^i + \bar{\mathbf{u}}_f^i$, where $\bar{\mathbf{u}}_f^i = \mathbf{S}^i \mathbf{q}_f^i$, one can write Eq. 84 as

$$\begin{aligned}\mathbf{m}_{\theta\theta}^i &= \int_{V^i} \rho^i \left[\bar{\mathbf{u}}_o^i + \bar{\mathbf{u}}_f^i\right]^{\mathrm{T}} \left[\bar{\mathbf{u}}_o^i + \bar{\mathbf{u}}_f^i\right] dV^i \\ &= \int_{V^i} \rho^i \left[\bar{\mathbf{u}}_o^{i\mathrm{T}} \bar{\mathbf{u}}_o^i + 2\bar{\mathbf{u}}_o^{i\mathrm{T}} \bar{\mathbf{u}}_f^i + \bar{\mathbf{u}}_f^{i\mathrm{T}} \bar{\mathbf{u}}_f^i\right] dV^i \\ &= \left(m_{\theta\theta}^i\right)_{rr} + \left(m_{\theta\theta}^i\right)_{rf} + \left(m_{\theta\theta}^i\right)_{ff}\end{aligned} \tag{5.85}$$

in which the submatrix $\mathbf{m}_{\theta\theta}^i$ reduces to a scalar that can be written as the sum of three components. The first component, $(m_{\theta\theta}^i)_{rr}$, can be recognized as the mass moment of inertia, in the undeformed state, of the body about the origin of the body reference. This scalar component can be written as

$$\begin{aligned}\left(m_{\theta\theta}^i\right)_{rr} &= \int_{V^i} \rho^i \bar{\mathbf{u}}_o^{i\mathrm{T}} \bar{\mathbf{u}}_o^i dV^i = \int_{V^i} \rho^i \begin{bmatrix} x_1^i & x_2^i \end{bmatrix} \begin{bmatrix} x_1^i \\ x_2^i \end{bmatrix} dV^i \\ &= \int_{V^i} \rho^i \left[\left(x_1^i\right)^2 + \left(x_2^i\right)^2\right] dV^i\end{aligned} \tag{5.86}$$

Clearly, this integral has a constant value and does not depend on the body deformation. The last two scalar components, $(m_{\theta\theta}^i)_{rf}$ and $(m_{\theta\theta}^i)_{ff}$, of Eq. 85 represent the change in the mass moment of inertia of the body due to deformation. These two components are evaluated according to

$$\left(m^i_{\theta\theta}\right)_{rf} = 2\int_{V^i} \rho^i \bar{\mathbf{u}}^{i\mathrm{T}}_o \bar{\mathbf{u}}^i_f \, dV^i = 2\left[\int_{V^i} \rho^i \bar{\mathbf{u}}^{i\mathrm{T}}_o \mathbf{S}^i \, dV^i\right]\mathbf{q}^i_f \tag{5.87}$$

$$\begin{aligned}\left(m^i_{\theta\theta}\right)_{ff} &= \int_{V^i} \rho^i \bar{\mathbf{u}}^{i\mathrm{T}}_f \bar{\mathbf{u}}^i_f \, dV^i = \int_{V^i} \rho^i \mathbf{q}^{i\mathrm{T}}_f \mathbf{S}^{i\mathrm{T}} \mathbf{S}^i \mathbf{q}^i_f \, dV^i \\ &= \mathbf{q}^{i\mathrm{T}}_f \left[\int_{V^i} \rho^i \mathbf{S}^{i\mathrm{T}} \mathbf{S}^i \, dV^i\right]\mathbf{q}^i_f \end{aligned} \tag{5.88}$$

If we use the definition of Eq. 46f, Eq. 88 can be written in the following form:

$$\left(m^i_{\theta\theta}\right)_{ff} = \mathbf{q}^{i\mathrm{T}}_f \mathbf{m}^i_{ff} \mathbf{q}^i_f \tag{5.89}$$

Note that the two scalars $(m^i_{\theta\theta})_{rf}$ and $(m^i_{\theta\theta})_{ff}$ depend on the elastic deformation of the body. Finally, we write Eq. 46e as

$$\mathbf{m}^i_{\theta f} = \int_{V^i} \rho^i \mathbf{B}^{i\mathrm{T}} \mathbf{A}^i \mathbf{S}^i \, dV^i = \int_{V^i} \rho^i \bar{\mathbf{u}}^{i\mathrm{T}} \mathbf{A}^{i\mathrm{T}}_\theta \mathbf{A}^i \mathbf{S}^i \, dV^i \tag{5.90}$$

It can be shown that the product $\mathbf{A}^{i\mathrm{T}}_\theta \mathbf{A}^i$ is a skew symmetric matrix defined as

$$\tilde{\mathbf{I}} = \mathbf{A}^{i\mathrm{T}}_\theta \mathbf{A}^i = \begin{bmatrix} 0 & 1 \\ -1 & 0 \end{bmatrix} \tag{5.91}$$

Substituting Eq. 91 into Eq. 90 and writing $\bar{\mathbf{u}}^i$ in a more explicit form yields

$$\begin{aligned}\mathbf{m}^i_{\theta f} &= \int_{V^i} \rho^i \left[\bar{\mathbf{u}}^i_o + \bar{\mathbf{u}}^i_f\right]^{\mathrm{T}} \tilde{\mathbf{I}} \mathbf{S}^i \, dV^i \\ &= \int_{V^i} \rho^i \bar{\mathbf{u}}^{i\mathrm{T}}_o \tilde{\mathbf{I}} \mathbf{S}^i \, dV^i + \mathbf{q}^{i\mathrm{T}}_f \tilde{\mathbf{S}}^i \end{aligned} \tag{5.92}$$

where the constant skew symmetric matrix $\tilde{\mathbf{S}}^i$ is defined as

$$\tilde{\mathbf{S}}^i = \int_{V^i} \rho^i \mathbf{S}^{i\mathrm{T}} \tilde{\mathbf{I}} \mathbf{S}^i \, dV^i = \int_{V^i} \rho^i \left[\mathbf{S}^{i\mathrm{T}}_1 \mathbf{S}^i_2 - \mathbf{S}^{i\mathrm{T}}_2 \mathbf{S}^i_1\right] dV^i \tag{5.93}$$

in which $\mathbf{S}^i_1$ and $\mathbf{S}^i_2$ are the rows of the shape function $\mathbf{S}^i$. One may also observe that the submatrix $\mathbf{m}^i_{\theta f}$ consists of two parts; the first part is constant, while the second part depends on the elastic coordinates of the body.

We conclude that, to completely describe the inertia properties of the deformable body in plane motion, a set of inertia shape integrals is required. These integrals, which depend on the assumed displacement field, are

$$\mathbf{I}^i_1 = \int_{V^i} \rho^i \left[x^i_1 \quad x^i_2\right]^{\mathrm{T}} dV^i, \quad I^i_{kl} = \int_{V^i} \rho^i x^i_k x^i_l \, dV^i, \quad k, l = 1, 2 \tag{5.94}$$

$$\bar{\mathbf{I}}^i_{kl} = \int_{V^i} \rho^i x^i_k \mathbf{S}^i_l \, dV^i, \quad \bar{\mathbf{S}}^i = \int_{V^i} \rho^i \mathbf{S}^i \, dV^i, \quad \bar{\mathbf{S}}^i_{kl} = \int_{V^i} \rho^i \mathbf{S}^{i\mathrm{T}}_k \mathbf{S}^i_l \, dV^i, \quad k, l = 1, 2 \tag{5.95}$$

where the constant matrix $\mathbf{m}^i_{ff}$ can be written in terms of these integrals as

$$\mathbf{m}^i_{ff} = \int_{V^i} \rho^i \mathbf{S}^{i\mathrm{T}} \mathbf{S}^i \, dV^i = \int_{V^i} \rho^i \left[\mathbf{S}^{i\mathrm{T}}_1 \mathbf{S}^i_1 + \mathbf{S}^{i\mathrm{T}}_2 \mathbf{S}^i_2\right] dV^i = \bar{\mathbf{S}}^i_{11} + \bar{\mathbf{S}}^i_{22} \tag{5.96}$$

If the body is rigid, only the integrals of Eq. 94 are required. On the other hand, if the large rigid body displacement is not permitted, which is the case in linear structural systems, only the constant matrix of Eq. 96 is required.

Example 5.3 For the deformable beam given in Example 1, the inertia shape integral $\bar{\mathbf{S}}^i$ of Eq. 95 is given by (since we have only one body, the superscript i is omitted for simplicity)

$$\bar{\mathbf{S}} = \int_V \rho \mathbf{S}\, dV$$
$$= \int_V \rho \begin{bmatrix} \xi & 0 \\ 0 & 3(\xi)^2 - 2(\xi)^3 \end{bmatrix} dV$$

where ρ is the mass density of the beam material; $\xi = (x/l)$; l is the length of the beam; and V is the volume. If the beam is assumed to have a constant cross-sectional area a, then

$$dV = a\, dx = al\frac{dx}{l} = V\, d\xi$$

Assuming that the mass density ρ is constant, we have

$$\rho V\, d\xi = m\, d\xi$$

where m is the total mass of the beam. The matrix $\bar{\mathbf{S}}$ can then be written as

$$\bar{\mathbf{S}} = \int_0^1 m \begin{bmatrix} \xi & 0 \\ 0 & 3(\xi)^2 - 2(\xi)^3 \end{bmatrix} d\xi = \frac{m}{2}\begin{bmatrix} 1 & 0 \\ 0 & 1 \end{bmatrix}$$

The skew symmetric matrix $\tilde{\mathbf{S}}^i$ of Eq. 93 is

$$\tilde{\mathbf{S}} = \int_V \rho \mathbf{S}^{\mathrm{T}} \tilde{\mathbf{I}} \mathbf{S}\, dV = \int_0^1 m \mathbf{S}^{\mathrm{T}} \tilde{\mathbf{I}} \mathbf{S}\, d\xi = m \begin{bmatrix} 0 & \frac{7}{20} \\ -\frac{7}{20} & 0 \end{bmatrix}$$

The constant matrix $\mathbf{m}_{ff}$ can be evaluated by using Eq. 96 as

$$\mathbf{m}_{ff} = \int_V \rho \mathbf{S}^{\mathrm{T}} \mathbf{S}\, dV = m \int_0^1 \mathbf{S}^{\mathrm{T}} \mathbf{S}\, d\xi = m \begin{bmatrix} \frac{1}{3} & 0 \\ 0 & \frac{13}{35} \end{bmatrix}$$

Since the location of an arbitrary point on the beam is

$$\bar{\mathbf{u}}_o = [x \quad 0]^{\mathrm{T}} = l\begin{bmatrix} \frac{x}{l} & 0 \end{bmatrix}^{\mathrm{T}} = l\,[\xi \quad 0]^{\mathrm{T}}$$

one has the following:

$$\mathbf{I}_1 = \int_V \rho \bar{\mathbf{u}}_o\, dV = ml \int_0^1 [\xi \quad 0]^{\mathrm{T}}\, d\xi = \frac{ml}{2}[1 \quad 0]^{\mathrm{T}}$$

$$\int_V \rho \bar{\mathbf{u}}_o^{\mathrm{T}} \mathbf{S}\, dV = \frac{ml}{3}[1 \quad 0]$$

$$\int_V \bar{\mathbf{u}}_o^{\mathrm{T}} \tilde{\mathbf{I}} \mathbf{S}\, dV = ml \begin{bmatrix} 0 & \frac{7}{20} \end{bmatrix}^{\mathrm{T}}$$

$$(m_{\theta\theta})_{rr} = \int_V \rho \bar{\mathbf{u}}_o^{\mathrm{T}} \bar{\mathbf{u}}_o\, dV = m(l)^2 \int_0^1 [\xi \quad 0] \begin{bmatrix} \xi \\ 0 \end{bmatrix} d\xi = \frac{m(l)^2}{3}$$

which is the mass moment of inertia of the beam, in the undeformed state, about the $\mathbf{X}_3$ axis.

Let the mass of the beam be 1.236 kg and the length 0.5 m, and let at a given instant of time the vector of beam coordinates be given by

$$\begin{aligned} \mathbf{q} &= \begin{bmatrix} \mathbf{q}_r^{\mathrm{T}} & \mathbf{q}_f^{\mathrm{T}} \end{bmatrix}^{\mathrm{T}} = [R_1 \quad R_2 \quad \theta \quad q_{f1} \quad q_{f2}]^{\mathrm{T}} \\ &= [1.0 \quad 0.5 \quad 30^\circ \quad 0.001 \quad 0.01]^{\mathrm{T}} \end{aligned}$$

The components of the mass matrix of Eq. 45 can be evaluated as follows. By use of Eq. 78, the matrix $\mathbf{m}_{RR}$ associated with the translational coordinates is given by

$$\mathbf{m}_{RR} = \begin{bmatrix} m & 0 \\ 0 & m \end{bmatrix} = \begin{bmatrix} 1.236 & 0 \\ 0 & 1.236 \end{bmatrix}$$

The vector $\mathbf{I}_1$ of Eq. 82, in this case, is

$$\mathbf{I}_1 = \int_V \rho \bar{\mathbf{u}}_o\, dV = \frac{ml}{2} [1 \quad 0]^{\mathrm{T}} = [0.309 \quad 0]^{\mathrm{T}}$$

and

$$\bar{\mathbf{S}}\mathbf{q}_f = \frac{m}{2} \begin{bmatrix} 1 & 0 \\ 0 & 1 \end{bmatrix} \begin{bmatrix} q_{f1} \\ q_{f2} \end{bmatrix} = \begin{bmatrix} 0.000618 \\ 0.00618 \end{bmatrix}$$

that is,

$$\mathbf{I}_1 + \bar{\mathbf{S}}\mathbf{q}_f = \begin{bmatrix} 0.309 \\ 0 \end{bmatrix} + \begin{bmatrix} 0.000618 \\ 0.00618 \end{bmatrix} = \begin{bmatrix} 0.30962 \\ 0.00618 \end{bmatrix}$$

Since the matrix $\mathbf{A}_\theta$ in this case is given by

$$\mathbf{A}_\theta = \begin{bmatrix} -\sin\theta & -\cos\theta \\ \cos\theta & -\sin\theta \end{bmatrix} = \begin{bmatrix} -0.5 & -0.866 \\ 0.866 & -0.5 \end{bmatrix}$$

the matrix $\mathbf{m}_{R\theta}$ of Eq. 81 is given by

$$\mathbf{m}_{R\theta} = \mathbf{A}_\theta \left[\mathbf{I}_1 + \bar{\mathbf{S}}\mathbf{q}_f\right] = \begin{bmatrix} -0.5 & -0.866 \\ 0.866 & -0.5 \end{bmatrix} \begin{bmatrix} 0.30962 \\ 0.00618 \end{bmatrix} = \begin{bmatrix} -0.1602 \\ 0.2650 \end{bmatrix}$$

For the value of $\theta = 30^\circ$, the transformation matrix $\mathbf{A}$ is

$$\mathbf{A} = \begin{bmatrix} \cos\theta & -\sin\theta \\ \sin\theta & \cos\theta \end{bmatrix} = \begin{bmatrix} 0.866 & -0.5 \\ 0.500 & 0.866 \end{bmatrix}$$

and the matrix $\mathbf{m}_{Rf}$ of Eq. 83 is given by

$$\begin{aligned} \mathbf{m}_{Rf} = \mathbf{A}\bar{\mathbf{S}} &= \begin{bmatrix} 0.866 & -0.5 \\ 0.500 & 0.866 \end{bmatrix} \begin{bmatrix} 1 & 0 \\ 0 & 1 \end{bmatrix} \left(\frac{1.236}{2}\right) \\ &= \begin{bmatrix} 0.5352 & -0.309 \\ 0.309 & 0.5352 \end{bmatrix} \end{aligned}$$

The scalars $(m_{\theta\theta})_{rf}$ and $(m_{\theta\theta})_{ff}$ of Eqs. 87 and 88 are given by

$$(m_{\theta\theta})_{rf} = 2\left[\int_V \rho\bar{\mathbf{u}}_o^{\mathrm{T}}\mathbf{S}\,dV\right]\mathbf{q}_f = 2\frac{ml}{3}[1 \quad 0]\begin{bmatrix} q_{f1} \\ q_{f2} \end{bmatrix}$$

$$= \frac{2ml}{3}q_{f1} = \frac{2(1.236)(0.5)}{3}(0.001) = 4.12 \times 10^{-4}$$

$$(m_{\theta\theta})_{ff} = \mathbf{q}_f^{\mathrm{T}}\mathbf{m}_{ff}\mathbf{q}_f = \frac{1.236}{2}[0.001 \quad 0.01]\begin{bmatrix} \frac{1}{3} & 0 \\ 0 & \frac{13}{35} \end{bmatrix}\begin{bmatrix} 0.001 \\ 0.01 \end{bmatrix}$$

$$= 2.316 \times 10^{-5}$$

Therefore, $\mathbf{m}_{\theta\theta}$ in Eq. 45 is given by

$$\mathbf{m}_{\theta\theta} = (m_{\theta\theta})_{rr} + (m_{\theta\theta})_{rf} + (m_{\theta\theta})_{ff}$$

$$= \frac{m(l)^2}{3} + (4.12) \times 10^{-4} + 2.316 \times 10^{-5}$$

$$= \frac{1.236(0.5)^2}{3} + (4.12) \times 10^{-4} + 2.316 \times 10^{-5} = 0.10306$$

Using Eq. 92, we can evaluate the matrix $\mathbf{m}_{\theta f}$ as

$$\mathbf{m}_{\theta f} = \int_V \rho\bar{\mathbf{u}}_0^{\mathrm{T}}\tilde{\mathbf{I}}\mathbf{S}\,dV + \mathbf{q}_f^{\mathrm{T}}\tilde{\mathbf{S}}$$

$$= ml\left[0 \quad \tfrac{7}{20}\right] + \left[q_{f1} \quad q_{f2}\right](m)\begin{bmatrix} 0 & \frac{7}{20} \\ -\frac{7}{20} & 0 \end{bmatrix}$$

$$= (1.236)(0.5)\left[0 \quad \tfrac{7}{20}\right] + 1.236\,[0.001 \quad 0.01]\begin{bmatrix} 0 & \frac{7}{20} \\ -\frac{7}{20} & 0 \end{bmatrix}$$

$$= \left[-4.326 \times 10^{-3} \quad 0.2167\right]$$

Finally, the matrix $\mathbf{m}_{ff}$ is, in this example, the 2×2 matrix given by

$$\mathbf{m}_{ff} = 1.236\begin{bmatrix} \frac{1}{3} & 0 \\ 0 & \frac{13}{35} \end{bmatrix} = \begin{bmatrix} 0.412 & 0 \\ 0 & 0.4591 \end{bmatrix}$$

Therefore, the mass matrix of the beam at this instant of time is given by

$$\mathbf{M} = \begin{bmatrix} \mathbf{m}_{RR} & \mathbf{m}_{R\theta} & \mathbf{m}_{Rf} \\ & \mathbf{m}_{\theta\theta} & \mathbf{m}_{\theta f} \\ \text{symmetric} & & \mathbf{m}_{ff} \end{bmatrix}$$

$$= \begin{bmatrix} 1.236 & 0 & -0.1602 & 0.5352 & -0.309 \\ & 1.236 & 0.2650 & 0.309 & 0.5352 \\ & & 0.1034 & -4.326 \times 100^{-3} & 0.2167 \\ & & & 0.412 & 0 \\ & \text{symmetric} & & & 0.4591 \end{bmatrix}$$

Lumped Masses In this section, the kinetic energy of the deformable body was developed in terms of a finite set of coordinates. This was achieved by assuming the deformation shape using the body shape functions that depend on the spatial coordinates. Therefore, the deformation at any point on the body can be obtained by specifying the coordinates of this point in the body shape function. This approach leads to what is called *consistent mass* formulation. Another approach that is also used to formulate the dynamic equations of deformable bodies is based on using *lumped mass* techniques. In the lumped mass formulation the interest is focused on the displacement of selected *grid points* on the deformable body. Instead of using shape functions, a set of shape vectors are used to describe the relative motion between these grid points. These shape vectors can be assumed or can be determined experimentally. They can also be the mode shapes of vibration of the deformable body. In the lumped mass formulation the total mass of the body is distributed among the grid points. By increasing the number of the grid points more accurate results can be obtained.

In the remainder of this section we develop the inertia properties of deformable bodies that undergo finite rotations using a lumped mass technique. This development leads to a set of inertia shape matrices similar to the ones that appeared in the consistent mass formulation. These matrices are required in order to evaluate the coupling between the reference motion and elastic deformation.

As pointed out earlier, in the lumped mass formulation, the motion of the deformable body is identified by a set of shape vectors that describe the displacement of selected grid points. This set of shape vectors should be linearly independent and should contain low-frequency modes of vibration of the body analyzed. In this section, grid point displacements are expressed in terms of the elastic generalized coordinates of the deformable body. The deformation vector of a grid point j on body i can be written as

$$\bar{\mathbf{u}}_f^{ij} = \mathbf{N}^{ij}\mathbf{q}_f^i, \quad j = 1, 2, \ldots, n_j$$

where $\bar{\mathbf{u}}_f^{ij}$ is the vector of elastic deformation at the grid point j, $\mathbf{q}_f^i$ is the vector of elastic coordinates of body i, $\mathbf{N}^{ij}$ is a partition of the assumed shape matrix associated with the displacements of the grid point j, and n_j is the total number of the grid points. The partition $\mathbf{N}^{ij}$ is $3 \times n_f$ matrix, where n_f is the total number of elastic coordinates of body i. As pointed out earlier, the body shape matrix can be determined experimentally by using modal testing or numerically by first using the finite-element method to discretize the deformable body and then solve for the eigenvalue problem.

The global position of the grid point j can be written as

$$\mathbf{r}^{ij} = \mathbf{R}^i + \mathbf{A}^i(\bar{\mathbf{u}}_o^{ij} + \bar{\mathbf{u}}_f^{ij})$$

where $\mathbf{R}^i$ is the set of Cartesian coordinates that define the location of the origin of the body reference, $\mathbf{A}^i$ is the transformation matrix from the local coordinate system to the global coordinate system, $\bar{\mathbf{u}}_o^{ij}$ is the position of the grid point j in the undeformed state, and $\bar{\mathbf{u}}_f^{ij}$ is the deformation vector. Differentiating the preceding equation with respect to time yields

$$\dot{\mathbf{r}}^{ij} = \dot{\mathbf{R}}^i + \dot{\mathbf{A}}^i\bar{\mathbf{u}}^{ij} + \mathbf{A}^i\mathbf{N}^{ij}\dot{\mathbf{q}}_f^i$$

where

$$\bar{\mathbf{u}}^{ij} = \bar{\mathbf{u}}_o^{ij} + \bar{\mathbf{u}}_f^{ij}$$

The kinetic energy of the deformable body i can then be defined as

$$T^i = \sum_{j=1}^{n_j} T^{ij}$$

where T^{ij} is the kinetic energy of the grid point j defined as

$$T^{ij} = \frac{1}{2} m^{ij} \dot{\mathbf{r}}^{ij\mathrm{T}} \dot{\mathbf{r}}^{ij}$$

in which m^{ij} is the mass of the grid point j.

Using the preceding two equations and following a procedure similar to the one described in the preceding sections, we can write the kinetic energy of the deformable body i, based on a lumped mass model, as

$$T^i = \frac{1}{2} [\dot{\mathbf{R}}^{i\mathrm{T}} \quad \dot{\boldsymbol{\theta}}^{i\mathrm{T}} \quad \dot{\mathbf{q}}_f^{i\mathrm{T}}] \begin{bmatrix} \mathbf{m}_{RR}^i & & \text{symmetric} \\ \mathbf{m}_{\theta R}^i & \mathbf{m}_{\theta\theta}^i & \\ \mathbf{m}_{fR}^i & \mathbf{m}_{f\theta}^i & \mathbf{m}_{ff}^i \end{bmatrix} \begin{bmatrix} \dot{\mathbf{R}}^i \\ \dot{\boldsymbol{\theta}}^i \\ \dot{\mathbf{q}}_f^i \end{bmatrix}$$

where $\boldsymbol{\theta}^i$ is the vector of rotation coordinates of the body reference and

$$\mathbf{m}_{RR}^i = \sum_{j=1}^{n_j} m^{ij} \mathbf{I}, \quad \mathbf{m}_{\theta R}^i = \sum_{j=1}^{n_j} m^{ij} \mathbf{B}^{ij\mathrm{T}}$$

$$\mathbf{m}_{\theta\theta}^i = \sum_{j=1}^{n_j} m^{ij} \mathbf{B}^{ij\mathrm{T}} \mathbf{B}^{ij}, \quad \mathbf{m}_{fR}^i = \sum_{j=1}^{n_j} m^{ij} \mathbf{N}^{ij\mathrm{T}} \mathbf{A}^{i\mathrm{T}}$$

$$\mathbf{m}_{f\theta}^i = \sum_{j=1}^{n_j} m^{ij} \mathbf{N}^{ij\mathrm{T}} \mathbf{A}^{i\mathrm{T}} \mathbf{B}^{ij}, \quad \mathbf{m}_{ff}^i = \sum_{j=1}^{n_j} m^{ij} \mathbf{N}^{ij\mathrm{T}} \mathbf{N}^{ij}$$

in which $\mathbf{I}$ is the identity matrix and $\mathbf{B}^{ij}$ is the matrix whose columns are defined as

$$\mathrm{Col}\,(\mathbf{B}^{ij})_k = \frac{\partial}{\partial \theta_k^i} (\mathbf{A}^i \bar{\mathbf{u}}^{ij})$$

The physical interpretation of the components of the mass matrix obtained using the lumped mass formulation is similar to the interpretation given in the case of the consistent mass approach.

5.3 GENERALIZED FORCES

In this section, we develop expressions for the generalized forces associated with the generalized coordinates of the deformable body i in the multibody system. We consider the elastic forces arising from the body deformation and also externally applied forces as well as restoring forces due to elastic and dissipating elements such as springs and dampers.

Generalized Elastic Forces In this section, we consider a linear isotropic material. The more general case of nonlinear elastic, orthotropic materials can also be formulated by changing the form of the body stiffness matrix.

In the preceding chapter, it was shown that the virtual work due to the elastic forces can be written as

$$\delta W_s^i = -\int_{V^i} \boldsymbol{\sigma}^{i\mathrm{T}} \delta \boldsymbol{\varepsilon}^i \, dV^i \tag{5.97}$$

where $\boldsymbol{\sigma}^i$ and $\boldsymbol{\varepsilon}^i$ are, respectively, the stress and strain vectors, and δW_s^i is the virtual work of the elastic forces. Since the rigid body motion corresponds to the case of constant strains and since we defined the deformation with respect to the body reference, there is no loss of generality in writing the strain displacement relations in the following form:

$$\boldsymbol{\varepsilon}^i = \mathbf{D}^i \bar{\mathbf{u}}_f^i \tag{5.98}$$

where $\mathbf{D}^i$ is a differential operator defined in the preceding chapter and $\bar{\mathbf{u}}_f^i$ is the deformation vector. In terms of the elastic generalized coordinates of body i, one may write Eq. 98 as

$$\boldsymbol{\varepsilon}^i = \mathbf{D}^i \mathbf{S}^i \mathbf{q}_f^i \tag{5.99}$$

For a linear isotropic material, the constitutive equations relating the stress and strains can be written as

$$\boldsymbol{\sigma}^i = \mathbf{E}^i \boldsymbol{\varepsilon}^i \tag{5.100}$$

where $\mathbf{E}^i$ is the symmetric matrix of elastic coefficients. Substituting Eq. 99 into Eq. 100 yields

$$\boldsymbol{\sigma}^i = \mathbf{E}^i \mathbf{D}^i \mathbf{S}^i \mathbf{q}_f^i \tag{5.101}$$

in which the stress vector is written in terms of the elastic generalized coordinates of body i. Substituting Eqs. 99 and 101 into Eq. 97 yields

$$\delta W_s^i = -\int_{V^i} \mathbf{q}_f^{i\mathrm{T}} (\mathbf{D}^i \mathbf{S}^i)^{\mathrm{T}} \mathbf{E}^i \mathbf{D}^i \mathbf{S}^i \delta \mathbf{q}_f^i \, dV^i \tag{5.102}$$

where the symmetry of the matrix of elastic coefficients is used. Because the vector $\mathbf{q}_f^i$ depends only on time, Eq. 102 can be written as

$$\delta W_s^i = -\mathbf{q}_f^{i\mathrm{T}} \left[\int_{V^i} (\mathbf{D}^i \mathbf{S}^i)^{\mathrm{T}} \mathbf{E}^i \mathbf{D}^i \mathbf{S}^i \, dV^i \right] \delta \mathbf{q}_f^i \tag{5.103}$$

One may write Eq. 103 as

$$\delta W_s^i = -\mathbf{q}_f^{i\mathrm{T}} \mathbf{K}_{ff}^i \delta \mathbf{q}_f^i \tag{5.104}$$

where $\mathbf{K}_{ff}^i$ is the symmetric positive definite *stiffness matrix* associated with the elastic coordinates of body i in the multibody system. This matrix is defined as

$$\mathbf{K}^i_{ff} = \int_{V^i} (\mathbf{D}^i\mathbf{S}^i)^{\mathrm{T}}\mathbf{E}^i\mathbf{D}^i\mathbf{S}^i dV^i \tag{5.105}$$

For convenience, we rewrite the virtual work of Eq. 104 as

$$\delta W^i_s = -\begin{bmatrix}\mathbf{R}^{i\mathrm{T}} & \theta^{i\mathrm{T}} & \mathbf{q}^{i\mathrm{T}}_f\end{bmatrix}\begin{bmatrix}\mathbf{0} & \mathbf{0} & \mathbf{0}\\ \mathbf{0} & \mathbf{0} & \mathbf{0}\\ \mathbf{0} & \mathbf{0} & \mathbf{K}^i_{ff}\end{bmatrix}\begin{bmatrix}\delta\mathbf{R}^i\\ \delta\theta^i\\ \delta\mathbf{q}^i_f\end{bmatrix} \tag{5.106}$$

The strain energy can also be used to define the stiffness matrix of Eq. 105. This is demonstrated in the following example.

Example 5.4 Neglecting the shear deformation and using the assumptions of *Euler–Bernoulli beam theory*, the strain energy of the beam presented in Example 1 can be written as

$$U = \frac{1}{2}\int_0^l [EI(u''_{f2})^2 + Ea(u'_{f1})^2]dx$$

where E is the modulus of elasticity of the beam, I is the second moment of area, a is the cross-sectional area, l is the beam length, u_{f1} and u_{f2} are, respectively, the axial and transverse displacements, and ($'$) denotes differentiation with respect to the spatial coordinate. The preceding strain energy expression can be written in a matrix form as

$$U = \frac{1}{2}\int_0^l [u'_{f1} \quad u''_{f2}]\begin{bmatrix}Ea & 0\\ 0 & EI\end{bmatrix}\begin{bmatrix}u'_{f1}\\ u''_{f2}\end{bmatrix}dx$$

The assumed displacement field is given by

$$\mathbf{u}_f = \begin{bmatrix}u_{f1}\\ u_{f2}\end{bmatrix} = \begin{bmatrix}\xi & 0\\ 0 & 3(\xi)^2 - 2(\xi)^3\end{bmatrix}\begin{bmatrix}q_{f1}\\ q_{f2}\end{bmatrix}$$

It follows that

$$\begin{bmatrix}u'_{f1}\\ u''_{f2}\end{bmatrix} = \begin{bmatrix}\frac{1}{l} & 0\\ 0 & \frac{1}{(l)^2}(6-12\xi)\end{bmatrix}\begin{bmatrix}q_{f1}\\ q_{f2}\end{bmatrix}$$

where $\xi = (x/l)$. Substituting the preceding equation into the strain energy expression, one gets

$$U = \frac{1}{2}\int_0^l [q_{f1} \quad q_{f2}]\begin{bmatrix}\frac{1}{l} & 0\\ 0 & \frac{(6-12\xi)}{(l)^2}\end{bmatrix}\begin{bmatrix}Ea & 0\\ 0 & EI\end{bmatrix}\begin{bmatrix}\frac{1}{l} & 0\\ 0 & \frac{(6-12\xi)}{(l)^2}\end{bmatrix}\begin{bmatrix}q_{f1}\\ q_{f2}\end{bmatrix}dx$$

Since q_{f1} and q_{f2} are only time-dependent, and $dx = l(dx/l) = ld\xi$, one can write U in the following form:

$$U = \frac{1}{2}\mathbf{q}_f^{\mathrm{T}}\mathbf{K}_{ff}\mathbf{q}_f \tag{5.107}$$

where $\mathbf{q}_f = [q_{f1} \quad q_{f2}]^{\mathrm{T}}$ and $\mathbf{K}_{ff}$ is defined as

$$\mathbf{K}_{ff} = \int_0^1 l \begin{bmatrix} \frac{1}{l} & 0 \\ 0 & \frac{(6-12\xi)}{(l)^2} \end{bmatrix} \begin{bmatrix} Ea & 0 \\ 0 & EI \end{bmatrix} \begin{bmatrix} \frac{1}{l} & 0 \\ 0 & \frac{(6-12\xi)}{(l)^2} \end{bmatrix} d\xi$$

$$= \begin{bmatrix} \frac{Ea}{l} & 0 \\ 0 & \frac{12EI}{(l)^3} \end{bmatrix} \tag{5.108}$$

Generalized External Forces The virtual work of all external forces acting on body i in the multibody system can be written in compact form as

$$\delta W_e^i = \mathbf{Q}_e^{i\mathrm{T}} \delta \mathbf{q}^i \tag{5.109}$$

where $\mathbf{Q}_e^i$ is the vector of generalized forces associated with the ith body generalized coordinates. In a partitioned form, the virtual work can be written as

$$\delta W_e^i = [\mathbf{Q}_R^{i\mathrm{T}} \quad \mathbf{Q}_\theta^{i\mathrm{T}} \quad \mathbf{Q}_f^{i\mathrm{T}}] \begin{bmatrix} \delta \mathbf{R}^i \\ \delta \boldsymbol{\theta}^i \\ \delta \mathbf{q}_f^i \end{bmatrix} \tag{5.110}$$

where $\mathbf{Q}_R^i$ and $\mathbf{Q}_\theta^i$ are the generalized forces associated, respectively, with the translational and rotational coordinates of the selected body reference, and $\mathbf{Q}_f^i$ is the vector of generalized forces associated with the elastic generalized coordinates of body i. In Eq. 109 or 110, the generalized forces may depend on the system generalized coordinates, velocities, and possibly on time.

Example 5.5 In Fig. 3, the force $\mathbf{F}^i(\mathbf{q}^i, t)$ acts at point P of the deformable body i. The force vector $\mathbf{F}^i(\mathbf{q}^i, t)$ has three components, which are defined in the global coordinate system and denoted as F_1^i, F_2^i, and F_3^i, that is,

$$\mathbf{F}^i = [F_1^i \quad F_2^i \quad F_3^i]^{\mathrm{T}}$$

The virtual work of the force $\mathbf{F}^i$ is defined as

$$\delta W_e^i = \mathbf{F}^{i\mathrm{T}} \delta \mathbf{r}_P^i$$

where $\mathbf{r}_P^i$ is the global position vector of point P and is defined as

$$\mathbf{r}_P^i = \mathbf{R}^i + \mathbf{A}^i \bar{\mathbf{u}}^i$$

where $\bar{\mathbf{u}}^i$ is the local position of point P with respect to the body coordinate system. The virtual change $\delta \mathbf{r}_P^i$ is then defined as

$$\delta \mathbf{r}_P^i = \delta \mathbf{R}^i + \frac{\partial}{\partial \boldsymbol{\theta}^i} [\mathbf{A}^i \bar{\mathbf{u}}^i] \delta \boldsymbol{\theta}^i + \mathbf{A}^i \mathbf{S}^i \delta \mathbf{q}_f^i = \delta \mathbf{R}^i + \mathbf{B}^i \delta \boldsymbol{\theta}^i + \mathbf{A}^i \mathbf{S}^i \delta \mathbf{q}_f^i$$

where $\mathbf{B}^i$ is the matrix whose columns are the partial derivatives of the vector $\mathbf{A}^i \bar{\mathbf{u}}^i$ with respect to the reference rotational coordinates. The vector $\delta \mathbf{r}_P^i$ can be written in a partitioned form as

$$\delta \mathbf{r}_P^i = [\mathbf{I} \quad \mathbf{B}^i \quad \mathbf{A}^i \mathbf{S}^i] \begin{bmatrix} \delta \mathbf{R}^i \\ \delta \boldsymbol{\theta}^i \\ \delta \mathbf{q}_f^i \end{bmatrix}$$

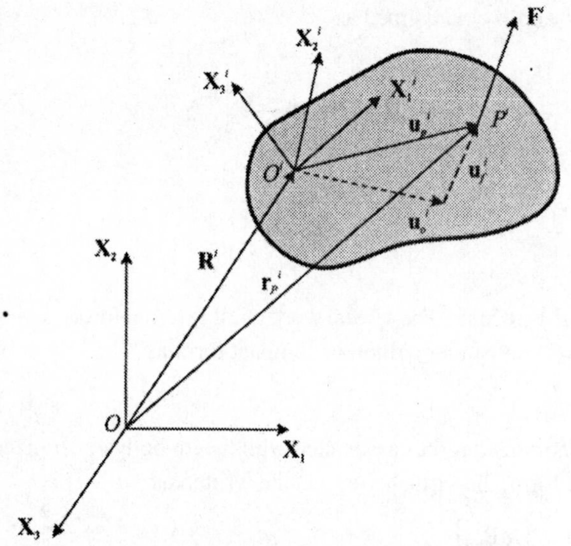

Figure 5.3 Generalized forces of the deformable body.

where the shape matrix $\mathbf{S}^i$ and the matrix $\mathbf{B}^i$ are defined at point P. Thus the virtual work δW_e^i is defined as

$$\delta W_e^i = \mathbf{F}^{i\mathrm{T}}[\mathbf{I} \quad \mathbf{B}^i \quad \mathbf{A}^i\mathbf{S}^i] \begin{bmatrix} \delta\mathbf{R}^i \\ \delta\boldsymbol{\theta}^i \\ \delta\mathbf{q}_f^i \end{bmatrix}$$

Equivalently, one can write δW_e^i as

$$\delta W_e^i = [\mathbf{Q}_R^{i\mathrm{T}} \quad \mathbf{Q}_\theta^{i\mathrm{T}} \quad \mathbf{Q}_f^{i\mathrm{T}}] \begin{bmatrix} \delta\mathbf{R}^i \\ \delta\boldsymbol{\theta}^i \\ \delta\mathbf{q}_f^i \end{bmatrix}$$

where the generalized forces $\mathbf{Q}_R^i$, $\mathbf{Q}_\theta^i$, and $\mathbf{Q}_f^i$ can be recognized as

$$\mathbf{Q}_R^{i\mathrm{T}} = \mathbf{F}^{i\mathrm{T}}, \qquad \mathbf{Q}_\theta^{i\mathrm{T}} = \mathbf{F}^{i\mathrm{T}}\mathbf{B}^i, \qquad \mathbf{Q}_f^{i\mathrm{T}} = \mathbf{F}^{i\mathrm{T}}\mathbf{A}^i\mathbf{S}^i$$

Example 5.6 Figure 4 shows a spring–damper–actuator element attached between bodies i and j in the multibody system. The attachment point on body i is P^i, while the attachment point on body j is P^j. The spring stiffness is assumed to be k, the damping coefficient is c, and the actuator force is F_a. The undeformed length of the spring is l_o. The relative position of point P^i with respect to point P^j can be expressed as

$$\mathbf{l} = \mathbf{R}^i + \mathbf{A}^i\bar{\mathbf{u}}^i - \mathbf{R}^j - \mathbf{A}^j\bar{\mathbf{u}}^j \tag{5.111}$$

where $\bar{\mathbf{u}}^i$ and $\bar{\mathbf{u}}^j$ are, respectively, the local position vectors of points P^i and P^j.

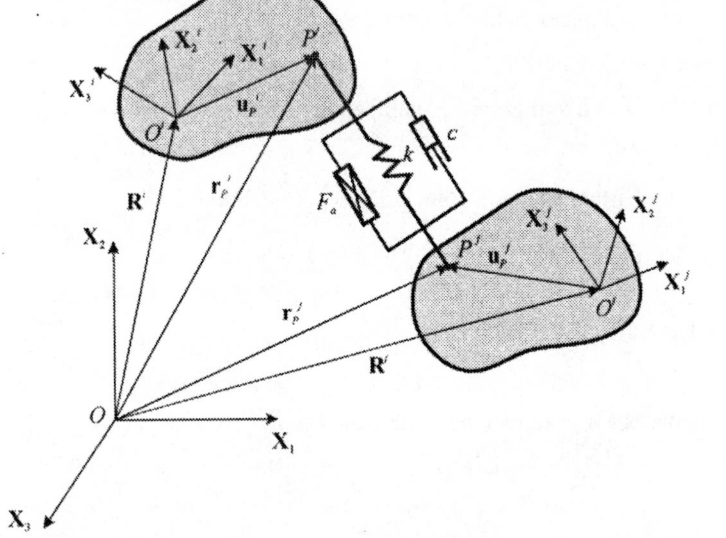

Figure 5.4 Spring–damper–actuator force element.

The vector $\mathbf{l}$ of Eq. 111 has the three components, l_1, l_2, and l_3, that is,

$$\mathbf{l} = \begin{bmatrix} l_1 & l_2 & l_3 \end{bmatrix}^{\mathrm{T}}$$

The square of length of the spring can be written as

$$\mathbf{l}^{\mathrm{T}}\mathbf{l} = (l_1)^2 + (l_2)^2 + (l_3)^2$$

We define the current length of the spring as

$$l = \sqrt{\mathbf{l}^{\mathrm{T}}\mathbf{l}} = \sqrt{(l_1)^2 + (l_2)^2 + (l_3)^2}$$

One can verify that the rate of change of the spring length can be written as

$$\dot{l} = \frac{1}{l}\mathbf{l}^{\mathrm{T}}\dot{\mathbf{l}} = \hat{\mathbf{l}}^{\mathrm{T}}\dot{\mathbf{l}}$$

where $\hat{\mathbf{l}}$ is a unit vector along the line joining points P^i and P^j and $\dot{\mathbf{l}}$ is defined as the time derivative of $\mathbf{l}$ of Eq. 111, that is,

$$\dot{\mathbf{l}} = \dot{\mathbf{R}}^i + \mathbf{B}^i\dot{\theta}^i + \mathbf{A}^i\mathbf{S}^i\dot{\mathbf{q}}_f^i - \dot{\mathbf{R}}^j - \mathbf{B}^j\dot{\theta}^j - \mathbf{A}^j\mathbf{S}^j\dot{\mathbf{q}}_f^i$$

where $\mathbf{B}^i$ and $\mathbf{B}^j$ are evaluated, respectively, at points P^i and P^j, and $\mathbf{S}^i$ and $\mathbf{S}^j$ are, respectively, the values of the shape functions at points P^i and P^j. The vector $\dot{\mathbf{l}}$ can also be written as

$$\dot{\mathbf{l}} = \mathbf{L}^i\dot{\mathbf{q}}^i - \mathbf{L}^j\dot{\mathbf{q}}^j$$

where $\mathbf{L}^i$ and $\mathbf{L}^j$ are defined as

$$\mathbf{L}^i = \begin{bmatrix} \mathbf{I} & \mathbf{B}^i & \mathbf{A}^i\mathbf{S}^i \end{bmatrix} \tag{5.112}$$

$$\mathbf{L}^j = \begin{bmatrix} \mathbf{I} & \mathbf{B}^j & \mathbf{A}^j\mathbf{S}^j \end{bmatrix} \tag{5.113}$$

Having defined the spring length and its time derivative, one can write the force along the spring–damper–actuator element as

$$F_s = k(l - l_o) + c\dot{l} + F_a$$

The virtual work due to this spring–damper–actuator force can be written as

$$\delta W_e = -F_s \delta l$$

where δl is the virtual change in the spring length defined as

$$\delta l = \frac{1}{l}\mathbf{l}^{\mathrm{T}}\delta\mathbf{l} = \hat{\mathbf{l}}^{\mathrm{T}}\delta\mathbf{l}$$

and $\delta\mathbf{l}$ is derived by using Eq. 111 as

$$\delta\mathbf{l} = \mathbf{L}^i\delta\mathbf{q}^i - \mathbf{L}^j\delta\mathbf{q}^j$$

The virtual work δW_e can then be written as

$$\begin{aligned}\delta W_e &= -F_s\hat{\mathbf{l}}^{\mathrm{T}}[\mathbf{L}^i\delta\mathbf{q}^i - \mathbf{L}^j\delta\mathbf{q}^j]\\ &= -F_s\hat{\mathbf{l}}^{\mathrm{T}}\mathbf{L}^i\delta\mathbf{q}^i + F_s\hat{\mathbf{l}}^{\mathrm{T}}\mathbf{L}^j\delta\mathbf{q}^j\end{aligned}$$

This equation can be written in the following simple form

$$\delta W_e = \mathbf{Q}^{i\mathrm{T}}\delta\mathbf{q}^i + \mathbf{Q}^{j\mathrm{T}}\delta\mathbf{q}^j$$

where the generalized force vectors $\mathbf{Q}^i$ and $\mathbf{Q}^j$ are defined as

$$\mathbf{Q}^{i\mathrm{T}} = -F_s\hat{\mathbf{l}}^{\mathrm{T}}\mathbf{L}^i, \qquad \mathbf{Q}^{j\mathrm{T}} = F_s\hat{\mathbf{l}}^{\mathrm{T}}\mathbf{L}^j$$

For body i, if we write $\mathbf{Q}^i$ in the partitioned form

$$\mathbf{Q}^i = \left[\mathbf{Q}_R^{i\mathrm{T}} \quad \mathbf{Q}_\theta^{i\mathrm{T}} \quad \mathbf{Q}_f^{i\mathrm{T}}\right]^{\mathrm{T}},$$

then

$$\mathbf{Q}_R^{i\mathrm{T}} = -F_s\hat{\mathbf{l}}^{\mathrm{T}}, \qquad \mathbf{Q}_\theta^{i\mathrm{T}} = -F_s\hat{\mathbf{l}}^{\mathrm{T}}\mathbf{B}^i, \qquad \mathbf{Q}_f^{i\mathrm{T}} = -F_s\hat{\mathbf{l}}^{\mathrm{T}}\mathbf{A}^i\mathbf{S}^i$$

Similarly,

$$\mathbf{Q}_R^{j\mathrm{T}} = F_s\hat{\mathbf{l}}^{\mathrm{T}}, \qquad \mathbf{Q}_\theta^{j\mathrm{T}} = F_s\hat{\mathbf{l}}^{\mathrm{T}}\mathbf{B}^j, \qquad \mathbf{Q}_f^{j\mathrm{T}} = F_s\hat{\mathbf{l}}^{\mathrm{T}}\mathbf{A}^j\mathbf{S}^j$$

One may observe that these generalized forces depend on both the reference motion as well as the elastic deformation of the two bodies.

5.4 KINEMATIC CONSTRAINTS

In multibody systems, the system coordinates are not independent because of the specified motion trajectories as well as mechanical joints such as universal, prismatic, and revolute joints. These kinematic constraints can be introduced to the dynamic formulation by using a set of nonlinear algebraic constraint equations that depend on the system generalized coordinates and possibly on time. One can write the vector of all kinematic constraint functions as

$$\mathbf{C}(\mathbf{q}, t) = \mathbf{0} \tag{5.114}$$

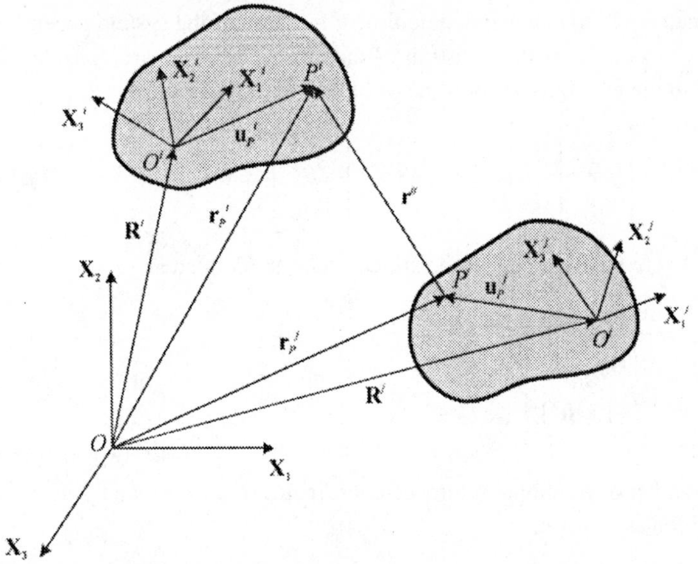

Figure 5.5 Constraints between deformable bodies.

where $\mathbf{q} = [\mathbf{q}^{1\mathrm{T}}\, \mathbf{q}^{2\mathrm{T}} \cdots \mathbf{q}^{n_b\mathrm{T}}]^{\mathrm{T}}$ is the total vector of system generalized coordinates, t is time, $\mathbf{C} = [C_1\, C_2 \cdots C_{n_c}]^{\mathrm{T}}$ is the vector of linearly independent constraint functions, and n_c is the number of constraint equations. For example, we may consider the two-body system shown in Fig. 5; one may require that the motion of point P^i on body i relative to point P^j on body j be specified, that is,

$$\mathbf{r}^{ij} = \mathbf{f}(t) \tag{5.115}$$

where $\mathbf{f}(t)$ is a time-dependent vector function and $\mathbf{r}^{ij}$ is the position vector of point P^i relative to point P^j. This relative position vector can be written as

$$\mathbf{r}^{ij} = \mathbf{R}^i + \mathbf{A}^i\bar{\mathbf{u}}^i - \mathbf{R}^j - \mathbf{A}^j\bar{\mathbf{u}}^j \tag{5.116}$$

Using Eq. 115 and writing $\bar{\mathbf{u}}^i$ and $\bar{\mathbf{u}}^j$ in a more explicit form, one obtains

$$\mathbf{R}^i + \mathbf{A}^i(\bar{\mathbf{u}}_o^i + \mathbf{S}^i\mathbf{q}_f^i) - \mathbf{R}^j - \mathbf{A}^j(\bar{\mathbf{u}}_o^j + \mathbf{S}^j\mathbf{q}_f^j) = \mathbf{f}(t) \tag{5.117}$$

where $\bar{\mathbf{u}}_o^i$ and $\bar{\mathbf{u}}_o^j$ are, respectively, the positions of points P^i and P^j in the undeformed state and $\mathbf{S}^i$ and $\mathbf{S}^j$ are the shape functions of the two bodies evaluated at points P^i and P^j, respectively. One may note that Eq. 115 or, alternatively, Eq. 117 represents three scalar equations that depend on both the reference and elastic coordinates of the two bodies as well as time. If the function $\mathbf{f}(t)$ is equal to zero, that is, points P^i and P^j coincide, Eq. 117 yields the algebraic constraint equations that describe the spherical joint in the spatial analysis and the revolute joint in the planar analysis.

For a virtual change in the system generalized coordinates, Eq. 114 yields

$$\mathbf{C}_{\mathbf{q}}\delta\mathbf{q} = \mathbf{0} \tag{5.118}$$

where $\mathbf{C}_{\mathbf{q}}$ is the constraint Jacobian matrix that has a full row rank because the constraint functions are assumed to be linearly independent. In multibody systems, the Jacobian matrix $\mathbf{C_q}$ is, in general, a nonlinear function of the system generalized coordinates. For example, if the constraint functions of Eq. 117 are considered, a virtual change in the generalized coordinates of bodies i and j yields

$$[\mathbf{I} \quad \mathbf{B}^i \quad \mathbf{A}^i\mathbf{S}^i]\begin{bmatrix}\delta\mathbf{R}^i\\ \delta\boldsymbol{\theta}^i\\ \delta\mathbf{q}_f^i\end{bmatrix} - [\mathbf{I} \quad \mathbf{B}^j \quad \mathbf{A}^j\mathbf{S}^j]\begin{bmatrix}\delta\mathbf{R}^j\\ \delta\boldsymbol{\theta}^j\\ \delta\mathbf{q}_f^j\end{bmatrix} = \mathbf{0} \tag{5.119}$$

where $\mathbf{B}^i$ and $\mathbf{B}^j$ are the matrices whose kth columns are defined as

$$\left.\begin{aligned}\text{Col}\,(\mathbf{B}^i)_k &= \frac{\partial}{\partial\theta_k^i}[\mathbf{A}^i\bar{\mathbf{u}}^i]\\ \text{Col}\,(\mathbf{B}^j)_k &= \frac{\partial}{\partial\theta_k^j}[\mathbf{A}^j\bar{\mathbf{u}}^j]\end{aligned}\right\}, \quad k = 1, 2, \ldots, n_r \tag{5.120}$$

and n_r is the number of rotational reference coordinates of bodies i and j. Equation 119 can be written as

$$\mathbf{L}^i\delta\mathbf{q}^i - \mathbf{L}^j\delta\mathbf{q}^j = \mathbf{0} \tag{5.121}$$

where $\mathbf{q}^i$ and $\mathbf{q}^j$ are, respectively, the vectors of the generalized coordinates of bodies i and j, and $\mathbf{L}^i$ and $\mathbf{L}^j$ are as defined in Eqs. 112 and 113. Equation 121 can be written in a matrix form as

$$[\mathbf{L}^i \quad -\mathbf{L}^j]\begin{bmatrix}\delta\mathbf{q}^i\\ \delta\mathbf{q}^j\end{bmatrix} = \mathbf{0}$$

where the Jacobian matrix in this case can be recognized as

$$\mathbf{C_q} = [\mathbf{L}^i \quad -\mathbf{L}^j]$$

Note that in this simple two-body example, the Jacobian matrix is a nonlinear function of the coordinates of the two bodies.

We proceed one step further and differentiate Eq. 114 with respect to time. This yields

$$\mathbf{C_q}\dot{\mathbf{q}} = -\mathbf{C}_t \tag{5.122}$$

where $\mathbf{C}_t$ is the partial derivative of the vector of constraint functions with respect to time. If the constraint functions do not depend explicitly on time, the vector $\mathbf{C}_t$ vanishes. Equation 122 is a kinematic equation that relates the generalized velocities of the multibody system. To obtain a solution for this equation, the vector $\mathbf{C}_t$ must lie in the column space of the Jacobian matrix. For the previous example of the two-body system, one can show that Eq. 122 yields

$$\mathbf{L}^i\dot{\mathbf{q}}^i - \mathbf{L}^j\dot{\mathbf{q}}^j = \mathbf{f}_t(t)$$

where $\mathbf{f}_t(t)$ is the partial derivative of $\mathbf{f}$ with respect to time.

To obtain the kinematic equations relating the accelerations in the flexible multibody system, we differentiate Eq. 122 with respect to time to yield

$$\mathbf{C_q}\ddot{\mathbf{q}} = -[\mathbf{C}_{tt} + (\mathbf{C_q}\dot{\mathbf{q}})_{\mathbf{q}}\dot{\mathbf{q}} + 2\mathbf{C}_{\mathbf{q}t}\dot{\mathbf{q}}] \tag{5.123}$$

This equation will be used in subsequent sections when the numerical solution of the nonlinear dynamic equations of motion of the flexible multibody system is discussed.

Intermediate Joint Coordinate Systems The formulation of the kinematic constraints of some joints such as revolute and cylindrical joints in the three-dimensional analysis may require introducing body fixed *intermediate joint coordinate systems* in addition to the deformable body coordinate systems. Figure 6 depicts two deformable bodies i and j, which are connected by a revolute joint. The coordinate system of body i is denoted as $\mathbf{X}_1^i\mathbf{X}_2^i\mathbf{X}_3^i$, while the coordinate system of body j is denoted as $\mathbf{X}_1^j\mathbf{X}_2^j\mathbf{X}_3^j$. As previously pointed out, the two body coordinate systems need not be rigidly attached to a material point on the two deformable bodies. To describe the joint between the two deformable bodies, one may introduce two body fixed intermediate joint coordinate systems. The first one is $\mathbf{X}_1^{iJ}\mathbf{X}_2^{iJ}\mathbf{X}_3^{iJ}$, which is rigidly attached to a joint definition point P^i on body i. The second coordinate system $\mathbf{X}_1^{jJ}\mathbf{X}_2^{jJ}\mathbf{X}_3^{jJ}$ is rigidly attached to a joint definition point P^j on body j as shown in Fig. 6. In the case of small deformation analysis, the orientation of the intermediate joint coordinate systems with respect to their body coordinate systems can be described using the infinitesimal rotation matrices $\mathbf{A}_s^i$ and $\mathbf{A}_s^j$. Let $\mathbf{v}_1^{iJ}$ and $\mathbf{v}^{jJ}$ be two constant vectors defined along the joint axis in the intermediate joint coordinate systems of bodies i and j, respectively. Using the constant vector $\mathbf{v}_1^{iJ}$, a systematic procedure can be used to define, on body i, two constant vectors $\mathbf{v}_2^{iJ}$ and $\mathbf{v}_3^{iJ}$, which are perpendicular to $\mathbf{v}_1^{iJ}$ (Shabana 1994a). The vectors $\mathbf{v}_1^{iJ}$, $\mathbf{v}_2^{iJ}$, and $\mathbf{v}_3^{iJ}$ and the vector

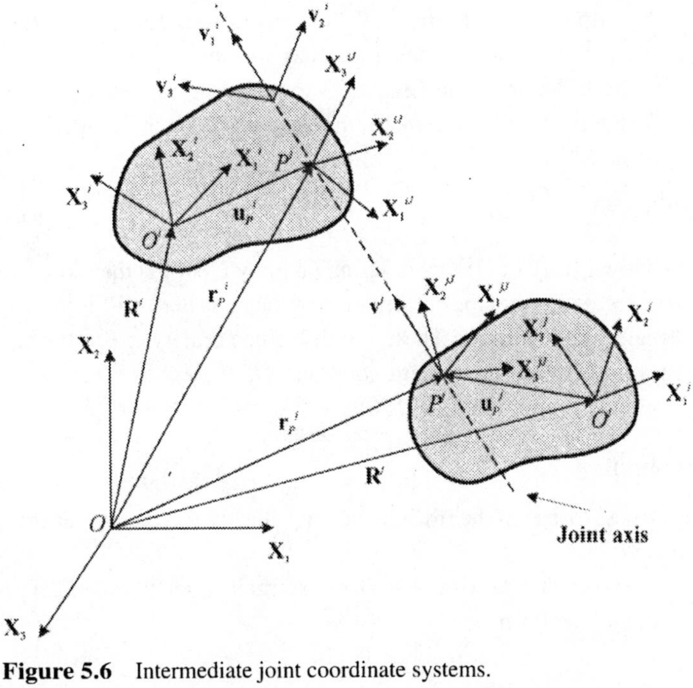

Figure 5.6 Intermediate joint coordinate systems.

$\mathbf{v}^{jJ}$ can be defined in the global coordinate system as

$$\mathbf{v}_k^i = \mathbf{A}^i \mathbf{A}_s^i \mathbf{v}_k^{iJ}, \quad k = 1, 2, 3$$

$$\mathbf{v}^j = \mathbf{A}^j \mathbf{A}_s^j \mathbf{v}^{jJ}$$

Note that these global vectors depend on the finite rotations of the body coordinate systems and the infinitesimal rotations of the intermediate joint coordinate systems. Using these vectors and the global position vectors of the joint definition points P^i and P^j, the kinematic constraint equations of the revolute joint in the three dimensional analysis can be written as

$$\mathbf{r}_P^i - \mathbf{r}_P^j = \text{constant}, \quad \mathbf{v}^{j^{\mathrm{T}}} \mathbf{v}_1^i = 0, \quad \mathbf{v}^{j\mathrm{T}} \mathbf{v}_2^i = 0$$

These are five scalar nonlinear algebraic equations that define the kinematic constraints of the revolute joint in the three-dimensional analysis. Note that by introducing the intermediate joint coordinate systems, the same form of the constraint equations used in rigid body dynamics (Shabana 1994a) can also be used to define the constraints between deformable bodies. In the case of deformable bodies, however, the constraint equations depend on the deformations as well as the reference motion of the deformable bodies. Using a procedure similar to the one described for formulating the revolute joint constraints, other types of joint constraints between deformable bodies can be formulated. This is left to the reader as an exercise.

5.5 EQUATIONS OF MOTION

Having determined, in the preceding sections, the kinetic energy of the deformable body i, the virtual work of the internal and external forces, and the kinematic constraints that describe mechanical joints as well as specified trajectories, one can use Lagrange's equation developed in Chapter 3 to write the system equations of motion of body i in the multibody system. To this end, we write the virtual work of the forces acting on body i as

$$\delta W^i = \delta W_s^i + \delta W_e^i \tag{5.124}$$

where δW^i is the virtual work of all forces acting on body i, δW_s^i is the virtual work of the elastic forces resulting from the deformation of the body, and δW_e^i is the virtual work due to externally applied forces. These forces include gravity effect, spring and damping forces acting between the system components, and control forces. It was shown in the preceding sections that (Eq. 106)

$$\delta W_s^i = -\mathbf{q}^{i\mathrm{T}} \mathbf{K}^i \delta \mathbf{q}^i \tag{5.125}$$

where $\mathbf{K}^i$ is the stiffness matrix of the ith body and $\mathbf{q}^i$ is the total vector of generalized coordinates of body i.

It has also been shown that the virtual work of externally applied forces δW_e^i can, in general, be written in the form

$$\delta W_e^i = \mathbf{Q}_e^{i\mathrm{T}} \delta \mathbf{q}^i \tag{5.126}$$

where $\mathbf{Q}_e^i$ is the vector of generalized forces associated with the generalized coordinates of body i. Equations 124–126 lead to

$$\delta W^i = -\mathbf{q}^{i\mathrm{T}}\mathbf{K}^i\delta\mathbf{q}^i + \mathbf{Q}_e^{i\mathrm{T}}\delta\mathbf{q}^i \tag{5.127}$$

This can be written as

$$\delta W^i = \mathbf{Q}^{i\mathrm{T}}\delta\mathbf{q}^i \tag{5.128}$$

where $\mathbf{Q}^i$ is the vector of generalized forces associated with the coordinates of body i and given by

$$\mathbf{Q}^i = -\mathbf{K}^i\mathbf{q}^i + \mathbf{Q}_e^i \tag{5.129}$$

For body i in the multibody system, Lagrange's equation takes the form

$$\frac{d}{dt}\left(\frac{\partial T^i}{\partial \dot{\mathbf{q}}^i}\right)^{\mathrm{T}} - \left(\frac{\partial T^i}{\partial \mathbf{q}^i}\right)^{\mathrm{T}} + \mathbf{C}_{\mathbf{q}^i}^{\mathrm{T}}\boldsymbol{\lambda} = \mathbf{Q}^i \tag{5.130}$$

where T^i is the kinetic energy of body i, $\mathbf{C}_{\mathbf{q}^i}$ is the constraint Jacobian matrix, and $\boldsymbol{\lambda}$ is the vector of Lagrange multipliers. Using the general expression of the kinetic energy of Eq. 42, we can write the first two terms on the left-hand side of Eq. 130 as

$$\frac{d}{dt}\left(\frac{\partial T^i}{\partial \dot{\mathbf{q}}^i}\right)^{\mathrm{T}} - \left(\frac{\partial T^i}{\partial \mathbf{q}^i}\right)^{\mathrm{T}} = \mathbf{M}^i\ddot{\mathbf{q}}^i + \dot{\mathbf{M}}^i\dot{\mathbf{q}}^i - \left[\frac{\partial}{\partial \mathbf{q}^i}\left(\frac{1}{2}\dot{\mathbf{q}}^{i\mathrm{T}}\mathbf{M}^i\dot{\mathbf{q}}^i\right)\right]^{\mathrm{T}} \tag{5.131}$$

We may define $\mathbf{Q}_v^i$ to be

$$\mathbf{Q}_v^i = -\dot{\mathbf{M}}^i\dot{\mathbf{q}}^i + \frac{1}{2}\left[\frac{\partial}{\partial \mathbf{q}^i}(\dot{\mathbf{q}}^{i\mathrm{T}}\mathbf{M}^i\dot{\mathbf{q}}^i)\right]^{\mathrm{T}}$$

where $\mathbf{Q}_v^i$ is a quadratic velocity vector resulting from the differentiation of the kinetic energy with respect to time and with respect to the body coordinates. This quadratic velocity vector contains the gyroscopic and Coriolis force components. Equation 131 can then be written as

$$\frac{d}{dt}\left(\frac{\partial T^i}{\partial \dot{\mathbf{q}}^i}\right)^{\mathrm{T}} - \left(\frac{\partial T^i}{\partial \mathbf{q}^i}\right)^{\mathrm{T}} = \mathbf{M}^i\ddot{\mathbf{q}}^i - \mathbf{Q}_v^i \tag{5.132}$$

With the use of this equation and Eq. 129, Eq. 130 leads to

$$\mathbf{M}^i\ddot{\mathbf{q}}^i + \mathbf{K}^i\mathbf{q}^i + \mathbf{C}_{\mathbf{q}^i}^{\mathrm{T}}\boldsymbol{\lambda} = \mathbf{Q}_e^i + \mathbf{Q}_v^i, \quad i = 1, 2 \ldots, n_b \tag{5.133a}$$

where n_b is the total number of bodies in the multibody system. Equation 133a can be written in a partitioned matrix form as

$$\begin{bmatrix} \mathbf{m}_{RR}^i & \mathbf{m}_{R\theta}^i & \mathbf{m}_{Rf}^i \\ & \mathbf{m}_{\theta\theta}^i & \mathbf{m}_{\theta f}^i \\ \text{symmetric} & & \mathbf{m}_{ff}^i \end{bmatrix} \begin{bmatrix} \ddot{\mathbf{R}}^i \\ \ddot{\boldsymbol{\theta}}^i \\ \ddot{\mathbf{q}}_f^i \end{bmatrix} + \begin{bmatrix} \mathbf{0} & \mathbf{0} & \mathbf{0} \\ \mathbf{0} & \mathbf{0} & \mathbf{0} \\ \mathbf{0} & \mathbf{0} & \mathbf{K}_{ff}^i \end{bmatrix} \begin{bmatrix} \mathbf{R}^i \\ \boldsymbol{\theta}^i \\ \mathbf{q}_f^i \end{bmatrix} + \begin{bmatrix} \mathbf{C}_{\mathbf{R}^i}^{\mathrm{T}} \\ \mathbf{C}_{\boldsymbol{\theta}^i}^{\mathrm{T}} \\ \mathbf{C}_{\mathbf{q}_f^i}^{\mathrm{T}} \end{bmatrix} \boldsymbol{\lambda}$$

$$= \begin{bmatrix} (\mathbf{Q}_e^i)_R \\ (\mathbf{Q}_e^i)_\theta \\ (\mathbf{Q}_e^i)_f \end{bmatrix} + \begin{bmatrix} (\mathbf{Q}_v^i)_R \\ (\mathbf{Q}_v^i)_\theta \\ (\mathbf{Q}_v^i)_f \end{bmatrix}, \quad i = 1, 2, \ldots, n_b \tag{5.133b}$$

Equation 133 is a system of second-order differential equations whose solution has to satisfy the algebraic constraint equations describing mechanical joints in the multibody system as well as specified trajectories and can be written in the following vector form:

$$\mathbf{C}(\mathbf{q}, t) = \mathbf{0} \tag{5.134}$$

Equations 133 and 134 are a mixed system of differential and algebraic equations that have to be solved simultaneously. In large-scale multibody systems, a closed-form solution of such a system of equations is difficult to obtain. Therefore, one usually resorts to numerical methods. Some of the numerical techniques used in solving this system of equations are discussed in later sections.

In a general multibody system consisting of interconnected rigid and deformable components, the number of coordinates and accordingly the number of differential and algebraic equations can be quite large. The objective of solving these equations is to determine the system coordinates, velocities, and accelerations as well as the vector of Lagrange multipliers that can be used to determine the generalized reaction forces given by $\mathbf{C}_{\mathbf{q}^i}^{\mathrm{T}}\boldsymbol{\lambda}$. By use of Lagrange multipliers, the multibody system equations of motion are formulated in terms of both the independent and dependent variables. The advantage of using this approach appears when general-purpose computer programs are developed for the dynamic analysis of large-scale multibody systems. In this case, general forcing functions and constraint equations that may depend on the system-dependent and independent coordinates and velocities can be introduced to the dynamic formulation in a straightforward manner. It is important, however, to point out that if the system of constraint equations is holonomic or nonholonomic, the use of Lagrange multipliers is not necessary. An alternate approach, discussed in Chapter 3, is to identify a set of dependent variables and write these variables in terms of the independent ones using the generalized coordinate partitioning of the constraint Jacobian matrix. In this case the system differential equations of motion can be written in terms of the independent variables only. These equations can be integrated forward in time by using direct numerical integration methods. The solution of the independent differential equations defines the independent variables. Dependent variables are determined by using the kinematic relations.

Quadratic Velocity Vector In the preceding sections, we discussed in detail the components of the mass matrix and pointed out the nonlinearities that arise because of the coupling between the rigid body motion and elastic deformation. We have also outlined methods for evaluating the stiffness matrix, the Jacobian matrix, and the generalized force vector. Using Eq. 131, one can also show that the quadratic velocity vector $\mathbf{Q}_v^i$ can be defined as

$$\mathbf{Q}_v^i = \left[(\mathbf{Q}_v^i)_R^{\mathrm{T}} \ (\mathbf{Q}_v^i)_\theta^{\mathrm{T}} \ (\mathbf{Q}_v^i)_f^{\mathrm{T}}\right]^{\mathrm{T}} \tag{5.135}$$

where in the *planar analysis*, the components of this vector are defined in terms of the inertia shape integrals of Eqs. 94 and 95 as

$$\left.\begin{aligned}(\mathbf{Q}_v^i)_R &= (\dot{\theta}^i)^2\mathbf{A}^i(\bar{\mathbf{S}}^i\mathbf{q}_f^i + \mathbf{I}_1^i) - 2\dot{\theta}^i\mathbf{A}_\theta^i\bar{\mathbf{S}}^i\dot{\mathbf{q}}_f^i\\ (\mathbf{Q}_v^i)_\theta &= -2\dot{\theta}^i\dot{\mathbf{q}}_f^{i\mathrm{T}}(\mathbf{m}_{ff}^i\mathbf{q}_f^i + \bar{\mathbf{I}}_o^i)\\ (\mathbf{Q}_v^i)_f &= (\dot{\theta}^i)^2(\mathbf{m}_{ff}^i\mathbf{q}_f^i + \bar{\mathbf{I}}_o^i) + 2\dot{\theta}^i\tilde{\mathbf{S}}^i\dot{\mathbf{q}}_f^i\end{aligned}\right\} \tag{5.136}$$

where $\mathbf{A}^i$ and $\mathbf{A}_\theta^i$ are, respectively, defined by Eqs. 5a and 80, and θ^i is the rotation angle of the body reference about the axis of rotation. The matrices $\mathbf{I}_1^i$, $\bar{\mathbf{S}}^i$, $\tilde{\mathbf{S}}^i$, and $\mathbf{m}_{ff}^i$ are defined in Eqs. 82, 95, 93, and 96, respectively. The vector $\bar{\mathbf{I}}_o^i$ is defined as

$$\bar{\mathbf{I}}_o^i = \int_{V^i} \rho^i\mathbf{S}^{i\mathrm{T}}\bar{\mathbf{u}}_o^i\,dV^i = (\bar{\mathbf{I}}_{11}^i + \bar{\mathbf{I}}_{22}^i)^{\mathrm{T}} \tag{5.137}$$

in which ρ^i and V^i are, respectively, the mass density and volume of body i, $\mathbf{S}^i$ is the ith body shape matrix, and $\bar{\mathbf{u}}_o^i = [x_1^i \;\; x_2^i]^{\mathrm{T}}$ is the local position of an arbitrary point on the body in the undeformed state. To simplify the term $(\mathbf{Q}_v^i)_\theta$ of Eq. 136, advantage is taken of the fact that

$$\dot{\mathbf{q}}_f^{i\mathrm{T}}\tilde{\mathbf{S}}^i\dot{\mathbf{q}}_f^i = 0$$

because $\tilde{\mathbf{S}}^i$ is a skew symmetric matrix.

In the *three-dimensional analysis* the components of the vector $\mathbf{Q}_v^i$ of Eq. 135 are defined as

$$\left.\begin{aligned}(\mathbf{Q}_v^i)_R &= -\mathbf{A}^i[(\tilde{\bar{\omega}}^i)^2\bar{\mathbf{S}}_t^i + 2\tilde{\bar{\omega}}^i\bar{\mathbf{S}}^i\dot{\mathbf{q}}_f^i]\\ (\mathbf{Q}_v^i)_\theta &= -2\dot{\bar{\mathbf{G}}}^{i\mathrm{T}}\bar{\mathbf{I}}_{\theta\theta}^i\bar{\omega}^i - 2\dot{\bar{\mathbf{G}}}^{i\mathrm{T}}\bar{\mathbf{I}}_{\theta f}^i\dot{\mathbf{q}}_f^i - \bar{\mathbf{G}}^{i\mathrm{T}}\dot{\bar{\mathbf{I}}}_{\theta\theta}^i\bar{\omega}^i\\ (\mathbf{Q}_v^i)_f &= -\int_{V^i}\rho^i\{\mathbf{S}^{i\mathrm{T}}[(\tilde{\bar{\omega}}^i)^2\bar{\mathbf{u}}^i + 2\tilde{\bar{\omega}}^i\dot{\bar{\mathbf{u}}}_f^i]\}dV^i\end{aligned}\right\} \tag{5.138}$$

where $\bar{\mathbf{S}}_t^i$, $\bar{\mathbf{S}}^i$, $\bar{\mathbf{I}}_{\theta\theta}^i$, and $\bar{\mathbf{I}}_{\theta f}^i$ are defined, respectively, by Eqs. 59, 63, 69, and 73. Clearly, the quadratic velocity vector that includes the effect of Coriolis and centrifugal forces is a nonlinear function of the system generalized coordinates and velocities. One can also obtain this vector by using the acceleration vector of Eq. 38 and the expression for the virtual work of the inertia forces, which is defined as

$$\delta W_i^i = \int_{V^i}\rho^i\delta\mathbf{r}^{i\mathrm{T}}\ddot{\mathbf{r}}^i dV^i \tag{5.139}$$

This also leads to the definition of the mass matrix as well as the Coriolis and centrifugal forces.

Generalized Newton–Euler Equations Equation 133a is expressed in terms of the generalized coordinates of the deformable bodies. The use of the generalized rotational coordinates, however, is not convenient in developing recursive formulations for multibody systems. In this section, an alternate form for the dynamic equation of motion of deformable multibody system is presented in terms of the angular acceleration vector of the deformable body reference.

In the case of Euler parameters, it can be shown that the relationship between the angular acceleration vector and the time derivative of Euler parameters is given by

$$\bar{\alpha}^i = \bar{\mathbf{G}}^i \ddot{\theta}^i$$

where $\bar{\alpha}^i$ is the angular acceleration vector of the reference of the deformable body i defined in the body coordinate system, $\ddot{\theta}^i$ is the time derivative of Euler parameters, and $\bar{\mathbf{G}}^i$ is a matrix that depends linearly on Euler parameters. In Chapter 3, it is shown how the preceding equation can be used to obtain Newton–Euler equations for rigid bodies that undergo finite rotations. Using a similar procedure, it is left to the reader to show that by relaxing the assumptions of Newton–Euler equations for the rigid bodies, the use of the preceding equation and Eq. 133b leads to the following *generalized Newton–Euler equations* for the unconstrained motion of the deformable body i that undergoes large reference displacements:

$$\begin{bmatrix} \mathbf{m}^i_{RR} & \mathbf{A}^i\bar{\tilde{\mathbf{S}}}^{iT}_t & \mathbf{A}^i\bar{\mathbf{S}}^i \\ & \bar{\mathbf{I}}^i_{\theta\theta} & \bar{\mathbf{I}}^i_{\theta f} \\ \text{symmetric} & & \mathbf{m}^i_{ff} \end{bmatrix} \begin{bmatrix} \ddot{\mathbf{R}}^i \\ \bar{\alpha}^i \\ \ddot{\mathbf{q}}^i_f \end{bmatrix} = \begin{bmatrix} (\mathbf{Q}^i_e)_R \\ (\mathbf{Q}^i_e)_\alpha \\ (\mathbf{Q}^i_e)_f - \mathbf{K}^i_{ff}\mathbf{q}^i_f \end{bmatrix} + \begin{bmatrix} (\mathbf{Q}^i_v)_R \\ (\mathbf{Q}^i_v)_\alpha \\ (\mathbf{Q}^i_v)_f \end{bmatrix} \tag{5.140}$$

where $\mathbf{A}^i$ is the orthogonal rotation matrix; $\mathbf{m}^i_{RR}$, $\bar{\tilde{\mathbf{S}}}^i_t$, $\bar{\mathbf{S}}^i$, $\bar{\mathbf{I}}^i_{\theta\theta}$, $\bar{\mathbf{I}}^i_{\theta f}$, and $\mathbf{m}^i_{ff}$ are defined, respectively, by Eqs. 52, 59, 63, 69, 73, and 46f; $(\mathbf{Q}^i_e)_\alpha$ is the vector of actual moments; and $(\mathbf{Q}^i_v)_\alpha$ is the quadratic velocity vector associated with rotation of the deformable body i and is given by

$$(\mathbf{Q}^i_v)_\alpha = -\bar{\omega}^i \times (\bar{\mathbf{I}}^i_{\theta\theta}\bar{\omega}^i) - \dot{\bar{\mathbf{I}}}^i_{\theta\theta}\bar{\omega}^i - \bar{\omega}^i \times (\bar{\mathbf{I}}^i_{\theta f}\dot{\mathbf{q}}^i_f) \tag{5.141}$$

The other components of the generalized Newton–Euler equations are the same as previously defined in this section.

One can show that by using the appropriate assumptions, the generalized Newton-Euler equations can be reduced to Newton–Euler equations used in the dynamic analysis of rigid bodies. For instance, if we assume that the body i is rigid, the generalized Newton–Euler equations reduce to

$$\begin{bmatrix} \mathbf{m}^i_{RR} & \mathbf{A}^i\bar{\tilde{\mathbf{S}}}^{iT}_t \\ \text{symmetric} & \bar{\mathbf{I}}^i_{\theta\theta} \end{bmatrix} \begin{bmatrix} \ddot{\mathbf{R}}^i \\ \bar{\alpha}^i \end{bmatrix} = \begin{bmatrix} (\mathbf{Q}^i_e)_R \\ (\mathbf{Q}^i_e)_\alpha \end{bmatrix} + \begin{bmatrix} (\mathbf{Q}^i_v)_R \\ (\mathbf{Q}^i_v)_\alpha \end{bmatrix}$$

Furthermore, if we assume that the origin of the rigid body reference is attached to the center of mass of the body, one has

$$\bar{\tilde{\mathbf{S}}}^i_t = \mathbf{0}$$

and the preceding equations reduce to

$$\begin{bmatrix} \mathbf{m}^i_{RR} & \mathbf{0} \\ \mathbf{0} & \bar{\mathbf{I}}^i_{\theta\theta} \end{bmatrix} \begin{bmatrix} \ddot{\mathbf{R}}^i \\ \bar{\alpha}^i \end{bmatrix} = \begin{bmatrix} (\mathbf{Q}^i_e)_R \\ (\mathbf{Q}^i_e)_\alpha \end{bmatrix} + \begin{bmatrix} \mathbf{0} \\ -\bar{\omega}^i \times (\bar{\mathbf{I}}^i_{\theta\theta}\bar{\omega}^i) \end{bmatrix}$$

which are the same Newton–Euler equations obtained in Chapter 3.

The use of the generalized Newton–Euler equations may prove useful in developing recursive formulations for multibody systems consisting of inter-connected rigid and deformable bodies. One needs only to develop space-independent kinematic relationships in which the absolute coordinates are expressed in terms of the joint variables. A set of *body-fixed intermediate joint coordinate systems* may be introduced at the joint definition points, for the convenience of describing the large relative displacements between deformable bodies.

5.6 COUPLING BETWEEN REFERENCE AND ELASTIC DISPLACEMENTS

In the floating frame of reference formulation described in this chapter, the configuration of each deformable body in the multibody system is identified by using two coupled sets of generalized coordinates: reference and elastic coordinates. Reference coordinates define the location and orientation of the deformable body reference, while elastic coordinates define the deformation of the body with respect to the body coordinate system. The use of the mixed set of reference and elastic coordinates in the floating frame of reference formulation leads to a highly nonlinear mass matrix as a result of the inertia coupling between the reference and the elastic displacements.

Inertia Shape Integrals It is also shown in this chapter that even though the mass matrix is highly nonlinear, the deformable body inertia can be defined in terms of a set of inertia shape integrals that depend on the assumed displacement field. These shape integrals are

$$\mathbf{I}_1^i = \int_{V^i} \rho^i \bar{\mathbf{u}}_o^i \, dV^i, \qquad I_{kl}^i = \int_{V^i} \rho^i x_k^i x_l^i \, dV^i, \quad k, l = 1, 2, 3 \tag{5.142}$$

$$\bar{\mathbf{S}}^i = \int_{V^i} \rho^i \mathbf{S}^i \, dV^i, \qquad \bar{\mathbf{S}}_{kl}^i = \int_{V^i} \rho^i \mathbf{S}_k^{i\mathrm{T}} \mathbf{S}_l^i \, dV^i, \quad k, l = 1, 2, 3 \tag{5.143}$$

$$\bar{\mathbf{I}}_{kl}^i = \int_{V^i} \rho^i x_k^i \mathbf{S}_l^i \, dV^i, \quad k, l = 1, 2, 3 \tag{5.144}$$

where $\bar{\mathbf{u}}_o^i = [x_1^i \;\; x_2^i \;\; x_3^i]^{\mathrm{T}}$ is the undeformed position of an arbitrary point on the deformable body i in the multibody system; ρ^i and V^i are, respectively, the mass density and volume of body i; and $\mathbf{S}_k^i$ is the kth row in the body shape function $\mathbf{S}^i$. In the special case of rigid body analysis the shape integrals are given by Eq. 142 only; these are the same shape integrals defined in Chapter 3. On the other hand, in the case of structural systems, wherein the reference motion is not allowed, the constant mass matrix $\mathbf{m}_{ff}^i$ that appears in the linear dynamics of the structural systems can be obtained by using the shape integrals of Eq. 143 as

$$\mathbf{m}_{ff}^i = \int_{V^i} \rho^i \mathbf{S}^{i\mathrm{T}} \mathbf{S}^i \, dV^i = \bar{\mathbf{S}}_{11}^i + \bar{\mathbf{S}}_{22}^i + \bar{\mathbf{S}}_{33}^i \tag{5.145}$$

The identification of the shape integrals of Eqs. 142–144 is useful in the computational mechanics of multibody systems consisting of interconnected rigid and deformable bodies. A general computational algorithm for multibody systems that contain rigid and structural components can be developed. In this computational algorithm the shape integrals of Eqs. 142–144 can be evaluated only once in advance for the dynamic analysis. Furthermore, the structures of the nonlinear mass matrix and the quadratic velocity vector that contains the Coriolis and gyroscopic force components remain the same when lumped mass techniques are used. One only needs to express the inertia shape integrals of Eqs. 142–144 in their lumped mass form. In this case the shape integrals are given by

$$
\begin{aligned}
\mathbf{I}_1^i &= \sum_{j=1}^{n_j} m^{ij}\bar{\mathbf{u}}_o^{ij}, \qquad I_{kl}^i = \sum_{j=1}^{n_j} m^{ij} x_k^{ij} x_l^{ij}, \quad k, l = 1, 2, 3 \\
\bar{\mathbf{S}}^i &= \sum_{j=1}^{n_j} m^{ij}\mathbf{N}^{ij}, \qquad \bar{\mathbf{S}}_{kl}^i = \sum_{j=1}^{n_j} m^{ij}\mathbf{N}_k^{ij\mathrm{T}}\mathbf{N}_l^{ij}, \quad k, l = 1, 2, 3 \\
\bar{\mathbf{I}}_{kl}^i &= \sum_{j=1}^{n_j} m^{ij} x_k^{ij}\mathbf{N}_l^{ij}, \quad k, l = 1, 2, 3
\end{aligned}
$$

where $\bar{\mathbf{u}}_o^{ij} = [x_1^{ij}\ \ x_2^{ij}\ \ x_3^{ij}]^{\mathrm{T}}$ is the undeformed position vector of the grid point j, m^{ij} is the mass of the grid point j, $\mathbf{N}_k^{ij}$ is the kth row of the matrix $\mathbf{N}^{ij}$ defined in Section 2 of this chapter, and n_j is the total number of the grid points.

Linear Theory of Elastodynamics The dynamics of multibody systems with deformable components has been a subject of interest in many different fields such as machine design and aerospace, for they represent many industrial and technological applications such as robotic manipulators, vehicle systems, and space structures. It was seen, however, that the dynamic equations of such systems are highly nonlinear because of the finite rotation of the deformable body reference. A solution strategy that has been used in the past is to treat the multibody system first as a collection of rigid bodies. General-purpose multi-rigid-body computer programs can then be used to solve for the inertia and reaction forces. These inertia and reaction forces obtained from the rigid body analysis are then introduced to a linear elasticity problem to solve for the deformations of the bodies in the multibody systems. The total motion of a body is then obtained by superimposing the small elastic deformation on the gross rigid body motion. This approach is usually referred to as the *linear theory of elastodynamics*. In this approach, rigid body motion and elastic deformation are not solved for simultaneously. Furthermore, the effect of the elastic deformation on rigid body motion is neglected. This assumption, however, may not be valid when high-speed, lightweight mechanical systems are considered. The effect of the coupling between the elastic deformation and the gross rigid body motion can be significant.

To understand the dynamic formulation based on the linear theory of elastodynamics, we write Eq. 133a in the following partitioned form:

$$
\begin{bmatrix} \mathbf{m}_{rr}^i & \mathbf{m}_{rf}^i \\ \mathbf{m}_{fr}^i & \mathbf{m}_{ff}^i \end{bmatrix}
\begin{bmatrix} \ddot{\mathbf{q}}_r^i \\ \ddot{\mathbf{q}}_f^i \end{bmatrix}
+ \begin{bmatrix} \mathbf{0} & \mathbf{0} \\ \mathbf{0} & \mathbf{K}_{ff}^i \end{bmatrix}
\begin{bmatrix} \mathbf{q}_r^i \\ \mathbf{q}_f^i \end{bmatrix}
= \begin{bmatrix} \bar{\mathbf{Q}}_r^i \\ \bar{\mathbf{Q}}_f^i \end{bmatrix}
\tag{5.146}
$$

where $\mathbf{q}_r^i = [\mathbf{R}^{i\mathrm{T}} \ \boldsymbol{\theta}^{i\mathrm{T}}]^\mathrm{T}$ is the vector of reference coordinates of body i; subscripts r and f refer, respectively, to reference and elastic coordinates; and $\bar{\mathbf{Q}}^i$ is the vector of generalized forces, including the external forces, reaction forces, and the quadratic velocity force vector $\mathbf{Q}_v^i$ of Eq. 133a, that is,

$$\bar{\mathbf{Q}}^i = \mathbf{Q}_e^i + \mathbf{Q}_v^i - \mathbf{C}_{\mathbf{q}^i}^\mathrm{T} \boldsymbol{\lambda}$$

Equation 146 yields the following two matrix equations:

$$\mathbf{m}_{rr}^i \ddot{\mathbf{q}}_r^i + \mathbf{m}_{rf}^i \ddot{\mathbf{q}}_f^i = \bar{\mathbf{Q}}_r^i \tag{5.147}$$

$$\mathbf{m}_{fr}^i \ddot{\mathbf{q}}_r^i + \mathbf{m}_{ff}^i \ddot{\mathbf{q}}_f^i + \mathbf{K}_{ff}^i \mathbf{q}_f^i = \bar{\mathbf{Q}}_f^i \tag{5.148}$$

In the linear theory of elastodynamics, the term $\mathbf{m}_{rf}^i \ddot{\mathbf{q}}_f^i$ in Eq. 147 is neglected. Furthermore, the matrix $\mathbf{m}_{rr}^i$ and the vector $\bar{\mathbf{Q}}_r^i$ are assumed not to depend on the elastic deformation of the body. Using these assumptions, one can write Eqs. 147 and 148 as

$$\mathbf{m}_{rr}^i \ddot{\mathbf{q}}_r^i = \bar{\mathbf{Q}}_r^i \tag{5.149}$$

$$\mathbf{m}_{ff}^i \ddot{\mathbf{q}}_f^i + \mathbf{K}_{ff}^i \mathbf{q}_f^i = \bar{\mathbf{Q}}_f^i - \mathbf{m}_{fr}^i \ddot{\mathbf{q}}_r^i \tag{5.150}$$

Equation 149 can be solved for the reference coordinates, velocities, and accelerations using rigid multibody computer programs. The information obtained from solving Eq. 149 can then be substituted into Eq. 150 in order to obtain a linear structural problem. Equation 150 can then be solved for the vector $\mathbf{q}_f^i$ by using any of the existing linear structural dynamics programs.

Deformable Body Axes As shown in this chapter, the nonlinear inertia coupling between the reference motion and the elastic deformation depends on the finite reference rotations as well as the elastic deformation of the body. Many attempts have been made in the past to simplify these coupling terms by a proper selection of the deformable body axes. In many of these investigations, deformable body references that satisfy the *mean axis conditions* were chosen. These conditions are obtained by minimizing the relative kinetic energy of the deformable body with respect to an observer stationed on the body. Applying the deformable body mean axis conditions leads to a weak coupling between the reference motion and the elastic deformation. It is important, however, to point out that, while any coordinate system can be selected for the deformable body, the deformation modes resulting from the application of the mean axis conditions may be suitable only in the dynamic analysis of specific applications (Shabana 1996a).

Shape Functions One of the questions that remain unanswered is how to select the appropriate displacement field or the shape functions for the deformable bodies, especially when the body has complex geometry or nonlinear boundary conditions. The finite-element method can be used to alleviate many of the problems of the Rayleigh–Ritz method. In the following chapter, a finite element floating frame of reference formulation is discussed. This formulation can be used to model deformable bodies with complex geometrical shapes. It can also be used to obtain exact modeling of the rigid body dynamics. The formulation presented in Chapter 6 shows that, by using finite-element shape functions that can describe an arbitrary rigid body translation, the number of inertia shape integrals required to evaluate the nonlinear terms that represent the coupling between the reference and elastic displacements can be significantly reduced. As described in the following chapter, the component mode synthesis method can also be used to transform the inertia shape integrals to the modal space, thereby eliminating many of the degrees of freedom associated with high-frequency modes of vibration. By evaluating the inertia shape integrals using the finite element method, the structure of the equations of motion developed in this chapter can be used in developing general purpose flexible multibody computer programs that can model deformable bodies that have complex geometrical shapes.

5.7 APPLICATION TO A MULTIBODY SYSTEM

The planar slider crank mechanism shown in Fig. 7 consists of four bodies: the fixed link (ground), denoted as body 1, the crankshaft *OA* (body 2), the connecting rod *AB* (body 3), and the slider block (body 4) at *B*. All bodies are assumed to be rigid except the flexible connecting rod *AB*, which has length l, mass density ρ, and modulus of elasticity E. The connecting rod *AB* is connected to the crankshaft and the slider block by means of revolute joints. For simplicity, the superscript that indicates the body number will be omitted in the discussion that follows with the understanding that all vectors and matrices are associated with the deformable connecting rod.

Assumed Displacement Field The displacement field of the connecting rod is assumed to be

$$\bar{\mathbf{u}}_f = \mathbf{S}_b \bar{\mathbf{q}}_f$$

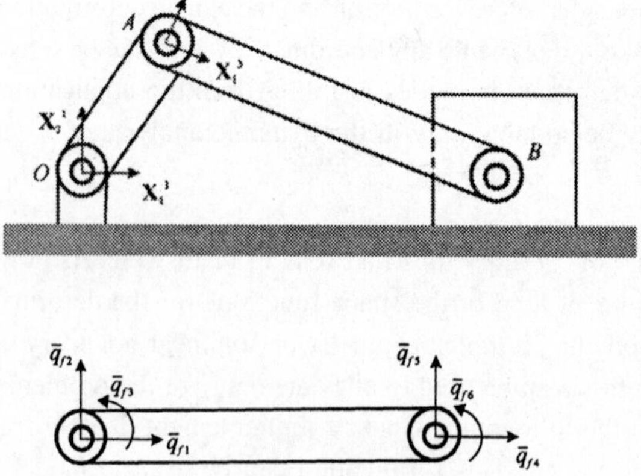

Figure 5.7 Planar slider crank mechanism.

where $\bar{\mathbf{u}}_f = [\bar{u}_{f1} \; \bar{u}_{f2}]^{\mathrm{T}}$ is the displacement vector, $\mathbf{S}_b$ is the shape matrix defined as

$$\mathbf{S}_b = \begin{bmatrix} 1-\xi & 0 & 0 & \xi & 0 & 0 \\ 0 & 1-3(\xi)^2+2(\xi)^3 & l[\xi-2(\xi)^2+(\xi)^3] & 0 & 3(\xi)^2-2(\xi)^3 & l[(\xi)^3-(\xi)^2] \end{bmatrix}$$

$\xi = (x/l)$, and the vector of elastic coordinates $\bar{\mathbf{q}}_f$ is

$$\bar{\mathbf{q}}_f = \begin{bmatrix} \bar{q}_{f1} & \bar{q}_{f2} & \bar{q}_{f3} & \bar{q}_{f4} & \bar{q}_{f5} & \bar{q}_{f6} \end{bmatrix}^{\mathrm{T}}$$

in which $\bar{q}_{f1}$ and $\bar{q}_{f4}$ are the axial displacements of the endpoints, $\bar{q}_{f2}$ and $\bar{q}_{f5}$ are the transverse displacements of the endpoints, and $\bar{q}_{f3}$ and $\bar{q}_{f6}$ are the slopes at the endpoints. One may observe that the assumed displacement field contains the *rigid body modes*. To define a unique displacement field with respect to the connecting rod reference whose origin is selected to be at point A, a set of *reference conditions* has to be imposed. It is clear that the connecting rod can be modeled as a simply supported beam. The simply supported end conditions imply that

$$\bar{u}_{f1}(0) = 0, \qquad \bar{u}_{f2}(0) = 0, \qquad \bar{u}_{f2}(l) = 0$$

which implies that

$$\bar{q}_{f1} = 0, \qquad \bar{q}_{f2} = 0, \qquad \bar{q}_{f5} = 0$$

Therefore, one can define the set of coordinates

$$\mathbf{q}_f = [\bar{q}_{f3} \quad \bar{q}_{f4} \quad \bar{q}_{f6}]^{\mathrm{T}}$$

and write $\bar{\mathbf{q}}_f$ in terms of $\mathbf{q}_f$ as

$$\bar{\mathbf{q}}_f = \mathbf{B}_r \mathbf{q}_f$$

or in a more explicit form as

$$\begin{bmatrix} \bar{q}_{f1} \\ \bar{q}_{f2} \\ \bar{q}_{f3} \\ \bar{q}_{f4} \\ \bar{q}_{f5} \\ \bar{q}_{f6} \end{bmatrix} = \begin{bmatrix} 0 & 0 & 0 \\ 0 & 0 & 0 \\ 1 & 0 & 0 \\ 0 & 1 & 0 \\ 0 & 0 & 0 \\ 0 & 0 & 1 \end{bmatrix} \begin{bmatrix} \bar{q}_{f3} \\ \bar{q}_{f4} \\ \bar{q}_{f6} \end{bmatrix}$$

in which the matrix $\mathbf{B}_r$ of the reference conditions is recognized as

$$\mathbf{B}_r = \begin{bmatrix} 0 & 0 & 0 \\ 0 & 0 & 0 \\ 1 & 0 & 0 \\ 0 & 1 & 0 \\ 0 & 0 & 0 \\ 0 & 0 & 1 \end{bmatrix}$$

One can then write the displacement field $\bar{\mathbf{u}}_f$ as

$$\bar{\mathbf{u}}_f = \mathbf{S}_b \bar{\mathbf{q}}_f = \mathbf{S}_b \mathbf{B}_r \mathbf{q}_f = \mathbf{S} \mathbf{q}_f$$

where the shape matrix $\mathbf{S}$ is defined as

$$\mathbf{S} = \mathbf{S}_b\mathbf{B}_r = \begin{bmatrix} 0 & \xi & 0 \\ l(\xi - 2(\xi)^2 + (\xi)^3) & 0 & l((\xi)^3 - (\xi)^2) \end{bmatrix}$$

This shape matrix defines a unique displacement field with respect to the body reference.

Mass Matrix of the Connecting Rod Let the vector $\mathbf{q}_r = [R_1 \quad R_2 \quad \theta]^{\mathrm{T}}$ denote the set of reference coordinates that define the location and orientation of the connecting rod reference shown in Fig. 7. According to this generalized coordinate partitioning, the mass matrix of the flexible connecting rod can be written as

$$\mathbf{M} = \begin{bmatrix} \mathbf{m}_{RR} & \mathbf{m}_{R\theta} & \mathbf{m}_{Rf} \\ & \mathbf{m}_{\theta\theta} & \mathbf{m}_{\theta f} \\ \text{symmetric} & & \mathbf{m}_{ff} \end{bmatrix}$$

where

$$\mathbf{m}_{RR} = \begin{bmatrix} m & 0 \\ 0 & m \end{bmatrix}$$

and m is the total mass of the connecting rod. The vector $\mathbf{I}_1$ and the matrix $\bar{\mathbf{S}}$ of Eq. 82 are given by

$$\mathbf{I}_1 = \int_V \rho \bar{\mathbf{u}}_o \, dV = \int_0^l \rho \begin{bmatrix} x \\ 0 \end{bmatrix} a \, dx = \begin{bmatrix} \frac{ml}{2} \\ 0 \end{bmatrix}$$

$$\bar{\mathbf{S}} = \int_V \rho \mathbf{S} \, dV = \int_0^l \rho a \mathbf{S} \, dx = \int_0^1 \rho \, a\mathbf{S}l \, d\xi$$

$$= m \int_0^1 \mathbf{S} \, d\xi = \frac{m}{12} \begin{bmatrix} 0 & 6 & 0 \\ l & 0 & -l \end{bmatrix}$$

where a is the cross-sectional area of the connecting rod. Let the vector of elastic coordinates be $\mathbf{q}_f = [q_{f1} \, q_{f2} \, q_{f3}]^{\mathrm{T}} = [\bar{q}_{f3} \, \bar{q}_{f4} \, \bar{q}_{f6}]^{\mathrm{T}}$, then the vector $\bar{\mathbf{S}}\mathbf{q}_f$ can be written as

$$\bar{\mathbf{S}}\mathbf{q}_f = \frac{m}{12} \begin{bmatrix} 0 & 6 & 0 \\ l & 0 & -l \end{bmatrix} \begin{bmatrix} q_{f1} \\ q_{f2} \\ q_{f3} \end{bmatrix} = \frac{m}{12} \begin{bmatrix} 6q_{f2} \\ l(q_{f1} - q_{f3}) \end{bmatrix}$$

The term $\mathbf{m}_{R\theta}$ of the mass matrix is then given by

$$\mathbf{m}_{R\theta} = \mathbf{A}_\theta(\mathbf{I}_1 + \bar{\mathbf{S}}\mathbf{q}_f)$$

where $\mathbf{A}_\theta$ is the partial derivative of the planar transformation $\mathbf{A}$. The matrices $\mathbf{A}$ and $\mathbf{A}_\theta$ are given by

$$\mathbf{A} = \begin{bmatrix} \cos\theta & -\sin\theta \\ \sin\theta & \cos\theta \end{bmatrix}, \quad \mathbf{A}_\theta = \begin{bmatrix} -\sin\theta & -\cos\theta \\ \cos\theta & -\sin\theta \end{bmatrix}$$

and accordingly

$$\mathbf{m}_{R\theta} = \begin{bmatrix} -\sin\theta & -\cos\theta \\ \cos\theta & -\sin\theta \end{bmatrix} \left\{ \begin{bmatrix} \frac{ml}{2} \\ 0 \end{bmatrix} + \begin{bmatrix} \frac{mq_{f2}}{2} \\ \frac{lm(q_{f1}-q_{f3})}{12} \end{bmatrix} \right\}$$

$$= \frac{m}{12} \begin{bmatrix} -6(l+q_{f2})\sin\theta - l(q_{f1}-q_{f3})\cos\theta \\ 6(l+q_{f2})\cos\theta - l(q_{f1}-q_{f3})\sin\theta \end{bmatrix}$$

The inertia coupling between the reference translation and the elastic deformation can be evaluated by using Eq. 83, that is,

$$\mathbf{m}_{Rf} = \mathbf{A}\bar{\mathbf{S}} = \frac{m}{12} \begin{bmatrix} \cos\theta & -\sin\theta \\ \sin\theta & \cos\theta \end{bmatrix} \begin{bmatrix} 0 & 6 & 0 \\ l & 0 & -l \end{bmatrix}$$

$$= \frac{m}{12} \begin{bmatrix} -l\sin\theta & 6\cos\theta & l\sin\theta \\ l\cos\theta & 6\sin\theta & -l\cos\theta \end{bmatrix}$$

Using Eq. 86, one has

$$(m_{\theta\theta})_{rr} = \int_V \rho\bar{\mathbf{u}}_o^{\mathrm{T}}\bar{\mathbf{u}}_o\,dV = \int_0^l \rho[x \quad 0]\begin{bmatrix} x \\ 0 \end{bmatrix} a\,dx = \frac{m(l)^2}{3}$$

which is the mass moment of inertia about point A in the undeformed state. From Eq. 87, one has

$$(m_{\theta\theta})_{rf} = 2\int_V \rho\bar{\mathbf{u}}_o^{\mathrm{T}}\bar{\mathbf{u}}_f\,dV = 2\int_0^l \rho\bar{\mathbf{u}}_o^{\mathrm{T}}\bar{\mathbf{u}}_f a\,dx = 2\int_0^1 \rho a\bar{\mathbf{u}}_o^{\mathrm{T}}\bar{\mathbf{u}}_f l\,d\xi$$

$$= 2\int_0^1 m[x \quad 0]\begin{bmatrix} 0 & \xi & 0 \\ l[\xi - 2(\xi)^2 + (\xi)^3] & 0 & l[(\xi)^3 - (\xi)^2] \end{bmatrix}\begin{bmatrix} q_{f1} \\ q_{f2} \\ q_{f3} \end{bmatrix} d\xi$$

$$= 2ml\int_0^1 (\xi)^2 q_{f2}\,d\xi = \frac{2}{3}mlq_{f2}$$

The matrix $\mathbf{m}_{ff}$ that appears in Eq. 89 is given by

$$\mathbf{m}_{ff} = \int_V \rho\mathbf{S}^{\mathrm{T}}\mathbf{S}\,dV = m\begin{bmatrix} \frac{(l)^2}{105} & & \text{symmetric} \\ 0 & \frac{1}{3} & \\ \frac{-(l)^2}{140} & 0 & \frac{(l)^2}{105} \end{bmatrix}$$

Hence, the scalar $(m_{\theta\theta})_{ff}$ of Eq. 89 is given by

$$(m_{\theta\theta})_{ff} = \mathbf{q}_f^{\mathrm{T}}\mathbf{m}_{ff}\mathbf{q}_f$$

$$= m[q_{f1} \quad q_{f2} \quad q_{f3}]\begin{bmatrix} \frac{(l)^2}{105} & 0 & \frac{-(l)^2}{140} \\ 0 & \frac{1}{3} & 0 \\ \frac{-(l)^2}{140} & 0 & \frac{(l)^2}{105} \end{bmatrix}\begin{bmatrix} q_{f1} \\ q_{f2} \\ q_{f3} \end{bmatrix}$$

$$= m\left[\frac{(l)^2}{105}(q_{f1})^2 + \frac{1}{3}(q_{f2})^2 + \frac{(l)^2}{105}(q_{f3})^2 - \frac{(l)^2}{70}q_{f1}q_{f3}\right]$$

It follows that the mass moment of inertia about point A is

$$\begin{aligned}\mathbf{m}_{\theta\theta} &= (m_{\theta\theta})_{rr} + (m_{\theta\theta})_{rf} + (m_{\theta\theta})_{ff}\\ &= m\left[\frac{(l)^2}{3} + \frac{2l}{3}q_{f2} + \frac{(l)^2}{105}(q_{f1})^2 + \frac{1}{3}(q_{f2})^2 + \frac{(l)^2}{105}(q_{f3})^2 - \frac{(l)^2}{70}q_{f1}q_{f3}\right]\end{aligned}$$

Clearly, $\mathbf{m}_{\theta\theta}$ depends on the elastic coordinates of the connecting rod. Let $\tilde{\mathbf{I}}$ be the skew symmetric matrix

$$\tilde{\mathbf{I}} = \begin{bmatrix} 0 & 1\\ -1 & 0\end{bmatrix}$$

The first integral in Eq. 92 can then be written as

$$\begin{aligned}\int_V \rho\bar{\mathbf{u}}_o^{\mathrm{T}}\tilde{\mathbf{I}}\mathbf{S}\,dV &= \int_0^1 a(l)^2[0 \quad \xi]\begin{bmatrix} 0 & \xi & 0\\ l[\xi - 2(\xi)^2 + (\xi)^3] & 0 & l[(\xi)^3 - (\xi)^2]\end{bmatrix}d\xi\\ &= m(l)^2\begin{bmatrix}\frac{1}{30} & 0 & -\frac{1}{20}\end{bmatrix}\end{aligned}$$

One can also verify that the skew symmetric matrix of Eq. 93 is

$$\tilde{\mathbf{S}} = \int_V \rho\mathbf{S}^{\mathrm{T}}\tilde{\mathbf{I}}\mathbf{S}\,dV = \frac{ml}{60}\begin{bmatrix} 0 & -2 & 0\\ 2 & 0 & -3\\ 0 & 3 & 0\end{bmatrix}$$

that is,

$$\mathbf{q}_f^{\mathrm{T}}\tilde{\mathbf{S}} = \frac{ml}{60}[2q_{f2} \quad (3q_{f3} - 2q_{f1}) \quad -3q_{f2}]$$

The matrix $\mathbf{m}_{\theta f}$ of Eq. 92 is then given by

$$\begin{aligned}\mathbf{m}_{\theta f} &= \int_V \rho\bar{\mathbf{u}}_o^{\mathrm{T}}\tilde{\mathbf{I}}\mathbf{S}\,dV + \mathbf{q}_f^{\mathrm{T}}\tilde{\mathbf{S}}\\ &= m(l)^2\begin{bmatrix}\frac{1}{30} & 0 & -\frac{1}{20}\end{bmatrix} + \frac{ml}{60}[2q_{f2} \quad (3q_{f3} - 2q_{f1}) \quad -3q_{f2}]\\ &= \begin{bmatrix}\frac{ml}{30}(l + q_{f2}) & \frac{ml}{60}(3q_{f3} - 2q_{f1}) & -\frac{ml}{20}(l + q_{f2})\end{bmatrix}\end{aligned}$$

This completes the evaluation of the components of the mass matrix of the flexible connecting rod.

Elastic Forces To develop expressions for the generalized elastic forces associated with the elastic coordinates of the connecting rod, one needs to evaluate the stiffness matrix. To this end, the following strain energy expression is used:

$$U = \frac{1}{2}\int_0^l [EI(\bar{u}''_{f2})^2 + Ea(\bar{u}'_{f1})^2]dx$$

where I is the second moment of area of the connecting rod; $\bar{u}_{f1}$ and $\bar{u}_{f2}$ are, respectively, the axial and transverse displacements; and $(\,')$ denotes the partial derivative with respect to the spatial coordinate x. This strain energy expression can be written in a matrix form as

$$U = \frac{1}{2}\int_0^l \begin{bmatrix} \bar{u}'_{f1} & \bar{u}''_{f2} \end{bmatrix} \begin{bmatrix} Ea & 0 \\ 0 & EI \end{bmatrix} \begin{bmatrix} \bar{u}'_{f1} \\ \bar{u}''_{f2} \end{bmatrix} dx$$

and since $\bar{\mathbf{u}}_f = \mathbf{S}\mathbf{q}_f$, one has

$$\bar{u}'_{f1} = \begin{bmatrix} 0 & \frac{1}{l} & 0 \end{bmatrix} \begin{bmatrix} q_{f1} \\ q_{f2} \\ q_{f3} \end{bmatrix}$$

and

$$\bar{u}''_{f2} = \begin{bmatrix} l\left(\frac{4}{(l)^2} - \frac{6x}{(l)^3}\right) & 0 & l\left(\frac{6x}{(l)^3} - \frac{2}{(l)^2}\right) \end{bmatrix} \begin{bmatrix} q_{f1} \\ q_{f2} \\ q_{f3} \end{bmatrix}$$

that is,

$$\begin{bmatrix} \bar{u}'_{f1} \\ \bar{u}''_{f2} \end{bmatrix} = \begin{bmatrix} 0 & \frac{1}{l} & 0 \\ \left(\frac{4}{l} - \frac{6x}{(l)^2}\right) & 0 & \frac{6x}{(l)^2} - \frac{2}{l} \end{bmatrix} \begin{bmatrix} q_{f1} \\ q_{f2} \\ q_{f3} \end{bmatrix}$$

Substituting this in the strain energy expression and keeping in mind that the vector of elastic coordinates $\mathbf{q}_f$ depends only on time, one can verify that

$$U = \frac{1}{2}\mathbf{q}_f^{\mathrm{T}}\mathbf{K}_{ff}\mathbf{q}_f$$

where the stiffness matrix $\mathbf{K}_{ff}$ is defined as

$$\mathbf{K}_{ff} = \begin{bmatrix} \frac{4EI}{l} & & \text{symmetric} \\ 0 & \frac{Ea}{l} & \\ \frac{2EI}{l} & 0 & \frac{4EI}{l} \end{bmatrix}$$

The strain energy expression can also be written as

$$U = \frac{1}{2}\begin{bmatrix} \mathbf{q}_r^{\mathrm{T}} & \mathbf{q}_f^{\mathrm{T}} \end{bmatrix} \begin{bmatrix} \mathbf{0} & \mathbf{0} \\ \mathbf{0} & \mathbf{K}_{ff} \end{bmatrix} \begin{bmatrix} \mathbf{q}_r \\ \mathbf{q}_f \end{bmatrix} = \frac{1}{2}\mathbf{q}^{\mathrm{T}}\mathbf{K}\mathbf{q}$$

where $\mathbf{q}_r$ is the vector of reference coordinates, $\mathbf{q}_r = [R_1\ R_2\ \theta]^{\mathrm{T}}$, $\mathbf{q} = [\mathbf{q}_r^{\mathrm{T}}\ \mathbf{q}_f^{\mathrm{T}}]^{\mathrm{T}}$ is the total vector of coordinates of the connecting rod, and the stiffness matrix $\mathbf{K}$ is

$$\mathbf{K} = \begin{bmatrix} 0 & & & & \text{symmetric} & \\ 0 & 0 & & & & \\ 0 & 0 & 0 & & & \\ 0 & 0 & 0 & \frac{4EI}{l} & & \\ 0 & 0 & 0 & 0 & \frac{Ea}{l} & \\ 0 & 0 & 0 & \frac{2EI}{l} & 0 & \frac{4EI}{l} \end{bmatrix}$$

External Forces The only external force acting on the flexible connecting rod is the weight mg, where g is the gravity constant. Reaction forces of the constraints such as revolute joints can be introduced to the dynamic formulation by using the vector of Lagrange multipliers. The weight of the connecting rod acts vertically; therefore, one can define the force vector $\mathbf{F}_e$ of the external forces as

$$\mathbf{F}_e = [0 \quad -mg]^{\mathrm{T}}$$

The virtual work due to this force is given by

$$\delta W_e = \mathbf{F}_e^{\mathrm{T}} \delta \mathbf{r}_c$$

where $\mathbf{r}_c$ is the global position vector of the center of mass of the connecting rod defined as

$$\mathbf{r}_c = \mathbf{R} + \mathbf{A}\bar{\mathbf{u}}$$

in which $\bar{\mathbf{u}}$ is the local position vector of the center of mass, which can be written as the sum of the two vectors

$$\bar{\mathbf{u}} = \bar{\mathbf{u}}_o + \bar{\mathbf{u}}_f$$

where $\bar{\mathbf{u}}_o$ is the undeformed position of the mass center defined in the connecting rod coordinate system and $\bar{\mathbf{u}}_f$ is the deformation vector. The vector $\bar{\mathbf{u}}_o$ is given by

$$\bar{\mathbf{u}}_o = \begin{bmatrix} \frac{l}{2} & 0 \end{bmatrix}^{\mathrm{T}}$$

and since, at the center of mass $\xi = 0.5$, one has

$$\bar{\mathbf{u}}_f = \mathbf{S}(\xi = 0.5)\mathbf{q}_f = \begin{bmatrix} 0 & 0.5 & 0 \\ l(0.125) & 0 & -(0.125)l \end{bmatrix} \begin{bmatrix} q_{f1} \\ q_{f2} \\ q_{f3} \end{bmatrix}$$

The virtual displacement $\delta \mathbf{r}_c$ is given by

$$\begin{aligned} \delta \mathbf{r}_c &= \delta \mathbf{R} + \mathbf{A}_\theta \bar{\mathbf{u}} \, \delta\theta + \mathbf{A}\mathbf{S}(\xi = 0.5) \, \delta \mathbf{q}_f \\ &= [\mathbf{I} \quad \mathbf{A}_\theta \bar{\mathbf{u}} \quad \mathbf{A}\mathbf{S}(\xi = 0.5)] \begin{bmatrix} \delta \mathbf{R} \\ \delta\theta \\ \delta \mathbf{q}_f \end{bmatrix} \end{aligned}$$

where $\mathbf{I}$ is a 2×2 identify matrix. The virtual work can then be written as

$$\delta W_e = \mathbf{F}_e^{\mathrm{T}} \delta \mathbf{r}_c = [0 \quad -mg][\mathbf{I} \quad \mathbf{A}_\theta \bar{\mathbf{u}} \quad \mathbf{A}\mathbf{S}(\xi = 0.5)] \begin{bmatrix} \delta \mathbf{R} \\ \delta\theta \\ \delta \mathbf{q}_f \end{bmatrix}$$

which can be written as

$$\begin{aligned} \delta W_e &= (\mathbf{Q}_e)_R^{\mathrm{T}} \delta \mathbf{R} + (\mathbf{Q}_e)_\theta^{\mathrm{T}} \delta\theta + (\mathbf{Q}_e)_f^{\mathrm{T}} \delta \mathbf{q}_f \\ &= [(\mathbf{Q}_e)_R^{\mathrm{T}} \quad (\mathbf{Q}_e)_\theta^{\mathrm{T}} \quad (\mathbf{Q}_e)_f^{\mathrm{T}}] \begin{bmatrix} \delta \mathbf{R} \\ \delta\theta \\ \delta \mathbf{q}_f \end{bmatrix} \end{aligned}$$

in which $(\mathbf{Q}_e)_R$, $(\mathbf{Q}_e)_\theta$, and $(\mathbf{Q}_e)_f$ are, respectively, the generalized forces associated with the reference translation, reference rotation, and the elastic deformation and are given by

$$(\mathbf{Q}_e)_R^{\mathrm{T}} = [0 \quad -mg]\mathbf{I} = [0 \quad -mg]$$

$$(\mathbf{Q}_e)_\theta^{\mathrm{T}} = [0 \quad -mg]\mathbf{A}_\theta\bar{\mathbf{u}}$$

$$= [0 \quad -mg]\begin{bmatrix} -\sin\theta & -\cos\theta \\ \cos\theta & -\sin\theta \end{bmatrix}\begin{bmatrix} \frac{l}{2} + 0.5q_{f2} \\ 0 + l(0.125)(q_{f1} - q_{f3}) \end{bmatrix}$$

$$= mg\{0.125l(q_{f1} - q_{f3})\sin\theta - 0.5(l + q_{f2})\cos\theta\}$$

$$(\mathbf{Q}_e)_f^{\mathrm{T}} = [0 \quad -mg]\mathbf{A}\mathbf{S}(\xi = 0.5)$$

$$= [0 \quad -mg]\begin{bmatrix} \cos\theta & -\sin\theta \\ \sin\theta & \cos\theta \end{bmatrix}\begin{bmatrix} 0 & 0.5 & 0 \\ l(0.125) & 0 & -(0.125)l \end{bmatrix}$$

$$= mg[-0.125l\cos\theta \quad -0.5\sin\theta \quad 0.125l\cos\theta]$$

Constraint Equations For the revolute joint at A, let $\mathbf{r}_A(t) = [r_1(t)\ r_2(t)]^{\mathrm{T}}$ be the global position vector of point A evaluated by using the coordinates of the crankshaft OA. The revolute joint constraints at A require that

$$\mathbf{R} + \mathbf{A}\bar{\mathbf{u}} = \mathbf{r}_A(t)$$

where $\bar{\mathbf{u}}$ is the local position of point A defined in the connecting rod coordinate system. Since point A coincides with the origin of the body reference of the connecting rod and as such $\xi = 0$, one can verify that $\bar{\mathbf{u}} = \mathbf{0}$; and the revolute joint constraints reduce to

$$\mathbf{R} = \mathbf{r}_A(t)$$

For the revolute joint at B, let $\mathbf{r}_B(t) = [x_4 \ 0]^{\mathrm{T}}$ be the global position vector of point B. The two algebraic constraint equations representing the revolute joint at B can be written as

$$\mathbf{R} + \mathbf{A}\bar{\mathbf{u}} = \mathbf{r}_B(t)$$

where, in this case, $\bar{\mathbf{u}}$ can be written as

$$\bar{\mathbf{u}} = \bar{\mathbf{u}}_o + \bar{\mathbf{u}}_f$$

and

$$\bar{\mathbf{u}}_o = [l \quad 0]^{\mathrm{T}}$$

Since at point B, $\xi = 1$, the deformation vector $\bar{\mathbf{u}}_f$ can be evaluated as

$$\bar{\mathbf{u}}_f = \mathbf{S}(\xi = 1)\mathbf{q}_f = \begin{bmatrix} 0 & 1 & 0 \\ 0 & 0 & 0 \end{bmatrix}\begin{bmatrix} q_{f1} \\ q_{f2} \\ q_{f3} \end{bmatrix} = \begin{bmatrix} q_{f2} \\ 0 \end{bmatrix}$$

The vector $\bar{\mathbf{u}}$ can then be defined as

$$\bar{\mathbf{u}} = \bar{\mathbf{u}}_o + \bar{\mathbf{u}}_f = \begin{bmatrix} l \\ 0 \end{bmatrix} + \begin{bmatrix} q_{f2} \\ 0 \end{bmatrix} = \begin{bmatrix} l + q_{f2} \\ 0 \end{bmatrix}$$

and the algebraic constraint equations of the revolute joint at B lead to

$$\mathbf{R} + \mathbf{A}\bar{\mathbf{u}} = \begin{bmatrix} R_1 \\ R_2 \end{bmatrix} + \begin{bmatrix} \cos\theta & -\sin\theta \\ \sin\theta & \cos\theta \end{bmatrix} \begin{bmatrix} l + q_{f2} \\ 0 \end{bmatrix} = \mathbf{r}_B(t) = \begin{bmatrix} x_4 \\ 0 \end{bmatrix}$$

that is,

$$R_1 + (l + q_{f2})\cos\theta = x_4$$
$$R_2 + (l + q_{f2})\sin\theta = 0$$

Quadratic Velocity Vector The quadratic velocity vector $(\mathbf{Q}_v)_R$ associated with the reference translation and given in Eq. 136 can be evaluated once the inertia shape integrals $\bar{\mathbf{S}}$ and $\mathbf{I}_1$ are determined. These quantities that were evaluated earlier can be used to yield the following:

$$\bar{\mathbf{S}}\mathbf{q}_f + \mathbf{I}_1 = \frac{m}{2} \begin{bmatrix} l + q_{f2} \\ \frac{q_{f1} - q_{f3}}{6} \end{bmatrix}$$

It follows that

$$(\dot{\theta})^2 \mathbf{A}(\bar{\mathbf{S}}\mathbf{q}_f + \mathbf{I}_1) = \frac{m(\dot{\theta})^2}{2} \begin{bmatrix} \cos\theta & -\sin\theta \\ \sin\theta & \cos\theta \end{bmatrix} \begin{bmatrix} l + q_{f2} \\ \frac{q_{f1} - q_{f3}}{6} \end{bmatrix}$$
$$= \frac{m(\dot{\theta})^2}{2} \begin{bmatrix} (l + q_{f2})\cos\theta - \frac{(q_{f1} - q_{f3})\sin\theta}{6} \\ (l + q_{f2})\sin\theta + \frac{(q_{f1} - q_{f3})\cos\theta}{6} \end{bmatrix}$$

One can verify that

$$\bar{\mathbf{S}}\dot{\mathbf{q}}_f = \frac{m}{12} \begin{bmatrix} 6\dot{q}_{f2} \\ l(\dot{q}_{f1} - \dot{q}_{f3}) \end{bmatrix}$$

and therefore

$$2\dot{\theta}\mathbf{A}_\theta \bar{\mathbf{S}}\dot{\mathbf{q}}_f = \frac{m\dot{\theta}}{6} \begin{bmatrix} -\sin\theta & -\cos\theta \\ \cos\theta & -\sin\theta \end{bmatrix} \begin{bmatrix} 6\dot{q}_{f2} \\ \dot{q}_{f1} - \dot{q}_{f3} \end{bmatrix}$$
$$= \frac{m\dot{\theta}}{6} \begin{bmatrix} -6\dot{q}_{f2}\sin\theta - (\dot{q}_{f1} - \dot{q}_{f3})\cos\theta \\ 6\dot{q}_{f2}\cos\theta - (\dot{q}_{f1} - \dot{q}_{f3})\sin\theta \end{bmatrix}$$

Therefore, the velocity vector $(\mathbf{Q}_v)_R$ is given by

$$(\mathbf{Q}_v)_R = (\dot{\theta})^2 \mathbf{A}(\bar{\mathbf{S}}\mathbf{q}_f + \mathbf{I}_1) - 2\dot{\theta}\mathbf{A}_\theta \bar{\mathbf{S}}\dot{\mathbf{q}}_f$$
$$= \frac{m(\dot{\theta})^2}{2} \begin{bmatrix} (l + q_{f2})\cos\theta - \frac{(q_{f1} - q_{f3})\sin\theta}{6} \\ (l + q_{f2})\sin\theta + \frac{(q_{f1} - q_{f3})\cos\theta}{6} \end{bmatrix}$$
$$- \frac{m\dot{\theta}}{6} \begin{bmatrix} -6\dot{q}_{f2}\sin\theta - (\dot{q}_{f1} - \dot{q}_{f3})\cos\theta \\ 6\dot{q}_{f2}\cos\theta - (\dot{q}_{f1} - \dot{q}_{f3})\sin\theta \end{bmatrix}$$

The quadratic velocity vector $(\mathbf{Q}_v)_\theta$ associated with the rotational coordinates of the body reference is given in Eq. 136 as

$$(\mathbf{Q}_v)_\theta = -2\dot{\theta}\dot{\mathbf{q}}_f^{\mathrm{T}}(\mathbf{m}_{ff}\mathbf{q}_f + \bar{\mathbf{I}}_o)$$

where one can verify that

$$\dot{\mathbf{q}}_f^{\mathrm{T}}\bar{\mathbf{I}}_o = \dot{\mathbf{q}}_f^{\mathrm{T}} \int_V \rho(\bar{\mathbf{u}}_o^{\mathrm{T}}\tilde{\mathbf{S}})^{\mathrm{T}} dV = \frac{ml}{3}\dot{q}_{f2}$$

and

$$\dot{\mathbf{q}}_f^{\mathrm{T}}\mathbf{m}_{ff}\mathbf{q}_f = m\left\{\frac{(l)^2}{105}q_{f1}\dot{q}_{f1} + \frac{1}{3}q_{f2}\dot{q}_{f2} + \frac{(l)^2}{105}q_{f3}\dot{q}_{f3} - \frac{(l)^2}{140}q_{f1}\dot{q}_{f3} - \frac{(l)^2}{140}\dot{q}_{f1}q_{f3}\right\}$$

It follows that

$$(\mathbf{Q}_v)_\theta = -2\dot{\theta}m\left[\frac{l}{3}\dot{q}_{f2} + \frac{(l)^2}{105}q_{f1}\dot{q}_{f1} + \frac{1}{3}q_{f2}\dot{q}_{f2} + \frac{(l)^2}{105}q_{f3}\dot{q}_{f3} - \frac{(l)^2}{140}q_{f1}\dot{q}_{f3} - \frac{(l)^2}{140}\dot{q}_{f1}q_{f3}\right]$$

The quadratic velocity vector $(\mathbf{Q}_v)_f$ associated with the elastic coordinates, also given in Eq. 136, is defined as

$$(\mathbf{Q}_v)_f = (\dot{\theta})^2(\mathbf{m}_{ff}\mathbf{q}_f + \bar{\mathbf{I}}_o) + 2\dot{\theta}\bar{\tilde{\mathbf{S}}}\dot{\mathbf{q}}_f$$

in which

$$\mathbf{m}_{ff}\mathbf{q}_f + \bar{\mathbf{I}}_o = m\begin{bmatrix} (l)^2\left(\frac{q_{f1}}{105} - \frac{q_{f3}}{140}\right) \\ \frac{q_{f2}}{3} \\ (l)^2\left(\frac{q_{f3}}{105} - \frac{q_{f1}}{140}\right)\end{bmatrix} + \begin{bmatrix} 0 \\ \frac{ml}{3} \\ 0\end{bmatrix} = m\begin{bmatrix} (l)^2\left(\frac{q_{f1}}{105} - \frac{q_{f3}}{140}\right) \\ \frac{q_{f2}+l}{3} \\ (l)^2\left(\frac{q_{f3}}{105} - \frac{q_{f1}}{140}\right)\end{bmatrix}$$

The vector $2\bar{\tilde{\mathbf{S}}}\dot{\mathbf{q}}_f$ is given by

$$2\bar{\tilde{\mathbf{S}}}\dot{\mathbf{q}}_f = -\frac{ml}{30}[2\dot{q}_{f2} \quad (3\dot{q}_{f3} - 2\dot{q}_{f1}) \quad -3\dot{q}_{f2}]^{\mathrm{T}}$$

Thus, the vector $(\mathbf{Q}_v)_f$ is

$$(\mathbf{Q}_v)_f = \dot{\theta}m\begin{bmatrix} \dot{\theta}(l)^2\left(\frac{q_{f1}}{105} - \frac{q_{f3}}{140}\right) - \frac{l\dot{q}_{f2}}{15} \\ \frac{\dot{\theta}}{3}(q_{f2} + l) - \frac{l}{30}(3\dot{q}_{f3} - 2\dot{q}_{f1}) \\ \dot{\theta}(l)^2\left(\frac{q_{f3}}{105} - \frac{q_{f1}}{140}\right) + \frac{1}{10}\dot{q}_{f2}\end{bmatrix}$$

5.8 USE OF INDEPENDENT COORDINATES

In the following few sections, computational algorithms for the solution of the dynamic equations of motion of multibody systems consisting of interconnected rigid and deformable bodies are discussed. We first consider the case in which the dynamic equations are formulated in terms of the independent coordinates (system degrees of freedom). The computational algorithms in this case, which also include *recursive formulations* that employ relative joint coordinates, are much simpler than the computational algorithms based on dynamic equations formulated in terms of the dependent and independent coordinates and in which nonlinear constraint equations are adjoined to the system differential equations of motion by using the vector of Lagrange multipliers. The use of the independent coordinates leads to a set of second-order differential equations, while the use of the multipliers leads to a coupled set of differential and algebraic equations that have to be solved simultaneously. Computational algorithms based on the Lagrangian formulation with the multipliers will also be discussed in later sections of this chapter.

In this section, a computational algorithm based on the dynamic formulation for multibody systems that employs a minimum set of differential equations is presented. This minimum set of differential equations can be obtained by using two different approaches. In the first approach the system degrees of freedom can be identified from the start and used to define the configuration of the multibody system. The degrees of freedom may be a set of *joint variables* that represent relative translational and rotational displacements between the multibody system components. This approach leads, in many applications, to recursive kinematic and dynamic equations expressed in terms of the independent degrees of freedom. In the other approach, however, the system kinematic equations are formulated in terms of both the independent and dependent coordinates. These kinematic relationships, such as nonlinear algebraic constraint equations that represent mechanical joints in the system, can be used to identify a set of independent coordinates. The dependent coordinates can be written in terms of the independent coordinates using the kinematic equations. These kinematic equations can then be used to develop a minimum set of independent differential equations of motion expressed in terms of the independent variables that in this case may or may not represent joint degrees of freedom.

Let $\mathbf{q}_i$ be the total vector of independent coordinates or the degrees of freedom of the multibody system consisting of interconnected rigid and deformable bodies. If the dependent coordinates are eliminated from the dynamic equations, the motion of the multibody system, as will be demonstrated in Section 10, is governed by only a set of second-order differential equations of motion that can be written in the following matrix form:

$$\mathbf{M}_{ii}\ddot{\mathbf{q}}_i + \mathbf{K}_{ii}\mathbf{q}_i = \mathbf{Q}_{ei} + \mathbf{Q}_{vi} \tag{5.151}$$

where $\mathbf{q}_i$ is the vector of system independent coordinates, $\mathbf{M}_{ii}$ is assumed to be a *positive definite* mass matrix associated with the independent coordinates, $\mathbf{K}_{ii}$ is the system stiffness matrix, $\mathbf{Q}_{ei}$ is the vector of generalized external forces associated with the independent coordinates, and $\mathbf{Q}_{vi}$ is the quadratic velocity vector that includes the gyroscopic and Coriolis force components. Clearly, the constraint forces do not appear in Eq. 151, since the dependent coordinates are eliminated (Shabana 1994a). The preceding equation can be written as a linear system of algebraic equations in the accelerations as follows:

$$\mathbf{M}_{ii}\ddot{\mathbf{q}}_i = \mathbf{Q}_{ei} + \mathbf{Q}_{vi} - \mathbf{K}_{ii}\mathbf{q}_i \tag{5.152}$$

Since $\mathbf{M}_{ii}$ is assumed to be a positive definite matrix, this equation can be solved for the vector of accelerations as follows:

$$\ddot{\mathbf{q}}_i = \mathbf{M}_{ii}^{-1}[\mathbf{Q}_{ei} + \mathbf{Q}_{vi} - \mathbf{K}_{ii}\mathbf{q}_i] = \ddot{\mathbf{q}}_i(\mathbf{q}_i, \dot{\mathbf{q}}_i) \tag{5.153}$$

One may then form the state vectors

$$\mathbf{y} = \begin{bmatrix} \mathbf{q}_i \\ \dot{\mathbf{q}}_i \end{bmatrix}, \qquad \dot{\mathbf{y}} = \begin{bmatrix} \dot{\mathbf{q}}_i \\ \ddot{\mathbf{q}}_i \end{bmatrix} = \mathbf{f}(\mathbf{y}, t) \tag{5.154}$$

Given a set of initial conditions $\mathbf{y}_o = [\mathbf{q}_{io}^{\mathrm{T}}\ \dot{\mathbf{q}}_{io}^{\mathrm{T}}]^{\mathrm{T}}$, one can obtain the acceleration vector from Eq. 153. This acceleration vector and the initial coordinates and velocities are sufficient to define the function $\dot{\mathbf{y}} = \mathbf{f}(\mathbf{y}, t)$, which is required for the numerical solution of the differential equations. The solution vector $\mathbf{y}_1$ at time $t_1 = t_o + h$, where h is the time step, can then be obtained and used to evaluate the accelerations at the new point in time t_1. The vector $\mathbf{y}_1$ as well as the acceleration vector can be used to define the function $\dot{\mathbf{y}}_1 = \mathbf{f}(\mathbf{y}_1, t_1)$, which can be used to advance the numerical integration another step to reach a new point in time t_2. This process continues until the end of the desired simulation time. The computational algorithm is shown schematically in Fig. 8 and proceeds in the following routine:

1. The inertia shape integrals that depend on the assumed displacement field and appear in the differential equations of motion for the multibody system consisting of interconnected rigid and deformable bodies are first identified and evaluated only once in advance for the dynamic analysis. These shape integrals can be evaluated in a preprocessor structural dynamic computer program.
2. An accurate estimate for the initial value for the vectors of independent coordinates $\mathbf{q}_i$ and independent velocities $\dot{\mathbf{q}}_i$ or equivalently $\mathbf{y} = [\mathbf{q}_i^{\mathrm{T}}\ \dot{\mathbf{q}}_i^{\mathrm{T}}]^{\mathrm{T}}$ must be provided. These vectors include both the rigid body and deformation variables. An accurate estimate for the initial value of the state vector $\mathbf{y}$ may require that we perform static analysis. A computational algorithm for the static analysis of multibody systems is discussed in the sections that follow.
3. The state vector as well as the inertia shape integrals can be used for the automatic generation of the multibody system equations of motion. In terms of these quantities, the mass matrix $\mathbf{M}_{ii}$ as well as the right-hand side of Eq. 152, which includes external, gyroscopic, and Coriolis forces, as well as the elastic forces, can be evaluated in a straightforward manner. Note that both the mass matrix and the right-hand side of Eq. 152 are nonlinear functions of the state variables; therefore, they have to be iteratively updated during the dynamic simulation.

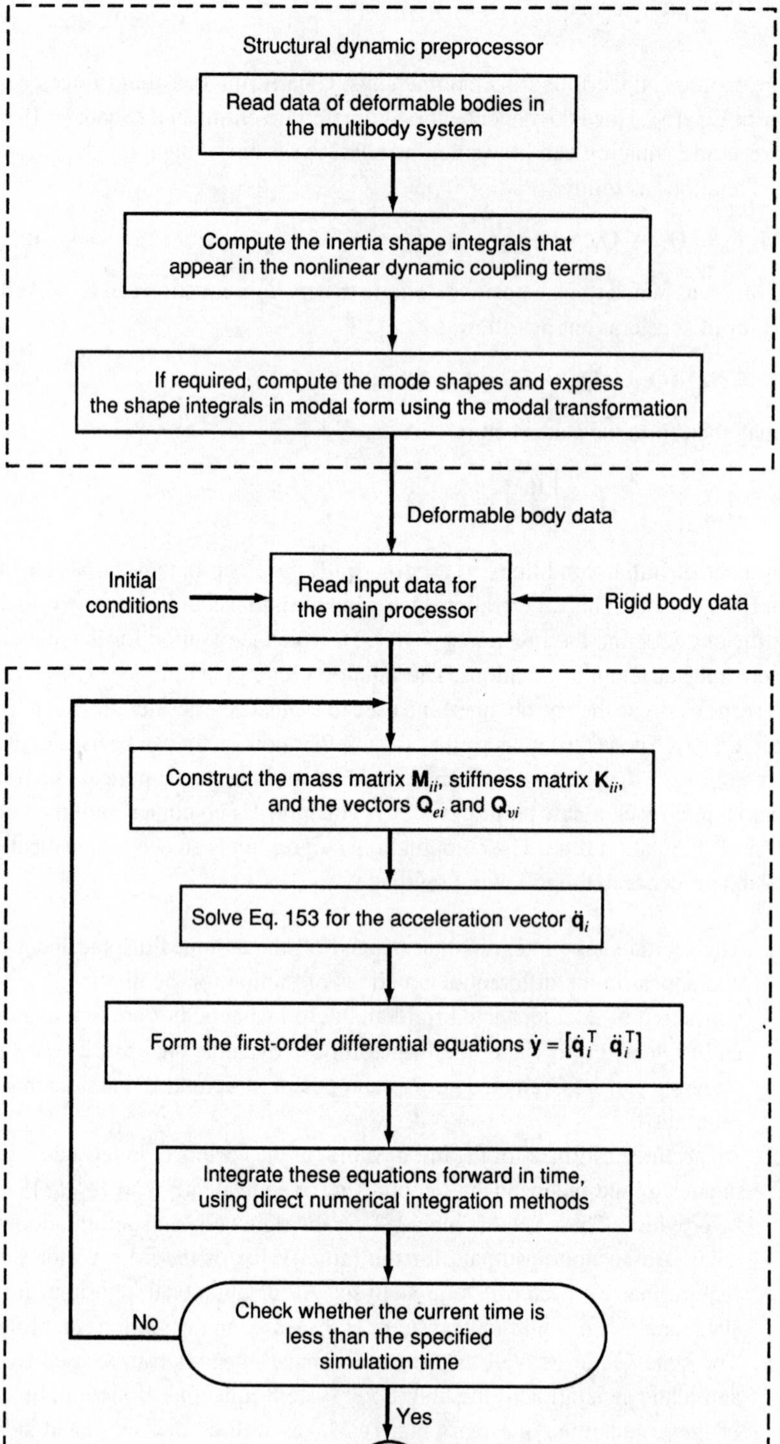

Figure 5.8 Computational algorithm.

4. Even though Eq. 152 is a highly nonlinear differential equation of motion in the coordinates and velocities, it can be considered as a linear system of algebraic equations in the acceleration vector $\ddot{\mathbf{q}}_i$. Therefore, solving these equations to obtain the acceleration vector is straightforward.
5. Since the acceleration vector $\ddot{\mathbf{q}}_i$ has been determined at this point in time and the velocity vector $\dot{\mathbf{q}}_i$ is assumed to be known, the system of first-order differential equations at this point in time can be formed as

$$\dot{\mathbf{y}} = \mathbf{f}(\mathbf{y}, t)$$

6. Using the initial conditions and the vector function $\mathbf{f}$ that is known at this stage, the system differential equations can be integrated forward in time using a direct numerical integration method. The solution of these differential equations defines the state vector at the end of the specified time step. This solution can be used to advance the integration in order to obtain the solution at the end of the second time step. This process continues until the end of the simulation time is reached.

5.9 DYNAMIC EQUATIONS WITH MULTIPLIERS

The kinematic constraints that describe mechanical joints as well as specified trajectories in the multibody system consisting of interconnected rigid and deformable components can be formulated by using a set of nonlinear algebraic constraint equations. This vector of equations defined by Eq. 134 is shown here for convenience:

$$\mathbf{C}(\mathbf{q}, t) = \mathbf{0} \tag{5.155}$$

where $\mathbf{C} = [C_1\ C_2\ \cdots\ C_{n_c}]^{\mathrm{T}}$ is the vector of linearly independent constraint equations, t is time, and $\mathbf{q}$ is the total vector of system generalized coordinates that can be written in a partitioned form as

$$\mathbf{q} = \begin{bmatrix}\mathbf{q}_r^{\mathrm{T}} & \mathbf{q}_f^{\mathrm{T}}\end{bmatrix}^{\mathrm{T}} \tag{5.156}$$

where the subscripts r and f refer, respectively, to reference and flexible (elastic) coordinates, and $\mathbf{q}_r$ and $\mathbf{q}_f$ are, respectively, the vectors of the system reference and elastic coordinates.

Equation 155 can then be written in terms of reference and elastic coordinates as follows:

$$\mathbf{C}(\mathbf{q}_r, \mathbf{q}_f, t) = \mathbf{0} \tag{5.157}$$

In the following $\mathbf{q}_f$ represents the vector of generalized elastic coordinates that can be introduced by using the finite-element method, Rayleigh–Ritz methods, or a set of experimentally identified data (Shabana 1986). This vector can be a set of physical or modal elastic coordinates.

Using Eq. 133a, the general system differential equations of motion of the multibody system can be written in a matrix form as

$$\mathbf{M}\ddot{\mathbf{q}} + \mathbf{K}\mathbf{q} + \mathbf{C}_{\mathbf{q}}^{\mathrm{T}}\boldsymbol{\lambda} = \mathbf{Q}_e + \mathbf{Q}_v \tag{5.158}$$

where $\mathbf{M}$ and $\mathbf{K}$ are, respectively, the system mass and stiffness matrices, $\mathbf{C_q}$ is the constraint Jacobian matrix, λ is the vector of Lagrange multipliers, $\mathbf{Q}_e$ is the vector of generalized externally applied forces, and $\mathbf{Q}_v$ is the quadratic velocity vector that contains the gyroscopic as well as Coriolis components and results from differentiating the kinetic energy with respect to time and with respect to the system generalized coordinates. According to the generalized coordinate partitioning of Eq. 156, Eq. 158 can be written in a more explicit form as

$$\begin{bmatrix} \mathbf{m}_{rr} & \mathbf{m}_{rf} \\ \text{symmetric} & \mathbf{m}_{ff} \end{bmatrix} \begin{bmatrix} \ddot{\mathbf{q}}_r \\ \ddot{\mathbf{q}}_f \end{bmatrix} + \begin{bmatrix} \mathbf{0} & \mathbf{0} \\ \mathbf{0} & \mathbf{K}_{ff} \end{bmatrix} \begin{bmatrix} \mathbf{q}_r \\ \mathbf{q}_f \end{bmatrix} + \begin{bmatrix} \mathbf{C}^{\mathrm{T}}_{\mathbf{q}_r} \\ \mathbf{C}^{\mathrm{T}}_{\mathbf{q}_f} \end{bmatrix} \lambda = \begin{bmatrix} (\mathbf{Q}_r)_e \\ (\mathbf{Q}_f)_e \end{bmatrix} + \begin{bmatrix} (\mathbf{Q}_r)_v \\ (\mathbf{Q}_f)_v \end{bmatrix} \tag{5.159}$$

A systematic approach for solving for the acceleration vector $\ddot{\mathbf{q}}$ and the generalized reaction forces $\mathbf{C}^{\mathrm{T}}_{\mathbf{q}}\lambda$ is discussed in the following paragraphs.

Accelerations and Lagrange Multipliers Differentiating Eq. 155 with respect to time yields

$$\mathbf{C_q}\dot{\mathbf{q}} + \mathbf{C}_t = \mathbf{0} \tag{5.160}$$

where $\dot{\mathbf{q}}$ is the velocity vector and $\mathbf{C}_t$ is the partial derivative of the vector of constraints with respect to time. According to the generalized coordinate partitioning of Eq. 156, Eq. 160 can be written in a more explicit form as

$$\mathbf{C}_{\mathbf{q}_r}\dot{\mathbf{q}}_r + \mathbf{C}_{\mathbf{q}_f}\dot{\mathbf{q}}_f + \mathbf{C}_t = \mathbf{0} \tag{5.161}$$

that is,

$$[\mathbf{C}_{\mathbf{q}_r} \quad \mathbf{C}_{\mathbf{q}_f}] \begin{bmatrix} \dot{\mathbf{q}}_r \\ \dot{\mathbf{q}}_f \end{bmatrix} + \mathbf{C}_t = \mathbf{0} \tag{5.162}$$

Differentiating Eq. 160 with respect to time yields

$$\mathbf{C_q}\ddot{\mathbf{q}} = -\mathbf{C}_{tt} - 2\mathbf{C}_{\mathbf{q}t}\dot{\mathbf{q}} - (\mathbf{C_q}\dot{\mathbf{q}})_{\mathbf{q}}\dot{\mathbf{q}} \tag{5.163}$$

that is,

$$[\mathbf{C}_{\mathbf{q}_r} \quad \mathbf{C}_{\mathbf{q}_f}] \begin{bmatrix} \ddot{\mathbf{q}}_r \\ \ddot{\mathbf{q}}_f \end{bmatrix} = \mathbf{Q}_c \tag{5.164}$$

where $\mathbf{Q}_c$ is a vector that depends on the reference and elastic coordinates and velocities and possibly on time. This vector is defined as

$$\mathbf{Q}_c = -\mathbf{C}_{tt} - 2\mathbf{C}_{\mathbf{q}t}\dot{\mathbf{q}} - (\mathbf{C_q}\dot{\mathbf{q}})_{\mathbf{q}}\dot{\mathbf{q}} \tag{5.165}$$

One may now combine Eqs. 158 and 163 and write the matrix equation

$$\begin{bmatrix} \mathbf{M} & \mathbf{C}^{\mathrm{T}}_{\mathbf{q}} \\ \text{symmetric} & \mathbf{0} \end{bmatrix} \begin{bmatrix} \ddot{\mathbf{q}} \\ \lambda \end{bmatrix} = \begin{bmatrix} \mathbf{Q}_e + \mathbf{Q}_v - \mathbf{K}\mathbf{q} \\ \mathbf{Q}_c \end{bmatrix} \tag{5.166}$$

or, alternatively, combine Eqs. 159 and 164 to obtain a more explicit form of Eq. 166 as

$$\begin{bmatrix} \mathbf{m}_{rr} & \mathbf{m}_{rf} & \mathbf{C}_{\mathbf{q}_r}^{\mathrm{T}} \\ & \mathbf{m}_{ff} & \mathbf{C}_{\mathbf{q}_f}^{\mathrm{T}} \\ \text{symmetric} & & \mathbf{0} \end{bmatrix} \begin{bmatrix} \ddot{\mathbf{q}}_r \\ \ddot{\mathbf{q}}_f \\ \boldsymbol{\lambda} \end{bmatrix} = \begin{bmatrix} (\mathbf{Q}_r)_e + (\mathbf{Q}_r)_v \\ (\mathbf{Q}_f)_e + (\mathbf{Q}_f)_v - \mathbf{K}_{ff}\mathbf{q}_f \\ \mathbf{Q}_c \end{bmatrix} \tag{5.167}$$

Equation 166 or its explicit form of Eq. 167 can be considered as a system of linear algebraic equations that can be solved for the accelerations and the vector of Lagrange multipliers. This system of equations can be written as

$$\mathbf{M}_\lambda \mathbf{q}_\lambda = \bar{\mathbf{F}} \tag{5.168}$$

where

$$\mathbf{q}_\lambda = [\ddot{\mathbf{q}}^{\mathrm{T}} \quad \boldsymbol{\lambda}^{\mathrm{T}}]^{\mathrm{T}}$$

$$\mathbf{M}_\lambda = \begin{bmatrix} \mathbf{M} & \mathbf{C}_{\mathbf{q}}^{\mathrm{T}} \\ \mathbf{C}_{\mathbf{q}} & \mathbf{0} \end{bmatrix}$$

and

$$\bar{\mathbf{F}} = \begin{bmatrix} \mathbf{Q}_e + \mathbf{Q}_v - \mathbf{K}\mathbf{q} \\ \mathbf{Q}_c \end{bmatrix} = \begin{bmatrix} \bar{\mathbf{F}}_q \\ \bar{\mathbf{F}}_\lambda \end{bmatrix}$$

For a physically meaningful system, the coefficient matrix $\mathbf{M}_\lambda$ of Eq. 168 is nonsingular. Therefore, one can solve for the vector $\mathbf{q}_\lambda$ and write

$$\mathbf{q}_\lambda = \bar{\mathbf{M}}\bar{\mathbf{F}} \tag{5.169}$$

where the matrix $\bar{\mathbf{M}}$ is the inverse of the matrix $\mathbf{M}_\lambda$, that is

$$\bar{\mathbf{M}} = \bar{\mathbf{M}}(\mathbf{q}) = \mathbf{M}_\lambda^{-1} \tag{5.170}$$

We may write Eq. 169 in a partitioned form as

$$\begin{bmatrix} \ddot{\mathbf{q}} \\ \boldsymbol{\lambda} \end{bmatrix} = \begin{bmatrix} \bar{\mathbf{M}}_{qq} & \bar{\mathbf{M}}_{q\lambda} \\ \bar{\mathbf{M}}_{\lambda q} & \bar{\mathbf{M}}_{\lambda\lambda} \end{bmatrix} \begin{bmatrix} \bar{\mathbf{F}}_q \\ \bar{\mathbf{F}}_\lambda \end{bmatrix} \tag{5.171}$$

from which the acceleration vector $\ddot{\mathbf{q}}$ and the vector of Lagrange multipliers can be written as

$$\ddot{\mathbf{q}} = \bar{\mathbf{M}}_{qq}\bar{\mathbf{F}}_q + \bar{\mathbf{M}}_{q\lambda}\bar{\mathbf{F}}_\lambda \tag{5.172}$$

$$\boldsymbol{\lambda} = \bar{\mathbf{M}}_{\lambda q}\bar{\mathbf{F}}_q + \bar{\mathbf{M}}_{\lambda\lambda}\bar{\mathbf{F}}_\lambda \tag{5.173}$$

Having determined the vector of Lagrange multipliers $\boldsymbol{\lambda}$, one can evaluate the generalized reaction force vector $\mathbf{C}_{\mathbf{q}}^{\mathrm{T}}\boldsymbol{\lambda}$.

Note that in the floating frame of reference formulation, the matrix $\mathbf{M}_\lambda$ of Eq. 168 depends on the system generalized coordinates. Thus, its inverse $\bar{\mathbf{M}}$ of Eq. 169 and the submatrices $\bar{\mathbf{M}}_{qq}$, $\bar{\mathbf{M}}_{q\lambda}$, $\bar{\mathbf{M}}_{\lambda q}$, and $\bar{\mathbf{M}}_{\lambda\lambda}$ are also functions of the system generalized coordinates. Furthermore, we know from the development of the preceding sections that the vector $\bar{\mathbf{F}}$ of Eq. 169 depends on the system generalized coordinates and velocities as well as on time. Therefore, it is expected that the acceleration vector of Eq. 172 also depends on the system generalized coordinates, velocities, and time. In a large-scale mechanical system, it is quite difficult to have a closed-form solution for the acceleration vector in terms of the generalized coordinates, velocities, and time. Therefore, one may write the vector $\ddot{\mathbf{q}}$ in the following simplified form:

$$\ddot{\mathbf{q}} = \mathbf{f}(\mathbf{q},\ \dot{\mathbf{q}},\ t) \tag{5.174}$$

where $\mathbf{f}$ is a vector function whose numerical values are determined by Eq. 172 as the result of solving the system of algebraic equations of Eq. 168. The vector function $\mathbf{f}$ is then defined as

$$\mathbf{f}(\mathbf{q},\ \dot{\mathbf{q}},\ t) = \bar{\mathbf{M}}_{qq}\bar{\mathbf{F}}_q + \bar{\mathbf{M}}_{q\lambda}\bar{\mathbf{F}}_\lambda \tag{5.175}$$

Numerical Procedures If the expression for the acceleration functions is simple such that they can be integrated in a closed form, the constraint equations of Eq. 155 are automatically satisfied since the accelerations are obtained from the solution of Eq. 167, which includes the effect of the constraint forces. Another approach is to identify a set of independent coordinates and integrate only the independent components of the accelerations associated with these independent coordinates. Having determined these independent coordinates and velocities in a closed form, one can then use the kinematic constraint equations to determine the dependent coordinates and velocities. One is seldom able, however, to obtain simple expressions for the accelerations in multibody system dynamics since the governing equations are highly nonlinear. One must then resort to direct numerical integration methods to obtain an approximate solution. The error in the state vector obtained by using the direct numerical methods will certainly depend on the numerical integration technique used. Furthermore, the kinematic constraint equations are not automatically satisfied because the solution obtained is not exact. The error in the system coordinates will lead to a violation in the kinematic constraint equations.

On the basis of the Lagrangian formulation with the multipliers, several numerical schemes have been used in some of the existing computer programs developed for the nonlinear dynamic analysis of multibody systems. Some of these numerical schemes are summarized below.

1. Perhaps the simplest approach is to assume that the numerical integration routine is close to perfect and hope that the error resulting from the numerical integration is small such that the violations in the kinematic constraint equations can be neglected. In this case the total vector of system accelerations can be integrated forward in time to determine the system coordinates and velocities. This numerical scheme is simple and provides acceptable solutions in many engineering applications. In many other applications, however, the accumulation of the error of the numerical integration increases with time, and this, in turn, leads to violations in the kinematic constraint relationships.

2. Another approach is to identify the system-independent coordinates. This can be achieved by using the generalized coordinate partitioning of the constraint Jacobian matrix. Only independent state equations are integrated forward in time by use of a direct numerical integration method. The solutions of the state equations define the system-independent generalized coordinates and velocities. Dependent generalized coordinates and velocities are then determined by using the kinematic constraint equations. In this scheme, since the constraint relationships are used to determine the dependent coordinates, there is no violation in the kinematic constraint equations. This, however, does not mean that the solution obtained is exact. It is clear that in this approach, dependent coordinates must be adjusted according to the error in the independent coordinates resulting from the numerical integration, and even though the kinematic relationships are not violated, the error in the dependent coordinates will be at least of the same order as the error in the independent coordinates as the result of using approximate direct numerical integration methods. This approach was proposed by Wehage (1980).

3. A third scheme is to integrate all the state equations associated with both the dependent and independent coordinates and then adjust the dependent coordinates by applying the kinematic constraint equations. The integration of all coordinates may provide a better estimate for the dependent variables; therefore, fewer iterations will be required in solving numerically the nonlinear kinematic constraint equations for the dependent variables. Many of the accurate predictor–corrector multistep numerical integration routines, however, use the history of the coordinates and velocities in evaluating the coefficients of the time-dependent polynomials used to advance the numerical integration (Shampine and Gordon 1975). Furthermore, many of these routines have sophisticated error-control schemes that can detect any change in the time history of the state variables. Adjusting the dependent coordinates may be a source of disturbance to the numerical integration routine. In many applications this leads to numerical problems.

It is clear that the preceding numerical schemes provide only approximate solutions for the state of the multibody system. These three numerical integration schemes are implemented in the general-purpose computer program DAMS (Dynamic Analysis of Multibody Systems) (Shabana 1985). In this computer program the system nonlinear differential and algebraic constraint equations are computer-generated and integrated forward in time. In the following sections we select the second approach proposed by Wehage (1980) for more detailed discussion. Computational algorithms for the static, kinematic, and dynamic analysis will also be discussed.

5.10 GENERALIZED COORDINATE PARTITIONING

The dynamics of multibody systems consisting of interconnected rigid and deformable bodies can be described by the coupled set of differential and algebraic equations given by Eqs. 155 and 158. Many techniques are available in the literature for the numerical solution of a mixed set of differential and algebraic equations. In this section, however, we outline the technique proposed by Wehage (1980) and discuss the use of *Wehage's algorithm* for solving the dynamic equations of multibody systems consisting of interconnected rigid and deformable bodies.

Dependent and Independent Coordinates In the preceding section, we outlined a method for obtaining the acceleration vector. As pointed out earlier, the acceleration vector is a vector function of the system generalized coordinates, velocities, and time. This functional relationship represents n second-order differential equations, where n is the number of system generalized coordinates. The solutions of these equations, however, are not independent because of the kinematic constraints that describe mechanical joints as well as specified trajectories. One has to identify a set of independent coordinates and the associated set of differential equations that can be integrated forward in time in order to define the independent variables. Dependent coordinates (variables) can then be determined by using the kinematic relations.

For a virtual change $\delta\mathbf{q}$ in the system generalized coordinates, Eq. 155 yields

$$\mathbf{C_q}\delta\mathbf{q} = \mathbf{0} \tag{5.176}$$

The Jacobian matrix $\mathbf{C_q}$ is an $n_c \times n$ matrix where $n_c < n$. Since the constraint equations are assumed to be linearly independent, the Jacobian $\mathbf{C_q}$ has a full row rank; therefore, one must be able to identify $(n - n_c)$ independent coordinates and write the vector of generalized coordinates $\mathbf{q}$ as

$$\mathbf{q} = [\mathbf{q}_i^{\mathrm{T}} \quad \mathbf{q}_d^{\mathrm{T}}]^{\mathrm{T}} \tag{5.177}$$

here $\mathbf{q}_i$ and $\mathbf{q}_d$ are, respectively, the vectors of independent and dependent coordinates. Both the dependent and independent vectors of coordinates can be a mixed set of reference and elastic coordinates. According to the generalized coordinate partitioning of Eq. 177, Eq. 176 can be written as

$$\mathbf{C}_{\mathbf{q}_i}\delta\mathbf{q}_i + \mathbf{C}_{\mathbf{q}_d}\delta\mathbf{q}_d = \mathbf{0}$$

or

$$\mathbf{C}_{\mathbf{q}_d}\delta\mathbf{q}_d = -\mathbf{C}_{\mathbf{q}_i}\delta\mathbf{q}_i \tag{5.178}$$

where $\mathbf{C}_{\mathbf{q}_d}$ is an $n_c \times n_c$ matrix and $\mathbf{C}_{\mathbf{q}_i}$ is an $n_c \times (n - n_c)$ matrix. Because the constraint equations are linearly independent, one, in general, must be able to select the independent coordinates such that the matrix $\mathbf{C}_{\mathbf{q}_d}$ is nonsingular. Equation 178 implies that the virtual change in the dependent coordinates can be written in terms of the virtual change of the independent ones, that is,

$$\delta\mathbf{q}_d = \mathbf{C}_{di}\delta\mathbf{q}_i \tag{5.179}$$

where $\mathbf{C}_{di}$ is the matrix

$$\mathbf{C}_{di} = -\mathbf{C}_{\mathbf{q}_d}^{-1}\mathbf{C}_{\mathbf{q}_i} \tag{5.180}$$

In like manner, the first derivative of the constraint equations with respect to time given by Eq. 160 can be written as

$$\mathbf{C}_{\mathbf{q}_i}\dot{\mathbf{q}}_i + \mathbf{C}_{\mathbf{q}_d}\dot{\mathbf{q}}_d + \mathbf{C}_t = \mathbf{0}$$

This equation can be used to write the dependent velocities in terms of the independent ones as follows:

$$\dot{\mathbf{q}}_d = -\mathbf{C}_{\mathbf{q}_d}^{-1}[\mathbf{C}_{\mathbf{q}_i}\dot{\mathbf{q}}_i + \mathbf{C}_t]$$

or

$$\dot{\mathbf{q}}_d = \mathbf{C}_{di}\dot{\mathbf{q}}_i - \mathbf{C}_{\mathbf{q}_d}^{-1}\mathbf{C}_t \tag{5.181}$$

where the matrix $\mathbf{C}_{di}$ is given by Eq. 180.

In a large-scale flexible multibody system, the identification of the independent and dependent coordinates may be a very difficult task. It is, therefore, more appropriate to use numerical techniques that take advantage of the numerical structure of the Jacobian matrix to identify the independent coordinates. Such numerical algorithms are available in the literature, and the interested reader should consult texts on the subject of numerical methods.

Having identified the independent and dependent sets of coordinates, one may now define the following state vector:

$$\mathbf{y} = \begin{bmatrix}\mathbf{q}_i^{\mathrm{T}} & \dot{\mathbf{q}}_i^{\mathrm{T}}\end{bmatrix}^{\mathrm{T}} \tag{5.182}$$

The associated independent state equations can now be defined as

$$\dot{\mathbf{y}} = \begin{bmatrix}\dot{\mathbf{q}}_i^{\mathrm{T}} & \ddot{\mathbf{q}}_i^{\mathrm{T}}\end{bmatrix}^{\mathrm{T}} = \mathbf{g}(\mathbf{q}, \dot{\mathbf{q}}, t) \tag{5.183}$$

Given a set of initial conditions, one can integrate the state equations forward in time using a direct numerical integration method. The solution of the state equations defines the independent coordinates and velocities. Dependent coordinates and velocities can then be determined, respectively, using the kinematic relations of Eqs. 155 and 181. The independent and dependent accelerations are determined by solving Eq. 168.

Embedding Techniques The solution of Eq. 168 is not the only approach for determining the independent accelerations. An alternate approach is to use Eq. 158 and the kinematic relationships developed in this section to obtain a minimum number of differential equations expressed in terms of the independent accelerations.

Using the generalized coordinate partitioning of Eq. 177, Eqs. 163 and 165 yield

$$\mathbf{C}_{\mathbf{q}_i}\ddot{\mathbf{q}}_i + \mathbf{C}_{\mathbf{q}_d}\ddot{\mathbf{q}}_d = \mathbf{Q}_c$$

That is,

$$\ddot{\mathbf{q}}_d = -\mathbf{C}_{\mathbf{q}_d}^{-1}\mathbf{C}_{\mathbf{q}_i}\ddot{\mathbf{q}}_i + \mathbf{C}_{\mathbf{q}_d}^{-1}\mathbf{Q}_c$$

The total vector of system accelerations can then be written in terms of the independent accelerations as

$$\ddot{\mathbf{q}} = \begin{bmatrix}\ddot{\mathbf{q}}_i \\ \ddot{\mathbf{q}}_d\end{bmatrix} = \begin{bmatrix}\mathbf{I} \\ -\mathbf{C}_{\mathbf{q}_d}^{-1}\mathbf{C}_{\mathbf{q}_i}\end{bmatrix}\ddot{\mathbf{q}}_i + \begin{bmatrix}\mathbf{0} \\ \mathbf{C}_{\mathbf{q}_d}^{-1}\mathbf{Q}_c\end{bmatrix}$$

where $\mathbf{I}$ in this equation is an identity matrix whose dimension is equal to the number of independent coordinates. The preceding equations can be rewritten in a more compact form as

$$\ddot{\mathbf{q}} = \mathbf{B}_{di}\ddot{\mathbf{q}}_i + \bar{\mathbf{Q}}_c$$

where

$$\mathbf{B}_{di} = \begin{bmatrix}\mathbf{I} \\ -\mathbf{C}_{\mathbf{q}_d}^{-1}\mathbf{C}_{\mathbf{q}_i}\end{bmatrix}, \qquad \bar{\mathbf{Q}}_c = \begin{bmatrix}\mathbf{0} \\ \mathbf{C}_{\mathbf{q}_d}^{-1}\mathbf{Q}_c\end{bmatrix}$$

Assuming that the accelerations and coordinates in Eq. 158 are rearranged according to the partitioning of Eq. 177, the vector of accelerations $\ddot{\mathbf{q}}$ can be substituted into

Eq. 158 to yield

$$\mathbf{M}(\mathbf{B}_{di}\ddot{\mathbf{q}}_i + \bar{\mathbf{Q}}_c) + \mathbf{C}_{\mathbf{q}}^{\mathrm{T}}\boldsymbol{\lambda} = \mathbf{Q}_e + \mathbf{Q}_v - \mathbf{K}\mathbf{q}$$

Premultiplying this equation by the transpose of the matrix $\mathbf{B}_{di}$ and rearranging terms, one obtains

$$\mathbf{B}_{di}^{\mathrm{T}}\mathbf{M}\mathbf{B}_{di}\ddot{\mathbf{q}}_i + \mathbf{B}_{di}^{\mathrm{T}}\mathbf{C}_{\mathbf{q}}^{\mathrm{T}}\boldsymbol{\lambda} = \mathbf{B}_{di}^{\mathrm{T}}(\mathbf{Q}_e + \mathbf{Q}_v - \mathbf{K}\mathbf{q}) - \mathbf{B}_{di}^{\mathrm{T}}\mathbf{M}\bar{\mathbf{Q}}_c$$

which can be written as

$$\mathbf{M}_{ii}\ddot{\mathbf{q}}_i + \mathbf{B}_{di}^{\mathrm{T}}\mathbf{C}_{\mathbf{q}}^{\mathrm{T}}\boldsymbol{\lambda} = \mathbf{Q}_i \tag{5.184}$$

where $\mathbf{M}_{ii}$ and $\mathbf{Q}_i$ are, respectively, the mass matrix and force vector associated with the independent coordinates. The matrix $\mathbf{M}_{ii}$ and the vector $\mathbf{Q}_i$ are defined as

$$\mathbf{M}_{ii} = \mathbf{B}_{di}^{\mathrm{T}}\mathbf{M}\mathbf{B}_{di}$$
$$\mathbf{Q}_i = \mathbf{B}_{di}^{\mathrm{T}}(\mathbf{Q}_e + \mathbf{Q}_v - \mathbf{K}\mathbf{q}) - \mathbf{B}_{di}^{\mathrm{T}}\mathbf{M}\bar{\mathbf{Q}}_c$$

One can show that $\mathbf{B}_{di}^{\mathrm{T}}\mathbf{C}_{\mathbf{q}}^{\mathrm{T}}$ is a null matrix. To this end, we write

$$\mathbf{B}_{di}^{\mathrm{T}}\mathbf{C}_{\mathbf{q}}^{\mathrm{T}} = [\mathbf{I} \quad -(\mathbf{C}_{\mathbf{q}_d}^{-1}\mathbf{C}_{\mathbf{q}_i})^{\mathrm{T}}]\begin{bmatrix}\mathbf{C}_{\mathbf{q}_i}^{\mathrm{T}} \\ \mathbf{C}_{\mathbf{q}_d}^{\mathrm{T}}\end{bmatrix}$$
$$= [\mathbf{C}_{\mathbf{q}_i}^{\mathrm{T}} - (\mathbf{C}_{\mathbf{q}_d}^{-1}\mathbf{C}_{\mathbf{q}_i})\mathbf{C}_{\mathbf{q}_d}^{\mathrm{T}}]$$

Since the inverse of the transpose of a nonsingular matrix is equal to the transpose of the inverse, one has

$$\mathbf{B}_{di}^{\mathrm{T}}\mathbf{C}_{\mathbf{q}}^{\mathrm{T}} = [\mathbf{C}_{\mathbf{q}_i}^{\mathrm{T}} - \mathbf{C}_{\mathbf{q}_i}^{\mathrm{T}}(\mathbf{C}_{\mathbf{q}_d}^{\mathrm{T}})^{-1}\mathbf{C}_{\mathbf{q}_d}^{\mathrm{T}}] = \mathbf{C}_{\mathbf{q}_i}^{\mathrm{T}} - \mathbf{C}_{\mathbf{q}_i}^{\mathrm{T}} = \mathbf{0}$$

and accordingly, Eq. 184 reduces to

$$\mathbf{M}_{ii}\ddot{\mathbf{q}}_i = \mathbf{Q}_i \tag{5.185}$$

which indeed proves that the force of constraints can be eliminated when the differential equations are formulated in terms of the independent coordinates. This procedure of eliminating the constraint forces is called the *embedding technique*. The use of Eq. 184, however, for determining the independent accelerations is not as efficient compared with the use of Eq. 168 since $\mathbf{M}_{ii}$ of Eq. 184 is a full matrix while $\mathbf{M}_\lambda$ of Eq. 168 is a sparse matrix. Furthermore, the use of the embedding technique for determining the independent accelerations requires more matrix multiplications and inversions than the use of Eq. 168.

5.11 ORGANIZATION OF MULTIBODY COMPUTER PROGRAMS

In this section, we discuss some considerations that should be taken into account in developing a general computational scheme for the dynamic analysis of flexible multibody systems using the floating frame of reference formulation. The organization of the multibody computer programs as well as the functions of their modules will be explained.

Preprocessor It is clear from the development presented in this chapter that a set of inertia shape integrals is required for each deformable body to generate the inertia coupling between the reference motion and the elastic deformation of the body. These integrals can be evaluated only once in advance for the dynamic analysis using a structural analysis program. The inertia shape integrals can also be generated using the finite-element method as described in the following chapter. It is more computationally efficient to evaluate these integrals once in advance for the dynamic analysis and store them for use whenever they are needed. This can be done in a preprocessor structural analysis program called PREDAMS. This program has the capability of generating the inertia shape integrals of the deformable bodies using consistent or lumped masses. Furthermore, the preprocessor PREDAMS can be linked with any existing finite-element program to generate the shape integrals for some special elements that are not available in the library of the program.

Main Processor Having determined the intertia shape integrals of each deformable body in the preprocessor PREDAMS, one can input these matrices to the dynamic analysis program DAMS along with the description of the rigid components in the multibody system. The computer code DAMS, which has the capability of performing the two- and three-dimensional analysis of flexible multibody systems, is divided into four modules – the *Constraint Module* (CONMOD), the *Mass Module* (MASMOD), the *Force Module* (FRCMOD), and the *Numerical Module* (NUMMOD):

1. *CONMOD* (*Constraint Module*) To perform the kinematic and dynamic analysis, one has to evaluate the constraint functions of Eq. 155, the Jacobian matrix $\mathbf{C_q}$, the first time derivative of the constraint function $\mathbf{C}_t$ (Eq. 160), and the vector $\mathbf{Q}_c$ of Eq. 164. Evaluation of the constraint functions and the Jacobian matrix is necessary for the position analysis since in the DAMS program a *Newton–Raphson* algorithm is used to correct for constraint violations. According to this algorithm, a solution of Eq. 155 is obtained by solving iteratively the following nonlinear system of equations

$$\mathbf{C_q}\Delta\mathbf{q} = -\mathbf{C}(\mathbf{q}, t) \tag{5.186}$$

where $\Delta\mathbf{q}$ are the *Newton differences*. Equation 155 is then satisfied if the norm of the vector $\Delta\mathbf{q}$ or the norm of the vector $\mathbf{C}$ is small, that is,

$$|\Delta\mathbf{q}| < \varepsilon_1, \qquad |\mathbf{C}| < \varepsilon_2$$

where $|\;|$ denotes a selected norm and ε_1 and ε_2 are small numbers. The Jacobian matrix as well as the time derivative of the constraints with respect to time are required for the velocity analysis (see Eq. 181). The Jacobian matrix $\mathbf{C_q}$ and the vector $\mathbf{Q}_c$ are needed for the acceleration analysis (see Eq. 166 or 167). In the constraint module CONMOD, the Jacobian matrix $\mathbf{C_q}$, the vectors $\mathbf{C}$, $\mathbf{C}_t$, and $\mathbf{Q}_c$ are computer-generated for a set of standard joints that can be utilized by providing a standard set of input data. In the two-dimensional analysis, the following standard library constraints are available:

(a) Revolute joints between rigid bodies, between deformable bodies, and between rigid and deformable bodies.

(b) Rigid joints between rigid bodies, between deformable bodies, and between rigid and deformable bodies.
(c) Translational (prismatic) joints between rigid bodies.
(d) Constraints that fix the generalized coordinates with respect to time.
(e) Specified trajectories of arbitrary points on the rigid and/or deformable bodies.

In the three-dimensional analysis the following standard library kinematic constraints are available:

(a) Constraints that fix the generalized coordinates with respect to time.
(b) Specified trajectories of arbitrary points on the rigid and/or deformable bodies.
(c) Spherical joints between rigid bodies, between deformable bodies, and between rigid and deformable bodies.
(d) Revolute joints between rigid bodies, between deformable bodies, and between rigid and deformable bodies.
(e) Translational (prismatic) joints between rigid bodies.
(f) Cylindrical joints between rigid bodies.
(g) Rigid joints between rigid bodies, between deformable bodies, and between rigid and deformable bodies.

Any other kinematic constraints can be introduced by the user through a set of user subroutines.

2. *MASMOD* (*Mass Module*) In this module the mass matrix of each rigid and deformable body is constructed. These matrices are used to define the entire system mass matrix $\mathbf{M}$ of the multibody system consisting of interconnected rigid and deformable bodies. In addition to the system mass matrix, the quadratic velocity vector $\mathbf{Q}_v$ of Eq. 166 and the coefficient matrix $\mathbf{M}_\lambda$ of Eq. 168 are also evaluated in the mass module MASMOD.

3. *FRCMOD* (*Force Module*) In this module the generalized forces as well as the elastic forces of deformable bodies are evaluated and the generalized forces of spring–damper–actuator elements connecting two rigid bodies, two deformable bodies, or a rigid body and a deformable body in the system are automatically generated. This module calls user subroutines that allow the user to supply any generalized forcing functions that may depend on the system generalized coordinates and velocities and possibly on time.

4. *NUMMOD* (*Numerical Module*) This module has three main functions:

(a) The capacity to determine the rank and the independent rows and columns of a system of algebraic equations with a full (nonsparse) singular coefficient matrix. This function can be used to identify the multibody system independent coordinates.
(b) Solution of a system of algebraic equations with a sparse coefficient matrix. This function is used for the analysis of position, velocity, and acceleration.
(c) Direct numerical integration of a set of first-order differential equations. This function is used in the numerical integration of the state equations.

In the following section, we describe some of the numerical algorithms used in the DAMS program that consists of the four modules discussed above.

5.12 NUMERICAL ALGORITHMS

In this section, the numerical algorithms for the solution of coupled sets of differential and algebraic constraint equations that describe the dynamics of constrained mechanical systems are discussed. These numerical algorithms are implemented in the general-purpose computer program DAMS developed for the nonlinear dynamic analysis of general multibody systems consisting of interconnected rigid and deformable bodies. The program is capable of performing the following types of analyses: (1) static, (2) kinematic, (3) dynamic, and (4) static and dynamic. The analysis type can be chosen by setting the value of a flag called *IANL*. If IANL equals 1, the program performs static analysis; if IANL equals 2, the program performs dynamic analysis; if IANL equals 3, the program performs static and dynamic analysis; and if IANL equals 4, the program performs kinematic analysis, that is

$$\text{IANL} = \begin{cases} 1, \text{ static} \\ 2, \text{ dynamic} \\ 3, \text{ static and dynamic} \\ 4, \text{ kinematic} \end{cases}$$

In the following, the algorithms used in the DAMS program for static, kinematic, and dynamic analyses are discussed.

Static Analysis There are many numerical algorithms available in the literature for the *static analysis* of constrained mechanical systems. We, however, select one of these algorithms, which is implemented in the DAMS program, to discuss in this section. Before we provide an outline for this algorithm, we first discuss the governing equations used in this algorithm.

For the static analysis Eqs. 158 and 155 can be rewritten, respectively, as

$$\mathbf{K}\mathbf{q} + \mathbf{C}_{\mathbf{q}}^{\mathrm{T}}\boldsymbol{\lambda} = \mathbf{Q}_e \tag{5.187}$$

$$\mathbf{C}(\mathbf{q}) = \mathbf{0} \tag{5.188}$$

where the vector of constraint functions $\mathbf{C}$ depends only on the vector of system generalized coordinates. One may define $\mathbf{R}_e$ as

$$\mathbf{R}_e = \mathbf{Q}_e - \mathbf{K}\mathbf{q} \tag{5.189}$$

and write Eq.187 as

$$\mathbf{C}_{\mathbf{q}}^{\mathrm{T}}\boldsymbol{\lambda} - \mathbf{R}_e = \mathbf{0} \tag{5.190}$$

In the static analysis the vector $\mathbf{R}_e$ is a function of the system reference and elastic generalized coordinates $\mathbf{q}$, which can be written in a partitioned form as (see Eq. 177)

$$\mathbf{q} = \begin{bmatrix} \mathbf{q}_i^{\mathrm{T}} & \mathbf{q}_d^{\mathrm{T}} \end{bmatrix}^{\mathrm{T}} \tag{5.191}$$

where $\mathbf{q}_i$ and $\mathbf{q}_d$ are, respectively, the vectors of independent and dependent coordinates. For a virtual change $\delta\mathbf{q}$ in the system generalized coordinates, Eq. 190 leads to

$$(\mathbf{C}_\mathbf{q}^\mathrm{T}\lambda - \mathbf{R}_e)^\mathrm{T}\delta\mathbf{q} = \mathbf{0} \tag{5.192}$$

As explained in the preceding sections, one may use Eq. 188 to write the virtual change of the dependent coordinates in terms of the virtual change of the independent ones, that is,

$$\delta\mathbf{q}_d = \mathbf{C}_{di}\delta\mathbf{q}_i \tag{5.193}$$

where the matrix $\mathbf{C}_{di}$ is defined by Eq. 180. One can then write the vector $\delta\mathbf{q}$ as

$$\delta\mathbf{q} = \begin{bmatrix} \delta\mathbf{q}_i \\ \delta\mathbf{q}_d \end{bmatrix} = \begin{bmatrix} \mathbf{I} \\ \mathbf{C}_{di} \end{bmatrix} \delta\mathbf{q}_i = \mathbf{B}_{di}\delta\mathbf{q}_i \tag{5.194}$$

where $\mathbf{I}$ is an identity matrix and $\mathbf{B}_{di}$ is an $n \times (n - n_c)$ matrix, where n is the total number of the system generalized coordinates, and n_c is the number of the constraint equations. The matrix $\mathbf{B}_{di}$ is given by

$$\mathbf{B}_{di} = \begin{bmatrix} \mathbf{I} \\ \mathbf{C}_{di} \end{bmatrix} \tag{5.195}$$

Substituting Eq. 194 into Eq. 192 leads to

$$(\mathbf{C}_\mathbf{q}^\mathrm{T}\lambda - \mathbf{R}_e)^\mathrm{T}\mathbf{B}_{di}\delta\mathbf{q}_i = \mathbf{0} \tag{5.196}$$

Since the components of the vector $\delta\mathbf{q}_i$ are linearly independent, Eq. 196 yields

$$(\mathbf{C}_\mathbf{q}^\mathrm{T}\lambda - \mathbf{R}_e)^\mathrm{T}\mathbf{B}_{di} = \mathbf{0} \tag{5.197}$$

This is a system of $(n - n_c)$ nonlinear algebraic constraint equations whose solution determines the static equilibrium position of the multibody system consisting of interconnected rigid and deformable bodies. If a correct estimate is made for the system static configuration, Eq. 197 will be satisfied. For a large-scale flexible multibody system, an accurate estimate of the system configuration may be difficult. Therefore, one expects, by assuming a set of generalized coordinates $\mathbf{q}$, that Eq. 197 may be violated, that is,

$$(\mathbf{C}_\mathbf{q}^\mathrm{T}\lambda - \mathbf{R}_e)^\mathrm{T}\mathbf{B}_{di} = \bar{\mathbf{R}}^\mathrm{T} \tag{5.198}$$

Since, as shown in the preceding section, $\mathbf{C}_\mathbf{q}\mathbf{B}_{di}$ is the null matrix, Eq. 198 reduces to

$$-\mathbf{R}_e^\mathrm{T}\mathbf{B}_{di} = \bar{\mathbf{R}}^\mathrm{T} \tag{5.199}$$

where $\bar{\mathbf{R}}$ is called the vector of *residual forces* associated with the independent generalized coordinates. It is obvious that the vector $\bar{\mathbf{R}}$ depends on the assumed system configuration and the norm of this vector is small if the initial guess of the system configuration is close to the correct static configuration. In fact, Eq. 197 is satisfied if the vector $\bar{\mathbf{R}}$ is identically zero, that is

$$\bar{\mathbf{R}} = \mathbf{0} \tag{5.200}$$

The roots of this system of nonlinear homogeneous algebraic equations then determine the exact static equilibrium position of the multibody system. This system of $(n - n_c)$ equations can be solved numerically by using a Newton–Raphson algorithm, in which an iterative solution of the following system of algebraic equations is sought (Atkinson 1978).

$$\frac{\partial \bar{\mathbf{R}}}{\partial \mathbf{q}_i} \Delta \mathbf{q}_i = -\bar{\mathbf{R}} \tag{5.201}$$

where the vector $\Delta \mathbf{q}_i$ is the *vector* of *Newton differences*. Equation 201 is solved for the Newton differences, and the independent coordinates, the vector $\bar{\mathbf{R}}$ and the coefficient matrix $\partial \bar{\mathbf{R}}/\partial \mathbf{q}_i$ are iteratively updated. The roots of Eq. 200 are then obtained if the norm of the vector $\Delta \mathbf{q}_i$ or the vector $\bar{\mathbf{R}}$ is arbitrarily small, that is,

$$|\Delta \mathbf{q}_i| < \varepsilon_1 \quad \text{or} \quad |\bar{\mathbf{R}}| < \varepsilon_2 \tag{5.202}$$

where ε_1 and ε_2 are small numbers.

We are now in a position to outline the numerical algorithm implemented in the DAMS program for the static analysis of multibody systems consisting of interconnected rigid and deformable bodies. This computational algorithm is shown in Fig. 9 and proceeds in the following routine:

Step 1 The stiffness matrices of the deformable bodies are generated in the preprocessor PREDAMS. The stiffness matrices, an estimate of the static configuration of the system, and the rigid body information, together with the constrained mechanical system description, complete the required input data for a mechanical system of interconnected rigid and deformable bodies. These data are used as input to the main processor DAMS.

Step 2 Having set the flag IANL equal to one, the main processor DAMS calls for subroutine *STATIC*, which performs the static analysis in a routine described in the following steps.

Step 3 The constraint module (CONMOD) is called on to evaluate the constraint Jacobian matrix and check on constraint violations. The set of independent and dependent coordinates, denoted as $\mathbf{q}_i$ and $\mathbf{q}_d$, respectively, are then identified by calling the numerical module (NUMMOD). The independent coordinates are then fixed and the dependent ones are adjusted by solving Eq. 188 using a Newton–Raphson algorithm.

Step 4 Equation 199 is automatically generated by calling the constraint module (CONMOD) to evaluate the constraint Jacobian matrix and calling the force module (FRCMOD) to evaluate the vector $\mathbf{R}_e$, which contains the generalized external as well as stiffness forces. Equation 199 can then be used to define the residual force vector $\bar{\mathbf{R}}$.

Step 5 The numerical module (NUMMOD) is called on to solve Eq. 201 for Newton differences $\Delta \mathbf{q}_i$. The vector of independent coordinates is then updated by using Newton differences, that is,

$$\mathbf{q}_i = \mathbf{q}_i + \Delta \mathbf{q}_i \tag{5.203}$$

The norm of the vectors of Newton differences $\Delta \mathbf{q}_i$ and the residual force vector

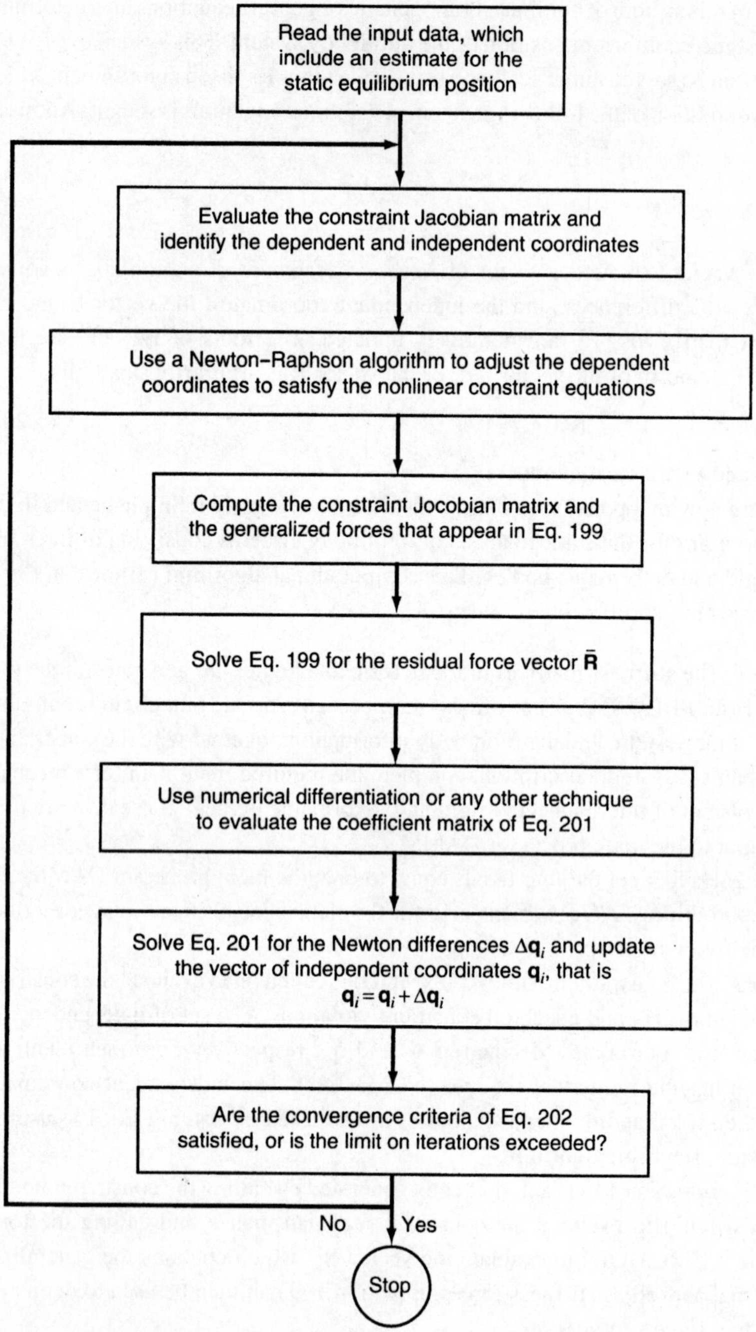

Figure 5.9 Computational algorithm for the static analysis.

$\bar{\mathbf{R}}$ are then checked according to Eq. 202. If Eq. 202 is satisfied, the dependent coordinates are adjusted by solving Eq. 188 using a Newton–Raphson algorithm. The new set of dependent and independent coordinates defines the static equilibrium position of the multibody system.

Step 6 If Eq. 202 is not satisfied, steps 3–5 are repeated.

Kinematic Analysis The *kinematic analysis* is an essential step in performing the dynamic analysis. Two cases of the kinematic analysis may be considered. In the first case, the number of the system generalized coordinates is equal to the number of the kinematic constraint equations, and accordingly there are no independent coordinates. In the second case the number of constraint equations is less than the number of system coordinates. In this case, the number of independent coordinates (degrees of freedom) is equal to $n - n_c$. In the following, we discuss the special case of kinematic analysis in which the number of constraint equations is equal to the number of generalized coordinates of the multibody system. The other case, in which the number of constraint equations is less than the number of system coordinates, will be discussed later when we consider the numerical algorithm for the dynamic analysis of multibody systems consisting of interconnected rigid and deformable bodies.

If the number of constraint functions is equal to the number of generalized coordinates, we have n constraint equations, which can be written in a vector form as

$$\mathbf{C}(\mathbf{q}, t) = \mathbf{0} \tag{5.204}$$

This is a system of n nonlinear algebraic constraint equations whose roots define the system generalized coordinate $\mathbf{q}$. For a specified value of the time t, one can use a Newton–Raphson algorithm to perform the *position analysis* by solving the following system of equations for the vector of Newton differences $\Delta\mathbf{q}$:

$$\mathbf{C_q}\Delta\mathbf{q} = -\mathbf{C} \tag{5.205}$$

where the Jacobian matrix $\mathbf{C_q}$ in this case is a square matrix. If the constraint equations are linearly independent, the Jacobian matrix $\mathbf{C_q}$ has a full row rank and thus is nonsingular. As pointed out earlier, the iterative solutions of Eq. 205 can be used to update the vector of system generalized coordinates, that is,

$$\mathbf{q} = \mathbf{q} + \Delta\mathbf{q} \tag{5.206}$$

A solution of Eq. 204 is obtained if the norm of the vector of Newton differences or the norm of the vector of constraint functions is less than a specified tolerance, that is,

$$|\Delta\mathbf{q}| < \varepsilon_1, \qquad |\mathbf{C}| < \varepsilon_2 \tag{5.207}$$

where ε_1 and ε_2 are specified tolerances.

In order to perform the *velocity analysis*, we differentiate Eq. 204 with respect to time. This gives

$$\mathbf{C_q}\dot{\mathbf{q}} + \mathbf{C}_t = \mathbf{0} \tag{5.208}$$

that is,

$$\mathbf{C_q}\dot{\mathbf{q}} = -\mathbf{C}_t \tag{5.209}$$

where $\mathbf{C}_t$ is the partial derivative of the vector $\mathbf{C}$ with respect to time. The vector $\mathbf{C}_t$ is a function of the system generalized coordinates and possibly time, that is,

$$\mathbf{C}_t = \mathbf{C}_t(\mathbf{q}, t) \tag{5.210}$$

Knowing $\mathbf{q}$ from the position analysis and by specifying the value of t, we can determine the vector $\mathbf{C}_t$. Because $\mathbf{C_q}$ is nonsingular, Eq. 209 can be solved for the velocity vector $\dot{\mathbf{q}}$. Having determined the vector of generalized coordinates $\mathbf{q}$ and the vector of generalized velocities $\dot{\mathbf{q}}$, one can proceed a step further to perform the *acceleration analysis*. This can be simply done in this case by solving Eq. 163, or by evaluating the coefficient matrix $\mathbf{M}_\lambda$ and the vector $\bar{\mathbf{F}}$ of Eq. 168, where $\mathbf{M}_\lambda$ depends on the system generalized coordinates and possibly on time, while $\bar{\mathbf{F}}$ depends on the system generalized coordinates, velocities, and possibly on time. The solution of Eq. 168 defines the acceleration vector as well as the vector of Lagrange multipliers. The vector of Lagrange multipliers λ can be used to determine the generalized reaction forces $\mathbf{C}_\mathbf{q}^{\mathrm{T}}\lambda$.

The procedure discussed above for the kinematic analysis of multibody systems has been implemented in the DAMS program. The user can perform the kinematic analysis of a general multibody system by setting the flag IANL equal to 4. The computational scheme for the kinematic analysis in the DAMS program is shown in Fig. 10 and proceeds in the following routine:

Step 1 A set of input data similar to the one described in step 1 of the computational algorithm for the static analysis must be supplied by the user.
Step 2 Having set the flag IANL equal to four, the main processor DAMS calls for subroutine *DYNAMC*. In the subroutine DYNAMC the time interval is divided into equal steps (subintervals) specified by the user. At the beginning of each subinterval, subroutine DYNAMC calls subroutine *F*, which performs the kinematic analysis in a routine described in the following steps.
Step 3 To perform the *position analysis*, a Newton–Raphson algorithm is used to solve Eq. 204 by using Eq. 205. Equation 205 can be constructed by calling CONMOD to evaluate the Jacobian matrix $\mathbf{C_q}$ and the vector of constraint functions $\mathbf{C}$. NUMMOD is then called on to solve Eq. 205 iteratively. A solution of Eq. 204 is obtained, if Eq. 207 is satisfied. The solution of Eq. 204 defines the total vector of system generalized coordinates that describe the configuration of the multibody system.
Step 4 Following determination of the vector $\mathbf{q}$, which describes the correct position of the system components, CONMOD is called on to evaluate the Jacobian matrix $\mathbf{C_q}$ and the vector $\mathbf{C}_t$ of Eq. 209. NUMMOD is then used to solve Eq. 209 for the velocity vector $\dot{\mathbf{q}}$.
Step 5 After determination of the vectors $\mathbf{q}$ and $\dot{\mathbf{q}}$, FRCMOD is called on to evaluate the vectors $\mathbf{Q}_e$ and $\mathbf{Kq}$ of Eq. 166. The vectors $\mathbf{Q}_e$ may depend on the system generalized coordinates, velocities, and possibly on time. CONMOD is

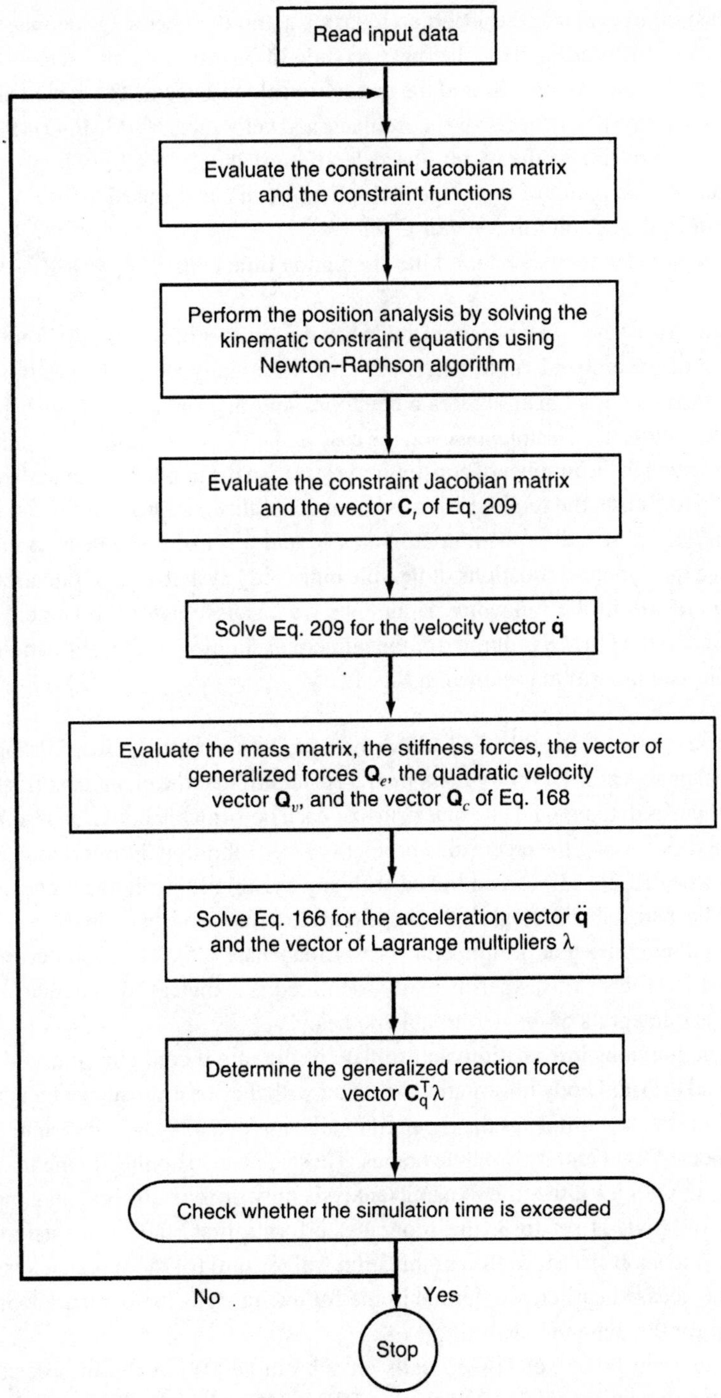

Figure 5.10 Computational algorithm for the kinematic analysis.

also called on to evaluate the Jacobian matrix $\mathbf{C_q}$ and the vector $\mathbf{Q}_c$ defined by Eq. 165. To construct Eq. 166, the mass module (MASMOD) is then called on to evaluate the mass matrix **M** and the quadratic velocity vector $\mathbf{Q}_v$, which also depends on the system generalized coordinates and velocities. NUMMOD is then used to solve Eq. 166 for the vector of accelerations $\ddot{\mathbf{q}}$ and the vector of Lagrange multipliers λ. The vector of Lagrange multipliers λ can then be used to determine the generalized reaction force vector $\mathbf{C}_\mathbf{q}^{\mathrm{T}}\lambda$.

Step 6 Steps 3–5 are repeated until the simulation time ends.

Dynamic Analysis We observed that when the number of constraints is equal to the number of generalized coordinates, the kinematic analysis requires only solutions of a system of nonlinear algebraic equations and there is no need for using numerical integration. This is not, however, the case in the *dynamic analysis* of multibody systems in which the number of constraints is less than the number of generalized coordinates. In this case, the total vector of system generalized coordinates $\mathbf{q}$ can be partitioned into a set of independent coordinates $\mathbf{q}_i$ and a set of dependent coordinates $\mathbf{q}_d$. Since the dynamic equations of flexible multibody systems are summarized in Sections 9 and 10, in the following paragraphs we discuss one of the numerical algorithms used in the DAMS program for the numerical solution of these equations. This computational algorithm is shown in Fig. 11.

Step 1 The preprocessor PREDAMS is employed to generate the inertia shape integrals that appear in the body mass matrix, in addition to the element stiffness matrix. If desired, the modal characteristics of each deformable body can also be determined by solving the eigenvalue problem of free vibration. Furthermore the preprocessor PREDAMS can be linked with any existing finite element code to evaluate the inertia shape integrals that appear in the deformable body mass matrix by using either consistent or lumped masses (Shabana 1985). The preprocessor PREDAMS can also use experimentally identified parameters to generate the inertia shape integrals of the deformable bodies.

Step 2 The previous information, an estimate of the initial configuration of the system, and the rigid body information, together with the constrained mechanical system description, completes the required input data for a mechanical system of interconnected rigid and deformable bodies. These data are supplied to the main processor DAMS for either the dynamic analysis only or for static and dynamic analysis. If IANL is set to 3, the program performs first static analysis and then dynamic analysis. Since the computational algorithm for the static analysis has been discussed earlier, we outline in the following steps the computational algorithm for the dynamic analysis.

Step 3 The main processor DAMS calls on subroutine DYNAMC in order to perform the dynamic analysis. Subroutine DYNAMC calls on subroutine F to perform the position, velocity, and acceleration analysis as outlined below.

Step 4 CONMOD is called on to evaluate the Jacobian matrix $\mathbf{C_q}$ and also to check on constraint violations. After evaluation of the constraint Jacobian matrix $\mathbf{C_q}$, NUMMOD is then used to identify the set of independent coordinates

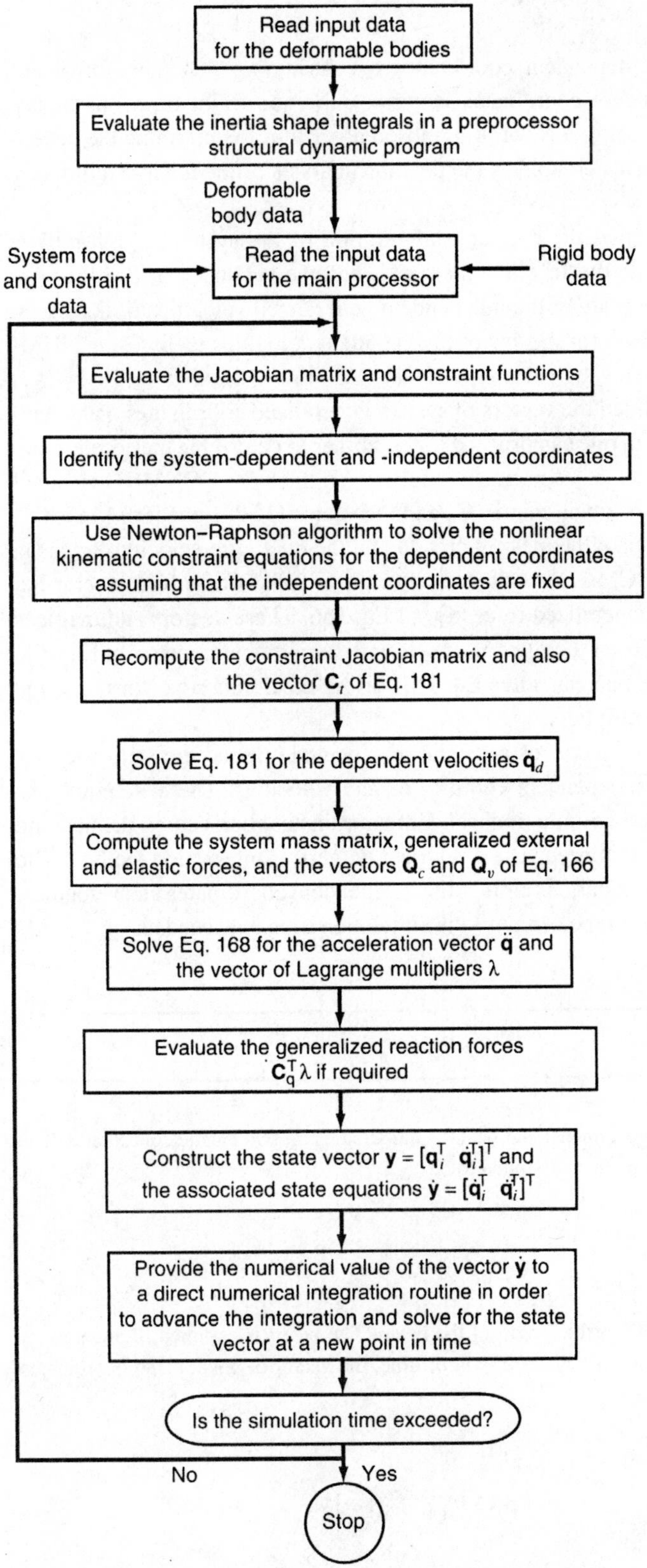

Figure 5.11 Computational algorithm for the dynamic analysis.

$\mathbf{q}_i$ and the set of dependent coordinates $\mathbf{q}_d$. Assuming that the estimate of the set of independent coordinates is correct, the constraint functions of Eq. 155 are solved by using a Newton–Raphson algorithm to determine the dependent coordinates. This completes the position analysis of the flexible multibody system.

Step 5 Having defined the correct configuration of the system, CONMOD is called on to evaluate the Jacobian matrix $\mathbf{C_q}$ and the vector $\mathbf{C}_t$ of Eq. 181. Assuming that the estimate of the independent generalized velocities $\dot{\mathbf{q}}_i$ is correct, Eq. 181 can be solved for the vector of dependent velocities $\dot{\mathbf{q}}_d$ by using NUMMOD. This completes the velocity analysis.

Step 6 Having defined the vectors of system generalized coordinates $\mathbf{q}$ and generalized velocities $\dot{\mathbf{q}}$, one can proceed a step further to determine the acceleration vector and the vector of Lagrange multipliers. To this end, CONMOD is called on to evaluate the Jacobian matrix $\mathbf{C_q}$ and the vector $\mathbf{Q}_c$ of Eq. 166. MASMOD is called on to evaluate the mass matrix $\mathbf{M}$ and the quadratic velocity vector $\mathbf{Q}_v$ of Eq. 166, and FRCMOD is called on to evaluate the stiffness force vector $\mathbf{Kq}$ and the vector of generalized forces $\mathbf{Q}_e$ of Eq. 166. These vectors and matrices can be used to construct the right-hand side and the coefficient matrix of Eq. 168. Using NUMMOD, one can solve Eq. 168 for the acceleration vector $\ddot{\mathbf{q}}$ and the vector Lagrange multipliers.

Step 7 Knowing the vector of accelerations, one can define the state vector $\mathbf{y}$ of Eq. 182 of the independent coordinates and velocities. The state equations (Eq. 183) associated with the independent coordinates then can be defined and integrated forward in time by use of a direct numerical integration method. The solution of the state equations defines the independent coordinates and velocities.

Step 8 Steps 4–7 are repeated until the simulation time is exceeded.

Problems

1. The dynamics of a two-dimensional beam is modeled using two elastic coordinates. The shape function of the beam is assumed to be

$$\mathbf{S}^i = \begin{bmatrix} \sin \pi\xi & 0 \\ 0 & \sin \pi\xi \end{bmatrix}^i$$

where $\xi = x_1/l$, and l is the length of the beam. The beam is assumed to undergo an arbitrary displacement. At a given instant of time, the vector of generalized coordinates of the beam is given by

$$\mathbf{q}^i = [R_1 \quad R_2 \quad \theta \quad q_{f1} \quad q_{f2}]^{i\mathrm{T}} = \left[3.0 \quad 2.0 \quad \frac{\pi}{2} \quad 0.5\times 10^{-3} \quad 1.0\times 10^{-3}\right]^{\mathrm{T}}$$

Determine the global position of the points $\xi = 0.5, 1.0$.

2. Determine the absolute velocities and accelerations of the points $\xi = 0.5, 1.0$ in Problem 1, if the vectors of generalized velocities and accelerations are given by

$$\dot{\mathbf{q}}^i = \begin{bmatrix} \dot{R}_1 & \dot{R}_2 & \dot{\theta} & \dot{q}_{f1} & \dot{q}_{f2} \end{bmatrix}^{i\mathrm{T}} = [0 \quad 0 \quad 50 \quad 5 \times 10^3 \quad 1.0 \times 10^2]^{\mathrm{T}}$$
$$\ddot{\mathbf{q}}^i = \begin{bmatrix} \ddot{R}_1 & \ddot{R}_2 & \ddot{\theta} & \ddot{q}_{f1} & \ddot{q}_{f2} \end{bmatrix}^{i\mathrm{T}} = [0 \quad 0 \quad 0 \quad 5 \times 10^4 \quad 1.0 \times 10^3]^{\mathrm{T}}$$

Determine also the absolute velocities and accelerations of the two points if the generalized elastic velocities and accelerations were equal to zero.

3. The dynamics of a two-dimensional beam is modeled using three elastic coordinates. The shape function of the beam is assumed to be

$$\mathbf{S}^i = \begin{bmatrix} \sin \pi\xi & 0 & 0 \\ 0 & \sin \pi\xi & \sin 2\pi\xi \end{bmatrix}^i$$

where $\xi = x_1/l$, and l is the length of the beam. At a given instant of time, the vector of generalized coordinates of the beam is given by

$$\mathbf{q}^i = \begin{bmatrix} R_1 & R_2 & \theta & q_{f1} & q_{f2} & q_{f3} \end{bmatrix}^{i\mathrm{T}}$$
$$= \begin{bmatrix} 3.0 & 2.0 & \dfrac{\pi}{2} & 0.5 \times 10^{-3} & 1.0 \times 10^{-3} & 1.0 \times 10^{-5} \end{bmatrix}^{\mathrm{T}}$$

Determine the global position vector of the points $\xi = 0.5, 1.0$. Compare the results obtained in this problem and the results obtained using the beam model described in Problem 1.

4. Determine the absolute velocities and accelerations of the points $\xi = 0.5, 1.0$ in Problem 3, if the vectors of generalized velocities and accelerations are given by

$$\dot{\mathbf{q}}^i = \begin{bmatrix} \dot{R}_1 & \dot{R}_2 & \dot{\theta} & \dot{q}_{f1} & \dot{q}_{f2} & \dot{q}_{f3} \end{bmatrix}^{i\mathrm{T}}$$
$$= [0 \quad 0 \quad 50 \quad 5 \times 10^3 \quad 1.0 \times 10^2 \quad 3.0 \times 10^4]^{\mathrm{T}}$$
$$\ddot{\mathbf{q}}^i = \begin{bmatrix} \ddot{R}_1 & \ddot{R}_2 & \ddot{\theta} & \ddot{q}_{f1} & \ddot{q}_{f2} & \ddot{q}_{f3} \end{bmatrix}^{i\mathrm{T}}$$
$$= [0 \quad 0 \quad 0 \quad 5 \times 10^4 \quad 1.0 \times 10^3 \quad 2.5 \times 10^5]^{\mathrm{T}}$$

Determine also the absolute velocities and accelerations if the reference velocities and accelerations were equal to zero.

5. The dynamics of a two-dimensional beam is modeled using three elastic coordinates. The shape function of the beam is assumed to be

$$\mathbf{S}^i = \begin{bmatrix} \xi & 0 & 0 \\ 0 & 3(\xi)^2 - 2(\xi)^3 & l((\xi)^3 - (\xi)^2) \end{bmatrix}^i$$

where $\xi = x_1/l$, and l is the length of the beam. At a given instant of time, the vector of the generalized coordinates of the beam is given by

$$\mathbf{q}^i = \begin{bmatrix} R_1 & R_2 & \theta & q_{f1} & q_{f2} & q_{f3} \end{bmatrix}^{i\mathrm{T}}$$
$$= \begin{bmatrix} 3.0 & 2.0 & \dfrac{\pi}{2} & 0.5 \times 10^{-3} & 1.0 \times 10^{-3} & 1.0 \times 10^{-5} \end{bmatrix}^{\mathrm{T}}$$

Determine the global position of the points $\xi = 0.5, 1.0$. Compare the results obtained in this problem and the results obtained using the beam model described in Problem 3.

6. Determine the absolute velocities and accelerations of the points $\xi = 0.5, 1.0$ in Problem 5, if the vectors of generalized velocities and accelerations are given by

$$\begin{aligned}\dot{\mathbf{q}}^i &= \left[\dot{R}_1 \quad \dot{R}_2 \quad \dot{\theta} \quad \dot{q}_{f1} \quad \dot{q}_{f2} \quad \dot{q}_{f3}\right]^{i\mathrm{T}} \\ &= [0 \quad 0 \quad 50 \quad 5\times 10^3 \quad 1.0\times 10^2 \quad 3.0\times 10^4]^{\mathrm{T}} \\ \ddot{\mathbf{q}}^i &= \left[\ddot{R}_1 \quad \ddot{R}_2 \quad \ddot{\theta} \quad \ddot{q}_{f1} \quad \ddot{q}_{f2} \quad \ddot{q}_{f3}\right]^{i\mathrm{T}} \\ &= [0 \quad 0 \quad 0 \quad 5\times 10^4 \quad 1.0\times 10^3 \quad 2.5\times 10^5]^{\mathrm{T}}\end{aligned}$$

Determine also the absolute velocities and accelerations if the reference velocities and accelerations were equal to zero.

7. Using the beam model described in Problem 1, determine the beam inertia shape integrals. Use these shape integrals to determine the mass matrix of the beam at the given instant of time.

8. Determine the inertia shape integrals of the beam model given in Problem 3. Use the vector of the generalized coordinates given in Problem 3 to evaluate the beam mass matrix.

9. Use the shape function and the vector of generalized coordinates given in Problem 5 to evaluate the mass matrix of the beam.

10. Three elastic coordinates are used to model the dynamics of a three-dimensional beam. The shape function of the beam is given by

$$\mathbf{S}^i = \begin{bmatrix} \sin \pi\xi & 0 & 0 \\ 0 & \sin \pi\xi & 0 \\ 0 & 0 & \sin \pi\xi \end{bmatrix}^i$$

where $\xi = x_1/l$, and l is the length of the beam. Determine the inertia shape integrals of the beam. Determine also the beam mass matrix if Euler parameters are used as the orientation coordinates.

11. Determine the nonlinear mass matrix of the beam model described in Problem 10 if Euler angles are used as the orientation coordinates.

12. Repeat Problem 10 if the following shape function is used:

$$\mathbf{S}^i = \begin{bmatrix} \xi & 0 & 0 \\ 0 & 3(\xi)^2 - 2(\xi)^3 & 0 \\ 0 & 0 & 3(\xi)^2 - 2(\xi)^3 \end{bmatrix}^i$$

where $\xi = x_1/l$, and l is the length of the beam.

13. Using Euler angles as the orientation coordinates, determine the nonlinear mass matrix of the three-dimensional beam using the shape function given in Problem 12.

14. Use the virtual work of the inertia forces of the elastic bodies to define the mass matrix in the case of planar motion.

15. In the case of three-dimensional motion, use the virtual work of the inertia forces to determine the mass matrix of the deformable bodies.

16. The force vector $\mathbf{F}^i = [2.5 \ -3.0]^{\mathrm{T}}$ N is assumed to act at the end of the beam described in Problem 3. Determine the generalized forces associated with the generalized coordinates

of the beam as the result of application of this force vector. Use the results obtained to demonstrate that the force vector in flexible body dynamics is not a sliding vector.

17. Repeat Problem 16 using the beam model described in Problem 5.

18. The three-dimensional force vector $\mathbf{F}^i = [2.5 \ \ -3.0 \ \ 9.0]^{\mathrm{T}}$ N is assumed to act at the end of the beam described in Problem 12. Determine the generalized forces associated with the generalized coordinates of the beam as the result of application of this force vector. Use Euler angles as the orientation coordinates.

19. Formulate the generalized forces of a spring–damper–actuator element connecting two flexible bodies in the case of planar motion.

20. Determine the stiffness matrix of the beam described in Problem 1.

21. Determine the stiffness matrix of the beam described in Problem 3. Compare the results with the results obtained using the beam model described in Problem 1.

22. Using the shape function defined in Problem 5, determine the beam stiffness matrix.

23. Formulate the constraint equations of the cylindrical joint between two deformable bodies. Obtain also the Jacobian matrix of the cylindrical joint constraints.

24. Formulate the nonlinear constraints equations that describe a universal joint between two deformable bodies. Formulate also the constraint Jacobian matrix of this joint.

25. Use the virtual work of the inertia forces to define the generalized centrifugal and Coriolis inertia forces of deformable bodies in planar motion.

26. Derive the expression for the generalized centrifugal and Coriolis forces of deformable bodies in the three-dimensional analysis using the virtual work of the inertia forces.

27. Provide the detailed derivation of the generalized Newton–Euler equations of deformable bodies.

28. Discuss the effect of selecting the deformable body coordinate system on the nonlinear inertia coupling between the reference and the elastic displacements.

29. Discuss the numerical algorithms used in the computer solution of flexible multibody equations. Discuss the basic differences between these algorithms and the algorithms used in the computer solution of rigid multibody equations.

6 FINITE-ELEMENT FORMULATION

In the classical finite-element formulation for beams and plates, infinitesimal rotations are used as nodal coordinates. As a result, beams and plates are not considered as *isoparametric elements*. Rigid body motion of these non-isoparametric elements does not result in zero strains and exact modeling of the rigid body inertia using these elements cannot be obtained. In this chapter, a formulation for the large reference displacement and small deformation analysis of deformable bodies using nonisoparametric finite elements is presented. This formulation, in which infinitesimal rotations are used as nodal coordinates, leads to exact modeling of the rigid body dynamics and results in zero strains under an arbitrary rigid body motion. It is crucial in this formulation that the assumed displacement field of the element can describe an arbitrary rigid body translation. Using this property and an *intermediate element coordinate system*, a concept similar to the *parallel axis theorem* used in rigid body dynamics can be applied to obtain an exact modeling of the rigid body inertia for deformable bodies that have complex geometrical shapes.

To develop a finite-element formulation for deformable bodies in multibody systems, the assumed displacement field of the finite element is first discussed and some important concepts that are fundamental in understanding large rotation problems in particular and the dynamics of constrained deformable bodies in general are introduced. In Section 2, the gross rigid body motion of the finite element is described using a set of *reference coordinates* that describe the gross rigid body translational and rotational displacements of a selected deformable body reference. To define a unique displacement field, the rigid body modes of the element shape functions have to be eliminated by using a set of *reference conditions*. These conditions, which define the nature of the deformable body axes, can be introduced with a set of linear algebraic equations. The general displacement of the finite element in a deformable body in the multibody system can then be described by using a coupled set of body reference coordinates and element *nodal elastic coordinates*. These coordinates are used in Sections 3 and 4 to develop the inertia and stiffness characteristics of the finite elements that undergo large translational and rotational displacements. The mass and stiffness matrices of the deformable body in the multibody system are then obtained by assembling the mass and stiffness matrices of the finite elements through the use of a standard finite-element procedure. As shown in this chapter, the use of the finite-element method can significantly reduce the number of inertia shape integrals required to formulate the nonlinear inertia terms that represent the dynamic coupling between the reference motion and the elastic deformation. Furthermore, the mass and stiffness matrices that appear in linear structural dynamics can be extracted from the nonlinear formulation presented in this chapter by considering the special case in which the large reference rotations of the deformable bodies are not permitted.

The general development of the inertia and stiffness characteristics will be exemplified using planar and spatial examples discussed, respectively, in Sections 5 and 6, wherein the mass and stiffness matrices of two- and three-dimensional beam elements are derived. Since the finite-element discretization of complex structures results in a large number of nodal coordinates, in Section 7 of this chapter, component mode synthesis techniques that are frequently employed to reduce the number of coordinates are briefly discussed. Before concluding this chapter, we discuss the computer implementation of the nonlinear formulation presented in this chapter.

6.1 ELEMENT SHAPE FUNCTIONS

In the previous chapter classical approximation methods such as Rayleigh–Ritz methods are used to describe the shape of deformation of the deformable bodies in the multibody systems. The finite-element method can be viewed as a special case of the Rayleigh–Ritz method wherein the deformable body is divided into small regions called *elements*. The deformable body is separated by imaginary lines or surfaces into a number of finite elements that are assumed to be interconnected at nodal points on their boundaries. The displacements of the selected nodal points are the basic unknowns of the problem. A piecewise fit is then used to uniquely describe the shape within each element. As pointed out by Cook (1981), the use of the Rayleigh–Ritz method has two undesirable properties. First, the assumed displacement fields cannot be used immediately. They must be adjusted to match the boundary conditions. Second, the time-dependent coordinates lack an obvious physical meaning. These undesirable features are avoided in the standard finite-element formulation by using coordinates that describe displacements, slopes, and curvatures at selected nodal points on the deformable body. Between these selected nodal points the displacement field within the element can be adequately described by using interpolating polynomials.

Nodal Coordinates The concept of interpolation is to select a function $f(x)$ from a given class of functions in such a way that the function passes through the given data points, which in this case are the nodal points. Therefore, exact and interpolated curves match at the endpoints (the nodes) but may differ elsewhere. We consider, for example, the two-dimensional beam element shown in Fig. 1. One may describe the

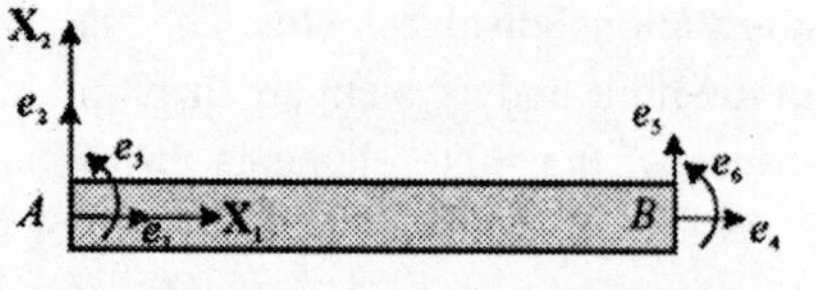

Figure 6.1 Two-dimensional beam.

displacement field within the element by the following polynomials:

$$w_1 = a_0 + a_1 x_1$$
$$w_2 = a_2 + a_3 x_1 + a_4 (x_1)^2 + a_5 (x_1)^3$$

or in a matrix form

$$\mathbf{w} = \mathbf{X}\mathbf{a} \tag{6.1}$$

where $\mathbf{X}$ is the matrix

$$\mathbf{X} = \begin{bmatrix} 1 & x_1 & 0 & 0 & 0 & 0 \\ 0 & 0 & 1 & x_1 & (x_1)^2 & (x_1)^3 \end{bmatrix}$$

and $\mathbf{a}$ is the time-dependent vector whose components are given by

$$\mathbf{a} = [a_0 \quad a_1 \quad a_2 \quad a_3 \quad a_4 \quad a_5]^{\mathrm{T}}$$

In the static analysis $\mathbf{a}$ is a constant vector, while in dynamics $\mathbf{a}$ is a function of time.

In Eq. 1 a linear interpolation is used to describe the *axial* displacement, while a cubic function is used to describe the *transverse* displacement. The coefficients $a_0, a_1, \ldots, a_5$ are determined by applying end conditions. In this particular example, we assume that the element has two nodal points at A and B, where A is located at $x_1 = 0$, B is located at $x_1 = l$, and l is the length of the element. We further assume that each nodal point has three degrees of freedom: two translational coordinates in x_1 and x_2 directions, respectively, and the third one describing the slope at the nodal point. Therefore, the total number of coordinates for this element is 6, denoted as e_1, e_2, e_3, e_4, e_5, and e_6, and can be written in a vector form as

$$\mathbf{e} = [e_1 \quad e_2 \quad e_3 \quad e_4 \quad e_5 \quad e_6]^{\mathrm{T}} \tag{6.2}$$

where e_1, e_2, e_4, and e_5 are the translational coordinates as shown in Fig. 1, while e_3 and e_6 are the slopes at the nodal points A and B, respectively. To determine the coefficients a_i, $i = 0, 1, \ldots, 5$, in Eq. 1 we impose the following end conditions:

$$\begin{aligned} w_1(0) &= e_1, \quad & w_2(0) &= e_2, \quad & w_2'(0) &= e_3 \\ w_1(l) &= e_4, \quad & w_2(l) &= e_5, \quad & w_2'(l) &= e_6 \end{aligned} \tag{6.3}$$

where ($'$) denotes partial differentiation with respect to the spatial coordinate x_1.

Using the representation of Eq. 1 and the end conditions of Eq. 3, one can write

$$\mathbf{e} = \bar{\mathbf{X}}\mathbf{a} \tag{6.4}$$

where the matrix $\bar{\mathbf{X}}$ is defined as

$$\bar{\mathbf{X}} = \begin{bmatrix} 1 & 0 & 0 & 0 & 0 & 0 \\ 0 & 0 & 1 & 0 & 0 & 0 \\ 0 & 0 & 0 & 1 & 0 & 0 \\ 1 & l & 0 & 0 & 0 & 0 \\ 0 & 0 & 1 & l & (l)^2 & (l)^3 \\ 0 & 0 & 0 & 1 & 2l & 3(l)^2 \end{bmatrix} \tag{6.5}$$

Therefore, the vector of the coefficients $\mathbf{a}$ can be determined in terms of the nodal coordinates $\mathbf{e}$ as

$$\mathbf{a} = \bar{\mathbf{X}}^{-1}\mathbf{e} \tag{6.6}$$

where $\bar{\mathbf{X}}^{-1}$ is the inverse of $\bar{\mathbf{X}}$. Using Eqs. 1 and 6, the displacement field of the beam element can be written in terms of the nodal coordinates as

$$\mathbf{w} = \mathbf{X}\bar{\mathbf{X}}^{-1}\mathbf{e}$$

or

$$\mathbf{w} = \mathbf{S}\mathbf{e} \tag{6.7}$$

where $\mathbf{S}$ is denoted as the *element shape function* and defined as

$$\mathbf{S} = \mathbf{X}\bar{\mathbf{X}}^{-1} \tag{6.8}$$

Using Eq. 5, the space-dependent shape function of the beam element is defined as

$$\mathbf{S} = \begin{bmatrix} 1-\xi & 0 & 0 & \xi & 0 & 0 \\ 0 & 1-3(\xi)^2+2(\xi)^3 & l(\xi-2(\xi)^2+(\xi)^3) & 0 & 3(\xi)^2-2(\xi)^3 & l[(\xi)^3-(\xi)^2] \end{bmatrix} \tag{6.9}$$

where

$$\xi = \frac{x_1}{l} \tag{6.10}$$

Note that at $x_1 = 0$ the elements of the shape function matrix associated with the coordinates of the second node are equal to zero, while at $x_1 = l$, the elements of the shape function matrix associated with the coordinates of the first node are equal to zero.

The procedure outlined above for writing the displacement field in terms of the nodal coordinates of the element is general and can be used for any type of element with any type of nodal coordinate. This procedure also applies for the planar analysis as well as the spatial analysis.

Rigid Body Modes The assumed displacement field has to satisfy the convergence requirements that guarantee that the exact solution will be approached when the number of elements increases. As pointed out by Cook (1981), the convergence conditions require that the displacement field within the element be continuous and also that the element be able to assume the state of constant strain. The continuity condition is easily met by describing the displacement field with the use of polynomials. The case of constant strains can be achieved if the derivatives of the displacement in the strain energy expression used are able to assume constant values. Element shape functions should also account for rigid body modes; that is, when the nodal coordinates correspond to rigid body motion, the strain and nodal forces must be equal to zero. The rigid body modes can be considered as a special case of the constant strain requirements in which the strain is equal to zero. For instance, in the previous example of the two-dimensional beam element, if in Eq. 7, $e_1 = e_4 = c_1$, where c_1 is a constant and $e_2 = e_3 = e_5 = e_6 = 0$, one can verify that this corresponds to a rigid body translation of the element in the x_1 direction. In a similar manner, if $e_2 = e_5 = c_2$ and $e_1 = e_3 = e_4 = e_6 = 0$, one can verify that this case corresponds to rigid body translation of the element in the x_2 direction. Therefore, the conventional shape function of the beam element defined by Eq. 9 can describe an arbitrary rigid body translation. This shape function, however, cannot be used to describe an arbitrary rigid body rotation if infinitesimal rotations instead of slopes are used as nodal coordinates. This fact is demonstrated in Chapter 7 where the problem of describing the finite rotations of the elements using the nodal coordinates is discussed.

Using the fact that the element shape function can describe an arbitrary rigid body translation, the vector of the nodal coordinates of the finite element can be written in any coordinate system whose axes are parallel to the axes of the element coordinate systems as

$$\mathbf{e} = \mathbf{e}_o + \mathbf{e}_f \tag{6.11}$$

where $\mathbf{e}_o$ represents the values of the coordinates in the undeformed state and $\mathbf{e}_f$ is the vector of elastic nodal coordinates associated with the deformation of the element.

Coordinate Systems In a general multibody system, a deformable body is normally divided into more than one element. To avoid any confusion in the notation, for an element j on a deformable body i, we may write Eq. 7 in the form

$$\mathbf{w}^{ij} = \mathbf{S}^{ij}\mathbf{e}^{ij} \tag{6.12}$$

where the superscript i refers to the body number in the multibody system and the superscript j refers to the element number in the finite-element discretization of the deformable body i. In a similar manner, we may write Eq. 11 as

$$\mathbf{e}^{ij} = \mathbf{e}_o^{ij} + \mathbf{e}_f^{ij} \tag{6.13}$$

Because the element shape function can describe an arbitrary rigid body translation, one can write Eq. 12 with respect to any coordinate system that is initially parallel to the element coordinate system. For instance, in Fig. 2, $\mathbf{X}_1^{ij}\mathbf{X}_2^{ij}\mathbf{X}_3^{ij}$ is the *element coordinate system* that translates and rotates with the element; that is, the origin of this coordinate system is rigidly attached to a point on the element. The $\mathbf{X}_1^i\mathbf{X}_2^i\mathbf{X}_3^i$ system is a selected *body coordinate system* that need not be rigidly attached to a point on the body (Shabana 1996a). The $\mathbf{X}_{i1}^{ij}\mathbf{X}_{i2}^{ij}\mathbf{X}_{i3}^{ij}$ system is an *intermediate element* coordinate system whose origin is rigidly attached to the origin of the body $\mathbf{X}_1^i\mathbf{X}_2^i\mathbf{X}_3^i$ coordinate system. The coordinate system $\mathbf{X}_{i1}^{ij}\mathbf{X}_{i2}^{ij}\mathbf{X}_{i3}^{ij}$ is assumed to have a fixed orientation

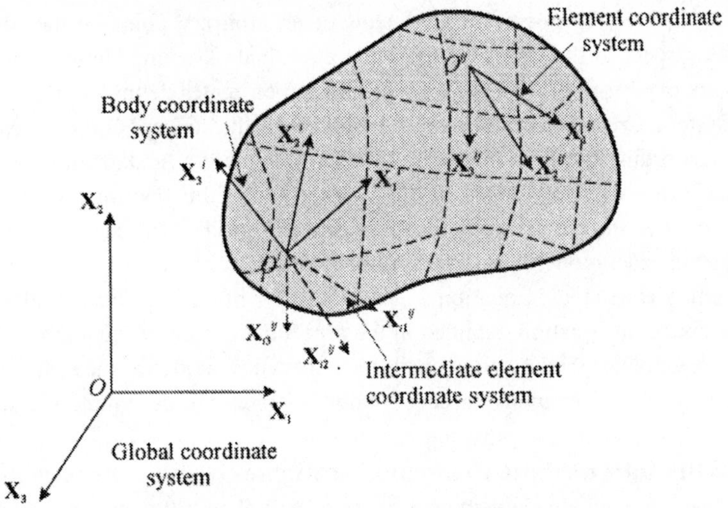

Figure 6.2 Finite-element coordinate systems.

with respect to the body coordinate system; that is, the $\mathbf{X}_{i1}^{ij}\mathbf{X}_{i2}^{ij}\mathbf{X}_{i3}^{ij}$ coordinate system translates and rotates with the body reference. Furthermore, it is assumed that the orientations of the axes $\mathbf{X}_{i1}^{ij}\mathbf{X}_{i2}^{ij}\mathbf{X}_{i3}^{ij}$ are selected in such a manner that they are initially parallel to the axes of the element coordinate system $\mathbf{X}_1^{ij}\mathbf{X}_2^{ij}\mathbf{X}_3^{ij}$. Therefore, Eqs. 12 and 13 can be used to define the configuration of the element ij with respect to the $\mathbf{X}_{i1}^{ij}\mathbf{X}_{i2}^{ij}\mathbf{X}_{i3}^{ij}$ system with the understanding that $\mathbf{e}^{ij}$ is replaced by $\mathbf{e}_i^{ij}$, that is,

$$\mathbf{w}_i^{ij} = \mathbf{S}^{ij}\mathbf{e}_i^{ij} \tag{6.14}$$

where $\mathbf{w}_i^{ij}$ is the assumed displacement field and $\mathbf{e}_i^{ij}$ are the nodal coordinates of the element ij. Both $\mathbf{w}_i^{ij}$ and $\mathbf{e}_i^{ij}$ are defined with respect to the $\mathbf{X}_{i1}^{ij}\mathbf{X}_{i2}^{ij}\mathbf{X}_{i3}^{ij}$ intermediate element coordinate system. Because the coordinate system $\mathbf{X}_{i1}^{ij}\mathbf{X}_{i2}^{ij}\mathbf{X}_{i3}^{ij}$ has a fixed orientation with respect to the body reference, one may define the vector of nodal coordinates $\mathbf{e}_i^{ij}$ in the body coordinate system as

$$\mathbf{e}_i^{ij} = \bar{\mathbf{C}}^{ij}\mathbf{q}_n^{ij} \tag{6.15}$$

where $\bar{\mathbf{C}}^{ij}$ is an orthogonal constant transformation matrix and $\mathbf{q}_n^{ij}$ is the vector of nodal coordinates of the element ij defined with respect to the coordinate system of the body i. It follows that the displacement vector $\bar{\mathbf{u}}^{ij}$ can be defined in the ith body coordinate system as

$$\bar{\mathbf{u}}^{ij} = \mathbf{C}^{ij}\mathbf{w}_i^{ij} = \mathbf{C}^{ij}\mathbf{S}^{ij}\mathbf{e}_i^{ij} \tag{6.16}$$

In the two-dimensional analysis $\mathbf{C}^{ij}$ is a 2×2 transformation matrix, while in three-dimensional analysis $\mathbf{C}^{ij}$ is a 3×3 matrix. The constant transformation $\bar{\mathbf{C}}^{ij}$ has a dimension that is equal to the number of nodal coordinates of the element. We will elaborate more on these transformations in subsequent sections.

Substituting Eq. 15 into Eq. 16 yields

$$\bar{\mathbf{u}}^{ij} = \mathbf{C}^{ij}\mathbf{S}^{ij}\bar{\mathbf{C}}^{ij}\mathbf{q}_n^{ij} \tag{6.17}$$

This equation defines the position coordinates of an arbitrary point on the finite element with respect to the origin of the body coordinate system. These position coordinates are expressed in terms of a set of nodal coordinates defined in the body coordinate system. Crucial to developing this equation is the concept of the intermediate element coordinate system (Shabana 1982), which plays a fundamental role in the nonlinear formulation developed in this chapter. Using this coordinate system, a concept similar to the parallel axis theorem used in rigid body dynamics can be applied to obtain exact modeling of the rigid body inertia for components that have complex geometrical shapes. Equation 17 can also be written in a form similar to the kinematic position equation obtained in the preceding chapter. By using Eq. 13, the position vector of an arbitrary point on the element can be written as the sum of the position vector in the undeformed reference configuration plus the deformation vector.

Role of the Intermediate Element Coordinate System As previously pointed out, the intermediate element coordinate system plays a fundamental role in the nonlinear formulation presented in this chapter (Shabana 1982). The use of this coordinate system with a shape function that can describe an arbitrary rigid body translation leads to an exact modeling of the rigid body kinematics. Without the use of this coordinate system, exact modeling of the rigid body kinematics cannot be obtained when conventional beam and plate element shape functions are used. As demonstrated in the following chapter, the use of the infinitesimal rotations as nodal coordinates in the case of beam and plate elements leads to a linearization of the rigid body kinematics when the large rotation of the element is described using the element nodal coordinates. By using the intermediate element coordinate system, this problem can be solved since Eq. 16 or Eq. 17 when used with the body reference coordinates leads to exact modeling of the rigid body kinematics. To demonstrate the fundamental role played by the intermediate element coordinate system, the two-dimensional beam element shown in Fig. 3 is considered. In the undeformed reference configuration, the location of the origin of the element coordinate system with respect to the intermediate element coordinate system is defined by the coordinates d_1^{ij} and d_2^{ij}. The orientation of the element and the intermediate element coordinate system with respect to the body coordinate system in the undeformed reference configuration is defined by the angle β^{ij}. Since in the undeformed reference configuration, the deformation of the element is equal to zero, the vector of nodal coordinates of the element used in Eq. 16 as defined in the intermediate element coordinate system is given by

$$\mathbf{e}_i^{ij} = [d_1 \quad d_2 \quad 0 \quad d_1 + l \quad d_2 \quad 0]^{ij^{\mathrm{T}}}$$

Using this vector of nodal coordinates and the shape function of Eq. 9, it can be shown that

$$\mathbf{S}^{ij}\mathbf{e}_i^{ij} = \begin{bmatrix} x_1 + d_1 \\ d_2 \end{bmatrix}^{ij}$$

which is the exact location of an arbitrary point on the element obtained here using the element shape function and the vector of nodal coordinates. The preceding equation defines the location of an arbitrary point with respect to the origin of the intermediate

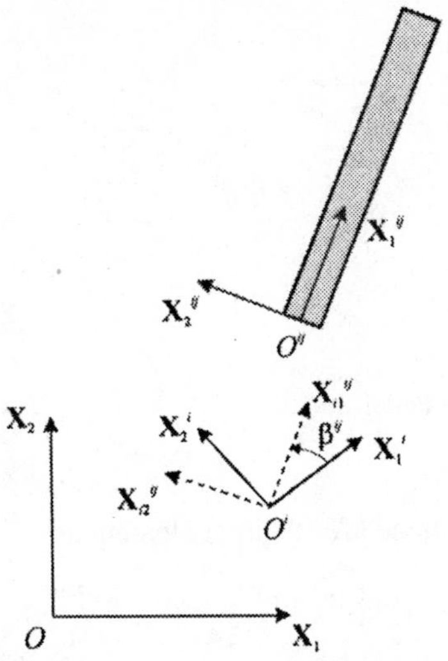

Figure 6.3 Two-dimensional beam element.

element coordinate system that is rigidly attached to the origin of the body coordinate system. The orientation of the element coordinate system with respect to the body coordinate system is defined by the transformation matrix $\mathbf{C}^{ij}$ used in Eq. 16. This matrix is given by

$$\mathbf{C}^{ij} = \begin{bmatrix} \cos\beta & -\sin\beta \\ \sin\beta & \cos\beta \end{bmatrix}^{ij}$$

As a consequence, Eq. 16 defines the exact location of an arbitrary point on the element in the body coordinate system. Using a similar procedure, it can be demonstrated that the use of the intermediate element coordinate system in the three-dimensional analysis leads to exact modeling of the rigid body kinematics.

Connectivity Conditions The coordinate system $\mathbf{X}_1^i\mathbf{X}_2^i\mathbf{X}_3^i$ of body i represents a unique standard for all elements of this body and as such serves to express the connectivity between these elements. Let $\mathbf{q}_n^i$ be the total vector of nodal coordinates of body i resulting from the finite-element discretization. Then the vector of element nodal coordinates can be written in terms of the nodal coordinates of the body as

$$\mathbf{q}_n^{ij} = \mathbf{B}_1^{ij}\mathbf{q}_n^i \tag{6.18}$$

where $\mathbf{B}_1^{ij}$ is a constant *Boolean transformation* whose elements are either zeros or ones and serves to express the connectivity of this element. For instance, consider the example shown in Fig. 4 where body i is divided into two beam elements that are rigidly connected at node 2. In this example, two coordinates are assumed for each node, and therefore the vector $\mathbf{q}_n^{ij}$ contains four elements, that is,

$$\mathbf{q}_n^{i1} = [e_1^{i1} \quad e_2^{i1} \quad e_3^{i1} \quad e_4^{i1}]^{\mathrm{T}} \tag{6.19}$$

$$\mathbf{q}_n^{i2} = [e_1^{i2} \quad e_2^{i2} \quad e_3^{i2} \quad e_4^{i2}]^{\mathrm{T}} \tag{6.20}$$

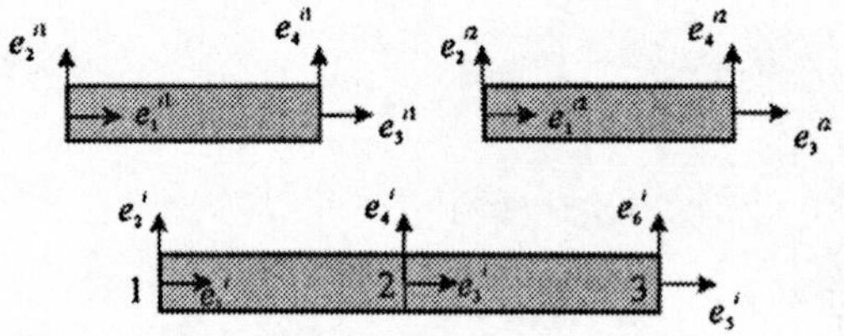

Figure 6.4 Element connectivity.

For the assembled body, however, the vector $\mathbf{q}_n^i$ is defined as

$$\mathbf{q}_n^i = [e_1^i \quad e_2^i \quad e_3^i \quad e_4^i \quad e_5^i \quad e_6^i]^{\mathrm{T}} \tag{6.21}$$

where the transformation of Eq. 18 can be recognized for the first element as

$$\mathbf{B}_1^{i1} = \begin{bmatrix} 1 & 0 & 0 & 0 & 0 & 0 \\ 0 & 1 & 0 & 0 & 0 & 0 \\ 0 & 0 & 1 & 0 & 0 & 0 \\ 0 & 0 & 0 & 1 & 0 & 0 \end{bmatrix} \tag{6.22}$$

and for the second element as

$$\mathbf{B}_1^{i2} = \begin{bmatrix} 0 & 0 & 1 & 0 & 0 & 0 \\ 0 & 0 & 0 & 1 & 0 & 0 \\ 0 & 0 & 0 & 0 & 1 & 0 \\ 0 & 0 & 0 & 0 & 0 & 1 \end{bmatrix} \tag{6.23}$$

Using Eq. 18, one can then write the displacement of element ij of Eq. 17 in terms of the nodal coordinates of body i as

$$\bar{\mathbf{u}}^{ij} = \mathbf{C}^{ij}\mathbf{S}^{ij}\bar{\mathbf{C}}^{ij}\mathbf{B}_1^{ij}\mathbf{q}_n^i \tag{6.24}$$

or in a compact form as

$$\bar{\mathbf{u}}^{ij} = \mathbf{N}^{ij}\mathbf{q}_n^i \tag{6.25}$$

where $\mathbf{N}^{ij}$ is the space-dependent matrix defined as

$$\mathbf{N}^{ij} = \mathbf{C}^{ij}\mathbf{S}^{ij}\bar{\mathbf{C}}^{ij}\mathbf{B}_1^{ij} \tag{6.26}$$

It can be observed that the displacement representation of Eq. 24 contains the rigid body modes, and accordingly the rigid body motion of the elements with respect to the body reference is allowed. It is convenient in the multibody system dynamics, however, to eliminate the rigid body modes of the shape functions in order to define a unique displacement field with respect to the body reference. The rigid body motion is described using a set of absolute reference coordinates that define the location and orientation of the selected body reference.

6.2 REFERENCE CONDITIONS

In the preceding section, it is shown how the displacement field of the element can be written in terms of the total vector of nodal coordinates of the body. It is also

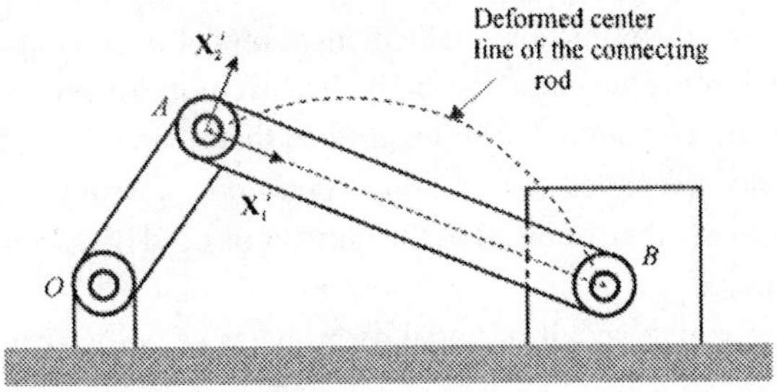

Figure 6.5 Planar slider crank mechanism.

pointed out that this representation contains the rigid body modes of the element. Our intention, however, as indicated earlier is to describe the rigid body motion by use of a coupled set of absolute Cartesian and rotational coordinates that, respectively, describe the location of the origin and orientation of the body reference. In so doing, many technical difficulties associated with the description of large rotations of the elements using translation and infinitesimal rotation nodal coordinates can be avoided. To define a unique displacement field, one then has to eliminate the rigid body modes associated with the element shape functions. This can be achieved by imposing a set of reference conditions that, in turn, define the nature of the body reference. These reference conditions are not a unique set, yet they have to be consistent with the kinematic constraints imposed on the boundary of the deformable body (Shabana 1996a).

As an illustration of the concept to be introduced in this section, we consider the slider crank mechanism shown in Fig. 5. The link OA is the crankshaft, AB is the connecting rod, and the mass concentrated at B is the slider block. We may consider the flexibility of the connecting rod and divide it into a set of finite beam elements. We consider the special case in which the connecting rod is divided into two beam elements and accordingly the vector of nodal coordinates is given by Eq. 21. The connecting rod is connected to the crankshaft and the slider block by pin joints at the ends A and B, respectively. The assumed displacement field must, therefore, assume the shape of a simply supported beam. The simply supported end conditions imply that the axial and transverse deformations at the endpoint A as well as the transverse deformation at the endpoint B vanish. These conditions are sufficient to eliminate the rigid body modes in the assumed displacement field. In fact, they define the way the displacement is measured with respect to the body reference and accordingly define the nature of this reference. In the slider crank mechanism shown in the figure, it is obvious that the $\mathbf{X}_1^i$ axis of the body reference has to pass through points A and B. This, however, does not imply that the origin of the body reference is rigidly attached to the connecting rod because the end conditions at A do not include the slope at this point. In fact, the origin of the body reference of the connecting rod is not rigidly attached to a specific point, and this results in a floating frame of reference. This is a significant difference between the kinematics of rigid and deformable bodies. In the rigid body analysis there is no difference between the reference and the body kinematics. The rigid body configuration is completely defined by the motion of its reference. It is often desirable to rigidly attach the origin of the reference to the center of mass of the rigid body to simplify the mathematical model by decoupling the translational and rotational coordinates of the body. This is not, however, the case when deformable bodies are considered. The conditions that define the nature of the deformable body reference are called the *reference conditions*. The number of reference conditions must be greater than or equal to the number of rigid body modes in the assumed displacement field.

In the finite-element analysis, the vector of nodal coordinates of body i can be written as

$$\mathbf{q}_n^i = \mathbf{q}_o^i + \bar{\mathbf{q}}_f^i \tag{6.27}$$

where $\mathbf{q}_o^i$ is the vector of nodal coordinates in the undeformed state and $\bar{\mathbf{q}}_f^i$ is the vector of nodal deformations. The reference conditions can be considered as a set of constraint equations relating the vector of nodal deformations. These reference conditions can then be used to write the vector of nodal elastic coordinates in terms of a new independent set of coordinates, that is,

$$\bar{\mathbf{q}}_f^i = \mathbf{B}_2^i \mathbf{q}_f^i \tag{6.28}$$

where $\mathbf{B}_2^i$ is a linear transformation that arises from imposing the reference conditions and $\mathbf{q}_f^i$ is the new vector of nodal deformations.

In the slider crank mechanism example, if the connecting rod is divided into two beam elements, the nodal coordinates are defined by Eq. 21. One can verify that by imposing reference conditions satisfying the simply supported end conditions the transformation $\mathbf{B}_2^i$ and the vector $\mathbf{q}_f^i$ are defined as

$$\mathbf{B}_2^i = \begin{bmatrix} 0 & 0 & 0 \\ 0 & 0 & 0 \\ 1 & 0 & 0 \\ 0 & 1 & 0 \\ 0 & 0 & 0 \\ 0 & 0 & 1 \end{bmatrix} \tag{6.29}$$

$$\mathbf{q}_f^{i\mathrm{T}} = [(e_f^i)_3 \quad (e_f^i)_4 \quad (e_f^i)_6] \tag{6.30}$$

Substituting Eq. 28 into Eq. 27 yields

$$\mathbf{q}_n^i = \mathbf{q}_o^i + \mathbf{B}_2^i \mathbf{q}_f^i \tag{6.31}$$

One may also observe that

$$\dot{\mathbf{q}}_n^i = \mathbf{B}_2^i \dot{\mathbf{q}}_f^i \tag{6.32}$$

Introducing the reference coordinates

$$\mathbf{q}_r^i = [\mathbf{R}^{i\mathrm{T}} \quad \boldsymbol{\theta}^{i\mathrm{T}}]^{\mathrm{T}} \tag{6.33}$$

that describe the location of the origin and the orientation of the body reference, one may uniquely define the position vector $\mathbf{r}^{ij}$ of an arbitrary point on element j of body i as

$$\mathbf{r}^{ij} = \mathbf{R}^i + \mathbf{A}^i \bar{\mathbf{u}}^{ij} \tag{6.34}$$

With the use of Eqs. 25 and 31, the above equation leads to (Shabana 1982)

$$\begin{aligned}\mathbf{r}^{ij} &= \mathbf{R}^i + \mathbf{A}^i\mathbf{N}^{ij}\mathbf{q}_n^i \\ &= \mathbf{R}^i + \mathbf{A}^i\mathbf{N}^{ij}(\mathbf{q}_o^i + \mathbf{B}_2^i\mathbf{q}_f^i)\end{aligned} \tag{6.35}$$

where the shape matrix $\mathbf{N}^{ij}$ is defined by Eq. 26. In Eq. 35, the position vector $\mathbf{r}^{ij}$ of an arbitrary point on element j of body i is written in terms of the reference and elastic coordinates of body i.

6.3 KINETIC ENERGY

In this section, an expression for the kinetic energy of body i is obtained by developing the kinetic energy of its elements. This leads to the definition of the nonlinear terms that represent the inertia coupling between the reference motion of the body and the elastic deformation of the elements. The inertia shape integrals required to develop these coupling terms will also be identified.

Kinetic Energy of Finite Elements Differentiating Eq. 34 with respect to time yields

$$\dot{\mathbf{r}}^{ij} = \dot{\mathbf{R}}^i + \mathbf{A}^i(\bar{\omega}^i \times \bar{\mathbf{u}}^{ij}) + \mathbf{A}^i\mathbf{N}^{ij}\mathbf{B}_2^i\dot{\mathbf{q}}_f^i \tag{6.36}$$

where $\bar{\omega}^i$ is the angular velocity vector defined in the local coordinate system. Recall that

$$\bar{\omega}^i \times \bar{\mathbf{u}}^{ij} = -\tilde{\bar{\mathbf{u}}}^{ij}\bar{\omega}^i \tag{6.37}$$

where $\tilde{\bar{\mathbf{u}}}^{ij}$ is the skew symmetric matrix defined as

$$\tilde{\bar{\mathbf{u}}}^{ij} = \begin{bmatrix} 0 & -\bar{u}_3^{ij} & \bar{u}_2^{ij} \\ \bar{u}_3^{ij} & 0 & -\bar{u}_1^{ij} \\ -\bar{u}_2^{ij} & \bar{u}_1^{ij} & 0 \end{bmatrix} \tag{6.38}$$

and $\bar{u}_1^{ij}$, $\bar{u}_2^{ij}$, and $\bar{u}_3^{ij}$ are the components of the vector $\bar{\mathbf{u}}^{ij}$ given by Eq. 25. The angular velocity vector $\bar{\omega}^i$ can be written in terms of the derivatives of the reference rotational coordinates of body i as

$$\bar{\omega}^i = \bar{\mathbf{G}}^i\dot{\boldsymbol{\theta}}^i \tag{6.39}$$

where $\bar{\mathbf{G}}^i$ is a matrix defined in Chapter 2 and $\boldsymbol{\theta}^i$ is the vector of rotational coordinates of the body reference.

Equations 37 and 39 yield

$$\bar{\omega}^i \times \bar{\mathbf{u}}^{ij} = -\tilde{\bar{\mathbf{u}}}^{ij}\bar{\mathbf{G}}^i\dot{\boldsymbol{\theta}}^i \tag{6.40}$$

Substituting this expression into Eq. 36 leads to

$$\dot{\mathbf{r}}^{ij} = \dot{\mathbf{R}}^i - \mathbf{A}^i\tilde{\bar{\mathbf{u}}}^{ij}\bar{\mathbf{G}}^i\dot{\boldsymbol{\theta}}^i + \mathbf{A}^i\mathbf{N}^{ij}\mathbf{B}_2^i\dot{\mathbf{q}}_f^i \tag{6.41}$$

which can be written in a partitioned form as

$$\dot{\mathbf{r}}^{ij} = \begin{bmatrix} \mathbf{I} & -\mathbf{A}^i\tilde{\bar{\mathbf{u}}}^{ij}\bar{\mathbf{G}}^i & \mathbf{A}^i\mathbf{N}^{ij}\mathbf{B}_2^i \end{bmatrix} \begin{bmatrix} \dot{\mathbf{R}}^i \\ \dot{\theta}^i \\ \dot{\mathbf{q}}_f^i \end{bmatrix} \tag{6.42}$$

where $\mathbf{I}$ is the 3×3 identity matrix.

The kinetic energy of element j of body i can be defined as

$$T^{ij} = \frac{1}{2}\int_{V^{ij}} \rho^{ij}\dot{\mathbf{r}}^{ij^T}\dot{\mathbf{r}}^{ij}\,dV^{ij} \tag{6.43}$$

where ρ^{ij} and V^{ij} are, respectively, the mass density and volume of the ijth element. Substituting Eq. 42 into Eq. 43 yields

$$T^{ij} = \frac{1}{2}\dot{\mathbf{q}}^{iT}\mathbf{M}^{ij}\dot{\mathbf{q}}^i \tag{6.44}$$

where $\mathbf{q}^i$ is the total vector of generalized coordinates of body i defined as

$$\mathbf{q}^i = [\mathbf{R}^{iT} \quad \theta^{iT} \quad \mathbf{q}_f^{iT}]^{\mathrm{T}} \tag{6.45}$$

and $\mathbf{M}^{ij}$ is the *mass matrix* of the element ij, which can be written according to the partition of Eq. 45 as

$$\mathbf{M}^{ij} = \int_{V^{ij}} \rho^{ij} \begin{bmatrix} \mathbf{I} & -\mathbf{A}^i\tilde{\bar{\mathbf{u}}}^{ij}\bar{\mathbf{G}}^i & \mathbf{A}^i\mathbf{N}^{ij}\mathbf{B}_2^i \\ & \bar{\mathbf{G}}^{iT}\tilde{\bar{\mathbf{u}}}^{ijT}\tilde{\bar{\mathbf{u}}}^{ij}\bar{\mathbf{G}}^i & \bar{\mathbf{G}}^{iT}\tilde{\bar{\mathbf{u}}}^{ij}\mathbf{N}^{ij}\mathbf{B}_2^i \\ \text{symmetric} & & \mathbf{B}_2^{iT}\mathbf{N}^{ijT}\mathbf{N}^{ij}\mathbf{B}_2^i \end{bmatrix} dV^{ij} \tag{6.46}$$

where the orthogonality of the transformation matrix has been used. This mass matrix can be written in a more simplified form as

$$\mathbf{M}^{ij} = \begin{bmatrix} \mathbf{m}_{RR} & \mathbf{m}_{R\theta} & \mathbf{m}_{Rf} \\ & \mathbf{m}_{\theta\theta} & \mathbf{m}_{\theta f} \\ \text{symmetric} & & \mathbf{m}_{ff} \end{bmatrix}^{ij} \tag{6.47}$$

Inertia Shape Integrals The first submatrix in Eq. 47 can be defined as

$$\mathbf{m}_{RR}^{ij} = \int_{V^{ij}} \rho^{ij}\mathbf{I}\,dV^{ij} = \begin{bmatrix} m^{ij} & 0 & 0 \\ 0 & m^{ij} & 0 \\ 0 & 0 & m^{ij} \end{bmatrix} = m^{ij}\mathbf{I} \tag{6.48}$$

in which m^{ij} is the mass of the element ij. The submatrix $\mathbf{m}_{RR}^{ij}$ is diagonal and constant.

The submatrix $\mathbf{m}_{R\theta}^{ij}$, which represents the inertia coupling between the translation and rotation of the body reference, is defined as

$$\mathbf{m}_{R\theta}^{ij} = -\int_{V^{ij}} \rho^{ij}\mathbf{A}^i\tilde{\bar{\mathbf{u}}}^{ij}\bar{\mathbf{G}}^i\,dV^{ij} = -\mathbf{A}^i\left[\int_{V^{ij}} \rho^{ij}\tilde{\bar{\mathbf{u}}}^{ij}\,dV^{ij}\right]\bar{\mathbf{G}}^i \tag{6.49}$$

Using the definition of $\bar{\mathbf{u}}^{ij}$ of Eq. 24 and the definition of the skew symmetric matrix of Eq. 38, it can be shown that the following integration is required to evaluate the matrix $\mathbf{m}_{R\theta}^{ij}$ of Eq. 49:

$$\bar{\mathbf{S}}^{ij} = \mathbf{C}^{ij}\left[\int_{V^{ij}} \rho^{ij}\mathbf{S}^{ij}\,dV^{ij}\right]\bar{\mathbf{C}}^{ij}\mathbf{B}_1^{ij} \tag{6.50}$$

The matrix $\bar{\mathbf{S}}^{ij}$, which depends on the assumed displacement field within the element, is constant. Furthermore, one can show that this matrix is also required for evaluating the matrix $\mathbf{m}_{Rf}^{ij}$, which represents the inertia coupling between the translation of the body reference and the elastic deformation of the element. To see this, we write $\mathbf{m}_{Rf}^{ij}$ as follows:

$$\mathbf{m}_{Rf}^{ij} = \int_{V^{ij}} \rho^{ij}\mathbf{A}^i\mathbf{N}^{ij}\mathbf{B}_2^i\,dV^{ij} = \mathbf{A}^i\left[\int_{V^{ij}} \rho^{ij}\mathbf{N}^{ij}\,dV^{ij}\right]\mathbf{B}_2^i \tag{6.51}$$

Using the definition of $\mathbf{N}^{ij}$ of Eq. 26 and the fact that the matrices $\mathbf{C}^{ij}$, $\bar{\mathbf{C}}^{ij}$, and $\mathbf{B}_2^{ij}$ are constant matrices, one can write Eq. 51 as

$$\mathbf{m}_{Rf}^{ij} = \mathbf{A}^i\mathbf{C}^{ij}\left[\int_{V^{ij}} \rho^{ij}\mathbf{S}^{ij}\,dV^{ij}\right]\bar{\mathbf{C}}^{ij}\mathbf{B}_1^{ij}\mathbf{B}_2^i$$

Using the definition of the matrix $\bar{\mathbf{S}}^{ij}$ of Eq. 50, we can write the matrix $\mathbf{m}_{Rf}^{ij}$ as

$$\mathbf{m}_{Rf}^{ij} = \mathbf{A}^i\bar{\mathbf{S}}^{ij}\mathbf{B}_2^i \tag{6.52}$$

The central term in the matrix of Eq. 47 can be written as

$$\begin{aligned}
\mathbf{m}_{\theta\theta}^{ij} &= \int_{V^{ij}} \rho^{ij}\bar{\mathbf{G}}^{i\mathrm{T}}\tilde{\bar{\mathbf{u}}}^{ij\mathrm{T}}\tilde{\bar{\mathbf{u}}}^{ij}\bar{\mathbf{G}}^i\,dV^{ij} \\
&= \bar{\mathbf{G}}^{i\mathrm{T}}\left[\int_{V^{ij}} \rho^{ij}\tilde{\bar{\mathbf{u}}}^{ij\mathrm{T}}\tilde{\bar{\mathbf{u}}}^{ij}\,dV^{ij}\right]\bar{\mathbf{G}}^i \\
&= \bar{\mathbf{G}}^{i\mathrm{T}}\bar{\mathbf{I}}_{\theta\theta}^{ij}\bar{\mathbf{G}}^i
\end{aligned} \tag{6.53}$$

where $\bar{\mathbf{I}}_{\theta\theta}^{ij}$ is the symmetric inertia tensor of the ijth element defined with respect to the body reference. The inertia tensor is then defined as

$$\bar{\mathbf{I}}_{\theta\theta}^{ij} = \int_{V^{ij}} \rho^{ij}\tilde{\bar{\mathbf{u}}}^{ij\mathrm{T}}\tilde{\bar{\mathbf{u}}}^{ij}\,dV^{ij} \tag{6.54}$$

The inertia tensor defined by the preceding equation depends on the elastic generalized coordinates of the element. This can be demonstrated by writing Eq. 54 in a more explicit form as

$$\bar{\mathbf{I}}_{\theta\theta}^{ij} = \begin{bmatrix} i_{11} & i_{12} & i_{13} \\ & i_{22} & i_{23} \\ \text{symmetric} & & i_{33} \end{bmatrix} \tag{6.55}$$

where the elements i_{kl} $(k, l = 1, 2, 3)$ can be defined by carrying out the matrix multiplication of Eq. 54. Using Eq. 38, the elements of the inertia tensor can be defined as follows:

$$\left.\begin{aligned}
i_{11} &= \int_{V^{ij}} \rho^{ij}[(\bar{u}_2^{ij})^2 + (\bar{u}_3^{ij})^2]dV^{ij} \\
i_{12} &= -\int_{V^{ij}} \rho^{ij}\bar{u}_2^{ij}\bar{u}_1^{ij}dV^{ij} \\
i_{13} &= -\int_{V^{ij}} \rho^{ij}\bar{u}_3^{ij}\bar{u}_1^{ij}dV^{ij} \\
i_{22} &= \int_{V^{ij}} \rho^{ij}[(\bar{u}_3^{ij})^2 + (\bar{u}_1^{ij})^2]dV^{ij} \\
i_{23} &= -\int_{V^{ij}} \rho^{ij}\bar{u}_3^{ij}\bar{u}_2^{ij}dV^{ij} \\
i_{33} &= \int_{V^{ij}} \rho^{ij}[(\bar{u}_1^{ij})^2 + (\bar{u}_2^{ij})^2]dV^{ij}
\end{aligned}\right\} \tag{6.56}$$

Using Eq. 25, the integrals of Eq. 56 can be written as follows:

$$\left.\begin{aligned}
i_{11} &= \mathbf{q}_n^{i\mathrm{T}}\left[\int_{V^{ij}} \rho^{ij}[\mathbf{N}_2^{ij\mathrm{T}}\mathbf{N}_2^{ij} + \mathbf{N}_3^{ij\mathrm{T}}\mathbf{N}_3^{ij}]dV^{ij}\right]\mathbf{q}_n^i \\
i_{12} &= -\mathbf{q}_n^{i\mathrm{T}}\left[\int_{V^{ij}} \rho^{ij}\mathbf{N}_2^{ij\mathrm{T}}\mathbf{N}_1^{ij}dV^{ij}\right]\mathbf{q}_n^i \\
i_{13} &= -\mathbf{q}_n^{i\mathrm{T}}\left[\int_{V^{ij}} \rho^{ij}\mathbf{N}_3^{ij\mathrm{T}}\mathbf{N}_1^{ij}dV^{ij}\right]\mathbf{q}_n^i \\
i_{22} &= \mathbf{q}_n^{i\mathrm{T}}\left[\int_{V^{ij}} \rho^{ij}[\mathbf{N}_3^{ij\mathrm{T}}\mathbf{N}_3^{ij} + \mathbf{N}_1^{ij\mathrm{T}}\mathbf{N}_1^{ij}]dV^{ij}\right]\mathbf{q}_n^i \\
i_{23} &= -\mathbf{q}_n^{i\mathrm{T}}\left[\int_{V^{ij}} \rho^{ij}\mathbf{N}_3^{ij\mathrm{T}}\mathbf{N}_2^{ij}dV^{ij}\right]\mathbf{q}_n^i \\
i_{33} &= \mathbf{q}_n^{i\mathrm{T}}\left[\int_{V^{ij}} \rho^{ij}[\mathbf{N}_2^{ij\mathrm{T}}\mathbf{N}_2^{ij} + \mathbf{N}_1^{ij\mathrm{T}}\mathbf{N}_1^{ij}]dV^{ij}\right]\mathbf{q}_n^i
\end{aligned}\right\} \tag{6.57}$$

where $\mathbf{N}_k^{ij}$ is the kth row of the matrix $\mathbf{N}^{ij}$ of Eq. 25, that is,

$$\mathbf{N}^{ij} = \begin{bmatrix} \mathbf{N}_1^{ij} \\ \mathbf{N}_2^{ij} \\ \mathbf{N}_3^{ij} \end{bmatrix} \tag{6.58}$$

Let $\mathbf{S}_1^{ij}$, $\mathbf{S}_2^{ij}$, and $\mathbf{S}_3^{ij}$ be the rows of the ijth element shape function, that is,

$$\mathbf{S}^{ij} = \begin{bmatrix} \mathbf{S}_1^{ij} \\ \mathbf{S}_2^{ij} \\ \mathbf{S}_3^{ij} \end{bmatrix} \tag{6.59}$$

Recalling that $\mathbf{N}^{ij}$ is given by

$$\mathbf{N}^{ij} = \mathbf{C}^{ij}\mathbf{S}^{ij}\bar{\mathbf{C}}^{ij}\mathbf{B}_1^{ij} \tag{6.60}$$

one can verify that the following six inertia shape integrals $\mathbf{S}_{kl}^{ij}$ are required to evaluate the inertia tensor of the finite element:

$$\mathbf{S}_{kl}^{ij} = \mathbf{B}_1^{ij\mathrm{T}}\bar{\mathbf{C}}^{ij\mathrm{T}}\left[\int_{V^{ij}} \rho^{ij}\mathbf{S}_k^{ij\mathrm{T}}\mathbf{S}_l^{ij}\,dV^{ij}\right]\bar{\mathbf{C}}^{ij}\mathbf{B}_1^{ij}, \quad k,l = 1,2,3 \tag{6.61}$$

In this equation, the transformations $\bar{\mathbf{C}}^{ij}$ and $\mathbf{B}_1^{ij}$ are as defined in Eq. 26.

The six constant matrices, which depend on the assumed displacement field, are also required to evaluate the matrix $\mathbf{m}_{\theta f}^{ij}$ (Eq. 47), which describes the inertia coupling between the rotation of the body reference and the deformation of the element. This can be demonstrated by writing $\mathbf{m}_{\theta f}^{ij}$ as follows:

$$\begin{aligned}\mathbf{m}_{\theta f}^{ij} &= \int_{V^{ij}} \rho^{ij}\bar{\mathbf{G}}^{i\mathrm{T}}\tilde{\bar{\mathbf{u}}}^{ij}\mathbf{N}^{ij}\mathbf{B}_2^i\,dV^{ij} \\ &= \bar{\mathbf{G}}^{i\mathrm{T}}\left[\int_{V^{ij}} \rho^{ij}\tilde{\bar{\mathbf{u}}}^{ij}\mathbf{N}^{ij}\,dV^{ij}\right]\mathbf{B}_2^i\end{aligned} \tag{6.62}$$

Writing $\mathbf{N}^{ij}$ in the partitioned form of Eq. 58, and using the definition of the skew symmetric matrix $\tilde{\bar{\mathbf{u}}}^{ij}$ of Eq. 38, we can write the matrix multiplication in the integrand of Eq. 62 as follows:

$$\tilde{\bar{\mathbf{u}}}^{ij}\mathbf{N}^{ij} = \begin{bmatrix} 0 & -\bar{u}_3^{ij} & \bar{u}_2^{ij} \\ -\bar{u}_3^{ij} & 0 & -\bar{u}_1^{ij} \\ -\bar{u}_2^{ij} & \bar{u}_1^{ij} & 0 \end{bmatrix}\begin{bmatrix}\mathbf{N}_1^{ij} \\ \mathbf{N}_2^{ij} \\ \mathbf{N}_3^{ij}\end{bmatrix} = \begin{bmatrix} \bar{u}_2^{ij}\mathbf{N}_3^{ij} - \bar{u}_3^{ij}\mathbf{N}_2^{ij} \\ \bar{u}_3^{ij}\mathbf{N}_1^{ij} - \bar{u}_1^{ij}\mathbf{N}_3^{ij} \\ \bar{u}_1^{ij}\mathbf{N}_2^{ij} - \bar{u}_2^{ij}\mathbf{N}_1^{ij} \end{bmatrix} \tag{6.63}$$

Substituting Eq. 25 into this equation leads to

$$\tilde{\bar{\mathbf{u}}}^{ij}\mathbf{N}^{ij} = \begin{bmatrix} \mathbf{q}_n^{i\mathrm{T}}(\mathbf{N}_2^{ij\mathrm{T}}\mathbf{N}_3^{ij} - \mathbf{N}_3^{ij\mathrm{T}}\mathbf{N}_2^{ij}) \\ \mathbf{q}_n^{i\mathrm{T}}(\mathbf{N}_3^{ij\mathrm{T}}\mathbf{N}_1^{ij} - \mathbf{N}_1^{ij\mathrm{T}}\mathbf{N}_3^{ij}) \\ \mathbf{q}_n^{i\mathrm{T}}(\mathbf{N}_1^{ij\mathrm{T}}\mathbf{N}_2^{ij} - \mathbf{N}_2^{ij\mathrm{T}}\mathbf{N}_1^{ij}) \end{bmatrix} \tag{6.64}$$

By substituting this equation into Eq. 62 and using the definition of $\mathbf{N}^{ij}$ given by Eqs. 26 and 58, one can write $\mathbf{m}_{\theta f}^{ij}$ of Eq. 62 as

$$\mathbf{m}_{\theta f}^{ij} = \bar{\mathbf{G}}^{i\mathrm{T}}\begin{bmatrix}\mathbf{q}_n^{i\mathrm{T}}\tilde{\mathbf{N}}_{23}^{ij} \\ \mathbf{q}_n^{i\mathrm{T}}\tilde{\mathbf{N}}_{31}^{ij} \\ \mathbf{q}_n^{i\mathrm{T}}\tilde{\mathbf{N}}_{12}^{ij}\end{bmatrix}\mathbf{B}_2^i \tag{6.65}$$

where $\tilde{\mathbf{N}}_{12}^{ij}$, $\tilde{\mathbf{N}}_{23}^{ij}$, and $\tilde{\mathbf{N}}_{31}^{ij}$ are constant skew symmetric matrices that can be expressed in the following form:

$$\left.\begin{aligned}\tilde{\mathbf{N}}_{12}^{ij} &= \mathbf{N}_{12}^{ij} - \mathbf{N}_{12}^{ij\mathrm{T}} \\ \tilde{\mathbf{N}}_{23}^{ij} &= \mathbf{N}_{23}^{ij} - \mathbf{N}_{23}^{ij\mathrm{T}} \\ \tilde{\mathbf{N}}_{31}^{ij} &= \mathbf{N}_{31}^{ij} - \mathbf{N}_{31}^{ij\mathrm{T}}\end{aligned}\right\} \tag{6.66}$$

in which

$$\tilde{\mathbf{N}}_{kl}^{ij} = \mathbf{N}_{kl}^{ij} - \mathbf{N}_{kl}^{ij\mathrm{T}} = \int_{V^{ij}} \rho^{ij}(\mathbf{N}_k^{ij\mathrm{T}}\mathbf{N}_l^{ij} - \mathbf{N}_l^{ij\mathrm{T}}\mathbf{N}_k^{ij})\,dV^{ij}, \quad k,l = 1,2,3$$

The inertia shape integrals $\mathbf{S}_{kl}^{ij}$ defined by Eq. 61 are also required to evaluate the skew symmetric matrices $\tilde{\mathbf{N}}_{kl}^{ij}$.

Finally, the last term in the element mass matrix of Eq. 47 is evaluated. It is clear that this term, denoted as $\mathbf{m}_{ff}^{ij}$, is a constant matrix and can be written as

$$\mathbf{m}_{ff}^{ij} = \int_{V^{ij}} \rho^{ij} \mathbf{B}_2^{i\mathrm{T}} \mathbf{N}^{ij\mathrm{T}} \mathbf{N}^{ij} \mathbf{B}_2^i \, dV^{ij}$$
$$= \mathbf{B}_2^{i\mathrm{T}} \left[\int_{V^{ij}} \rho^{ij} \mathbf{N}^{ij\mathrm{T}} \mathbf{N}^{ij} \, dV^{ij} \right] \mathbf{B}_2^i \tag{6.67}$$

Substituting for $\mathbf{N}^{ij}$ from Eq. 26, one obtains

$$\mathbf{m}_{ff}^{ij} = \mathbf{B}_2^{i\mathrm{T}} \mathbf{S}_{ff}^{ij} \mathbf{B}_2^i \tag{6.68}$$

where the symmetric matrix $\mathbf{S}_{ff}^{ij}$ is defined as

$$\mathbf{S}_{ff}^{ij} = \mathbf{B}_1^{ij\mathrm{T}} \bar{\mathbf{C}}^{ij\mathrm{T}} \left[\int_{V^{ij}} \rho^{ij} \mathbf{S}^{ij\mathrm{T}} \mathbf{S}^{ij} \, dV^{ij} \right] \bar{\mathbf{C}}^{ij} \mathbf{B}_1^{ij} \tag{6.69}$$

In terms of the inertia shape integrals of Eq. 61, the matrix $\mathbf{S}_{ff}^{ij}$ can be written as

$$\mathbf{S}_{ff}^{ij} = \mathbf{S}_{11}^{ij} + \mathbf{S}_{22}^{ij} + \mathbf{S}_{33}^{ij}$$

It follows that the matrix $\mathbf{m}_{ff}^{ij}$ can be written in terms of the inertia shape integrals as

$$\mathbf{m}_{ff}^{ij} = \mathbf{B}_2^{i\mathrm{T}} (\mathbf{S}_{11}^{ij} + \mathbf{S}_{22}^{ij} + \mathbf{S}_{33}^{ij}) \mathbf{B}_2^i$$

Even though the mass matrix of Eq. 47 depends on the rotation of the body reference as well as the elastic deformation of the element, it is clear that, to determine the mass matrix of the finite element ij in the three-dimensional analysis, seven inertia shape integrals are required. These are the matrix $\bar{\mathbf{S}}^{ij}$ of Eq. 50 and the six matrices $\mathbf{S}_{kl}^{ij}$ of Eq. 61.

Kinetic Energy of the Deformable Body The kinetic energy of body i can be determined by summing up the kinetic energies of its elements, that is,

$$T^i = \sum_{j=1}^{n_e} T^{ij} \tag{6.70}$$

where T^i is the kinetic energy of body i and n_e is the number of elements resulting from the finite-element discretization of body i. Substituting Eq. 44 into Eq. 70 yields

$$T^i = \frac{1}{2} \sum_{j=1}^{n_e} \dot{\mathbf{q}}^{i\mathrm{T}} \mathbf{M}^{ij} \dot{\mathbf{q}}^i$$
$$= \frac{1}{2} \dot{\mathbf{q}}^{i\mathrm{T}} \left[\sum_{j=1}^{n_e} \mathbf{M}^{ij} \right] \dot{\mathbf{q}}^i = \frac{1}{2} \dot{\mathbf{q}}^{i\mathrm{T}} \mathbf{M}^i \dot{\mathbf{q}}^i \tag{6.71}$$

where $\mathbf{M}^i$ is the mass matrix of body i defined as

$$\mathbf{M}^i = \sum_{j=1}^{n_e} \mathbf{M}^{ij} \tag{6.72}$$

which can be written in a partitioned form as

$$\mathbf{M}^i = \begin{bmatrix} \mathbf{m}^i_{RR} & \mathbf{m}^i_{R\theta} & \mathbf{m}^i_{Rf} \\ & \mathbf{m}^i_{\theta\theta} & \mathbf{m}^i_{\theta f} \\ \text{symmetric} & & \mathbf{m}^i_{ff} \end{bmatrix} \tag{6.73}$$

The components of the mass matrix of Eq. 73 can be evaluated as follows. The matrix $\mathbf{m}^i_{RR}$ is the sum of the element $\mathbf{m}^{ij}_{RR}$ matrices, that is,

$$\mathbf{m}^i_{RR} = \sum_{j=1}^{n_e} \mathbf{m}^{ij}_{RR} = \begin{bmatrix} m^i & 0 & 0 \\ 0 & m^i & 0 \\ 0 & 0 & m^i \end{bmatrix} \tag{6.74}$$

where m^i is the total mass of the body, that is,

$$m^i = \sum_{j=1}^{n_e} m^{ij} \tag{6.75}$$

and m^{ij} is the mass of the ijth element.

The matrix $\mathbf{m}^i_{R\theta}$ that represents the coupling between the translation and rotation of the body reference is defined as

$$\mathbf{m}^i_{R\theta} = \sum_{j=1}^{n_e} \mathbf{m}^{ij}_{R\theta} = -\mathbf{A}^i \left[\sum_{j=1}^{n_e} \int_{V^{ij}} \rho^{ij} \tilde{\bar{\mathbf{u}}}^{ij} \, dV^{ij} \right] \bar{\mathbf{G}}^i \tag{6.76}$$

Using the definition of $\bar{\mathbf{u}}^{ij}$ of Eq. 17 and the definition of the skew symmetric matrix $\tilde{\bar{\mathbf{u}}}^{ij}$ of Eq. 38, one can verify that the constant element matrices $\bar{\mathbf{S}}^{ij}$ of Eq. 50 are required to evaluate the matrix $\mathbf{m}^{ij}_{R\theta}$ of Eq. 76. These element matrices can be assembled to yield the following body matrix:

$$\bar{\mathbf{S}}^i = \sum_{j=1}^{n_e} \bar{\mathbf{S}}^{ij} \tag{6.77}$$

The matrix $\bar{\mathbf{S}}^i$ is also needed to evaluate the matrix $\mathbf{m}^i_{Rf}$ of Eq. 73. This matrix can be written using Eq. 52 as

$$\mathbf{m}^i_{Rf} = \sum_{j=1}^{n_e} \mathbf{m}^{ij}_{Rf} = \mathbf{A}^i \sum_{j=1}^{n_e} \bar{\mathbf{S}}^{ij} \mathbf{B}^i_2$$

which, on using Eq. 77, yields

$$\mathbf{m}^i_{Rf} = \mathbf{A}^i \bar{\mathbf{S}}^i \mathbf{B}^i_2 \tag{6.78}$$

With the use of Eq. 53, the central term $\mathbf{m}^i_{\theta\theta}$ of Eq. 73 can be written as

$$\begin{aligned} \mathbf{m}^i_{\theta\theta} &= \sum_{j=1}^{n_e} \mathbf{m}^{ij}_{\theta\theta} = \bar{\mathbf{G}}^{i\mathrm{T}} \left[\sum_{j=1}^{n_e} \bar{\mathbf{I}}^{ij}_{\theta\theta} \right] \bar{\mathbf{G}}^i \\ &= \bar{\mathbf{G}}^{i\mathrm{T}} \bar{\mathbf{I}}^i_{\theta\theta} \bar{\mathbf{G}}^i \end{aligned} \tag{6.79}$$

where $\bar{\mathbf{I}}^i_{\theta\theta}$ is the inertia tensor of the deformable body i defined in the body coordinate system. Using Eqs. 54–61, one can verify that the following square matrices are required to evaluate the inertia tensor $\bar{\mathbf{I}}^i_{\theta\theta}$:

$$\mathbf{S}^i_{kl} = \sum_{j=1}^{n_e} \mathbf{S}^{ij}_{kl}, \quad k, l = 1, 2, 3 \tag{6.80}$$

where the element matrices $\mathbf{S}^{ij}_{kl}$ are defined in Eq. 61. Similarly, the matrix $\mathbf{m}^i_{\theta f}$, which represents the inertia coupling between the reference rotations and the elastic deformation, can be written as

$$\mathbf{m}^i_{\theta f} = \sum_{j=1}^{n_e} \mathbf{m}^{ij}_{\theta f} \tag{6.81}$$

Substituting Eq. 65 into Eq. 81 yields

$$\mathbf{m}^i_{\theta f} = \bar{\mathbf{G}}^{i\mathrm{T}} \begin{bmatrix} \mathbf{q}^{i\mathrm{T}}_n \tilde{\mathbf{N}}^i_{23} \\ \mathbf{q}^{i\mathrm{T}}_n \tilde{\mathbf{N}}^i_{31} \\ \mathbf{q}^{i\mathrm{T}}_n \tilde{\mathbf{N}}^i_{12} \end{bmatrix} \mathbf{B}^i_2 \tag{6.82}$$

where the skew symmetric matrices $\tilde{\mathbf{N}}^i_{kl}$ are defined as

$$\tilde{\mathbf{N}}^i_{12} = \sum_{j=1}^{n_e} \tilde{\mathbf{N}}^{ij}_{12}, \quad \tilde{\mathbf{N}}^i_{23} = \sum_{j=1}^{n_e} \tilde{\mathbf{N}}^{ij}_{23}, \quad \tilde{\mathbf{N}}^i_{31} = \sum_{j=1}^{n_e} \tilde{\mathbf{N}}^{ij}_{31} \tag{6.83}$$

in which the skew symmetric matrices $\tilde{\mathbf{N}}^{ij}_{kl}$ can be evaluated by using the inertia shape integrals defined by Eq. 80.

Finally, the matrix $\mathbf{m}^i_{ff}$ associated with the elastic coordinates can, by using Eq. 68, be written as

$$\mathbf{m}^i_{ff} = \sum_{j=1}^{n_e} \mathbf{m}^{ij}_{ff} = \mathbf{B}^{i\mathrm{T}}_2 \mathbf{S}^i_{ff} \mathbf{B}^i_2 \tag{6.84}$$

where $\mathbf{S}^i_{ff}$ is the assembled matrix of the element $\mathbf{S}^{ij}_{ff}$ matrices of Eq. 69, that is

$$\mathbf{S}^i_{ff} = \sum_{j=1}^{n_e} \mathbf{S}^{ij}_{ff} \tag{6.85}$$

Planar Motion of the Deformable Body In the case of the planar motion of the deformable body in the $\mathbf{X}_1\mathbf{X}_2$ plane, the velocity vector $\dot{\mathbf{r}}^{ij}$ of Eq. 41 can be written as

$$\dot{\mathbf{r}}^{ij} = \dot{\mathbf{R}}^i + \mathbf{A}^i_\theta \bar{\mathbf{u}}^{ij} \dot{\theta}^i + \mathbf{A}^i \mathbf{N}^{ij} \mathbf{B}^i_2 \dot{\mathbf{q}}^i_f \tag{6.86}$$

where the two orthogonal matrices $\mathbf{A}^i$ and $\mathbf{A}^i_\theta$ are defined as

$$\mathbf{A}^i = \begin{bmatrix} \cos\theta^i & -\sin\theta^i \\ \sin\theta^i & \cos\theta^i \end{bmatrix}, \quad \mathbf{A}^i_\theta = \begin{bmatrix} -\sin\theta^i & -\cos\theta^i \\ \cos\theta^i & -\sin\theta^i \end{bmatrix} \tag{6.87}$$

in which θ^i is the angular rotation about the $\mathbf{X}_3$ axis. In Eq. 86, $\bar{\mathbf{u}}^{ij}$ is a two-dimensional vector. Using the definition of the kinetic energy of Eq. 43, one can show that the mass matrix of Eq. 44 is defined as

$$\mathbf{M}^{ij} = \int_{V^{ij}} \rho^{ij} \begin{bmatrix} \mathbf{I}_2 & \mathbf{A}^i_\theta \bar{\mathbf{u}}^{ij} & \mathbf{A}^i \mathbf{N}^{ij} \mathbf{B}^i_2 \\ & \bar{\mathbf{u}}^{ij\mathrm{T}} \bar{\mathbf{u}}^{ij} & \bar{\mathbf{u}}^{ij\mathrm{T}} \mathbf{A}^{i\mathrm{T}}_\theta \mathbf{A}^i \mathbf{N}^{ij} \mathbf{B}^i_2 \\ \text{symmetric} & & \mathbf{B}^{i\mathrm{T}}_2 \mathbf{N}^{ij\mathrm{T}} \mathbf{N}^{ij} \mathbf{B}^i_2 \end{bmatrix} dV^{ij} \tag{6.88}$$

where $\mathbf{I}_2$ is a 2×2 identity matrix, and the product $\mathbf{A}^{i\mathrm{T}}_\theta \mathbf{A}^i$ yields a skew symmetric matrix denoted as $\tilde{\mathbf{I}}$, where

$$\tilde{\mathbf{I}} = \mathbf{A}^{i\mathrm{T}}_\theta \mathbf{A}^i = \begin{bmatrix} 0 & 1 \\ -1 & 0 \end{bmatrix} \tag{6.89}$$

Following the procedure described in the preceding chapter, it can be shown that the following three element inertia shape integrals are required to evaluate the mass matrix of Eq. 88:

$$\bar{\mathbf{S}}^{ij} = \mathbf{C}^{ij} \left[\int_{V^{ij}} \rho^{ij} \mathbf{S}^{ij} dV^{ij} \right] \bar{\mathbf{C}}^{ij} \mathbf{B}^{ij}_1 \tag{6.90}$$

$$\tilde{\mathbf{S}}^{ij} = \mathbf{B}^{ij\mathrm{T}}_1 \bar{\mathbf{C}}^{ij\mathrm{T}} \left[\int_{V^{ij}} \rho^{ij} \mathbf{S}^{ij\mathrm{T}} \tilde{\mathbf{I}} \mathbf{S}^{ij} dV^{ij} \right] \bar{\mathbf{C}}^{ij} \mathbf{B}^{ij}_1 \tag{6.91}$$

$$\mathbf{S}^{ij}_{ff} = \mathbf{B}^{ij\mathrm{T}}_1 \bar{\mathbf{C}}^{ij} \left[\int_{V^{ij}} \rho^{ij} \mathbf{S}^{ij\mathrm{T}} \mathbf{S}^{ij} dV^{ij} \right] \bar{\mathbf{C}}^{ij} \mathbf{B}^{ij}_1 \tag{6.92}$$

Comparison between the finite-element formulation for the mass matrix of the deformable body presented in this section and the formulation presented in the preceding chapter reveals that the use of the finite-element method substantially reduces the number of inertia shape integrals that are required to formulate the nonlinear mass matrix of the deformable body that undergoes large translational and rotational displacements. This is mainly because, in the finite-element formulation, the position vector in the undeformed state can be expressed in terms of the element shape function that can describe an arbitrary rigid body translation. In the spatial analysis seven shape integrals, given by Eqs. 50 and 61, are defined for each finite element. In the planar analysis, three shape integrals, given by Eqs. 90–92, are defined for each finite element. It was also shown that the body shape integrals can be obtained by assembling the shape integrals of its elements by using a standard finite-element procedure.

6.4 GENERALIZED ELASTIC FORCES

In this section, a procedure for defining the stiffness matrix of the deformable body i assuming a linear isotropic material is outlined. For the jth element of the ith body, the virtual work of the elastic forces can be written as

$$\delta W^{ij}_s = -\int_{V^{ij}} \sigma^{ij\mathrm{T}} \delta \varepsilon^{ij} dV^{ij} \tag{6.93}$$

where $\boldsymbol{\sigma}^{ij}$ and $\boldsymbol{\varepsilon}^{ij}$ are, respectively, the stress and strain vectors. The constitutive equations of the element can be written in the following form:

$$\boldsymbol{\sigma}^{ij} = \mathbf{E}^{ij}\boldsymbol{\varepsilon}^{ij} \tag{6.94}$$

where $\mathbf{E}^{ij}$ is the matrix of elastic coefficients. The strain displacement relation can be written as

$$\boldsymbol{\varepsilon}^{ij} = \mathbf{D}^{ij}\bar{\mathbf{u}}_f^{ij} \tag{6.95}$$

where $\mathbf{D}^{ij}$ is a spatial differential operator relating strains and displacements.

Using Eqs. 25, 27, and 28, one can write the displacement vector $\bar{\mathbf{u}}_f^{ij}$ as

$$\bar{\mathbf{u}}_f^{ij} = \mathbf{N}^{ij}\mathbf{B}_2^i\mathbf{q}_f^i = \bar{\mathbf{N}}^{ij}\mathbf{q}_f^i \tag{6.96}$$

where $\mathbf{N}^{ij}$ is defined by Eq. 26 and the space-dependent matrix $\bar{\mathbf{N}}^{ij}$ is defined as

$$\bar{\mathbf{N}}^{ij} = \mathbf{N}^{ij}\mathbf{B}_2^i \tag{6.97}$$

Substituting Eq. 96 into Eq. 95 yields

$$\boldsymbol{\varepsilon}^{ij} = \mathbf{D}^{ij}\bar{\mathbf{N}}^{ij}\mathbf{q}_f^i \tag{6.98}$$

Substituting Eqs. 94 and 98 into Eq. 93 leads to

$$\delta W_s^{ij} = -\mathbf{q}_f^{i\mathrm{T}}\mathbf{K}_{ff}^{ij}\delta\mathbf{q}_f^i \tag{6.99}$$

where $\mathbf{K}_{ff}^{ij}$ is the *element stiffness matrix* defined as

$$\mathbf{K}_{ff}^{ij} = \int_{V^{ij}} (\mathbf{D}^{ij}\bar{\mathbf{N}}^{ij})^{\mathrm{T}}\mathbf{E}^{ij}(\mathbf{D}^{ij}\bar{\mathbf{N}}^{ij})\,dV^{ij} \tag{6.100}$$

The virtual work of elastic forces of body i can be written as

$$\delta W_s^i = \sum_{j=1}^{n_e} \delta W_s^{ij} \tag{6.101}$$

which, on substituting Eq. 99, yields

$$\delta W_s^i = -\mathbf{q}_f^{i\mathrm{T}}\mathbf{K}_{ff}^i\delta\mathbf{q}_f^i \tag{6.102}$$

where $\mathbf{K}_{ff}^i$ is the assembled stiffness matrix of body i, which is defined as

$$\mathbf{K}_{ff}^i = \sum_{j=1}^{n_e} \mathbf{K}_{ff}^{ij} \tag{6.103}$$

and the element stiffness matrix $\mathbf{K}_{ff}^{ij}$ is defined by Eq. 100.

According to the partitioning of the generalized coordinates of body i given by Eq. 45, the virtual work of Eq. 102 can be written in the following matrix form:

$$\delta W_s^i = -\begin{bmatrix}\mathbf{R}^{i\mathrm{T}} & \boldsymbol{\theta}^{i\mathrm{T}} & \mathbf{q}_f^{i\mathrm{T}}\end{bmatrix}\begin{bmatrix}\mathbf{0} & \mathbf{0} & \mathbf{0}\\ \mathbf{0} & \mathbf{0} & \mathbf{0}\\ \mathbf{0} & \mathbf{0} & \mathbf{K}_{ff}^i\end{bmatrix}\begin{bmatrix}\delta\mathbf{R}^i\\ \delta\boldsymbol{\theta}^i\\ \delta\mathbf{q}_f^i\end{bmatrix} \tag{6.104}$$

or in a compact matrix form as

$$\delta W_s^i = -\mathbf{q}^{i\mathrm{T}}\mathbf{K}^i\delta\mathbf{q}^i \tag{6.105}$$

where

$$\mathbf{K}^i = \begin{bmatrix} \mathbf{0} & \mathbf{0} & \mathbf{0} \\ \mathbf{0} & \mathbf{0} & \mathbf{0} \\ \mathbf{0} & \mathbf{0} & \mathbf{K}^i_{ff} \end{bmatrix} \tag{6.106}$$

Having defined the mass and stiffness matrices for the deformable body i, one can then substitute them into the differential equations given by Eq. 133 of the preceding chapter to obtain the differential equations of motion of the deformable body i in the multibody system. In the following two sections, we exemplify the preceding developments by use of two- and three-dimensional beam elements.

6.5 CHARACTERIZATION OF PLANAR ELASTIC SYSTEMS

As an illustration of the preceding development in the planar analysis, we consider the two-dimensional beam element shown in Fig. 6. The figure shows element j on a planar body i. The element $\mathbf{X}_1^{ij}$ axis forms an angle β^{ij} relative to the body $\mathbf{X}_1^i$ axis. The reference coordinates for this body are

$$\mathbf{q}_r^{i\mathrm{T}} = [\mathbf{R}^{i\mathrm{T}} \quad \theta^i] \tag{6.107}$$

This set of coordinates defines the location and orientation of the body reference relative to the $\mathbf{X}_1\mathbf{X}_2$ inertial frame. The elastic generalized coordinates are defined initially with respect to the intermediate element coordinate system $\mathbf{X}_{i1}^{ij}\mathbf{X}_{i2}^{ij}$, which is initially parallel to the element coordinate system and the origin of which is rigidly attached to the body coordinate system. This set of element elastic generalized coordinates

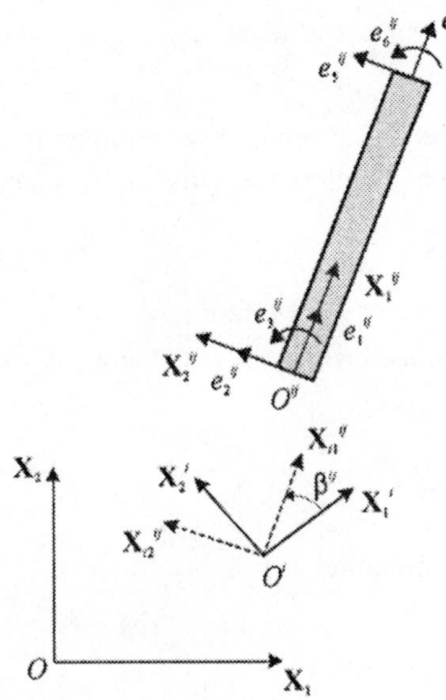

Figure 6.6 Two-dimensional beam element.

is denoted by $e^{ij}_{ik}(k = 1, \ldots, 6)$. These coordinates are the nodal coordinates and represent the location of the nodes and slopes of the element axis at these nodes. The location of an arbitrary point P^{ij} on element ij with respect to the $\mathbf{X}^{ij}_{i1}\mathbf{X}^{ij}_{i2}$ frame can be expressed as (Eq. 14)

$$\mathbf{w}^{ij}_i = \mathbf{S}^{ij}\mathbf{e}^{ij}_i \tag{6.108}$$

where $\mathbf{S}^{ij}$ is the element shape function, which is assumed to be

$$\mathbf{S}^{ij} = \begin{bmatrix} 1-\xi & 0 & 0 & \xi & 0 & 0 \\ 0 & 1-3(\xi)^2+2(\xi)^3 & l[\xi-2(\xi)^2+(\xi)^3] & 0 & 3(\xi)^2-2(\xi)^3 & l[(\xi)^3-(\xi)^2] \end{bmatrix}^{ij} \tag{6.109}$$

where the superscript ij on the major bracket indicates that all the elements inside this bracket are superscripted with ij, l^{ij} is the element length, and ξ^{ij} is the dimensionless parameter

$$\xi^{ij} = \frac{x^{ij}_1}{l^{ij}} \tag{6.110}$$

where x^{ij}_1 is the spatial coordinate along the element axis.

A transformation $\mathbf{C}^{ij}$ is employed to define $\mathbf{w}^{ij}_i$ with respect to the $\mathbf{X}^i_1\mathbf{X}^i_2$ frame as

$$\bar{\mathbf{u}}^{ij} = \mathbf{C}^{ij}\mathbf{w}^{ij}_i = \mathbf{C}^{ij}\mathbf{S}^{ij}\mathbf{e}^{ij}_i \tag{6.111}$$

where

$$\mathbf{C}^{ij} = \begin{bmatrix} \cos\beta^{ij} & -\sin\beta^{ij} \\ \sin\beta^{ij} & \cos\beta^{ij} \end{bmatrix} \tag{6.112}$$

Compatibility conditions between elements on a given body are simpler if the nodal coordinates are defined with respect to the body reference. This can be accomplished by the transformation

$$\mathbf{e}^{ij}_i = \bar{\mathbf{C}}^{ij}\mathbf{e}^{ij} \tag{6.113}$$

where $\mathbf{e}^{ij}$ is the set of nodal coordinates defined with respect to the $\mathbf{X}^i_1\mathbf{X}^i_2$ frame and

$$\bar{\mathbf{C}}^{ij} = \begin{bmatrix} \bar{\mathbf{C}}^{ij}_1 & \mathbf{0} \\ \mathbf{0} & \bar{\mathbf{C}}^{ij}_1 \end{bmatrix} \tag{6.114}$$

in which $\bar{\mathbf{C}}^{ij}_1$ is the 3×3 orthogonal transformation

$$\bar{\mathbf{C}}^{ij}_1 = \begin{bmatrix} \cos\beta^{ij} & \sin\beta^{ij} & 0 \\ -\sin\beta^{ij} & \cos\beta^{ij} & 0 \\ 0 & 0 & 1 \end{bmatrix} \tag{6.115}$$

Substituting Eq. 113 into Eq. 111 yields

$$\bar{\mathbf{u}}^{ij} = \mathbf{C}^{ij}\mathbf{S}^{ij}\bar{\mathbf{C}}^{ij}\mathbf{e}^{ij} \tag{6.116}$$

Recall that β^{ij} is a constant angle and accordingly the orthogonal transformation matrices $\mathbf{C}^{ij}$ and $\bar{\mathbf{C}}^{ij}$ are constant.

Without any loss of generality and to simplify the derivation, we consider the case of only one element. In this case one can write Eq. 116 as

$$\bar{\mathbf{u}}^{ij} = \mathbf{N}^{ij}\mathbf{e}^{ij} \tag{6.117}$$

where

$$\mathbf{N}^{ij} = \mathbf{C}^{ij}\mathbf{S}^{ij}\bar{\mathbf{C}}^{ij} \tag{6.118}$$

Element Mass Matrix In the following, the kinetic energy expression is used to develop the mass matrix of the two-dimensional beam element j on body i based on the assumed displacement field defined by Eq. 108.

The global position vector $\mathbf{r}^{ij}$ of an arbitrary point on element ij can then be expressed as

$$\mathbf{r}^{ij} = \mathbf{R}^i + \mathbf{A}^i\mathbf{N}^{ij}\mathbf{e}^{ij} \tag{6.119}$$

where $\mathbf{A}^i$ is the transformation matrix

$$\mathbf{A}^i = \begin{bmatrix} \cos\theta^i & -\sin\theta^i \\ \sin\theta^i & \cos\theta^i \end{bmatrix} \tag{6.120}$$

Differentiating Eq. 119 with respect to time gives

$$\dot{\mathbf{r}}^{ij} = \dot{\mathbf{R}}^i + \dot{\mathbf{A}}^i\mathbf{N}^{ij}\mathbf{e}^{ij} + \mathbf{A}^i\mathbf{N}^{ij}\dot{\mathbf{e}}^{ij} \tag{6.121}$$

where

$$\begin{aligned} \dot{\mathbf{A}}^i &= \dot{\theta}^i \begin{bmatrix} -\sin\theta^i & -\cos\theta^i \\ \cos\theta^i & -\sin\theta^i \end{bmatrix} \\ &= \dot{\theta}^i\mathbf{A}^i_\theta \end{aligned} \tag{6.122}$$

and $\mathbf{A}^i_\theta$ is the partial derivative of $\mathbf{A}^i$ with respect to the reference rotational degree of freedom θ^i. Substituting Eq. 122 into Eq. 121 and writing $\dot{\mathbf{r}}^{ij}$ in a partitioned form yields

$$\dot{\mathbf{r}}^{ij} = \begin{bmatrix} \mathbf{I} & \mathbf{A}^i_\theta\mathbf{N}^{ij}\mathbf{e}^{ij} & \mathbf{A}^i\mathbf{N}^{ij} \end{bmatrix} \begin{bmatrix} \dot{\mathbf{R}}^i \\ \dot{\theta}^i \\ \dot{\mathbf{e}}^{ij} \end{bmatrix} \tag{6.123}$$

The kinetic energy expression for element ij is given by

$$T^{ij} = \frac{1}{2}\int_{V^{ij}} \rho^{ij}\dot{\mathbf{r}}^{ij\mathrm{T}}\dot{\mathbf{r}}^{ij}\,dV^{ij} \tag{6.124}$$

$$= \frac{1}{2}\dot{\mathbf{q}}^{ij\mathrm{T}}\mathbf{M}^{ij}\dot{\mathbf{q}}^{ij} \tag{6.125}$$

where V^{ij} is the element volume, ρ^{ij} is the density of the element material, and $\mathbf{q}^{ij}$ and $\mathbf{M}^{ij}$ are defined as

$$\mathbf{q}^{ij} = [\mathbf{R}^{i\mathrm{T}} \quad \theta^i \quad \mathbf{e}^{ij\mathrm{T}}]^{\mathrm{T}} \tag{6.126}$$

$$\mathbf{M}^{ij} = \int_{V^{ij}} \rho^{ij} \begin{bmatrix} \mathbf{I}_2 & \mathbf{A}^i_\theta \bar{\mathbf{u}}^{ij} & \mathbf{A}^i \mathbf{N}^{ij} \\ & \bar{\mathbf{u}}^{ij\mathrm{T}}\bar{\mathbf{u}}^{ij} & \bar{\mathbf{u}}^{ij\mathrm{T}}\tilde{\mathbf{I}}\mathbf{N}^{ij} \\ \text{symmetric} & & \mathbf{N}^{ij\mathrm{T}}\mathbf{N}^{ij} \end{bmatrix} dV^{ij} \tag{6.127}$$

where the skew symmetric matrix $\tilde{\mathbf{I}}$ is defined in Eq. 89. In deriving the mass matrix of Eq. 127 the orthogonality of the matrices $\mathbf{A}^i$ and $\mathbf{A}^i_\theta$ is used. One may write Eq. 127 in a simpler form as

$$\mathbf{M}^{ij} = \begin{bmatrix} \mathbf{m}^{ij}_{RR} & \mathbf{m}^{ij}_{R\theta} & \mathbf{m}^{ij}_{Rf} \\ & \mathbf{m}^{ij}_{\theta\theta} & \mathbf{m}^{ij}_{\theta f} \\ \text{symmetric} & & \mathbf{m}^{ij}_{ff} \end{bmatrix} \tag{6.128}$$

Inertia Shape Integrals In the following, we define the components of the mass matrix of Eq. 128 and identify the inertia shape integrals that appear in the nonlinear terms that represent the coupling between the gross motion and the elastic deformation. The matrix $\mathbf{m}^{ij}_{RR}$ associated with the reference translation is given by

$$\mathbf{m}^{ij}_{RR} = \int_{V^{ij}} \rho^{ij}\mathbf{I}_2 dV^{ij} = \begin{bmatrix} m^{ij} & 0 \\ 0 & m^{ij} \end{bmatrix} \tag{6.129}$$

where m^{ij} is the mass of the ijth element. By using Eq. 118, one can write the matrix $\mathbf{m}^{ij}_{ff}$ associated with the elastic deformation of the element as

$$\mathbf{m}^{ij}_{ff} = \int_{V^{ij}} \rho^{ij}\mathbf{N}^{ij\mathrm{T}}\mathbf{N}^{ij} dV^{ij} = \bar{\mathbf{C}}^{ij^{\mathrm{T}}} \left[\int_{V^{ij}} \rho^{ij}\mathbf{S}^{ij\mathrm{T}}\mathbf{S}^{ij} dV^{ij} \right] \bar{\mathbf{C}}^{ij} \tag{6.130}$$

where the orthogonality of the transformation matrix $\mathbf{C}^{ij}$ of Eq. 112 is used. We may, then, write Eq. 130 in the following form:

$$\mathbf{m}^{ij}_{ff} = \bar{\mathbf{C}}^{ij^{\mathrm{T}}} \mathbf{S}^{ij}_{ff} \bar{\mathbf{C}}^{ij} \tag{6.131}$$

where, on using the shape function of Eq. 109, one can write the matrix $\mathbf{S}^{ij}_{ff}$ in a more explicit form as

$$\mathbf{S}^{ij}_{ff} = \int_{V^{ij}} \rho^{ij}\mathbf{S}^{ij\mathrm{T}}\mathbf{S}^{ij} dV^{ij} \tag{6.132}$$

or

$$\mathbf{S}^{ij}_{ff} = m^{ij} \begin{bmatrix} \frac{1}{3} & & & & & \\ 0 & \frac{13}{35} & & \text{symmetric} & & \\ 0 & \frac{11l}{210} & \frac{(l)^2}{105} & & & \\ \frac{1}{6} & 0 & 0 & \frac{1}{3} & & \\ 0 & \frac{9}{70} & \frac{13l}{420} & 0 & \frac{13}{35} & \\ 0 & -\frac{13l}{420} & -\frac{(l)^2}{140} & 0 & -\frac{11l}{210} & \frac{(l)^2}{105} \end{bmatrix} \tag{6.133}$$

Similarly

$$\mathbf{m}_{R\theta}^{ij} = \int_{V^{ij}} \rho^{ij} \mathbf{A}_{\theta}^{i} \bar{\mathbf{u}}^{ij} dV^{ij} = \mathbf{A}_{\theta}^{i} \int_{V^{ij}} \rho^{ij} \bar{\mathbf{u}}^{ij} dV^{ij}$$

Substituting Eq. 116 into the above equation, and using the fact that $\mathbf{C}^{ij}$ and $\bar{\mathbf{C}}^{ij}$ are constant matrices and $\mathbf{e}^{ij}$ depends only on time, one gets

$$\mathbf{m}_{R\theta}^{ij} = \mathbf{A}_{\theta}^{i} \mathbf{C}^{ij} \left\{ \int_{V^{ij}} \rho^{ij} \mathbf{S}^{ij} dV^{ij} \right\} \bar{\mathbf{C}}^{ij} \mathbf{e}^{ij} = \mathbf{A}_{\theta}^{i} \mathbf{C}^{ij} \bar{\mathbf{S}}^{ij} \bar{\mathbf{C}}^{ij} \mathbf{e}^{ij} \tag{6.134}$$

where the matrix $\bar{\mathbf{S}}^{ij}$ is defined as

$$\bar{\mathbf{S}}^{ij} = \frac{m^{ij}}{12} \begin{bmatrix} 6 & 0 & 0 & 6 & 0 & 0 \\ 0 & 6 & l^{ij} & 0 & 6 & -l^{ij} \end{bmatrix} \tag{6.135}$$

The matrix $\mathbf{m}_{Rf}^{ij}$, which represents the dynamic coupling between the reference translation and the elastic deformation of the element, can be expressed as

$$\begin{aligned} \mathbf{m}_{Rf}^{ij} &= \int_{V^{ij}} \rho^{ij} \mathbf{A}^{i} \mathbf{N}^{ij} dV^{ij} = \mathbf{A}^{i} \int_{V^{ij}} \rho^{ij} \mathbf{C}^{ij} \mathbf{S}^{ij} \bar{\mathbf{C}}^{ij} dV^{ij} \\ &= \mathbf{A}^{i} \mathbf{C}^{ij} \bar{\mathbf{S}}^{ij} \bar{\mathbf{C}}^{ij} \end{aligned} \tag{6.136}$$

where the matrix $\bar{\mathbf{S}}^{ij}$ is defined by Eq. 135.

The central term in the mass matrix of Eq. 128 is simply given by

$$\mathbf{m}_{\theta\theta}^{ij} = \int_{V^{ij}} \rho^{ij} \bar{\mathbf{u}}^{ij\mathrm{T}} \bar{\mathbf{u}}^{ij} dV^{ij} = \mathbf{e}^{ij\mathrm{T}} \mathbf{m}_{ff}^{ij} \mathbf{e}^{ij} \tag{6.137}$$

where the matrix $\mathbf{m}_{ff}^{ij}$ is given by Eq. 131. It is clear that $\mathbf{m}_{\theta\theta}^{ij}$, which reduces to a scalar in this case, is the mass moment of inertia of the element about the $\mathbf{X}_3^i$ axis of the body reference. This moment of inertia depends on the elastic coordinates of the elements. This can be demonstrated if we write the vector of nodal coordinates as

$$\mathbf{e}^{ij} = \mathbf{e}_o^{ij} + \mathbf{e}_f^{ij} \tag{6.138}$$

where $\mathbf{e}_o^{ij}$ is the nodal coordinates in the undeformed state and $\mathbf{e}_f^{ij}$ is the vector of deformation at the nodal points. Substituting Eq. 138 into Eq. 137 and using the symmetry of the matrix $\mathbf{m}_{ff}^{ij}$ leads to

$$\mathbf{m}_{\theta\theta}^{ij} = \mathbf{e}_o^{ij\mathrm{T}} \mathbf{m}_{ff}^{ij} \mathbf{e}_o^{ij} + 2\mathbf{e}_o^{ij\mathrm{T}} \mathbf{m}_{ff}^{ij} \mathbf{e}_f^{ij} + \mathbf{e}_f^{ij\mathrm{T}} \mathbf{m}_{ff}^{ij} \mathbf{e}_f^{ij} \tag{6.139}$$

where the first term in the right-hand side of the equation can be recognized as the mass moment of inertia if the element were rigid.

Finally, the matrix $\mathbf{m}_{\theta f}^{ij}$ that represents the coupling between the rotation of the body reference and the elastic deformation of the element can be written as

$$\mathbf{m}_{\theta f}^{ij} = \int_{V^{ij}} \rho^{ij} \bar{\mathbf{u}}^{ij\mathrm{T}} \tilde{\mathbf{I}} \mathbf{N}^{ij} dV^{ij}$$

which, with the use of Eqs. 117 and 118, yields

$$\mathbf{m}_{\theta f}^{ij} = \mathbf{e}^{ij\mathrm{T}}\bar{\mathbf{C}}^{ij\mathrm{T}}\tilde{\mathbf{S}}^{ij}\bar{\mathbf{C}}^{ij} \tag{6.140}$$

where the fact that $\mathbf{C}^{ij\mathrm{T}}\tilde{\mathbf{I}}\mathbf{C}^{ij} = \tilde{\mathbf{I}}$ is used and $\tilde{\mathbf{S}}^{ij}$ is the skew symmetric matrix defined as

$$\tilde{\mathbf{S}}^{ij} = \int_{V^{ij}} \rho^{ij}\mathbf{S}^{ij\mathrm{T}}\tilde{\mathbf{I}}\mathbf{S}^{ij}\,dV^{ij}$$

$$= \frac{m^{ij}}{60}\begin{bmatrix} 0 & 21 & 3l & 0 & 9 & -2l \\ -21 & 0 & 0 & -9 & 0 & 0 \\ -3l & 0 & 0 & -2l & 0 & 0 \\ 0 & 9 & 2l & 0 & 21 & -3l \\ -9 & 0 & 0 & -21 & 0 & 0 \\ 2l & 0 & 0 & 3l & 0 & 0 \end{bmatrix}^{ij} \tag{6.141}$$

Element Stiffness Matrix The stiffness matrix for the beam element j on body i can be developed by using the strain energy expression. Using the elementary beam theory and neglecting the shear deformation, the strain energy of element j on body i is given by

$$U^{ij} = \frac{1}{2}\int_0^{l^{ij}} \left[E^{ij}I^{ij}\left(u_2^{ij''}\right)^2 + E^{ij}a^{ij}\left(u_1^{ij'}\right)^2\right] dx_1^{ij} \tag{6.142}$$

where the primes indicate derivatives with respect to the spatial x_1^{ij} element coordinate, E^{ij} is the modulus of elasticity, I^{ij} is the second moment of area, and a^{ij} is the element cross-sectional area.

With the use of Eqs. 117 and 118, Eq. 142 becomes

$$U^{ij} = \frac{1}{2}\mathbf{e}_f^{ij\mathrm{T}}\bar{\mathbf{C}}^{ij\mathrm{T}}\bar{\mathbf{K}}_{ff}^{ij}\bar{\mathbf{C}}^{ij}\mathbf{e}_f^{ij} \tag{6.143}$$

where $\mathbf{e}_f^{ij}$ is defined in Eq. 138 and $\bar{\mathbf{K}}_{ff}^{ij}$ is the element stiffness matrix, which is symmetric and given by

$$\bar{\mathbf{K}}_{ff}^{ij} = \frac{E^{ij}I^{ij}}{l^{ij}}\begin{bmatrix} \frac{a}{I} & & & & & \\ 0 & \frac{12}{(l)^2} & & & & \\ 0 & \frac{6}{l} & 4 & & & \\ -\frac{a}{I} & 0 & 0 & \frac{a}{I} & & \\ 0 & -\frac{12}{(l)^2} & -\frac{6}{l} & 0 & \frac{12}{(l)^2} & \\ 0 & \frac{6}{l} & 2 & 0 & -\frac{6}{l} & 4 \end{bmatrix} \tag{6.144}$$

Equation 143 can be written in another form as

$$U^{ij} = \frac{1}{2}\mathbf{e}_f^{ij\mathrm{T}}\mathbf{K}_{ff}^{ij}\mathbf{e}_f^{ij} \tag{6.145}$$

where

$$\mathbf{K}_{ff}^{ij} = \bar{\mathbf{C}}^{ij\mathrm{T}}\bar{\mathbf{K}}_{ff}^{ij}\bar{\mathbf{C}}^{ij}$$

is the element stiffness matrix defined with respect to the body coordinate system.

6.6 CHARACTERIZATION OF SPATIAL ELASTIC SYSTEMS

Figure 7 shows a typical element j on body i. The element has 12 nodal coordinates that describe the translations and slopes of the two nodes. These nodal coordinates defined with respect to the $\mathbf{X}_{i1}^{ij}\mathbf{X}_{i2}^{ij}\mathbf{X}_{i3}^{ij}$ coordinate system are denoted by the vector $\mathbf{e}_i^{ij}$, that is,

$$\mathbf{e}_i^{ij} = \begin{bmatrix} e_{i1}^{ij} & e_{i2}^{ij} & \cdots & e_{i12}^{ij} \end{bmatrix}^{\mathrm{T}} \tag{6.146}$$

The location of the origin of the ith body reference with respect to the inertial frame is defined by the Cartesian coordinates $\mathbf{R}^i$. Let $\mathbf{w}_i^{ij} = [w_{i1}^{ij} \;\; w_{i2}^{ij} \;\; w_{i3}^{ij}]^{\mathrm{T}}$ locate an arbitrary point P^{ij} on element ij, relative to the $\mathbf{X}_{i1}^{ij}\mathbf{X}_{i2}^{ij}\mathbf{X}_{i3}^{ij}$ coordinate system, where w_{i1}^{ij}, w_{i2}^{ij}, and w_{i3}^{ij} are the $\mathbf{X}_{i1}^{ij}$, $\mathbf{X}_{i2}^{ij}$ and $\mathbf{X}_{i3}^{ij}$, components, respectively. Following the same procedure as in the previous section, $\bar{\mathbf{u}}^{ij}$ can be written as

$$\bar{\mathbf{u}}^{ij} = \mathbf{C}^{ij}\mathbf{S}^{ij}\bar{\mathbf{C}}^{ij}\mathbf{e}^{ij} \tag{6.147}$$

where $\mathbf{e}^{ij}$ is the vector of nodal coordinates of element ij defined in the body coordinate

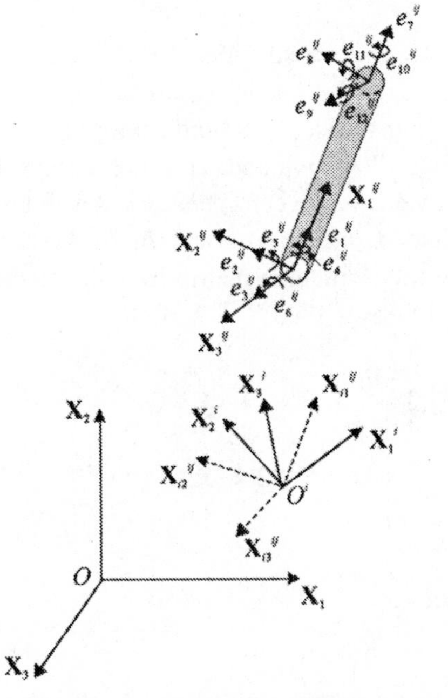

Figure 6.7 Three-dimensional beam element.

system and the shape function matrix $\mathbf{S}^{ij}$ is assumed to be

$$\mathbf{S}^{ij\mathrm{T}} = \begin{bmatrix} 1-\xi & 0 & 0 \\ 6[\xi-(\xi)^2]\eta & 1-3(\xi)^2+2(\xi)^3 & 0 \\ 6[\xi-(\xi)^2]\xi & 0 & 1-3(\xi)^2+2(\xi)^3 \\ 0 & -(1-\xi)l\zeta & -(1-\xi)l\eta \\ [1-4\xi+3(\xi)^2]l\zeta & 0 & [-\xi+2(\xi)^2-(\xi)^3]l \\ [-1+4\xi-3(\xi)^2]l\eta & [\xi-2(\xi)^2+(\xi)^3]l & 0 \\ \xi & 0 & 0 \\ 6[-\xi+(\xi)^2]\eta & 3(\xi)^2-2(\xi)^3 & 0 \\ 6[-\xi+(\xi)^2]\zeta & 0 & 3(\xi)^2-2(\xi)^3 \\ 0 & -l\xi\zeta & -l\xi\eta \\ [-2\xi+3(\xi)^2]l\zeta & 0 & [(\xi)^2-(\xi)^3]l \\ [2\xi-3(\xi)^2]l\eta & [-(\xi)^2+(\xi)^3]l & 0 \end{bmatrix}^{ij} \tag{6.148}$$

in which

$$\xi^{ij} = \frac{x_1^{ij}}{l^{ij}}, \qquad \eta^{ij} = \frac{x_2^{ij}}{l^{ij}}, \qquad \zeta^{ij} = \frac{x_3^{ij}}{l^{ij}}$$

and l^{ij} is the length of element ij and x_1^{ij}, x_2^{ij}, and x_3^{ij} are the spatial coordinates along the element axes.

To perform coordinate transformation from the element coordinate system to the body coordinate system, the rotation matrix $\mathbf{C}^{ij}$ is required. Direction cosines for the $\mathbf{X}_1^{ij}\mathbf{X}_2^{ij}\mathbf{X}_3^{ij}$ axes can be found directly by geometric considerations. An alternate approach involves successive rotations of axes. Let (a_1, a_2, a_3) and (b_1, b_2, b_3) be the locations of the nodes of element ij. The transformation matrix from the element axes to the body axes is given by (Gere and Weaver 1965)

$$\mathbf{C}^{ij} = \begin{bmatrix} c_1 & \frac{-c_1c_2}{\sqrt{(c_1)^2+(c_3)^2}} & \frac{-c_3}{\sqrt{(c_1)^2+(c_3)^2}} \\ c_2 & \sqrt{(c_1)^2+(c_3)^2} & 0 \\ c_3 & \frac{-c_2c_3}{\sqrt{(c_1)^2+(c_3)^2}} & \frac{c_1}{\sqrt{(c_1)^2+(c_3)^2}} \end{bmatrix}^{ij} \tag{6.149}$$

where

$$c_1^{ij} = \frac{b_1-a_1}{l^{ij}}, \qquad c_2^{ij} = \frac{b_2-a_2}{l^{ij}}, \qquad c_3^{ij} = \frac{b_3-a_3}{l^{ij}}$$

where the length of the element l^{ij} is given by

$$l^{ij} = \sqrt{(b_1 - a_1)^2 + (b_2 - a_2)^2 + (b_3 - a_3)^2} \tag{6.150}$$

It can be verified that the matrix $\mathbf{C}^{ij}$ is orthogonal, a property that is used throughout our development.

The preceding transformation $\mathbf{C}^{ij}$ is valid for all positions of the element, except when the element $\mathbf{X}_1^{ij}$ axis coincides with the body $\mathbf{X}_2^i$ axis. In this case, the transformation matrix $\mathbf{C}^{ij}$ is given by

$$\mathbf{C}^{ij} = \begin{bmatrix} 0 & -c_2 & 0 \\ c_2 & 0 & 0 \\ 0 & 0 & 1 \end{bmatrix}^{ij} \tag{6.151}$$

If the rotations at the nodes with respect to the body axes are infinitesimal, the same matrix $\mathbf{C}^{ij}$ of Eqs. 149 and 151 can be used to transform rotations from the element ij axes to the ith body coordinate system. That is, the matrix $\bar{\mathbf{C}}^{ij}$ of Eq. 147 is given by

$$\bar{\mathbf{C}}^{ij} = \begin{bmatrix} \mathbf{C}^{ij} & \mathbf{0} & \mathbf{0} & \mathbf{0} \\ \mathbf{0} & \mathbf{C}^{ij} & \mathbf{0} & \mathbf{0} \\ \mathbf{0} & \mathbf{0} & \mathbf{C}^{ij} & \mathbf{0} \\ \mathbf{0} & \mathbf{0} & \mathbf{0} & \mathbf{C}^{ij} \end{bmatrix}^{\mathrm{T}} \tag{6.152}$$

Mass Matrix Following the procedure described in the previous sections, it can be shown that the integrals that appear in the expression of the $\bar{\mathbf{S}}^{ij}$ and $\mathbf{S}_{kl}^{ij}$ matrices of Eqs. 50 and 61, respectively, are given by (Shabana 1982)

$$\int_{V^{ij}} \rho^{ij} \mathbf{S}^{ij\mathrm{T}} dV^{ij} = \begin{bmatrix} \frac{m}{2} & 0 & 0 \\ lQ_\eta & \frac{m}{2} & 0 \\ lQ_\zeta & 0 & \frac{m}{2} \\ 0 & -\left[\frac{(l)^2}{2}\right]Q_\zeta & \left[\frac{(l)^2}{2}\right]Q_\eta \\ 0 & 0 & \frac{-ml}{12} \\ 0 & \frac{ml}{12} & 0 \\ \frac{m}{2} & 0 & 0 \\ -Q_\eta l & \frac{m}{2} & 0 \\ -Q_\zeta l & 0 & \frac{m}{2} \\ 0 & -\left[\frac{(l)^2}{2}\right]Q_\zeta & -\left[\frac{(l)^2}{2}\right]Q_\eta \\ 0 & 0 & \frac{ml}{12} \\ 0 & \frac{-ml}{12} & 0 \end{bmatrix}^{ij} \tag{6.153}$$

$$\left[\int_V \rho \mathbf{S}_1^{\mathrm{T}} \mathbf{S}_1 \, dV\right]^{ij} =$$

$$\begin{bmatrix}
\frac{m}{3} & & & & & & & & & & & \\
\frac{Q_\eta l}{2} & \frac{6I_\zeta l}{5} & & & & & & & & & & \\
\frac{Q_\zeta l}{2} & \frac{6I_{\eta\zeta} l}{5} & \frac{6I_\eta l}{5} & & & & & & & & & \\
0 & 0 & 0 & 0 & & & & & & & & \\
\frac{Q_\zeta (l)^2}{12} & -\frac{I_{\eta\zeta}(l)^2}{10} & -\frac{I_\eta (l)^2}{10} & 0 & \frac{2I_\eta (l)^3}{15} & & & \text{symmetric} & & & & \\
-\frac{Q_\eta (l)^2}{12} & \frac{I_\zeta (l)^2}{10} & \frac{I_{\zeta\eta}(l)^2}{10} & 0 & -\frac{2I_{\eta\zeta}(l)^3}{15} & \frac{2I_\zeta (l)^3}{15} & & & & & & \\
\frac{m}{6} & \frac{lQ_\eta}{2} & \frac{Q_\zeta l}{2} & 0 & -\frac{Q_\zeta (l)^2}{12} & \frac{(l)^2 Q_\eta}{12} & \frac{m}{3} & & & & & \\
-\frac{Q_\eta l}{2} & -\frac{6I_\zeta l}{5} & -\frac{6I_{\eta\zeta} l}{5} & 0 & \frac{I_{\eta\zeta}(l)^2}{10} & -\frac{(l)^2 I_\zeta}{10} & -\frac{lQ_\eta}{2} & \frac{6lI_\zeta}{5} & & & & \\
-\frac{Q_\zeta l}{2} & -\frac{6I_{\eta\zeta} l}{5} & -\frac{6I_\eta l}{5} & 0 & \frac{I_\eta (l)^2}{10} & -\frac{(l)^2 I_{\eta\zeta}}{10} & -\frac{Q_\zeta l}{2} & \frac{6lI_{\eta\zeta}}{5} & \frac{6lI_\eta}{5} & & & \\
0 & 0 & 0 & 0 & 0 & 0 & 0 & 0 & 0 & 0 & & \\
-\frac{Q_\zeta (l)^2}{12} & -\frac{I_{\eta\zeta}(l)^2}{10} & -\frac{I_\eta (l)^2}{10} & 0 & -\frac{I_\eta (l)^3}{30} & \frac{(l)^3 I_{\eta\zeta}}{30} & \frac{Q_\zeta (l)^2}{12} & \frac{I_{\eta\zeta}(l)^2}{10} & \frac{I_\eta (l)^2}{10} & 0 & \frac{2I_\eta (l)^3}{15} & \\
\frac{Q_\eta (l)^2}{12} & \frac{I_\zeta (l)^2}{10} & \frac{I_{\zeta\eta}(l)^2}{10} & 0 & \frac{I_{\eta\zeta}(l)^3}{30} & -\frac{(l)^3 I_\zeta}{30} & -\frac{Q_\eta (l)^2}{12} & -\frac{I_\zeta (l)^2}{10} & -\frac{I_{\zeta\eta}(l)^2}{10} & 0 & -\frac{2I_{\eta\zeta}(l)^3}{15} & \frac{2I_\zeta (l)^3}{15}
\end{bmatrix}^{ij} \tag{6.154}$$

$$\left[\int_V \rho \mathbf{S}_2^{\mathrm{T}} \mathbf{S}_2 \, dV\right]^{ij} =
\begin{bmatrix}
0 & & & & & & & & & & & \\
0 & \frac{13m}{35} & & & & & & & & & & \\
0 & 0 & 0 & & & & & & \text{symmetric} & & & \\
0 & -\frac{7(l)^2}{20} Q_\zeta & 0 & \frac{(l)^3}{3} I_\eta & & & & & & & & \\
0 & 0 & 0 & 0 & 0 & & & & & & & \\
0 & \frac{11ml}{210} & 0 & -\frac{(l)^3}{20} Q_\zeta & 0 & \frac{m(l)^2}{105} & & & & & & \\
0 & 0 & 0 & 0 & 0 & 0 & 0 & & & & & \\
0 & \frac{9m}{70} & 0 & -\frac{3(l)^2}{20} Q_\zeta & 0 & \frac{13ml}{420} & 0 & \frac{13m}{35} & & & & \\
0 & 0 & 0 & 0 & 0 & 0 & 0 & 0 & 0 & & & \\
0 & -\frac{3(l)^2}{20} Q_\zeta & 0 & \frac{(l)^3}{6} I_\eta & 0 & -\frac{(l)^3}{30} Q_\zeta & 0 & -\frac{7(l)^2}{20} Q_\zeta & 0 & \frac{(l)^3}{3} I_\eta & & \\
0 & 0 & 0 & 0 & 0 & 0 & 0 & 0 & 0 & 0 & 0 & \\
0 & -\frac{13ml}{420} & 0 & \frac{(l)^3}{30} Q_\zeta & 0 & -\frac{m(l)^2}{140} & 0 & -\frac{11ml}{210} & 0 & \frac{(l)^3}{20} Q_\zeta & 0 & \frac{m(l)^2}{105}
\end{bmatrix}^{ij} \tag{6.155}$$

$$\left[\int_V \rho \mathbf{S}_3^{\mathrm{T}} \mathbf{S}_3 \, dV\right]^{ij} =
\begin{bmatrix}
0 & & & & & & & & & & & \\
0 & 0 & & & & & & & & & & \\
0 & 0 & \frac{13m}{35} & & & & \text{symmetric} & & & & & \\
0 & 0 & \frac{7(l)^2}{20} Q_\eta & \frac{(l)^3}{3} I_\zeta & & & & & & & & \\
0 & 0 & -\frac{11ml}{210} & -\frac{(l)^3}{20} Q_\eta & \frac{m(l)^2}{105} & & & & & & & \\
0 & 0 & 0 & 0 & 0 & 0 & & & & & & \\
0 & 0 & 0 & 0 & 0 & 0 & 0 & & & & & \\
0 & 0 & 0 & 0 & 0 & 0 & 0 & 0 & & & & \\
0 & 0 & \frac{9m}{70} & \frac{3(l)^2}{20} Q_\eta & -\frac{13ml}{420} & 0 & 0 & 0 & \frac{13m}{35} & & & \\
0 & 0 & \frac{3(l)^2}{20} Q_\eta & \frac{(l)^3}{6} I_\eta & -\frac{(l)^3}{30} Q_\eta & 0 & 0 & 0 & \frac{7(l)^2}{20} Q_\eta & \frac{(l)^3}{3} I_\zeta & & \\
0 & 0 & \frac{13ml}{420} & \frac{(l)^3}{30} Q_\eta & -\frac{m(l)^2}{140} & 0 & 0 & 0 & \frac{11ml}{210} & \frac{(l)^3}{20} Q_\eta & \frac{m(l)^2}{105} & \\
0 & 0 & 0 & 0 & 0 & 0 & 0 & 0 & 0 & 0 & 0 & 0
\end{bmatrix}^{ij} \tag{6.156}$$

$$
\left[\int_V \rho \mathbf{S}_1^{\mathrm{T}} \mathbf{S}_2 dV\right]^{ij} =
$$

$$
\begin{bmatrix}
0 & \frac{7m}{20} & 0 & -\frac{(l)^2}{3} Q_\zeta & 0 & \frac{ml}{20} & 0 & \frac{3m}{20} & 0 & -\frac{(l)^2}{6} Q_\zeta & 0 & -\frac{ml}{30} \\
0 & \frac{l}{2} Q_\eta & 0 & -\frac{(l)^2}{2} I_{\eta\zeta} & 0 & \frac{(l)^2}{10} Q_\eta & 0 & \frac{l}{2} Q_\eta & 0 & -\frac{(l)^2}{2} I_{\eta\zeta} & 0 & -\frac{(l)^2}{10} Q_\eta \\
0 & \frac{l}{2} Q_\zeta & 0 & -\frac{(l)^2}{2} I_\eta & 0 & \frac{(l)^2}{10} Q_\zeta & 0 & \frac{l}{2} Q_\zeta & 0 & -\frac{(l)^2}{2} I_\eta & 0 & -\frac{(l)^2}{10} Q_\zeta \\
0 & 0 & 0 & 0 & 0 & 0 & 0 & 0 & 0 & 0 & 0 & 0 \\
0 & \frac{(l)^2}{10} Q_\zeta & 0 & -\frac{(l)^3}{12} I_\eta & 0 & 0 & 0 & -\frac{(l)^2}{10} Q_\zeta & 0 & \frac{(l)^3}{12} I_\eta & 0 & \frac{(l)^3}{60} Q_\zeta \\
0 & -\frac{(l)^2}{10} Q_\eta & 0 & \frac{(l)^3}{12} I_{\eta\zeta} & 0 & 0 & 0 & \frac{(l)^2}{10} Q_\eta & 0 & -\frac{(l)^3}{12} I_{\eta\zeta} & 0 & -\frac{(l)^3}{60} Q_\eta \\
0 & \frac{3m}{20} & 0 & -\frac{(l)^2}{6} Q_\zeta & 0 & \frac{ml}{30} & 0 & \frac{7m}{20} & 0 & -\frac{(l)^2}{3} Q_\zeta & 0 & -\frac{ml}{20} \\
0 & -\frac{l}{2} Q_\eta & 0 & \frac{(l)^2}{2} I_{\eta\zeta} & 0 & -\frac{(l)^2}{10} Q_\eta & 0 & -\frac{l}{2} Q_\eta & 0 & \frac{(l)^2}{2} I_{\eta\zeta} & 0 & \frac{(l)^2}{10} Q_\eta \\
0 & -\frac{l}{2} Q_\zeta & 0 & \frac{(l)^2}{2} I_\eta & 0 & -\frac{(l)^2}{10} Q_\zeta & 0 & -\frac{l}{2} Q_\zeta & 0 & \frac{(l)^2}{2} I_\eta & 0 & \frac{(l)^2}{10} Q_\zeta \\
0 & 0 & 0 & 0 & 0 & 0 & 0 & 0 & 0 & 0 & 0 & 0 \\
0 & -\frac{(l)^2}{10} Q_\zeta & 0 & \frac{(l)^3}{12} I_\eta & 0 & -\frac{(l)^3}{60} Q_\zeta & 0 & \frac{(l)^2}{10} Q_\zeta & 0 & -\frac{(l)^3}{12} I_\eta & 0 & 0 \\
0 & \frac{(l)^2}{10} Q_\eta & 0 & -\frac{(l)^3}{12} I_{\eta\zeta} & 0 & \frac{(l)^3}{60} Q_\eta & 0 & -\frac{(l)^2}{10} Q_\eta & 0 & \frac{(l)^3}{12} I_{\eta\zeta} & 0 & 0
\end{bmatrix}^{ij} \tag{6.157}
$$

$$
\left[\int_V \rho \mathbf{S}_1^{\mathrm{T}} \mathbf{S}_3 dV\right]^{ij} =
$$

$$
\begin{bmatrix}
0 & 0 & \frac{7m}{20} & \frac{(l)^2}{3} Q_\eta & -\frac{ml}{20} & 0 & 0 & 0 & \frac{3m}{20} & \frac{(l)^2}{6} Q_\eta & -\frac{ml}{30} & 0 \\
0 & 0 & \frac{l}{2} Q_\eta & \frac{(l)^2}{2} I_\zeta & -\frac{(l)^2}{10} Q_\eta & 0 & 0 & 0 & \frac{l}{2} Q_\eta & \frac{(l)^2}{2} I_\zeta & \frac{(l)^2}{10} Q_\eta & 0 \\
0 & 0 & \frac{l}{2} Q_\zeta & \frac{(l)^2}{2} I_{\eta\zeta} & -\frac{(l)^2}{10} Q_\zeta & 0 & 0 & 0 & \frac{l}{2} Q_\zeta & \frac{(l)^2}{2} I_{\eta\zeta} & \frac{(l)^2}{10} Q_\zeta & 0 \\
0 & 0 & 0 & 0 & 0 & 0 & 0 & 0 & 0 & 0 & 0 & 0 \\
0 & 0 & \frac{(l)^2}{10} Q_\zeta & \frac{(l)^3}{12} I_{\zeta\eta} & 0 & 0 & 0 & 0 & -\frac{(l)^2}{10} Q_\zeta & -\frac{(l)^3}{12} I_{\eta\zeta} & -\frac{(l)^2}{60} Q_\zeta & 0 \\
0 & 0 & -\frac{(l)^2}{10} Q_\eta & -\frac{(l)^3}{12} I_\zeta & 0 & 0 & 0 & 0 & \frac{(l)^2}{10} Q_\eta & \frac{(l)^3}{12} I_\zeta & \frac{(l)^3}{60} Q_\eta & 0 \\
0 & 0 & \frac{3m}{20} & \frac{(l)^2}{6} Q_\eta & -\frac{ml}{30} & 0 & 0 & 0 & \frac{7m}{20} & \frac{(l)^2}{3} Q_\eta & \frac{ml}{20} & 0 \\
0 & 0 & -\frac{l}{2} Q_\eta & -\frac{(l)^2}{2} I_\zeta & \frac{(l)^2}{10} Q_\eta & 0 & 0 & 0 & -\frac{l}{2} Q_\eta & -\frac{(l)^2}{2} I_\zeta & -\frac{(l)^2}{10} Q_\eta & 0 \\
0 & 0 & -\frac{l}{2} Q_\zeta & -\frac{(l)^2}{2} I_{\eta\zeta} & \frac{(l)^2}{10} Q_\zeta & 0 & 0 & 0 & -\frac{l}{2} Q_\zeta & -\frac{(l)^2}{2} I_{\eta\zeta} & -\frac{(l)^2}{10} Q_\zeta & 0 \\
0 & 0 & 0 & 0 & 0 & 0 & 0 & 0 & 0 & 0 & 0 & 0 \\
0 & 0 & -\frac{(l)^2}{10} Q_\zeta & -\frac{(l)^3}{12} I_{\eta\zeta} & \frac{(l)^3}{60} Q_\zeta & 0 & 0 & 0 & \frac{(l)^2}{10} Q_\zeta & \frac{(l)^3}{12} I_{\eta\zeta} & 0 & 0 \\
0 & 0 & \frac{(l)^2}{10} Q_\eta & \frac{(l)^3}{12} I_\zeta & -\frac{(l)^3}{60} Q_\eta & 0 & 0 & 0 & -\frac{(l)^2}{10} Q_\eta & -\frac{(l)^3}{12} I_\zeta & 0 & 0
\end{bmatrix}^{ij} \tag{6.158}
$$

$$
\left[\int_V \rho \mathbf{S}_2^{\mathrm{T}} \mathbf{S}_3 dV\right]^{ij} =
$$

$$
\begin{bmatrix}
0 & 0 & 0 & 0 & 0 & 0 & 0 & 0 & 0 & 0 & 0 & 0 \\
0 & 0 & \frac{13m}{35} & \frac{7(l)^2}{20} Q_\eta & -\frac{11ml}{210} & 0 & 0 & 0 & \frac{9m}{70} & \frac{3(l)^2}{20} Q_\eta & \frac{13ml}{420} & 0 \\
0 & 0 & 0 & 0 & 0 & 0 & 0 & 0 & 0 & 0 & 0 & 0 \\
0 & 0 & -\frac{7(l)^2}{20} Q_\zeta & -\frac{(l)^3}{3} I_{\eta\zeta} & \frac{(l)^3}{20} Q_\zeta & 0 & 0 & 0 & -\frac{3(l)^2}{20} Q_\zeta & -\frac{(l)^3}{6} I_{\eta\zeta} & -\frac{(l)^3}{30} Q_\zeta & 0 \\
0 & 0 & 0 & 0 & 0 & 0 & 0 & 0 & 0 & 0 & 0 & 0 \\
0 & 0 & \frac{11ml}{210} & \frac{(l)^3}{20} Q_\eta & -\frac{m(l)^2}{105} & 0 & 0 & 0 & \frac{13ml}{420} & \frac{(l)^3}{30} Q_\eta & \frac{m(l)^2}{140} & 0 \\
0 & 0 & 0 & 0 & 0 & 0 & 0 & 0 & 0 & 0 & 0 & 0 \\
0 & 0 & \frac{9m}{70} & \frac{3(l)^2}{20} Q_\eta & -\frac{13ml}{420} & 0 & 0 & 0 & \frac{13m}{35} & \frac{7(l)^2}{20} Q_\eta & \frac{11ml}{210} & 0 \\
0 & 0 & 0 & 0 & 0 & 0 & 0 & 0 & 0 & 0 & 0 & 0 \\
0 & 0 & -\frac{3(l)^2}{20} Q_\zeta & -\frac{(l)^3}{6} I_{\eta\zeta} & \frac{(l)^3}{30} Q_\zeta & 0 & 0 & 0 & -\frac{7(l)^2}{20} Q_\zeta & -\frac{(l)^3}{3} I_{\eta\zeta} & \frac{(l)^3}{20} Q_\zeta & 0 \\
0 & 0 & 0 & 0 & 0 & 0 & 0 & 0 & 0 & 0 & 0 & 0 \\
0 & 0 & -\frac{13ml}{420} & -\frac{(l)^3}{30} Q_\eta & \frac{m(l)^2}{140} & 0 & 0 & 0 & -\frac{11ml}{210} & -\frac{(l)^2}{20} Q_\eta & -\frac{m(l)^2}{105} & 0
\end{bmatrix}^{ij} \tag{6.159}
$$

were ρ^{ij}, l^{ij}, and a^{ij} are, respectively, the mass density, length, and cross–sectional area of the element ij. The variables Q_η^{ij}, Q_ζ^{ij}, I_ζ^{ij}, I_η^{ij}, and $I_{\eta\zeta}^{ij}$ are defined as

$$
\begin{aligned}
Q_\eta^{ij} &= \left[\int_a \rho\eta\, da\right]^{ij}, \quad Q_\zeta^{ij} = \left[\int_a \rho\zeta\, da\right]^{ij} \\
I_\zeta^{ij} &= \left[\int_a \rho(\eta)^2 da\right]^{ij}, \quad I_\eta^{ij} = \left[\int_a \rho(\zeta)^2 da\right]^{ij} \\
I_{\eta\zeta}^{ij} &= \left[\int_a \rho\eta\zeta\, da\right]^{ij}
\end{aligned}
\tag{6.160}
$$

in which

$$
\xi^{ij} = \frac{x_1^{ij}}{l^{ij}}, \qquad \eta^{ij} = \frac{x_2^{ij}}{l^{ij}}, \qquad \zeta^{ij} = \frac{x_3^{ij}}{l^{ij}}
$$

Stiffness Matrix The stiffness matrix of the three-dimensional straight element of uniform cross-sectional area is given by (Przemineiecki 1968)

$$
\bar{\mathbf{K}}_{ff}^{ij} =
$$

$$
\begin{bmatrix}
\frac{Ea}{l} \\
0 & \frac{12EI_3}{(l)^3} \\
0 & 0 & \frac{12EI_2}{(l)^3} \\
0 & 0 & 0 & \frac{GJ_1}{l} & & & & \text{symmetric} \\
0 & 0 & -\frac{6EI_2}{(l)^2} & 0 & \frac{4EI_2}{l} \\
0 & \frac{6EI_3}{(l)^2} & 0 & 0 & 0 & \frac{4EI_3}{l} \\
-\frac{Ea}{l} & 0 & 0 & 0 & 0 & 0 & \frac{aE}{l} \\
0 & -\frac{12EI_3}{(l)^3} & 0 & 0 & 0 & -\frac{6EI}{(l)^2} & 0 & \frac{12EI_3}{(l)^3} \\
0 & 0 & -\frac{12EI_2}{(l)^3} & 0 & \frac{6EI_2}{(l)^2} & 0 & 0 & 0 & \frac{12EI_2}{(l)^3} \\
0 & 0 & 0 & -\frac{GJ_1}{l} & 0 & 0 & 0 & 0 & 0 & \frac{GJ}{l} \\
0 & 0 & -\frac{6EI_2}{(l)^2} & 0 & \frac{2EI_2}{l} & 0 & 0 & 0 & \frac{6EI}{(l)^2} & 0 & \frac{4EI_2}{l} \\
0 & \frac{6EI_3}{(l)^2} & 0 & 0 & 0 & \frac{2EI_3}{l} & 0 & -\frac{6EI_3}{(l)^2} & 0 & 0 & 0 & \frac{4EI_3}{l}
\end{bmatrix}^{ij}
\tag{6.161}
$$

where E^{ij} and G^{ij} are, respectively, the modulus of elasticity and the modulus of rigidity of element ij, and I_1^{ij}, I_2^{ij}, and I_3^{ij} are, respectively, the second moment of areas about the $\mathbf{X}_1^{ij}$, $\mathbf{X}_2^{ij}$, and $\mathbf{X}_3^{ij}$ element axes.

6.7 COORDINATE REDUCTION

Adequate representation of large-scale nonlinear mechanical systems using the finite-element method may require a large number of nodal coordinates. It is necessary to reduce this number of coordinates if a solution is to be obtained with a reasonable amount of computer time. *Substructuring* and *component mode synthesis* techniques have been used extensively in structural dynamics (Craig and Bampton 1968; Meirovitch 1997; Shabana 1997) to reduce the problem dimensionality. In many applications, the number of elastic coordinates is much larger than the number of reference coordinates, and therefore the problem dimension can be significantly decreased if insignificant elastic generalized coordinates are eliminated.

This section is devoted to an outline of the use of substructuring methods to reduce the number of elastic generalized coordinates in mechanical and structural systems in which the reference motion is coupled with the elastic deformation. Even though the component mode technique is considered in this section, application of the *condensation techniques* is straightforward and follows the same procedure, once the transformation matrix eliminating slave variables is identified.

As pointed out earlier, the problems addressed in this book differ from those commonly occurring in structural dynamics, in the sense that system components undergo finite rotations. This leads to inertia-variant systems, and accordingly the frequency spectrum is time-dependent. However, by imposing the appropriate reference conditions one can identify a transformation from the space of system nodal coordinates to the space of system generalized modal coordinates of lower dimension. By so doing, the motion of the elastic component can be identified by using three sets of modes: *rigid body, reference*, and *normal modes*. Rigid body modes describe translations and large angular rotations of a selected body reference. This set of modes is introduced using the Cartesian coordinates $\mathbf{R}^i$ and $\boldsymbol{\theta}^i$, which define the large translational and rotational displacements of the selected deformable body reference. Reference modes are the result of imposing the reference conditions and normal modes define the deformation of the body relative to the body reference. The normal modes defined in this section are introduced by using the *modal transformation*. The method developed in this section is based on solving the eigenvalue problem of the deformable bodies only once.

In the preceding chapter, it is shown that the equations of motion of the constrained body i in the multibody system can be written in a matrix form as

$$\mathbf{M}^i\ddot{\mathbf{q}}^i + \mathbf{K}^i\mathbf{q}^i = \mathbf{Q}_e^i + \mathbf{Q}_v^i - \mathbf{C}_{\mathbf{q}^i}^{\mathrm{T}}\boldsymbol{\lambda} \tag{6.162}$$

where $\mathbf{M}^i$ and $\mathbf{K}^i$ are, respectively, the symmetric mass and stiffness matrices of body i, $\mathbf{Q}_e^i$ is the vector of externally applied forces, $\mathbf{C}_{\mathbf{q}^i}$ is the constraint Jacobian matrix, $\boldsymbol{\lambda}$ is the vector of Lagrange multipliers, and $\mathbf{Q}_v^i$ is the quadratic velocity vector that arises from differentiating the kinetic energy with respect to time and with respect to the generalized coordinates of body i that can be written in a partitioned form as

$$\mathbf{q}^i = \left[\mathbf{q}_r^{i\mathrm{T}} \quad \mathbf{q}_f^{i\mathrm{T}}\right]^{\mathrm{T}} \tag{6.163}$$

where $\mathbf{q}_r^i = [\mathbf{R}^{i\mathrm{T}} \quad \boldsymbol{\theta}^{i\mathrm{T}}]^{\mathrm{T}}$ is the vector of reference coordinates and $\mathbf{q}_f^i$ is the vector of nodal coordinates resulting from the finite-element discretization. According to the generalized coordinate partitioning of Eq. 163, the equations of motion of the deformable body i, given by Eq. 162, can be written as

$$\begin{bmatrix} \mathbf{m}_{rr}^i & \mathbf{m}_{rf}^i \\ \mathbf{m}_{fr}^i & \mathbf{m}_{ff}^i \end{bmatrix} \begin{bmatrix} \ddot{\mathbf{q}}_r^i \\ \ddot{\mathbf{q}}_f^i \end{bmatrix} + \begin{bmatrix} \mathbf{0} & \mathbf{0} \\ \mathbf{0} & \mathbf{K}_{ff}^i \end{bmatrix} \begin{bmatrix} \mathbf{q}_r^i \\ \mathbf{q}_f^i \end{bmatrix} = \begin{bmatrix} (\mathbf{Q}_e^i)_r \\ (\mathbf{Q}_e^i)_f \end{bmatrix} + \begin{bmatrix} (\mathbf{Q}_v^i)_r \\ (\mathbf{Q}_v^i)_f \end{bmatrix} - \begin{bmatrix} \mathbf{C}_{\mathbf{q}_r^i}^{\mathrm{T}} \\ \mathbf{C}_{\mathbf{q}_f^i}^{\mathrm{T}} \end{bmatrix} \boldsymbol{\lambda} \tag{6.164}$$

where subscripts r and f refer, respectively, to reference and elastic coordinates, and $\mathbf{m}^i_{fr} = \mathbf{m}^{i\mathrm{T}}_{rf}$.

Modal Transformation If the body i is assumed to vibrate freely about a reference configuration, Eq. 164 yields

$$\mathbf{m}^i_{ff}\ddot{\mathbf{q}}^i_f + \mathbf{K}^i_{ff}\mathbf{q}^i_f = \mathbf{0} \tag{6.165}$$

The stiffness matrix $\mathbf{K}^i_{ff}$ is positive definite, because of imposing the reference conditions that define a unique displacement field. A trial solution for Eq. 165 is given by (Clough and Penzien 1975; Shabana 1997)

$$\mathbf{q}^i_f = \mathbf{a}^i e^{j\omega t} \tag{6.166}$$

The vector $\mathbf{a}^i$ represents maximum values, or amplitudes of vibratory motion, ω is the frequency, t is the time, and j is the complex operator defined as

$$j = \sqrt{-1}$$

Substituting Eq. 166 into Eq. 165 results in

$$-(\omega)^2\mathbf{m}^i_{ff}\mathbf{a}^i + \mathbf{K}^i_{ff}\mathbf{a}^i = \mathbf{0}$$

which can be written as

$$\mathbf{K}^i_{ff}\mathbf{a}^i = (\omega)^2\mathbf{m}^i_{ff}\mathbf{a}^i \tag{6.167}$$

Equation 167 is the *standard eigenvalue problem* that can be solved for a set of eigenvalues $(\omega_k)^2$ and the corresponding eigenvectors $\mathbf{a}^i_k$, $k = 1, 2, \ldots, n_f$, where n_f is the number of elastic nodal coordinates of the deformable body i. The eigenvectors are called the *normal modes* or the *mode shapes*. A reduced order model can be achieved by solving for only n_m mode shapes, where $n_m < n_f$. A coordinate transformation from the physical nodal coordinates to the modal elastic coordinates can be obtained as follows:

$$\mathbf{q}^i_f = \bar{\mathbf{B}}^i_m\mathbf{p}^i_f \tag{6.168}$$

where $\bar{\mathbf{B}}^i_m$ is the modal transformation matrix whose columns are the low-frequency n_m mode shapes. The vector $\mathbf{p}^i_f$ is the vector of modal coordinates. The n_m mode shapes should be selected such that a good approximation for the displaced shape can be obtained. With the use of Eq. 168, the reference and elastic generalized coordinates are written in terms of the reference and modal coordinates as

$$\begin{bmatrix} \mathbf{q}^i_r \\ \mathbf{q}^i_f \end{bmatrix} = \begin{bmatrix} \mathbf{I} & \mathbf{0} \\ \mathbf{0} & \bar{\mathbf{B}}^i_m \end{bmatrix} \begin{bmatrix} \mathbf{p}^i_r \\ \mathbf{p}^i_f \end{bmatrix} \tag{6.169}$$

or in compact form as

$$\mathbf{q}^i = \mathbf{B}^i_m\mathbf{p}^i$$

where $\mathbf{p}^i$ is the vector

$$\mathbf{p}^i = \begin{bmatrix} \mathbf{p}^{i\mathrm{T}}_r & \mathbf{p}^{i\mathrm{T}}_f \end{bmatrix}^{\mathrm{T}}$$

and the transformation $\mathbf{B}_m^i$ is defined as

$$\mathbf{B}_m^i = \begin{bmatrix} \mathbf{I} & \mathbf{0} \\ \mathbf{0} & \bar{\mathbf{B}}_m^i \end{bmatrix}$$

Dynamic Equations in Terms of the Modal Coordinates A transformation similar to Eq. 169 can be obtained if *condensation methods* are used. Therefore, the following steps are general, in the sense that they can be applied to all methods of substructuring. Substituting Eq. 169 into Eq. 164 and premultiplying by $\mathbf{B}_m^{i\mathrm{T}}$ yields the following system of equations written in terms of a coupled set of reference and modal elastic coordinates:

$$\begin{bmatrix} \bar{\mathbf{m}}_{rr}^i & \bar{\mathbf{m}}_{rf}^i \\ \bar{\mathbf{m}}_{fr}^i & \bar{\mathbf{m}}_{ff}^i \end{bmatrix} \begin{bmatrix} \ddot{\mathbf{p}}_r^i \\ \ddot{\mathbf{p}}_f^i \end{bmatrix} + \begin{bmatrix} \mathbf{0} & \mathbf{0} \\ \mathbf{0} & \bar{\mathbf{K}}_{ff}^i \end{bmatrix} \begin{bmatrix} \mathbf{p}_r^i \\ \mathbf{p}_f^i \end{bmatrix}$$

$$= \begin{bmatrix} (\bar{\mathbf{Q}}_e^i)_r \\ (\bar{\mathbf{Q}}_e^i)_f \end{bmatrix} + \begin{bmatrix} (\bar{\mathbf{Q}}_v^i)_r \\ (\bar{\mathbf{Q}}_v^i)_f \end{bmatrix} - \begin{bmatrix} \mathbf{C}_{\mathbf{p}_r^i}^{\mathrm{T}} \\ \mathbf{C}_{\mathbf{p}_f^i}^{\mathrm{T}} \end{bmatrix} \boldsymbol{\lambda} \qquad (6.170)$$

where

$$\bar{\mathbf{m}}_{rr}^i = \mathbf{m}_{rr}^i, \qquad \bar{\mathbf{m}}_{rf}^i = \bar{\mathbf{m}}_{fr}^{i\mathrm{T}} = \mathbf{m}_{rf}^i \bar{\mathbf{B}}_m^i$$

$$\bar{\mathbf{m}}_{ff}^i = \bar{\mathbf{B}}_m^{i\mathrm{T}} \mathbf{m}_{ff}^i \bar{\mathbf{B}}_m^i, \qquad \bar{\mathbf{K}}_{ff}^i = \bar{\mathbf{B}}_m^{i\mathrm{T}} \mathbf{K}_{ff}^i \bar{\mathbf{B}}_m^i$$

$$(\bar{\mathbf{Q}}_e^i)_r = (\mathbf{Q}_e^i)_r, \qquad (\bar{\mathbf{Q}}_e^i)_f = \bar{\mathbf{B}}_m^{i\mathrm{T}} (\mathbf{Q}_e^i)_f$$

$$(\bar{\mathbf{Q}}_v^i)_r = (\mathbf{Q}_v^i)_r, \qquad (\bar{\mathbf{Q}}_v^i)_f = \bar{\mathbf{B}}_m^{i\mathrm{T}} (\mathbf{Q}_v^i)_f$$

Note that the constraint Jacobian matrix must be evaluated by taking the partial derivative of the constraint equations with respect to the new set of coordinates $\mathbf{p}^i$. It is more convenient, however, to express these derivatives in terms of derivatives with respect to physical coordinates. This can be accomplished by using the following relation:

$$\mathbf{C}_{\mathbf{p}^i} = \frac{\partial \mathbf{C}}{\partial \mathbf{p}^i} = \frac{\partial \mathbf{C}}{\partial \mathbf{q}^i} \frac{\partial \mathbf{q}^i}{\partial \mathbf{p}^i} = \mathbf{C}_{\mathbf{q}^i} \mathbf{B}_m^i \qquad (6.171)$$

Therefore, the Jacobian matrices in Eq. 170 can be written as

$$\mathbf{C}_{\mathbf{p}_r^i} = \mathbf{C}_{\mathbf{q}_r^i}, \qquad \mathbf{C}_{\mathbf{p}_f^i} = \mathbf{C}_{\mathbf{q}_f^i} \bar{\mathbf{B}}_m^i \qquad (6.172)$$

where the modal transformation $\bar{\mathbf{B}}_m^i$ is defined by Eq. 168.

Remarks In this section a formal procedure is presented for reducing the number of elastic coordinates of deformable bodies in multibody systems. It is important, however, to point out that, in the computer implementation, this procedure is equivalent to transforming the inertia shape integrals to their modal form only once in advance for the dynamic analysis. This transformation can be carried out in a preprocessor computer program. One can show that, once the inertia shape integrals are expressed in their modal form, all the inertia forces including the Coriolis and centrifugal forces are automatically expressed in terms of the modal coordinates. That is, the same computer processor can be used for both cases of the physical and modal coordinates. In other words, the structure of the dynamic equations does not change by changing the set of coordinates as long as the inertia shape integrals are expressed in the proper form. Given the transformation matrix $\bar{\mathbf{B}}_m^i$ of Eq. 168, one can show that the inertia shape integrals of Eqs. 77 and 80 can be expressed in their modal form as

$$(\bar{\mathbf{S}}^i)_m = \bar{\mathbf{S}}^i \bar{\mathbf{B}}_m^i \tag{6.173}$$

$$\left(\mathbf{S}_{kl}^i\right)_m = \bar{\mathbf{B}}_m^{i\mathrm{T}} \mathbf{S}_{kl}^i \bar{\mathbf{B}}_m^i, \quad k, l = 1, 2, 3 \tag{6.174}$$

6.8 THE FLOATING FRAME OF REFERENCE AND LARGE DEFORMATION PROBLEM

In the finite-element floating frame of reference formulation presented in this chapter, a set of coordinate systems is introduced to obtain exact modeling of the rigid body inertia. In this formulation four coordinate systems are employed for the finite element:

1. A *global coordinate system* $\mathbf{X}_1\mathbf{X}_2\mathbf{X}_3$ is fixed in time and forms a single standard for the entire assembly of bodies, and as such serves to express the connectivity of all bodies in the system. Kinematic constraints that represent mechanical joints in the system such as revolute and prismatic joints are formulated in this coordinate system using a set of nonlinear algebraic constraint equations that depend on the system generalized coordinates and possibly on time.
2. A *body coordinate system* denoted as $\mathbf{X}_1^i\mathbf{X}_2^i\mathbf{X}_3^i$ forms a single standard for the entire assembly of elements in the body i and as such serves to express the connectivity of all the elements in this body. Compatibility conditions between elements are then defined by using a constant Boolean matrix. In the formulation presented in this and the preceding chapter, the body reference need not be rigidly attached to the body. Floating frames of reference are commonly employed to describe the motion of deformable bodies that undergo large angular rotations. The configuration of the body coordinate system is identified by using a set of reference coordinates that define the location and orientation of this rigid frame of reference.
3. For each element j on the deformable body i, the *element coordinate system* $\mathbf{X}_1^{ij}\mathbf{X}_2^{ij}\mathbf{X}_3^{ij}$ is rigidly attached to the element. This coordinate system translates and rotates with the element.
4. The fourth coordinate system is the *intermediate element coordinate system* $\mathbf{X}_{i1}^{ij}\mathbf{X}_{i2}^{ij}\mathbf{X}_{i3}^{ij}$ whose origin is rigidly attached to the origin of the body coordinate system. This coordinate system, which does not follow the deformation of the element, is initially oriented to be parallel to the element coordinate system.

With the use of these coordinate systems, the location of an arbitrary point on the element can be defined and used to develop the kinematic and dynamic equations of the elements that undergo large reference displacements. By defining the inertia shape integrals of the deformable body using the element shape function, a computer algorithm similar to the algorithm discussed in the preceding chapter can be used. In fact, the same main computer program can be used when classical approximation methods or the finite-element methods are used since in both cases the final forms of the dynamic equations of motion are the same. There is no need also to change the main processor when modal transformations, and/or consistent and lumped masses are used since in these cases one needs to change only the form of the inertia shape integrals.

The intermediate element coordinate system plays a fundamental role in the nonlinear formulation presented in this chapter. As previously pointed out, it is crucial in this formulation that the shape function of the finite element can describe an arbitrary rigid body translation, which is the case for most existing finite-element shape functions. Using this property of the shape function and the intermediate element coordinate system, one can develop a formulation, as demonstrated in this chapter, that leads to an exact modeling of the rigid body inertia. The concept of the intermediate element coordinate system is similar to the concept of the parallel axis theorem used in rigid body mechanics. Since the shape function can describe an arbitrary rigid body translation, exact modeling of the rigid body inertia in the intermediate element coordinate system can be obtained. Furthermore, since the intermediate element coordinate system has a constant orientation with respect to the body coordinate system, exact modeling of the rigid body inertia in the body coordinate system can be obtained using a constant transformation.

The intermediate element coordinate system is introduced to circumvent the problems associated with the description of large reference rotations using beam, plate, and shell elements. The conventional shape functions of these widely used elements cannot describe an arbitrary rigid body rotation since infinitesimal rotations are used as nodal coordinates. Using these elements, exact modeling of arbitrary rigid body rotations using the conventional element shape function and the vector of nodal coordinates cannot be obtained (Shabana 1996b). This is not, however, the case when other elements that employ only displacement coordinates are used. To demonstrate this fact, we consider the *rectangular element* shown in Fig. 8. The element has eight

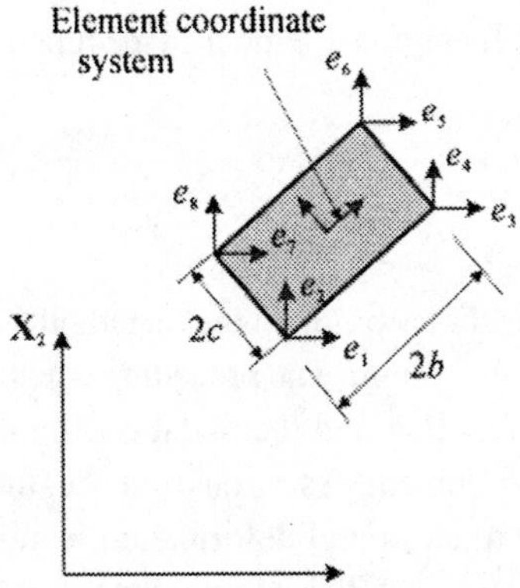

Figure 6.8 Rectangular element.

nodal coordinates that describe the displacements of the four nodes as shown in the figure. No rotations are used as nodal coordinates for this planar element. Dropping the superscripts that indicate the body and the element number for simplicity, the shape function of this element is defined as (Shabana 1997)

$$\mathbf{S} = \begin{bmatrix} N_1 & 0 & N_2 & 0 & N_3 & 0 & N_4 & 0 \\ 0 & N_1 & 0 & N_2 & 0 & N_3 & 0 & N_4 \end{bmatrix}$$

where

$$N_1 = \frac{1}{4bc}(b-x)(c-y), \qquad N_2 = \frac{1}{4bc}(b+x)(c-y)$$

$$N_3 = \frac{1}{4bc}(b+x)(c+y), \qquad N_4 = \frac{1}{4bc}(b-x)(c+y)$$

and $2b$ and $2c$ are the dimensions of the element as shown in Fig. 8. One can verify that

$$N_1 + N_2 + N_3 + N_4 = 1$$

If the rectangular element undergoes an arbitrary rigid body displacement defined by the vector

$$\mathbf{q}_r = [R_1 \quad R_2 \quad \theta]^{\mathrm{T}}$$

the vector of nodal coordinates of the rectangular element can be defined in the global coordinate system and can be written as

$$\mathbf{e} = \begin{bmatrix} e_1 \\ e_2 \\ e_3 \\ e_4 \\ e_5 \\ e_6 \\ e_7 \\ e_8 \end{bmatrix} = \begin{bmatrix} R_1 - b\cos\theta + c\sin\theta \\ R_2 - b\sin\theta - c\cos\theta \\ R_1 + b\cos\theta + c\sin\theta \\ R_2 + b\sin\theta - c\cos\theta \\ R_1 + b\cos\theta - c\sin\theta \\ R_2 + b\sin\theta + c\cos\theta \\ R_1 - b\cos\theta - c\sin\theta \\ R_2 - b\sin\theta + c\cos\theta \end{bmatrix}$$

Using this vector of the nodal coordinates and the rectangular element shape function, it can be shown that

$$\mathbf{Se} = \begin{bmatrix} R_1 + x_1\cos\theta - x_2\sin\theta \\ R_2 + x_1\sin\theta + x_2\cos\theta \end{bmatrix}$$

where x_1 and x_2 are the spatial coordinates of an arbitrary point on the rectangular element defined with respect to the element coordinate system. The preceding equation demonstrates that the rectangular element shape function and the nodal coordinates can describe an arbitrary rigid body motion of the element. As a consequence, there is no problem in using this element in the large rotation and deformation analysis of deformable bodies. In this case, there is no need to use the intermediate element coordinate system, since the element shape function and the nodal coordinates can be used to obtain exact modeling of the rigid body motion of the element. For this reason, elements such as the rectangular elements that do not use rotations as nodal coordinates are not the subject of extensive research because they do not suffer from the problems encountered when beams and plates are used in the large rotation problems. To obtain exact modeling of the rigid body inertia when infinitesimal rotations are used as nodal coordinates, the concept of the intermediate element coordinate system must be used. It is important, however, to point out that since infinitesimal rotations are used to describe the deformation of the element with respect to the deformable body coordinate system, the floating frame of reference formulation can be used only in the large reference rotation and small deformation problems. If this formulation is to be used in the large deformation problems, the deformable body must be divided into small elements, and each of these elements must be treated as a separate body. The elements can then be connected using nonlinear algebraic equations that describe the rigid joints between the elements. This approach, however, does not lead to an efficient procedure for the large deformation analysis of deformable bodies that undergo large rotations. A more elegant approach for the large deformation analysis of elements such as beams and plates is presented in the following chapter.

Problems

1. The displacement of a two-dimensional beam element is defined by the following interpolating polynomials

$$w_1 = a_0 + a_1 x_1, \qquad w_2 = a_2 + a_3 x_1$$

where a_0, a_1, a_2, and a_3 are the polynomial coefficients. Obtain the element shape function assuming that the element has two nodes, and each node has two translational degrees of freedom.

2. Show that the shape function obtained in the preceding problem can describe an arbitrary rigid body translation.

3. Discuss the relationship between the intermediate element coordinate system introduced in the nonlinear formulation presented in this chapter and the parallel axis theorem used in rigid body dynamics in the case of planar motion.

4. In the case of three-dimensional motion, discuss the relationship between the intermediate element coordinate system and the parallel axis theorem.

5. Using the shape function of Eq. 9 and the intermediate element coordinate system, determine the rigid body inertia matrix of an element displaced by a rigid body translation $\mathbf{c}$ and a rigid body rotation θ.

6. Repeat Problem 5 using the shape function obtained in Problem 1.

7. Using Eq. 13, show that the position vector of Eq. 17 can be written as the sum of the position vector of the arbitrary point on the element in the undeformed state plus the deformation vector.

8. The connecting rod of a slider crank mechanism is modeled using three beam elements, each defined by the shape function of Eq. 9. Obtain the Boolean matrix that describes the element connectivity. Obtain also the matrix of reference conditions if the beam is to be modeled as a simply supported beam.

9. The displacement field of a finite beam element is described using the following functions

$$w_1 = (1 - \xi)e_1 + \xi e_3, \qquad w_2 = (1 - \xi)e_2 + \xi e_4$$

where e_1 and e_2 are the translational coordinates of the first node, e_3 and e_4 are the translational coordinates of the second node, and $\xi = x/l$. Define the inertia shape integrals of this element.

10. Using the results obtained in Problem 9, define the mass matrix of the finite element. Identify the mass moment of inertia and discuss its dependence on the elastic coordinates.

11. Obtain the stiffness matrix of the finite beam element defined in Problem 9.

12. Show that the rigid body translational modes $\mathbf{B}_R^{ij}$ of an element that can describe an arbitrary rigid body translation must satisfy the following condition:

$$\mathbf{S}^{ij}\mathbf{B}_R^{ij} = \mathbf{I}$$

where $\mathbf{S}^{ij}$ is the element shape function and $\mathbf{I}$ is the identity matrix. Verify the preceding identity using the element shape function of Eq. 109 and the element shape function given in Problem 9.

13. Use the identity given in Problem 12 to show that two of the inertia shape integrals that appear in the planar analysis are related by the equation

$$\bar{\mathbf{S}}^{ij} = \mathbf{B}_R^{ij\mathrm{T}}\mathbf{m}_{ff}^{ij}$$

14. Derive the mass matrix of the beam element defined by the shape function of Eq. 109 when the body coordinate system is rigidly attached to the end of the beam (cantilever end conditions).

15. Derive the mass matrix of the beam element defined by the shape function of Eq. 109 when simply supported reference conditions are used.

16. Show that the identity given in Problem 12 holds when the shape function of the three-dimensional beam element defined in Eq. 148 is used. Show that the relationship given in Problem 13 also holds in the three-dimensional case.

7 THE LARGE DEFORMATION PROBLEM

There are two main concerns regarding the use of the classical finite-element formulations in the large deformation and rotation analysis of flexible multibody systems. First, in the classical finite-element literature on beams and plates, infinitesimal rotations are used as nodal coordinates. Such a use of coordinates does not lead to the exact modeling of a simple rigid body motion. Second, lumped mass techniques are used in many finite-element formulations and computer programs to describe the inertia of the deformable bodies. As will be demonstrated in this chapter, such a lumped mass representation of the inertia also does not lead to exact modeling of the equations of motion of the rigid bodies.

In the preceding chapter, a floating frame of reference formulation that uses classical finite-element methodologies is developed. This formulation, in which infinitesimal rotations can be considered as nodal coordinates, can be used only in the large reference displacement and small elastic deformation with respect to the flexible body reference. Using the concept of the intermediate element coordinate system, which is equivalent to the application of the parallel axis theorem used in rigid body dynamics, a nonlinear formulation that leads to exact modeling of the rigid body motion for elements whose coordinates are defined in terms of infinitesimal rotations can be developed. This floating frame of reference formulation also leads, in the case of lumped masses, to a nonlinear nondiagonal mass matrix as the result of the nonlinear inertia coupling between the reference motion and the elastic deformation.

In this chapter, an *absolute nodal coordinate formulation* that can be used in the large rotation and deformation analysis of flexible bodies that undergo arbitrary displacements is presented. In this formulation, no infinitesimal or finite rotations are used as the nodal coordinates; instead, absolute slopes and displacements at the nodal points are used as the element nodal coordinates. Crucial to the success of using this new formulation, however, is the use of a consistent mass approach. This is a necessary requirement that guarantees that exact modeling of the rigid body inertia can be obtained when the structures rotate as rigid bodies. In this chapter, the equivalence of the absolute nodal coordinate formulation and the floating frame of reference formulation that is widely used in flexible multibody simulations is established and used to shed more light on the basic differences between the use of infinitesimal rotations and the use of the slopes as nodal coordinates as well as the differences between the consistent and lumped mass approximations in flexible multibody dynamics.

The method presented in this chapter represents a departure from the classical finite-element formulations used in many engineering applications in the sense that not all the nodal coordinates have a physical meaning. These coordinates, however, can be systematically determined in the undeformed reference configuration using simple rigid body kinematics. The method presented in this chapter is also conceptually different from the floating frame of reference formulation discussed in the preceding two chapters since only absolute coordinates are used to define displacements and slopes at the nodes in a global inertial frame of reference.

The absolute nodal coordinate formulation is also different from the *mixed formulations* used in the finite-element literature. Despite the fact that displacement gradients are used as nodal coordinates in the mixed formulations, these formulations are often used in the framework of an incremental procedure, thus requiring the transformation of all the element matrices. The mixed formulations also suffer from serious limitations when flexible multibody applications are considered. This is mainly because many of these mixed formulations were developed for static problem applications, and in most cases the shape functions employed do not have a complete set of rigid body modes. As a consequence, exact modeling of the rigid body dynamics has not been a subject of research when the mixed formulations are considered. Therefore, it is not surprising that beams and plates are not considered in the classical finite-element literature as isoparametric elements.

7.1 BACKGROUND

The finite elements used in the static and dynamic analysis of mechanical and structural systems can be categorized into two main groups. The first group consists of elements that have only displacement coordinates. Most of these elements are of the *isoparametric type* since the locations of the material points on the elements as well as their displacements can be interpolated using the same shape functions. Examples of these isoparametric elements are the planar triangular and rectangular elements and the spatial solid and tetrahedral elements. These elements can be efficiently used in the large rotation and deformation analysis of flexible bodies since they lead to a constant mass matrix if the nodal coordinates are defined in a global inertial frame of reference.

The second group, on the other hand, consists of elements in which displacements as well as infinitesimal rotations are used as nodal coordinates. Examples of these elements are planar and spatial beam elements as well as plate and shell elements. It can be demonstrated, however, that the use of the infinitesimal rotations as nodal coordinates leads to a linearization of the equations of motion of the rigid body, and as a consequence, classical finite-element formulations do not describe an exact rigid body displacements (Shabana 1996b). Because of this fact, beams and plates are not considered in the finite-element literature as isoparametric elements, since an arbitrary rigid body motion of the element described in terms of infinitesimal nodal rotations does not result in zero strains. This problem can be circumvented if the nodal coordinates are expressed in terms of absolute nodal displacements and slopes. Using this new set of nodal coordinates, an absolute nodal coordinate formulation can be developed for the large deformation and rotation analysis of flexible structures that undergo an arbitrary reference displacement (Shabana 1996b,c). In this absolute nodal coordinate formulation, the nodal coordinates and slopes are defined in the inertial frame, and no infinitesimal or finite rotation coordinates are used in the kinematic description of the motion. Using this method, beams and plates can be treated as isoparametric elements and an arbitrary rigid body motion of these elements produces zero strain. The absolute nodal coordinate formulation leads to a constant mass matrix and a highly nonlinear stiffness matrix, and unlike existing incremental finite-element formulations, the absolute nodal coordinate formulation leads to exact modeling of the rigid body inertia when the structure rotates as a rigid body.

Absolute Coordinates In this chapter an absolute nodal coordinate formulation, which can be used in the large deformation analysis of flexible multibody systems, is presented and its relationship with the floating frame of reference formulation is established. In the absolute nodal coordinate formulation, the element nodal coordinates are defined in the inertial frame. These nodal coordinates are used with a *global shape function* that has a complete set of rigid body modes. Therefore, the global position vector of an arbitrary point on the element can be described using the global shape function and the absolute nodal coordinates as

$$\mathbf{r} = \mathbf{S}\mathbf{e} \tag{7.1}$$

where $\mathbf{S}$ is the global shape function and $\mathbf{e}$ is the vector of element nodal coordinates. In the absolute nodal coordinate formulation, no infinitesimal or finite rotations are used as nodal coordinates. The element coordinates are expressed in terms of nodal displacements and slopes that can be determined in the undeformed reference configuration using simple rigid body kinematics. Using this motion description, beams and plates can be treated as isoparametric elements without the need to introduce an orientation coordinate to describe the rigid body rotation of the deformable element. Introducing such an orientation coordinate leads to a redundant set of rigid body modes since the tangent and normal vectors at an arbitrary point on the deformed center line of the element can be obtained using the derivatives of the position vector with respect to the spatial coordinate. This fact can be demonstrated by considering the beam shown in Fig. 1. Using the Euler–Bernoulli beam assumptions, the orientation of the coordinate system defined by the tangent vector $\mathbf{t}$ and the normal vector $\mathbf{n}$ can be described in the inertial frame using the transformation matrix

$$\mathbf{A}_c = \begin{bmatrix} \cos\alpha & -\sin\alpha \\ \sin\alpha & \cos\alpha \end{bmatrix} \tag{7.2}$$

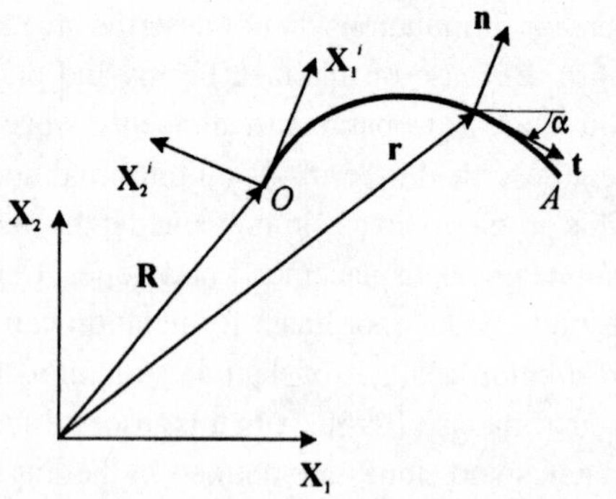

Figure 7.1 Absolute coordinates of the beam.

where

$$\cos\alpha = \frac{\frac{\partial r_1}{\partial x}}{\sqrt{\left(\frac{\partial r_1}{\partial x}\right)^2 + \left(\frac{\partial r_2}{\partial x}\right)^2}}, \qquad \sin\alpha = \frac{\frac{\partial r_2}{\partial x}}{\sqrt{\left(\frac{\partial r_1}{\partial x}\right)^2 + \left(\frac{\partial r_2}{\partial x}\right)^2}} \tag{7.3}$$

in which r_1 and r_2 are the components of the vector **r** that defines the global position vector of the arbitrary point as described by Eq. 1, and x is the coordinate of the point along the beam axis in the undeformed configuration. Using the *Frenet frame* whose orientation is defined in the inertial frame by the angle α, the position and orientation of the beam cross-section in Euler–Bernoulli beam theory can be uniquely defined using the vector **r** and the angle α.

Rigid Body Motion In the case of an arbitrary planar rigid body motion of the beam, the global position vector of an arbitrary point on the beam element can be written as

$$\mathbf{r} = \begin{bmatrix} r_1 \\ r_2 \end{bmatrix} = \begin{bmatrix} R_1 + x\cos\theta \\ R_2 + x\sin\theta \end{bmatrix} \tag{7.4}$$

where R_1 and R_2 are the global coordinates of the endpoint O, and θ in this case is the angle that defines the beam orientation as shown in Fig. 2. It follows that the slopes in the case of a rigid body motion are defined as

$$\frac{\partial r_1}{\partial x} = \cos\theta, \qquad \frac{\partial r_2}{\partial x} = \sin\theta \tag{7.5}$$

In this section, we consider cubic polynomials to define the elements of the vector **r**. It is justified to use the same representation for the elements of this vector since they are both defined in the inertial frame when the absolute nodal coordinate formulation is used. In this case, the global shape function is given by

$$\mathbf{S} = \begin{bmatrix} 1-3(\xi)^2+2(\xi)^3 & 0 & l(\xi-2(\xi)^2+(\xi)^3) & 0 & 3(\xi)^2-2(\xi)^3 & 0 & l((\xi)^3-(\xi)^2) & 0 \\ 0 & 1-3(\xi)^2+2(\xi)^3 & 0 & l(\xi-2(\xi)^2+(\xi)^3) & 0 & 3(\xi)^2-2(\xi)^3 & 0 & l((\xi)^3-(\xi)^2) \end{bmatrix} \tag{7.6}$$

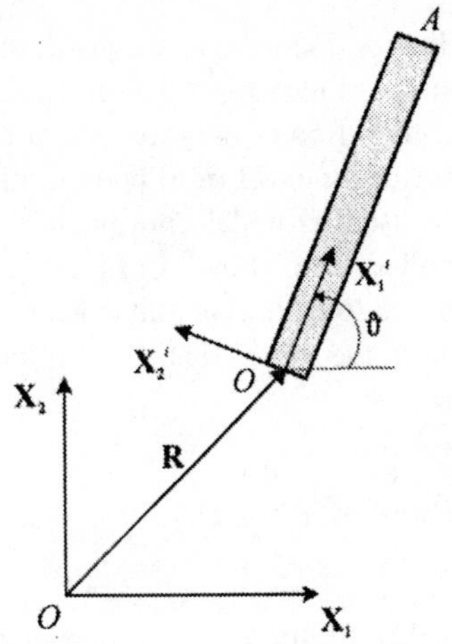

Figure 7.2 Rigid body motion.

and the vector of nodal coordinates is

$$\mathbf{e} = [e_1 \quad e_2 \quad e_3 \quad e_4 \quad e_5 \quad e_6 \quad e_7 \quad e_8]^{\mathrm{T}} \tag{7.7}$$

where $\xi = x/l$, l is the length of the element, and e_1, e_2, e_5, and e_6 are, respectively, the absolute coordinates of the nodes at O and A, and

$$e_3 = \frac{\partial r_1(x=0)}{\partial x}, \qquad e_4 = \frac{\partial r_2(x=0)}{\partial x}, \qquad e_7 = \frac{\partial r_1(x=l)}{\partial x}$$

$$e_8 = \frac{\partial r_2(x=l)}{\partial x} \tag{7.8}$$

Using the simple rigid body kinematic equations previously obtained in Eqs. 4 and 5, one can show that in the case of an arbitrary rigid body motion defined by the translations R_1 and R_2 of the endpoint O and the rotation defined by the angle θ, the vector of the nodal coordinates $\mathbf{e}$ can be written as

$$\mathbf{e} = [R_1 \quad R_2 \quad \cos\theta \quad \sin\theta \quad R_1 + l\cos\theta \quad R_2 + l\sin\theta \quad \cos\theta \quad \sin\theta]^{\mathrm{T}} \tag{7.9}$$

Using this vector of nodal coordinates, and the shape function of Eq. 6, it can be verified that

$$\mathbf{Se} = \begin{bmatrix} R_1 + x\cos\theta \\ R_2 + x\sin\theta \end{bmatrix} = \begin{bmatrix} r_1 \\ r_2 \end{bmatrix} = \mathbf{r} \tag{7.10}$$

which demonstrates that the element shape function of Eq. 6 and the vector of absolute nodal coordinates of Eq. 7 can describe an arbitrary rigid body motion.

Example 7.1 In the classical finite-element literature, the deformation of beams is defined with respect to a beam coordinate system. The axial displacement is interpolated using a linear polynomial, while the transverse displacement is interpolated using a cubic polynomial. Such a beam element is not considered an isoparametric element in the classical finite-element literature because infinitesimal rotations are used as nodal coordinates. While it is not justified to use such a beam element in the absolute nodal coordinate formulation because the displacement components are described using different polynomials, it can be demonstrated that such an element can describe an exact rigid body motion if slopes instead of infinitesimal rotations are used as nodal coordinates. To demonstrate this, we consider the uniform slender beam shown in Fig. 2. The coordinate system of this beam element is assumed to be initially attached to its left end, which is defined by point O as shown in the figure. The conventional shape function of this element is assumed to be

$$\mathbf{S} = \begin{bmatrix} 1-\xi & 0 & 0 & \xi & 0 & 0 \\ 0 & 1-3(\xi)^2+2(\xi)^3 & l(\xi-2(\xi)^2+(\xi)^3) & 0 & 3(\xi)^2-2(\xi)^3 & l((\xi)^3-(\xi)^2) \end{bmatrix} \tag{7.11}$$

where $\xi = x/l$ and l is the length of the beam. The vector of nodal coordinates associated with the shape function of Eq. 11 is

$$\mathbf{e} = [e_1 \quad e_2 \quad e_3 \quad e_4 \quad e_5 \quad e_6]^{\mathrm{T}} \tag{7.12}$$

where e_1 and e_2 are the translational coordinates at the node at O, e_4 and e_5 are the translational coordinates of the node at A, and e_3 and e_6 are the slopes at the two nodes. An arbitrary rigid body displacement of the beam is defined by the translation $\mathbf{R} = [R_1 \; R_2]^{\mathrm{T}}$ of the reference point O, and a rigid body rotation θ. As a result of this arbitrary rigid body displacement, the vector of nodal coordinates $\mathbf{e}$ can be defined in the global coordinate system using Eq. 4 as

$$\mathbf{e} = [R_1 \quad R_2 \quad \sin\theta \quad R_1 + l\cos\theta \quad R_2 + l\sin\theta \quad \sin\theta]^{\mathrm{T}} \tag{7.13}$$

Using Eqs. 11 and 13, it follows that

$$\mathbf{Se} = \begin{bmatrix} R_1 + x\cos\theta \\ R_2 + x\sin\theta \end{bmatrix} \tag{7.14}$$

which demonstrates that the element shape function and the nodal coordinates can describe an arbitrary rigid body displacement provided that the coordinates are defined in the global coordinate system and the slopes are defined in terms of trigonometric functions. Therefore, the conventional finite-element shape function can be used to obtain an exact modeling of the rigid body displacement (Shabana 1996b).

7.2 ABSOLUTE NODAL COORDINATE FORMULATION

Assuming that the shape function of the finite element can describe an arbitrary rigid body displacement, the global position vector of an arbitrary point on the element can be defined using Eq. 1. By differentiating this equation with respect to time, the absolute velocity vector can be defined. This velocity vector can be used to define the kinetic energy of the element as

$$T = \frac{1}{2}\dot{\mathbf{e}}^{\mathrm{T}}\mathbf{M}_a\dot{\mathbf{e}} \tag{7.15}$$

where $\mathbf{M}_a$ is the constant mass matrix of the element defined as

$$\mathbf{M}_a = \int_V \rho \mathbf{S}^{\mathrm{T}} \mathbf{S} \, dV \tag{7.16}$$

where ρ and V are the mass density and volume of the finite element, respectively. Note that this mass matrix is constant, and it is the same mass matrix that appears in linear structural dynamics.

Deformation While in the absolute nodal coordinate formulation the mass matrix takes a simple form, it can be shown using the kinematic description of Eq. 1 that the strain energy is a highly nonlinear function. Since in the absolute nodal coordinate formulation, the shape function can describe an arbitrary rigid body motion, the displacement of the element from a reference configuration defined by the vector of nodal coordinates $\mathbf{e}_o$ can be written as

$$\mathbf{u}_g = \mathbf{S}(\mathbf{e} - \mathbf{e}_o) \tag{7.17}$$

where $\mathbf{u}_g$ is the global displacement vector defined in an inertial frame. Using this vector, the matrix of displacement gradients can be systematically evaluated, and a continuum mechanics approach as described in Chapter 4 can be used to evaluate the elastic forces. This will lead to a simpler expression for the stiffness matrix as compared with the expression that might be obtained by using the classical beam and plate theories when the absolute nodal coordinate formulation is used. Furthermore, there is no need in this case to introduce any intermediate element coordinate system to define the deformation since the rigid body motion produces zero strains. However, when Eq. 17 is used in the large displacement analysis, a nonlinear strain-displacement relationship must be used to obtain an accurate solution. This may not be necessary in many applications when the classical beam and plate theories in which the deformations are defined in an element coordinate system are used. In this case, when small elements are used, the deformation in the element coordinate system remains small such that the use of a linear strain-displacement relationship can be justified. For this reason, we consider in the following discussion, as an example, an alternative that employs the beam theory to formulate the strain energy of the element in the absolute nodal coordinate formulation.

Beam Theory Classical beam and plate theories can also be used with the absolute nodal coordinate formulation to define the element stiffness matrix. This is demonstrated in this section using the two-dimensional classical beam theory. In this section, we consider only the case of small deformations for simplicity. In the case of large deformations, one needs only to change the form of the stiffness matrix, which is nonlinear even in the case of small deformations. If we select point O on the beam element as the reference point, the components of the relative displacement of an arbitrary point with respect to point O can be defined in the inertial coordinate system as

$$\mathbf{u} = \begin{bmatrix} u_1 \\ u_2 \end{bmatrix} = \begin{bmatrix} (\mathbf{S}_1 - \mathbf{S}_{1O})\mathbf{e} \\ (\mathbf{S}_2 - \mathbf{S}_{2O})\mathbf{e} \end{bmatrix} \tag{7.18}$$

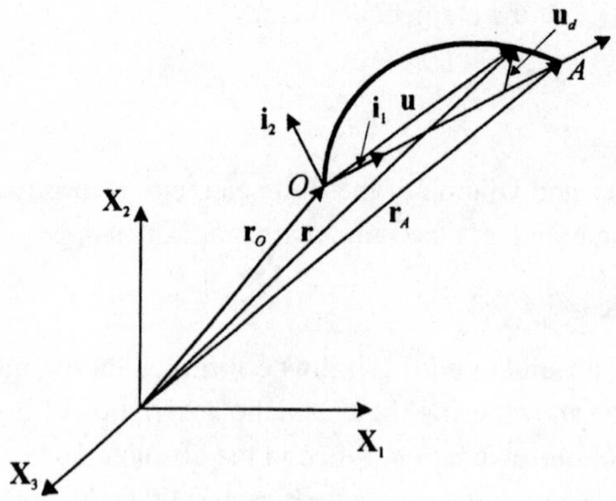

Figure 7.3 Beam deformation.

where $\mathbf{S}_1$ and $\mathbf{S}_2$ are the rows of the element shape function matrix, and $\mathbf{S}_{1O}$ and $\mathbf{S}_{2O}$ are the rows of the shape function matrix defined at the reference point O. To define the longitudinal and transverse displacements of the beam, one may first define the unit vector $\mathbf{i}_1$ along a selected beam axis as

$$\mathbf{i}_1 = [i_{11} \quad i_{12}]^{\mathrm{T}} = \frac{\mathbf{r}_A - \mathbf{r}_O}{|\mathbf{r}_A - \mathbf{r}_O|} \tag{7.19}$$

A unit vector $\mathbf{i}_2$ perpendicular to $\mathbf{i}_1$ can be obtained as

$$\mathbf{i}_2 = [i_{21} \quad i_{22}]^{\mathrm{T}} = \mathbf{i}_3 \times \mathbf{i}_1 \tag{7.20}$$

where $\mathbf{i}_3$ is a unit vector along the $\mathbf{X}_3$ axis. Then, the longitudinal and transverse deformations of the beam can be defined as shown in Fig. 3 as

$$\mathbf{u}_d = \begin{bmatrix} u_l \\ u_t \end{bmatrix} = \begin{bmatrix} \mathbf{u}^{\mathrm{T}}\mathbf{i}_1 - x \\ \mathbf{u}^{\mathrm{T}}\mathbf{i}_2 \end{bmatrix} = \begin{bmatrix} u_1 i_{11} + u_2 i_{12} - x \\ u_1 i_{21} + u_2 i_{22} \end{bmatrix} \tag{7.21}$$

If we assume a linear elastic model, a simple expression for the strain energy U can be written as

$$U = \frac{1}{2}\int_0^l \left(Ea\left(\frac{\partial u_l}{\partial x}\right)^2 + EI\left(\frac{\partial^2 u_t}{\partial x^2}\right)^2 \right) dx \tag{7.22}$$

where E is the modulus of elasticity, a is the cross-sectional area, and I is the second moment of area. Using the expressions for the deformation components given by Eq. 21, it can be shown that the strain energy of the finite element can be written as

$$U = \frac{1}{2}\mathbf{e}^{\mathrm{T}}\mathbf{K}_a\mathbf{e} \tag{7.23}$$

where $\mathbf{K}_a$ is the stiffness matrix of the element. This stiffness matrix is a highly nonlinear function of the element coordinates even in the case in which a linear elastic model is used. This matrix also includes stiffness coupling between the displacements of the element.

Isoparametric Property It can be shown that the use of the absolute nodal coordinate formulation produces zero deformation in the case of an arbitrary rigid body motion. Using Eq. 4, which describes an arbitrary rigid body displacement, it can be shown that unit vectors along and perpendicular to the element axis are defined using Eqs. 19 and 20 as

$$\mathbf{i}_1 = \begin{bmatrix} i_{11} \\ i_{12} \end{bmatrix} = \begin{bmatrix} \cos\theta \\ \sin\theta \end{bmatrix}, \quad \mathbf{i}_2 = \begin{bmatrix} i_{21} \\ i_{22} \end{bmatrix} = \begin{bmatrix} -\sin\theta \\ \cos\theta \end{bmatrix} \tag{7.24}$$

and

$$\mathbf{u} = \begin{bmatrix} u_1 \\ u_2 \end{bmatrix} = \begin{bmatrix} x\cos\theta \\ x\sin\theta \end{bmatrix} \tag{7.25}$$

The deformation gradients can then be evaluated as

$$\frac{\partial u_l}{\partial x} = \frac{\partial u_1}{\partial x} i_{11} + \frac{\partial u_2}{\partial x} i_{12} - 1 = \cos^2\theta + \sin^2\theta - 1 = 0 \tag{7.26}$$

$$\frac{\partial^2 u_t}{\partial x^2} = \frac{\partial^2 u_1}{\partial x^2} i_{21} + \frac{\partial^2 u_2}{\partial x^2} i_{22} = 0 \tag{7.27}$$

which show that, in the absolute nodal coordinate formulation, the deformation gradients remain zero under the rigid body motion, and consequently, the strain energy is equal to zero and the vector of elastic forces remains equal to zero. This result clearly demonstrates that the beam element can be treated in the absolute nodal coordinate formulation as an isoparametric element.

Example 7.2 It is the objective of this example to demonstrate that the linearization of the slopes and use of them as infinitesimal rotations does not produce zero strain under a rigid body motion. To demonstrate this, the conventional shape function of Eq. 11 is used. If the slopes are linearized, the vector of nodal coordinates of Eq. 13 reduces to

$$\mathbf{e} = [R_1 \quad R_2 \quad \theta \quad R_1 + l \quad R_2 + l\theta \quad \theta]^{\mathrm{T}} \tag{7.28}$$

which defines the following two unit vectors:

$$\mathbf{i}_1 = \frac{1}{\sqrt{1+(\theta)^2}} \begin{bmatrix} 1 \\ \theta \end{bmatrix}, \quad \mathbf{i}_2 = \frac{1}{\sqrt{1+(\theta)^2}} \begin{bmatrix} -\theta \\ 1 \end{bmatrix} \tag{7.29}$$

Using the preceding two equations and the shape function of Eq. 11, it can be shown that

$$\mathbf{u} = \begin{bmatrix} u_1 \\ u_2 \end{bmatrix} = \begin{bmatrix} x \\ x\theta \end{bmatrix} \tag{7.30}$$

and

$$\mathbf{u}_d = \begin{bmatrix} u_l \\ u_t \end{bmatrix} = x \begin{bmatrix} \sqrt{1+(\theta)^2} - 1 \\ 0 \end{bmatrix} \tag{7.31}$$

which shows that

$$\frac{\partial u_l}{\partial x} = \sqrt{1+(\theta)^2} - 1 \tag{7.32}$$

The preceding two equations show the error in the deformation and their gradients when the slopes are treated as infinitesimal rotations in the absolute nodal coordinate formulation. The conventional stiffness matrix used in linear structural analysis obtained using the shape function of Eq. 11 is

$$\mathbf{K}_l = \frac{EI}{l}\begin{bmatrix} \frac{a}{l} & 0 & 0 & -\frac{a}{l} & 0 & 0 \\ 0 & \frac{12}{(l)^2} & \frac{6}{l} & 0 & -\frac{12}{(l)^2} & \frac{6}{l} \\ 0 & \frac{6}{l} & 4 & 0 & -\frac{6}{l} & 2 \\ -\frac{a}{l} & 0 & 0 & \frac{a}{l} & 0 & 0 \\ 0 & -\frac{12}{(l)^2} & -\frac{6}{l} & 0 & \frac{12}{(l)^2} & -\frac{6}{l} \\ 0 & \frac{6}{l} & 2 & 0 & -\frac{6}{l} & 4 \end{bmatrix} \tag{7.33}$$

Note that when the vector of nodal coordinates is defined by Eq. 28, one obtains

$$\mathbf{K}_l\mathbf{e} \neq \mathbf{0} \tag{7.34}$$

which indicates that when the slopes are linearized the elastic forces are not equal to zero in the case of a rigid body displacement. The error that results in the inertia as the result of the linearization of the slope was evaluated and can be found in the literature (Shabana 1996b).

7.3 FORMULATION OF THE STIFFNESS MATRIX

In the *absolute nodal coordinate formulation* presented in this chapter, the mass matrix is constant, and it is the same matrix that appears in linear structural dynamics. The stiffness matrix, on the other hand, becomes a nonlinear function of time even in the case of linear elastic problems. It is clear from the analysis presented in the preceding section that the calculations of the stiffness matrix in the case of using linear strain-displacement relationships of beams requires the evaluation of the following stiffness integrals:

$$\left.\begin{aligned}
\mathbf{A}_{11} &= \frac{Ea}{l}\int_0^1 \left(\frac{\partial \mathbf{S}_1}{\partial \xi}\right)^{\mathrm{T}}\left(\frac{\partial \mathbf{S}_1}{\partial \xi}\right)d\xi, \quad \mathbf{A}_{12} = \frac{Ea}{l}\int_0^1 \left(\frac{\partial \mathbf{S}_1}{\partial \xi}\right)^{\mathrm{T}}\left(\frac{\partial \mathbf{S}_2}{\partial \xi}\right)d\xi \\
\mathbf{A}_{21} &= \frac{Ea}{l}\int_0^1 \left(\frac{\partial \mathbf{S}_2}{\partial \xi}\right)^{\mathrm{T}}\left(\frac{\partial \mathbf{S}_1}{\partial \xi}\right)d\xi, \quad \mathbf{A}_{22} = \frac{Ea}{l}\int_0^1 \left(\frac{\partial \mathbf{S}_2}{\partial \xi}\right)^{\mathrm{T}}\left(\frac{\partial \mathbf{S}_2}{\partial \xi}\right)d\xi \\
\mathbf{B}_{11} &= \frac{EI}{(l)^3}\int_0^1 \left(\frac{\partial^2 \mathbf{S}_1}{\partial \xi^2}\right)^{\mathrm{T}}\left(\frac{\partial^2 \mathbf{S}_1}{\partial \xi^2}\right)d\xi, \quad \mathbf{B}_{12} = \frac{EI}{(l)^3}\int_0^1 \left(\frac{\partial^2 \mathbf{S}_1}{\partial \xi^2}\right)^{\mathrm{T}}\left(\frac{\partial^2 \mathbf{S}_2}{\partial \xi^2}\right)d\xi \\
\mathbf{B}_{21} &= \frac{EI}{(l)^3}\int_0^1 \left(\frac{\partial^2 \mathbf{S}_2}{\partial \xi^2}\right)^{\mathrm{T}}\left(\frac{\partial^2 \mathbf{S}_1}{\partial \xi^2}\right)d\xi, \quad \mathbf{B}_{22} = \frac{EI}{(l)^3}\int_0^1 \left(\frac{\partial^2 \mathbf{S}_2}{\partial \xi^2}\right)^{\mathrm{T}}\left(\frac{\partial^2 \mathbf{S}_2}{\partial \xi^2}\right)d\xi \\
\mathbf{A}_1 &= Ea\int_0^1 \frac{\partial \mathbf{S}_1}{\partial \xi}\,d\xi, \quad \mathbf{A}_2 = Ea\int_0^1 \frac{\partial \mathbf{S}_2}{\partial \xi}\,d\xi
\end{aligned}\right\} \tag{7.35}$$

Using the element shape function, the explicit form of the matrices and the vectors given in Eq. 35 can be determined. These integrals can also be efficiently evaluated using symbolic manipulations. Despite the fact that in the case of large deformation analysis, the expression for the strain energy becomes more complex and more matrices must be evaluated to define the stiffness matrix, very little can be gained computationally by using the linear strain assumptions.

Using the stiffness integrals of Eq. 35, it can be shown that the strain energy of a beam element as defined by Eq. 22 can be written as

$$U = \frac{1}{2}\{\mathbf{e}^{\mathrm{T}}\mathbf{A}_{11}\mathbf{e}(i_{11})^2 + \mathbf{e}^{\mathrm{T}}(\mathbf{A}_{12} + \mathbf{A}_{21})\mathbf{e}i_{11}i_{12} + \mathbf{e}^{\mathrm{T}}\mathbf{A}_{22}\mathbf{e}(i_{12})^2 + \mathbf{e}^{\mathrm{T}}\mathbf{B}_{11}\mathbf{e}(i_{21})^2$$
$$+ \mathbf{e}^{\mathrm{T}}(\mathbf{B}_{12} + \mathbf{B}_{21})\mathbf{e}i_{21}i_{22} + \mathbf{e}^{\mathrm{T}}\mathbf{B}_{22}\mathbf{e}(i_{22})^2 - 2\mathbf{A}_1\mathbf{e}i_{11} - 2\mathbf{A}_2\mathbf{e}i_{12} + Eal\} \quad (7.36)$$

Using this expression for the strain energy, the vector of the element generalized elastic forces can be obtained from

$$\left(\frac{\partial U}{\partial \mathbf{e}}\right)^{\mathrm{T}} = \mathbf{A}_{11}\mathbf{e}(i_{11})^2 + (\mathbf{A}_{12} + \mathbf{A}_{21})\mathbf{e}i_{11}i_{12} + \mathbf{A}_{22}\mathbf{e}(i_{12})^2 + \mathbf{B}_{11}\mathbf{e}(i_{21})^2$$
$$+ (\mathbf{B}_{12} + \mathbf{B}_{21})\mathbf{e}i_{21}i_{22} + \mathbf{B}_{22}\mathbf{e}(i_{22})^2 - \mathbf{A}_1^{\mathrm{T}}i_{11} - \mathbf{A}_2^{\mathrm{T}}i_{12}$$
$$+ \mathbf{e}^{\mathrm{T}}\mathbf{A}_{11}\mathbf{e}i_{11}\left(\frac{\partial i_{11}}{\partial \mathbf{e}}\right)^{\mathrm{T}} + \frac{1}{2}\mathbf{e}^{\mathrm{T}}(\mathbf{A}_{12} + \mathbf{A}_{21})\mathbf{e}\left(i_{11}\left(\frac{\partial i_{12}}{\partial \mathbf{e}}\right)^{\mathrm{T}}\right.$$
$$\left. + i_{12}\left(\frac{\partial i_{11}}{\partial \mathbf{e}}\right)^{\mathrm{T}}\right) + \mathbf{e}^{\mathrm{T}}\mathbf{A}_{22}\mathbf{e}i_{12}\left(\frac{\partial i_{12}}{\partial \mathbf{e}}\right)^{\mathrm{T}} + \mathbf{e}^{\mathrm{T}}\mathbf{B}_{11}\mathbf{e}i_{21}\left(\frac{\partial i_{21}}{\partial \mathbf{e}}\right)^{\mathrm{T}}$$
$$+ \frac{1}{2}\mathbf{e}^{\mathrm{T}}(\mathbf{B}_{12} + \mathbf{B}_{21})\mathbf{e}\left(i_{21}\left(\frac{\partial i_{22}}{\partial \mathbf{e}}\right)^{\mathrm{T}} + i_{22}\left(\frac{\partial i_{21}}{\partial \mathbf{e}}\right)^{\mathrm{T}}\right)$$
$$+ \mathbf{e}^{\mathrm{T}}\mathbf{B}_{22}\mathbf{e}i_{22}\left(\frac{\partial i_{22}}{\partial \mathbf{e}}\right)^{\mathrm{T}} - \mathbf{A}_1\mathbf{e}\left(\frac{\partial i_{11}}{\partial \mathbf{e}}\right)^{\mathrm{T}} - \mathbf{A}_2\mathbf{e}\left(\frac{\partial i_{12}}{\partial \mathbf{e}}\right)^{\mathrm{T}} \quad (7.37)$$

This equation shows that when the rotation of the element is equal to zero, the stiffness coefficients reduce to the stiffness coefficients obtained using the linear structural analysis approach.

In general, a simpler expression for the strain energy can be obtained by eliminating the effect of the rigid body motion since such a motion has no effect on the elastic forces. This can be achieved by writing the vector of nodal coordinates $\mathbf{e}$ as

$$\mathbf{e} = \mathbf{e}_r + \mathbf{e}_f$$

where $\mathbf{e}_r$ is the vector of nodal coordinates resulting from the rigid body motion and $\mathbf{e}_f$ is the vector of nodal coordinates due to the deformation. Using the example of the coordinate system described in the preceding section, Eq. 19, and the coordinates associated with the shape function of Eq. 6, one can show that

$$\mathbf{e}_r = [e_1 \quad e_2 \quad i_{11} \quad i_{12} \quad e_1 + li_{11} \quad e_2 + li_{12} \quad i_{11} \quad i_{12}]^{\mathrm{T}}$$

In this case, the deformation vector of Eq. 21 can simply be written as

$$\mathbf{u}_d = \mathbf{A}_c^{\mathrm{T}}\mathbf{S}(\mathbf{e} - \mathbf{e}_r)$$

where $\mathbf{A}_c$ is the transformation matrix

$$\mathbf{A}_c = \begin{bmatrix} i_{11} & i_{21} \\ i_{12} & i_{22} \end{bmatrix}$$

It can be shown that the use of the deformation vector defined in the preceding equation leads to simpler expressions for the strain energy and the elastic forces.

Illustrative Example It can be shown using the cubic shape function of Eq. 6 that the explicit forms of the matrices and vectors that appear in Eq. 35 are as follows:

$$\mathbf{A}_{11} = \frac{Ea}{l}\begin{bmatrix} \frac{6}{5} & 0 & \frac{l}{10} & 0 & -\frac{6}{5} & 0 & \frac{l}{10} & 0 \\ 0 & 0 & 0 & 0 & 0 & 0 & 0 & 0 \\ \frac{l}{10} & 0 & \frac{2(l)^2}{15} & 0 & -\frac{l}{10} & 0 & -\frac{(l)^2}{30} & 0 \\ 0 & 0 & 0 & 0 & 0 & 0 & 0 & 0 \\ -\frac{6}{5} & 0 & -\frac{l}{10} & 0 & \frac{6}{5} & 0 & -\frac{l}{10} & 0 \\ 0 & 0 & 0 & 0 & 0 & 0 & 0 & 0 \\ \frac{l}{10} & 0 & -\frac{(l)^2}{30} & 0 & -\frac{l}{10} & 0 & \frac{2(l)^2}{15} & 0 \\ 0 & 0 & 0 & 0 & 0 & 0 & 0 & 0 \end{bmatrix} \tag{7.38}$$

$$\mathbf{A}_{22} = \frac{Ea}{l}\begin{bmatrix} 0 & 0 & 0 & 0 & 0 & 0 & 0 & 0 \\ 0 & \frac{6}{5} & 0 & \frac{l}{10} & 0 & -\frac{6}{5} & 0 & \frac{l}{10} \\ 0 & 0 & 0 & 0 & 0 & 0 & 0 & 0 \\ 0 & \frac{l}{10} & 0 & \frac{2(l)^2}{15} & 0 & -\frac{l}{10} & 0 & -\frac{(l)^2}{30} \\ 0 & 0 & 0 & 0 & 0 & 0 & 0 & 0 \\ 0 & -\frac{6}{5} & 0 & -\frac{l}{10} & 0 & \frac{6}{5} & 0 & -\frac{l}{10} \\ 0 & 0 & 0 & 0 & 0 & 0 & 0 & 0 \\ 0 & \frac{l}{10} & 0 & -\frac{(l)^2}{30} & 0 & -\frac{l}{10} & 0 & -\frac{2(l)^2}{15} \end{bmatrix} \tag{7.39}$$

$$\mathbf{A}_{12} + \mathbf{A}_{21} = \frac{Ea}{l}\begin{bmatrix} 0 & \frac{6}{5} & 0 & \frac{l}{10} & 0 & -\frac{6}{5} & 0 & \frac{l}{10} \\ \frac{6}{5} & 0 & \frac{l}{10} & 0 & -\frac{6}{5} & 0 & \frac{l}{10} & 0 \\ 0 & \frac{l}{10} & 0 & \frac{2(l)^2}{15} & 0 & -\frac{l}{10} & 0 & -\frac{(l)^2}{30} \\ \frac{l}{10} & 0 & \frac{2l^2}{15} & 0 & -\frac{l}{10} & 0 & -\frac{(l)^2}{30} & 0 \\ 0 & -\frac{6}{5} & 0 & -\frac{l}{10} & 0 & \frac{6}{5} & 0 & -\frac{l}{10} \\ -\frac{6}{5} & 0 & -\frac{l}{10} & 0 & \frac{6}{5} & 0 & -\frac{l}{10} & 0 \\ 0 & \frac{l}{10} & 0 & -\frac{(l)^2}{30} & 0 & -\frac{l}{10} & 0 & \frac{2(l)^2}{15} \\ \frac{l}{10} & 0 & -\frac{(l)^2}{30} & 0 & -\frac{l}{10} & 0 & \frac{2(l)^2}{15} & 0 \end{bmatrix} \tag{7.40}$$

$$\begin{aligned} \mathbf{A}_1 &= Ea[-1 \quad 0 \quad 0 \quad 0 \quad 1 \quad 0 \quad 0 \quad 0] \\ \mathbf{A}_2 &= Ea[0 \quad -1 \quad 0 \quad 0 \quad 0 \quad 1 \quad 0 \quad 0] \end{aligned} \tag{7.41}$$

$$\mathbf{B}_{11} = \frac{EI}{(l)^3}\begin{bmatrix} 12 & 0 & 6l & 0 & -12 & 0 & 6l & 0 \\ 0 & 0 & 0 & 0 & 0 & 0 & 0 & 0 \\ 6l & 0 & 4(l)^2 & 0 & -6l & 0 & -2(l)^2 & 0 \\ 0 & 0 & 0 & 0 & 0 & 0 & 0 & 0 \\ -12 & 0 & -6l & 0 & 12 & 0 & -6l & 0 \\ 0 & 0 & 0 & 0 & 0 & 0 & 0 & 0 \\ 6l & 0 & 2(l)^2 & 0 & -6l & 0 & 4(l)^2 & 0 \\ 0 & 0 & 0 & 0 & 0 & 0 & 0 & 0 \end{bmatrix} \tag{7.42}$$

$$\mathbf{B}_{22} = \frac{EI}{(l)^3}\begin{bmatrix} 0 & 0 & 0 & 0 & 0 & 0 & 0 & 0 \\ 0 & 12 & 0 & 6l & 0 & -12 & 0 & 6l \\ 0 & 0 & 0 & 0 & 0 & 0 & 0 & 0 \\ 0 & 6l & 0 & 4(l)^2 & 0 & -6l & 0 & 2(l)^2 \\ 0 & 0 & 0 & 0 & 0 & 0 & 0 & 0 \\ 0 & -12 & 0 & -6l & 0 & 12 & 0 & -6l \\ 0 & 0 & 0 & 0 & 0 & 0 & 0 & 0 \\ 0 & 6l & 0 & 2(l)^2 & 0 & -6l & 0 & 4(l)^2 \end{bmatrix} \tag{7.43}$$

$$\mathbf{B}_{12} + \mathbf{B}_{21} = \frac{EI}{(l)^3}\begin{bmatrix} 0 & 12 & 0 & 6l & 0 & -12 & 0 & 6l \\ 12 & 0 & 6l & 0 & -12 & 0 & 6l & 0 \\ 0 & 6l & 0 & 4(l)^2 & 0 & -6l & 0 & 2(l)^2 \\ 6l & 0 & 4l^2 & 0 & -6l & 0 & 2(l)^2 & 0 \\ 0 & -12 & 0 & -6l & 0 & 12 & 0 & -6l \\ -12 & 0 & -6l & 0 & 12 & 0 & -6l & 0 \\ 0 & 6l & 0 & 2(l)^2 & 0 & -6l & 0 & 4(l)^2 \\ 6l & 0 & 2(l)^2 & 0 & -6l & 0 & 4(l)^2 & 0 \end{bmatrix} \tag{7.44}$$

If we assume that one of the axes of a selected element coordinate system passes through points O and A, unit vectors along the axes of the element coordinate system can be expressed in terms of the nodal coordinates of the element as

$$\mathbf{i}_1 = \begin{bmatrix} i_{11} \\ i_{12} \end{bmatrix} = \frac{1}{(d)^{\frac{1}{2}}}\begin{bmatrix} e_5 - e_1 \\ e_6 - e_2 \end{bmatrix}, \quad \mathbf{i}_2 = \begin{bmatrix} i_{21} \\ i_{22} \end{bmatrix} = \frac{1}{(d)^{\frac{1}{2}}}\begin{bmatrix} -i_{12} \\ i_{11} \end{bmatrix} \tag{7.45}$$

where

$$d = (e_5 - e_1)^2 + (e_6 - e_2)^2 \tag{7.46}$$

The partial derivatives of the components of the unit vectors $\mathbf{i}_1$ and $\mathbf{i}_2$ with respect to the nodal coordinates of the element are given by

$$\begin{aligned} \frac{\partial i_{11}}{\partial \mathbf{e}} &= \frac{1}{(d)^{3/2}} \\ &\times [-(e_6 - e_2)^2 \quad (e_5 - e_1)(e_6 - e_2) \quad 0 \quad 0 \quad -(e_6 - e_2)^2 \quad -(e_5 - e_1)(e_6 - e_2) \quad 0 \quad 0] \\ \frac{\partial i_{12}}{\partial \mathbf{e}} &= \frac{1}{(d)^{3/2}} \\ &\times [(e_5 - e_1)(e_6 - e_2) \quad -(e_5 - e_1)^2 \quad 0 \quad 0 \quad -(e_5 - e_1)(e_6 - e_2) \quad (e_5 - e_1)^2 \quad 0 \quad 0] \\ \frac{\partial i_{21}}{\partial \mathbf{e}} &= -\frac{\partial i_{12}}{\partial \mathbf{e}}, \quad \frac{\partial i_{22}}{\partial \mathbf{e}} = \frac{\partial i_{11}}{\partial \mathbf{e}} \end{aligned} \tag{7.47}$$

The matrices and vectors presented in Eqs. 38–47 are evaluated using the cubic shape function of Eq. 6 and the associated nodal coordinates. If another shape function is used, another set of matrices and vectors, whose dimensions depend on the number of coordinates employed, must be evaluated.

7.4 EQUATIONS OF MOTION

As pointed out in Section 2, the use of the absolute nodal coordinate formulation leads to a constant mass matrix for the finite element. As a consequence, the centrifugal and Coriolis inertia forces are equal to zero. Furthermore, no coordinate transformation is required to determine the global mass matrix of the element. The elastic forces, on the other hand, are nonlinear in the coordinates and they have a more complex form as compared with the form of the elastic forces used in the floating frame of reference formulation. Because of the nonlinearity of the elastic forces in the absolute nodal coordinate formulation, little is to be gained from the use of the small strain assumptions.

Using the results obtained in the preceding two sections, one can show that the matrix equation of motion of the finite element in the case of the absolute nodal coordinate formulation takes the following form:

$$\mathbf{M}_a\ddot{\mathbf{e}} + \mathbf{K}_a\mathbf{e} = \mathbf{Q}_a \tag{7.48}$$

where $\mathbf{M}_a$ is the constant mass matrix, $\mathbf{K}_a$ is the nonlinear stiffness matrix, and $\mathbf{Q}_a$ is the vector of generalized nodal forces. Since the stiffness matrix is a nonlinear function of the element nodal coordinates, the preceding equation can be written as

$$\mathbf{M}_a\ddot{\mathbf{e}} = \mathbf{Q} \tag{7.49}$$

where

$$\mathbf{Q} = \mathbf{Q}_a - \mathbf{K}_a\mathbf{e} \tag{7.50}$$

Since the element mass matrix is constant, the equation of motion of the element can be written as

$$\ddot{\mathbf{e}} = \mathbf{M}_a^{-1}\mathbf{Q} = \mathbf{b}(\mathbf{e}, \dot{\mathbf{e}}, t) \tag{7.51}$$

where

$$\mathbf{b} = \mathbf{M}_a^{-1}\mathbf{Q} \tag{7.52}$$

Using the preceding element equations, connectivity conditions between the finite elements can be imposed and the equations of the elements can be assembled to obtain the equations of motion of the deformable bodies in the multibody system.

Generalized External Forces If a force $\mathbf{F}$ acts at an arbitrary point on the finite element, the virtual work of this force can be written as $\mathbf{F}^{\mathrm{T}}\,\delta\mathbf{r}$, where $\mathbf{r}$ is the global position vector of the point of application of the force. The virtual change in the vector $\mathbf{r}$ can be expressed in terms of the virtual changes in the absolute nodal coordinates, thereby defining the generalized forces associated with these absolute nodal coordinates.

In the case of an external moment acting at an arbitrary point on the finite element, one can, in general, define the orientation of the cross-section of the element at the point of application of the moment in terms of the slopes. To demonstrate the procedure for doing this, we consider the case of a planar beam element. Using Eqs. 2 and 3, the orientation of a coordinate system whose origin is rigidly attached to the point of the application of the moment can be defined using the following transformation matrix

$$\begin{bmatrix} \cos\alpha & -\sin\alpha \\ \sin\alpha & \cos\alpha \end{bmatrix} = \frac{1}{(d)^{\frac{1}{2}}} \begin{bmatrix} \frac{\partial r_1}{\partial x} & -\frac{\partial r_2}{\partial x} \\ \frac{\partial r_2}{\partial x} & \frac{\partial r_1}{\partial x} \end{bmatrix}$$

where

$$d = \left(\frac{\partial r_1}{\partial x}\right)^2 + \left(\frac{\partial r_2}{\partial x}\right)^2$$

Using the elements of the planar transformation matrix defined in the preceding equation, it is clear that

$$\sin\alpha = (d)^{-\frac{1}{2}}\left(\frac{\partial r_2}{\partial x}\right), \qquad \cos\alpha = (d)^{-\frac{1}{2}}\left(\frac{\partial r_1}{\partial x}\right)$$

Using these two equations, it can be verified that

$$\delta\alpha = \frac{r_1'\delta r_2' - r_2'\delta r_1'}{d}$$

where

$$r_i' = \frac{\partial r_i}{\partial x}, \qquad i = 1, 2$$

If the moment is applied at a node, the spatial derivatives of the displacements can be expressed in terms of the nodal coordinates of this node. For example, if the moment is applied at the second node of a beam element defined by the shape function of Eq. 6, then

$$\delta\alpha(x = l) = \frac{e_7\,\delta e_8 - e_8\,\delta e_7}{\sqrt{(e_7)^2 + (e_8)^2}}$$

In general, if a moment M is applied at an arbitrary point x on the beam, the virtual work of this moment can be written as $M\,\delta\alpha(x)$. This virtual work expression can be used to define the generalized forces associated with the absolute nodal coordinates of the element.

In the absolute nodal coordinate formulation, the generalized forces due to the spring–damper–actuator elements take a very simple form as compared with the form used in the floating frame of reference formulation. It is left to the reader as an exercise to derive the simple expressions of these forces when the absolute nodal coordinates are used.

7.5 RELATIONSHIP TO THE FLOATING FRAME OF REFERENCE FORMULATION

In the floating frame of reference formulation, the configuration of the body is described using a mixed set of absolute reference and local deformation coordinates.

As a consequence, not all coordinates represent absolute variables. The reference coordinates define the location and the orientation of a selected body coordinate system, and the deformation of the body is described using a set of local shape functions and a set of deformation coordinates defined in the body coordinate system. In the floating frame of reference formulation, it is assumed that there is no rigid body motion between the body and its coordinate system. As described in the preceding two chapters, the reference and deformation coordinates can be used to define the global position vector of an arbitrary point on the finite element of a deformable body as

$$\mathbf{r} = \mathbf{R} + \mathbf{A}\bar{\mathbf{u}} \tag{7.53}$$

where $\mathbf{R}$ defines the global position vector of the origin of the selected body coordinate system, $\mathbf{A}$ is the transformation matrix that defines the orientation of the selected body coordinate system with respect to the inertial frame, and $\bar{\mathbf{u}}$ is the local position vector of the arbitrary point defined with respect to the origin of the body coordinate system. In Eq. 53, the superscripts that indicate the element and the body numbers are dropped for simplicity. The local position vector $\bar{\mathbf{u}}$ may be represented in terms of *local shape functions* $\mathbf{S}_l$ as

$$\bar{\mathbf{u}} = \mathbf{S}_l \mathbf{q}_f \tag{7.54}$$

where $\mathbf{q}_f$ is the vector of time-dependent deformation coordinates that can also be used in the finite-element formulation to interpolate the local position as well as the deformation. Here the subscript l is used for the local shape function to distinguish it from the global shape function used in the absolute nodal coordinate formulation. This local shape function has the same meaning as the element shape function used in the preceding chapter. When the kinematic description of Eq. 53 is used, it is assumed that there is no rigid body motion between the element and its coordinate system. As a consequence, it is required that the local shape function matrix $\mathbf{S}_l$ contains no rigid body modes. Using Eqs. 53 and 54, the motion of the finite element can be described using the floating frame of reference formulation as

$$\mathbf{r} = \mathbf{R} + \mathbf{A}\mathbf{S}_l \mathbf{q}_f \tag{7.55}$$

where the vector $\mathbf{q}_f$ describes the local position and the deformation of an arbitrary point, and the vector

$$\mathbf{q}_r = \begin{bmatrix} \mathbf{R} \\ \boldsymbol{\theta} \end{bmatrix} \tag{7.56}$$

describes the reference motion. Therefore, the vector of generalized coordinates of the element used in the floating frame of reference formulation can be written in a partitioned form as

$$\mathbf{q} = \begin{bmatrix} \mathbf{R}^{\mathrm{T}} & \boldsymbol{\theta}^{\mathrm{T}} & \mathbf{q}_f^{\mathrm{T}} \end{bmatrix}^{\mathrm{T}} = \begin{bmatrix} \mathbf{q}_r^{\mathrm{T}} & \mathbf{q}_f^{\mathrm{T}} \end{bmatrix}^{\mathrm{T}} \tag{7.57}$$

Using Eq. 55 and the coordinate partitioning of Eq. 57, the mass matrix of the finite element in the case of the floating frame of reference can be written in a partitioned form, as demonstrated in the preceding chapter, as

$$\mathbf{M}_f = \begin{bmatrix} \mathbf{m}_{rr} & \mathbf{m}_{rf} \\ \mathbf{m}_{fr} & \mathbf{m}_{ff} \end{bmatrix} \tag{7.58}$$

As pointed out in the preceding chapter, this mass matrix is highly nonlinear in the coordinates $\mathbf{q} = [\mathbf{q}_r^{\mathrm{T}}\ \mathbf{q}_f^{\mathrm{T}}]^{\mathrm{T}}$ as the result of the dynamic coupling between the reference coordinates $\mathbf{q}_r$ and the deformation coordinates $\mathbf{q}_f$.

In the case of planar motion, one has

$$\mathbf{q}_r = \begin{bmatrix} R_1 \\ R_2 \\ \theta \end{bmatrix}, \qquad \mathbf{A} = \begin{bmatrix} \cos\theta & -\sin\theta \\ \sin\theta & \cos\theta \end{bmatrix} \tag{7.59}$$

where θ is the angle that defines the orientation of the body coordinate system. In this case of planar motion, it can be shown that the nonlinear mass matrix and the Coriolis and centrifugal forces of the finite element can be expressed in terms of the following constant inertia shape integrals:

$$\bar{\mathbf{S}} = \int_V \rho \mathbf{S}_l \, dV, \qquad \mathbf{m}_{ff} = \int_V \rho \mathbf{S}_l^{\mathrm{T}} \mathbf{S}_l \, dV, \qquad \tilde{\mathbf{S}} = \int_V \rho \mathbf{S}_l^{\mathrm{T}} \tilde{\mathbf{I}} \mathbf{S}_l \, dV \tag{7.60}$$

where ρ and V are the mass density and volume of the element, respectively, and

$$\tilde{\mathbf{I}} = \begin{bmatrix} 0 & 1 \\ -1 & 0 \end{bmatrix} \tag{7.61}$$

By establishing the relationship between the coordinates used in the floating frame of reference formulation and the coordinates used in the absolute nodal coordinate formulation, the nonlinear mass matrix of Eq. 58 can be obtained using the constant mass matrix of Eq. 16. Such a coordinate transformation can be used as the basis for developing a systematic procedure for evaluating the inertia shape integrals from the constant mass matrix that appears in the absolute nodal coordinate formulation. The use of such a transformation will be the subject of the following three sections.

7.6 COORDINATE TRANSFORMATION

In the absolute nodal coordinate formulation, beams and plates can be considered as isoparametric elements. Using this fact, the equivalence between the floating frame of reference formulation and the absolute nodal coordinate formulation can be demonstrated and used to examine the effect of using the consistent and lumped mass distribution on modeling the inertia of deformable bodies that undergo large reference displacements. To demonstrate the equivalence of the floating frame of reference formulation and the absolute nodal coordinate formulation, the relationship between the absolute and local slopes is first defined and then used to establish the relationship between the coordinates used in the two different formulations. In this chapter, a planar beam element will be used as an example, and cubic polynomials will be used to equally represent the displacement components of the element. The procedure developed in this section, however, can be applied to other interpolating functions as well as other planar and spatial element types, provided that the global shape function has a complete set of rigid body modes.

Slope Relationship Using Eq. 53, the global position vector of an arbitrary point on the planar beam element can be written as

$$\mathbf{r} = \begin{bmatrix} r_1 \\ r_2 \end{bmatrix} = \begin{bmatrix} R_1 + \bar{u}_1 \cos\theta - \bar{u}_2 \sin\theta \\ R_2 + \bar{u}_1 \sin\theta + \bar{u}_2 \cos\theta \end{bmatrix} \tag{7.62}$$

where $\bar{u}_1$ and $\bar{u}_2$ are the position coordinates of the arbitrary point defined with respect to the beam coordinate system. It follows in the case of a slender beam element that

$$\left.\begin{aligned} \frac{\partial r_1}{\partial x} &= \frac{\partial \bar{u}_1}{\partial x}\cos\theta - \frac{\partial \bar{u}_2}{\partial x}\sin\theta \\ \frac{\partial r_2}{\partial x} &= \frac{\partial \bar{u}_1}{\partial x}\sin\theta + \frac{\partial \bar{u}_2}{\partial x}\cos\theta \end{aligned}\right\} \tag{7.63}$$

This slope relationship plays a fundamental role in defining the relationship between the coordinates used in the absolute nodal coordinate formulation and the coordinates used in the floating frame of reference formulation, as will be demonstrated in this section.

In the case of an arbitrary rigid body motion, the global slopes of the beam can be obtained using Eq. 5 as

$$\begin{bmatrix} \frac{\partial r_1}{\partial x} \\ \frac{\partial r_2}{\partial x} \end{bmatrix} = \begin{bmatrix} \cos\theta \\ \sin\theta \end{bmatrix} \tag{7.64}$$

Using this equation and Eq. 63, it can be shown that, in the case of a rigid body motion, the local slopes are given by

$$\begin{bmatrix} \frac{\partial \bar{u}_1}{\partial x} \\ \frac{\partial \bar{u}_2}{\partial x} \end{bmatrix} = \begin{bmatrix} \cos\theta & \sin\theta \\ -\sin\theta & \cos\theta \end{bmatrix} \begin{bmatrix} \cos\theta \\ \sin\theta \end{bmatrix} = \begin{bmatrix} 1 \\ 0 \end{bmatrix} \tag{7.65}$$

which defines the unit tangent vector in the beam coordinate system.

Coordinate Transformation In the remainder of this section, we develop the relationship between the coordinates used in the floating frame of reference formulation and the coordinates used in the absolute nodal coordinate formulation. In the case of the absolute nodal coordinate formulation, we use the global element shape function defined by Eq. 6. In the floating frame of reference formulation, we consider as an example the case in which the origin of the beam coordinate system is rigidly attached to point O, as shown in Fig. 4. In this case, the local shape function can be obtained from the global shape function of Eq. 6 as

$$\mathbf{S}_l = \begin{bmatrix} l(\xi - 2(\xi)^2 + (\xi)^3) & 3(\xi)^2 - 2(\xi)^3 & 0 & l((\xi)^3 - (\xi)^2) & 0 \\ 0 & 0 & 3(\xi)^2 - 2(\xi)^3 & 0 & l((\xi)^3 - (\xi)^2) \end{bmatrix} \tag{7.66}$$

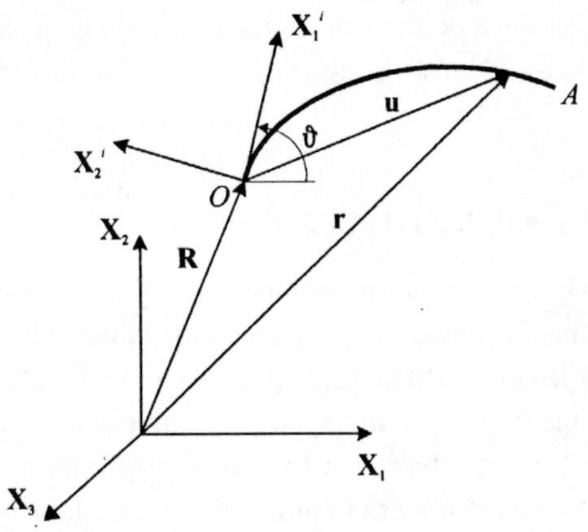

Figure 7.4 Floating frame of reference.

Note that this local shape function does not include any rigid body modes. The vector $\mathbf{q}_f$ in this case can be defined as

$$\mathbf{q}_f = [q_1 \quad q_2 \quad q_3 \quad q_4 \quad q_5]^{\mathrm{T}} \tag{7.67}$$

where q_2 and q_3 are the local coordinates of the node at A defined in the beam coordinate system and

$$q_1 = \frac{\partial \bar{u}_1(x = 0)}{\partial x}, \qquad q_4 = \frac{\partial \bar{u}_1(x = l)}{\partial x}, \qquad q_5 = \frac{\partial \bar{u}_2(x = l)}{\partial x} \tag{7.68}$$

Using Eq. 63, the vector $\mathbf{e}$ of Eq. 7 used in the absolute nodal coordinate formulation can be expressed in terms of the components of the vector

$$\mathbf{q} = [R_1 \quad R_2 \quad \theta \quad q_1 \quad q_2 \quad q_3 \quad q_4 \quad q_5]^{\mathrm{T}} \tag{7.69}$$

of the floating frame of reference formulation as

$$\mathbf{e} = \begin{bmatrix} e_1 \\ e_2 \\ e_3 \\ e_4 \\ e_5 \\ e_6 \\ e_7 \\ e_8 \end{bmatrix} \begin{bmatrix} R_1 \\ R_2 \\ q_1 \cos\theta \\ q_1 \sin\theta \\ R_1 + q_2 \cos\theta - q_3 \sin\theta \\ R_2 + q_2 \sin\theta + q_3 \cos\theta \\ q_4 \cos\theta - q_5 \sin\theta \\ q_4 \sin\theta + q_5 \cos\theta \end{bmatrix} \tag{7.70}$$

Using this vector and the global shape function of Eq. 6, it can be shown that

$$\mathbf{Se} = \mathbf{R} + \mathbf{A}\mathbf{S}_l\mathbf{q}_f = \mathbf{r} \tag{7.71}$$

This equation demonstrates the equivalence of the kinematic descriptions used in the floating frame of reference formulation and the absolute nodal coordinate formulation. Therefore, the coordinate transformation of Eq. 70 can be used to obtain the nonlinear mass matrix and the inertia shape integrals used in the floating frame of reference formulation from the constant mass matrix used in the absolute nodal coordinate formulation. This is demonstrated in the following section using the consistent mass formulation.

7.7 CONSISTENT MASS FORMULATION

Exact modeling of the rigid body motion can be obtained using the absolute nodal coordinate formulation only when a consistent mass approach is used. It will be demonstrated in this section that, when a consistent mass approach is employed, the nonlinear mass matrix and the inertia shape integrals of the floating frame of reference can be systematically obtained using the coordinate transformation presented in the preceding section. Equally important is that the inertia matrix of the rigid body can also be obtained using a similar transformation. In Section 9, it will be shown that this is not the case when lumped masses are used.

Using the coordinate partitioning of Eq. 57, it can be shown using the analysis presented in the preceding chapters that the mass matrix of the deformable beam element, in the floating frame of reference formulation, can be expressed in terms of the inertia shape integrals of Eq. 60 as

$$\mathbf{M}_f = \begin{bmatrix} m\mathbf{I} & \mathbf{A}_\theta \bar{\mathbf{S}} \mathbf{q}_f & \mathbf{A}\bar{\mathbf{S}} \\ & \mathbf{q}_f^{\mathrm{T}} \mathbf{m}_{ff} \mathbf{q}_f & \mathbf{q}_f^{\mathrm{T}} \tilde{\mathbf{S}} \\ \text{symmetric} & & \mathbf{m}_{ff} \end{bmatrix} \tag{7.72}$$

where $\mathbf{I}$ in this equation is a 2×2 identity matrix, m is the mass of the element, and $\mathbf{A}_\theta$ is the partial derivative of the transformation matrix $\mathbf{A}$ with respect to the orientation coordinate θ. The velocity transformation between the coordinates used in the two formulations can be written as

$$\dot{\mathbf{e}} = \mathbf{B}\dot{\mathbf{q}} = [\mathbf{B}_R \quad \mathbf{B}_\theta \quad \mathbf{B}_f] \begin{bmatrix} \dot{\mathbf{R}} \\ \dot{\theta} \\ \dot{\mathbf{q}}_f \end{bmatrix} \tag{7.73}$$

where $\mathbf{B}$ is a velocity transformation matrix. Let $\mathbf{M}_a$ be the mass matrix obtained using the absolute nodal coordinate formulation (Eq. 16); the mass matrix $\mathbf{M}_f$ that results from the use of the floating frame of reference formulation can be simply obtained as

$$\mathbf{M}_f = \mathbf{B}^{\mathrm{T}}\mathbf{M}_a\mathbf{B} = \begin{bmatrix} \mathbf{B}_R^{\mathrm{T}}\mathbf{M}_a\mathbf{B}_R & \mathbf{B}_R^{\mathrm{T}}\mathbf{M}_a\mathbf{B}_\theta & \mathbf{B}_R^{\mathrm{T}}\mathbf{M}_a\mathbf{B}_f \\ & \mathbf{B}_\theta^{\mathrm{T}}\mathbf{M}_a\mathbf{B}_\theta & \mathbf{B}_\theta^{\mathrm{T}}\mathbf{M}_a\mathbf{B}_f \\ \text{symmetric} & & \mathbf{B}_f^{\mathrm{T}}\mathbf{M}_a\mathbf{B}_f \end{bmatrix} \tag{7.74}$$

The inertia shape integrals of Eq. 60 can be obtained by comparing Eqs. 72 and 74. The use of this procedure shows that the nonlinear mass matrix and the inertia shape integrals of the floating frame of reference formulation can be systematically evaluated using the constant mass matrix $\mathbf{M}_a$ and the velocity transformation matrix $\mathbf{B}$ as demonstrated by the following example.

Cubic Interpolating Polynomials Using the local shape function of Eq. 66 and the definitions of the constant matrices given by Eq. 60, the inertia shape integrals that appear in the nonlinear mass matrix of the floating frame of reference formulation can be evaluated as

$$\bar{\mathbf{S}} = \int_V \rho\,\mathbf{S}\,dV = \frac{m}{12}\begin{bmatrix} l & 6 & 0 & -l & 0 \\ 0 & 0 & 6 & 0 & -l \end{bmatrix}$$

$$\mathbf{M}_{ff} = \int_V \rho\mathbf{S}^{\mathrm{T}}\mathbf{S}\,dV = m\begin{bmatrix} \frac{(l)^2}{105} & \frac{13l}{420} & 0 & -\frac{(l)^2}{140} & 0 \\ \frac{13l}{420} & \frac{13}{35} & 0 & -\frac{11l}{210} & 0 \\ 0 & 0 & \frac{13}{35} & 0 & -\frac{11l}{210} \\ -\frac{(l)^2}{140} & -\frac{11l}{210} & 0 & \frac{(l)^2}{105} & 0 \\ 0 & 0 & -\frac{11l}{210} & 0 & \frac{(l)^2}{105} \end{bmatrix} \quad (7.75)$$

$$\tilde{\mathbf{S}} = \int_V \rho\mathbf{S}^{\mathrm{T}}\tilde{\mathbf{I}}\mathbf{S}\,dV = m\begin{bmatrix} 0 & 0 & \frac{13l}{420} & 0 & -\frac{(l)^2}{140} \\ 0 & 0 & \frac{13}{35} & 0 & -\frac{11l}{210} \\ -\frac{13l}{420} & -\frac{13}{35} & 0 & \frac{11l}{210} & 0 \\ 0 & 0 & -\frac{11l}{210} & 0 & \frac{(l)^2}{105} \\ \frac{(l)^2}{140} & \frac{11l}{210} & 0 & -\frac{(l)^2}{105} & 0 \end{bmatrix}$$

Differentiating Eq. 70 with respect to time, one obtains

$$\dot{\mathbf{e}} = \begin{bmatrix} 1 & 0 & 0 & 0 & 0 & 0 & 0 & 0 \\ 0 & 1 & 0 & 0 & 0 & 0 & 0 & 0 \\ 0 & 0 & -q_1\sin\theta & \cos\theta & 0 & 0 & 0 & 0 \\ 0 & 0 & q_1\cos\theta & \sin\theta & 0 & 0 & 0 & 0 \\ 1 & 0 & -q_2\sin\theta - q_3\cos\theta. & 0 & \cos\theta & -\sin\theta & 0 & 0 \\ 0 & 1 & q_2\cos\theta - q_3\sin\theta & 0 & \sin\theta & \cos\theta & 0 & 0 \\ 0 & 0 & -q_4\sin\theta - q_5\cos\theta & 0 & 0 & 0 & \cos\theta & -\sin\theta \\ 0 & 0 & q_4\cos\theta - q_5\sin\theta & 0 & 0 & 0 & \sin\theta & \cos\theta \end{bmatrix}\begin{bmatrix} \dot{R}_1 \\ \dot{R}_2 \\ \dot{\theta} \\ \dot{q}_1 \\ \dot{q}_2 \\ \dot{q}_3 \\ \dot{q}_4 \\ \dot{q}_5 \end{bmatrix} \quad (7.76)$$

which defines the velocity transformation matrix $\mathbf{B}$ as

$$\mathbf{B} = \begin{bmatrix} 1 & 0 & 0 & 0 & 0 & 0 & 0 & 0 \\ 0 & 1 & 0 & 0 & 0 & 0 & 0 & 0 \\ 0 & 0 & -q_1\sin\theta & \cos\theta & 0 & 0 & 0 & 0 \\ 0 & 0 & q_1\cos\theta & \sin\theta & 0 & 0 & 0 & 0 \\ 1 & 0 & -q_2\sin\theta - q_3\cos\theta & 0 & \cos\theta & -\sin\theta & 0 & 0 \\ 0 & 1 & q_2\cos\theta - q_3\sin\theta & 0 & \sin\theta & \cos\theta & 0 & 0 \\ 0 & 0 & -q_4\sin\theta - q_5\cos\theta & 0 & 0 & 0 & \cos\theta & -\sin\theta \\ 0 & 0 & q_4\cos\theta - q_5\sin\theta & 0 & 0 & 0 & \sin\theta & \cos\theta \end{bmatrix} \quad (7.77)$$

The constant matrix $\mathbf{M}_a$ that appears in the absolute nodal coordinate formulation (see Eq. 16) can be obtained using the global shape function of Eq. 6 as

$$
\mathbf{M}_a = \int_V \rho \mathbf{S}^{\mathrm{T}} \mathbf{S} \, dV = m \begin{bmatrix}
\frac{13}{35} & 0 & \frac{11l}{210} & 0 & \frac{9}{70} & 0 & -\frac{13l}{420} & 0 \\
0 & \frac{13}{35} & 0 & \frac{11l}{210} & 0 & \frac{9}{70} & 0 & -\frac{13l}{420} \\
\frac{11l}{210} & 0 & \frac{(l)^2}{105} & 0 & \frac{13l}{420} & 0 & -\frac{(l)^2}{140} & 0 \\
0 & \frac{11l}{210} & 0 & \frac{(l)^2}{105} & 0 & \frac{13l}{420} & 0 & -\frac{(l)^2}{140} \\
\frac{9}{70} & 0 & \frac{13l}{420} & 0 & \frac{13}{35} & 0 & -\frac{11l}{210} & 0 \\
0 & \frac{9}{70} & 0 & \frac{13l}{420} & 0 & \frac{13}{35} & 0 & -\frac{11l}{210} \\
-\frac{13l}{420} & 0 & -\frac{(l)^2}{140} & 0 & -\frac{11l}{210} & 0 & \frac{(l)^2}{105} & 0 \\
0 & -\frac{13l}{420} & 0 & -\frac{(l)^2}{140} & 0 & -\frac{11l}{210} & 0 & \frac{(l)^2}{105}
\end{bmatrix} \tag{7.78}
$$

Using this matrix and Eq. 76, it can be shown that

$$
\mathbf{M}_f = \mathbf{B}^{\mathrm{T}} \mathbf{M}_a \mathbf{B}
$$

$$
= \begin{bmatrix}
m\mathbf{I} & \mathbf{A}_\theta \begin{bmatrix} \frac{ml}{12} & \frac{m}{2} & 0 & -\frac{ml}{12} & 0 \\ 0 & 0 & \frac{m}{2} & 0 & -\frac{ml}{12} \end{bmatrix} \mathbf{q}_f & \mathbf{A} \begin{bmatrix} \frac{ml}{12} & \frac{m}{2} & 0 & -\frac{ml}{12} & 0 \\ 0 & 0 & \frac{m}{2} & 0 & -\frac{ml}{12} \end{bmatrix} \\
 & m\mathbf{q}_f^{\mathrm{T}} \begin{bmatrix} \frac{(l)^2}{105} & \frac{13l}{420} & 0 & -\frac{(l)^2}{140} & 0 \\ \frac{13l}{420} & \frac{13}{35} & 0 & -\frac{11l}{210} & 0 \\ 0 & 0 & \frac{13}{35} & 0 & -\frac{11l}{210} \\ -\frac{(l)^2}{140} & -\frac{11l}{210} & 0 & \frac{(l)^2}{105} & 0 \\ 0 & 0 & -\frac{11l}{210} & 0 & \frac{(l)^2}{105} \end{bmatrix} \mathbf{q}_f & m\mathbf{q}_f^{\mathrm{T}} \begin{bmatrix} 0 & 0 & \frac{13l}{420} & 0 & -\frac{(l)^2}{140} \\ 0 & 0 & \frac{13}{35} & 0 & -\frac{11l}{210} \\ -\frac{13l}{420} & -\frac{13}{35} & 0 & \frac{11l}{210} & 0 \\ 0 & 0 & -\frac{11l}{210} & 0 & \frac{(l)^2}{105} \\ \frac{(l)^2}{140} & \frac{11l}{210} & 0 & -\frac{(l)^2}{105} & 0 \end{bmatrix} \\
 & \text{symmetric} & m \begin{bmatrix} \frac{(l)^2}{105} & \frac{13l}{420} & 0 & -\frac{(l)^2}{140} & 0 \\ \frac{13l}{420} & \frac{13}{35} & 0 & -\frac{11l}{210} & 0 \\ 0 & 0 & \frac{13}{35} & 0 & -\frac{11l}{210} \\ -\frac{(l)^2}{140} & -\frac{11l}{210} & 0 & \frac{(l)^2}{105} & 0 \\ 0 & 0 & -\frac{11l}{210} & 0 & \frac{(l)^2}{105} \end{bmatrix}
\end{bmatrix} \tag{7.79}
$$

Comparing this matrix with Eq. 72, the shape integrals presented in Eq. 75 can be easily identified, demonstrating the equivalence of the inertia forces used in the two formulations. This example also demonstrates that the nonlinear mass matrix and all the inertia shape integrals of the floating frame of reference formulation can be obtained from the constant consistent mass matrix used in linear structural dynamics.

Rigid Body Inertia In the case of the consistent mass formulation, exact modeling of the rigid body inertia of the beam can be obtained as a special case of the more general development presented in this section. In the case of a rigid body motion, one has

$$
q_1 = q_4 = 1, \quad q_2 = l, \quad q_3 = q_5 = 0 \tag{7.80}
$$

In this special case, the transformation of Eq. 76 reduces to

$$\dot{\mathbf{e}} = \mathbf{B}\dot{\mathbf{q}} = \begin{bmatrix} 1 & 0 & 0 \\ 0 & 1 & 0 \\ 0 & 0 & -\sin\theta \\ 0 & 0 & \cos\theta \\ 1 & 0 & -l\sin\theta \\ 0 & 1 & l\cos\theta \\ 0 & 0 & -\sin\theta \\ 0 & 0 & \cos\theta \end{bmatrix} \begin{bmatrix} \dot{R}_1 \\ \dot{R}_2 \\ \dot{\theta} \end{bmatrix} \tag{7.81}$$

Using the velocity transformation matrix in this equation and the mass matrix $\mathbf{M}_a$ of Eq. 78, it can be shown that in the case of a rigid body motion the mass matrix of the element reduces to

$$\mathbf{M}_f = \mathbf{B}^{\mathrm{T}}\mathbf{M}_a\mathbf{B} = m \begin{bmatrix} 1 & 0 & -\frac{l}{2}\sin\theta \\ 0 & 1 & \frac{l}{2}\cos\theta \\ -\frac{l}{2}\sin\theta & \frac{l}{2}\cos\theta & \frac{(l)^2}{3} \end{bmatrix} \tag{7.82}$$

which is the exact mass matrix of the rigid beam in the case of a noncentroidal beam coordinate system.

7.8 THE VELOCITY TRANSFORMATION MATRIX

Equation 71, which demonstrates the equivalence of the kinematic relationships used in the floating frame of reference formulation and the absolute nodal coordinate formulation, can be used to develop several interesting matrix identities. For example, by differentiating Eq. 71 with respect to time and using the partitioning of the velocity transformation matrix given by Eq. 73, one can show that the following simple relationships between the global and the local shape functions hold:

$$\mathbf{S}\mathbf{B}_R = \mathbf{I}, \qquad \mathbf{S}\mathbf{B}_\theta = \mathbf{A}_\theta\mathbf{S}_l\mathbf{q}_f, \qquad \mathbf{S}\mathbf{B}_f = \mathbf{A}\mathbf{S}_l \tag{7.83}$$

These relationships can be used to obtain other interesting matrix identities. For example, the first identity in Eq. 83 leads to

$$\mathbf{B}_R^{\mathrm{T}}\mathbf{S}^{\mathrm{T}}\mathbf{S}\mathbf{B}_R = \mathbf{I} \tag{7.84}$$

Multiplying this equation by the mass density ρ and integrating over the volume, one obtains

$$\mathbf{B}_R^{\mathrm{T}}\mathbf{M}_a\mathbf{B}_R = m\mathbf{I} \tag{7.85}$$

a relationship previously obtained in Eq. 79.

Equation 71 can also be used to obtain an alternative procedure for formulating the velocity transformation matrix that relates the derivatives of the coordinates used in the floating frame of reference formulation and the absolute nodal coordinate formulation. Differentiating this equation with respect to time, one obtains

$$\mathbf{S}\dot{\mathbf{e}} = \mathbf{L}\dot{\mathbf{q}} \tag{7.86}$$

where

$$\mathbf{L} = [\mathbf{I} \quad \mathbf{A}_\theta \mathbf{S}_l \mathbf{q}_f \quad \mathbf{A}\mathbf{S}_l], \qquad \mathbf{q} = \begin{bmatrix} \mathbf{R}^{\mathrm{T}} & \theta & \mathbf{q}_f^{\mathrm{T}} \end{bmatrix}^{\mathrm{T}} \tag{7.87}$$

Premultiplying Eq. 86 by the transpose of the global shape function **S** and integrating over the volume, it can be shown that

$$\mathbf{M}_a \dot{\mathbf{e}} = \mathbf{M}_B \dot{\mathbf{q}} \tag{7.88}$$

where

$$\mathbf{M}_B = \int_V \rho \mathbf{S}^{\mathrm{T}} \mathbf{L}\, dV \tag{7.89}$$

Since $\mathbf{M}_a$ is a positive definite matrix, one can write

$$\dot{\mathbf{e}} = \mathbf{M}_a^{-1} \mathbf{M}_B \dot{\mathbf{q}} = \mathbf{B} \dot{\mathbf{q}} \tag{7.90}$$

where **B** is the velocity transformation matrix defined as

$$\mathbf{B} = \mathbf{M}_a^{-1} \mathbf{M}_B \tag{7.91}$$

This equation defines the velocity transformation matrix **B** in terms of the inverse of the constant mass matrix $\mathbf{M}_a$.

7.9 LUMPED MASS FORMULATION

The analysis presented in the preceding sections demonstrates that the absolute nodal coordinate formulation leads to an exact modeling of the rigid body motion provided that the global element shape function has a complete set of rigid body modes and a consistent mass formulation is used to describe the element inertia. This is not, however, the case when the lumped mass approach is used as demonstrated in this section. First, some basic concepts in rigid body dynamics are reviewed.

Rigid Body Mechanics To demonstrate some of the modeling problems that result from the use of the lumped mass formulation, we consider the slender beam shown in Fig. 5a. The beam is assumed to have mass m, mass moment of inertia I_O about its end point O, length l, volume V, and mass density ρ. The motion of the beam can be described using three coordinates: two coordinates R_1 and R_2, which describe the global position vector of point O, and the angle θ, which defines the orientation of the beam with respect to the inertial frame. In this case of rigid body motion, the vector of the beam generalized coordinates is defined as

$$\mathbf{q} = [R_1 \quad R_2 \quad \theta]^{\mathrm{T}} \tag{7.92}$$

Using this set of generalized coordinates, one can show that the inertia matrix of the beam is defined in the case of rigid body motion by Eq. 82.

Another model for the beam is shown in Fig. 5b. In this model the distributed inertia of the beam is replaced by two concentrated masses, each equal to $m/2$, at the two ends of the beam (nodal points). Using this model, one can show that the inertia

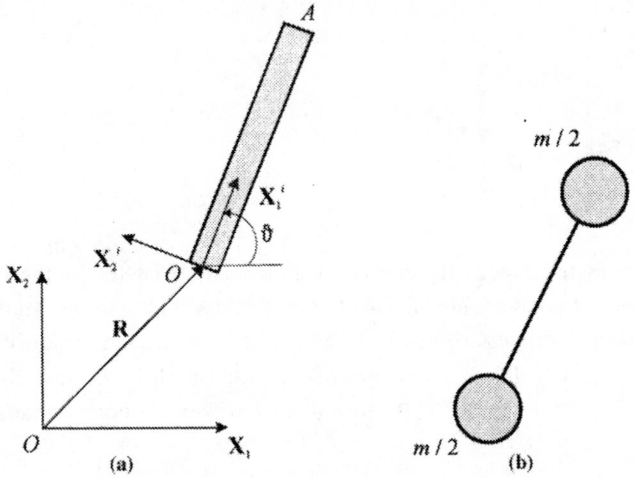

Figure 7.5 Lumped and consistent masses.

matrix of the rigid beam obtained using the lumped mass model is given by

$$\mathbf{M} = \begin{bmatrix} m & 0 & -\frac{ml}{2}\sin\theta \\ 0 & m & \frac{ml}{2}\cos\theta \\ -\frac{ml}{2}\sin\theta & \frac{ml}{2}\cos\theta & \frac{m(l)^2}{2} \end{bmatrix} \tag{7.93}$$

Comparing this mass matrix with the mass matrix previously obtained using the distributed inertia of the beam, it is clear that the lumped mass model shown in Fig. 5b does not lead to exact modeling of the mass matrix of the beam in the case of an arbitrary rigid body motion.

Use of Trigonometric Functions as Generalized Coordinates The equations of motion of the rigid beam can also be formulated in terms of $\sin\theta$ and $\cos\theta$ as coordinates instead of the angle θ. In this case, one has a redundant set of coordinates since the sine and cosine functions of an angle are not independent. Using the trigonometric functions as generalized coordinates, one can write the position vector of an arbitrary point on the slender rigid beam as

$$\mathbf{r} = \begin{bmatrix} r_1 \\ r_2 \end{bmatrix} = \begin{bmatrix} 1 & 0 & x & 0 \\ 0 & 1 & 0 & x \end{bmatrix} \begin{bmatrix} R_1 \\ R_2 \\ \cos\theta \\ \sin\theta \end{bmatrix} \tag{7.94}$$

where x is the spatial coordinate along the beam axis. Using this kinematic equation and a distributed mass model, it can be shown that the mass matrix associated with the new set of coordinates

$$\mathbf{q} = [R_1 \quad R_2 \quad \cos\theta \quad \sin\theta]^{\mathrm{T}} \tag{7.95}$$

is given by

$$\mathbf{M} = \begin{bmatrix} m & 0 & \frac{ml}{2} & 0 \\ 0 & m & 0 & \frac{ml}{2} \\ \frac{ml}{2} & 0 & \frac{m(l)^2}{3} & 0 \\ 0 & \frac{ml}{2} & 0 & \frac{m(l)^2}{3} \end{bmatrix} \tag{7.96}$$

Note that in this mass matrix, the rotary inertia coefficients associated with the sine and cosine function coordinates are equal. This fact will be used later in this section to define a lumped mass matrix associated with the coordinates used in the absolute nodal coordinate formulation. In the remainder of this section, the approximations that result from the use of the lumped mass formulation in flexible body dynamics will be examined.

Flexible Body Lumped Mass Matrix In the lumped mass approach, we assume, in the case of the absolute nodal coordinate formulation, a diagonal mass matrix in the form

$$\mathbf{M}_l = \begin{bmatrix} \frac{m}{2} & 0 & 0 & 0 & 0 & 0 & 0 & 0 \\ 0 & \frac{m}{2} & 0 & 0 & 0 & 0 & 0 & 0 \\ 0 & 0 & J_1 & 0 & 0 & 0 & 0 & 0 \\ 0 & 0 & 0 & J_1 & 0 & 0 & 0 & 0 \\ 0 & 0 & 0 & 0 & \frac{m}{2} & 0 & 0 & 0 \\ 0 & 0 & 0 & 0 & 0 & \frac{m}{2} & 0 & 0 \\ 0 & 0 & 0 & 0 & 0 & 0 & J_2 & 0 \\ 0 & 0 & 0 & 0 & 0 & 0 & 0 & J_2 \end{bmatrix} \tag{7.97}$$

where J_1 and J_2 are rotary inertia coefficients defined, respectively, at the first and second nodes of the beam element. Using the transformation of Eq. 77, it can be shown that the mass matrix obtained for the floating frame of reference formulation using the preceding diagonal mass matrix is given by

$$\mathbf{M}_f = \mathbf{B}^{\mathrm{T}}\mathbf{M}_l\mathbf{B}$$

$$= \begin{bmatrix} m\mathbf{I} & \mathbf{A}_\theta \begin{bmatrix} 0 & \frac{m}{2} & 0 & 0 & 0 \\ 0 & 0 & \frac{m}{2} & 0 & 0 \end{bmatrix} \mathbf{q}_f & \mathbf{A} \begin{bmatrix} 0 & \frac{m}{2} & 0 & 0 & 0 \\ 0 & 0 & \frac{m}{2} & 0 & 0 \end{bmatrix} \\ & m\mathbf{q}_f^{\mathrm{T}} \begin{bmatrix} J_1 & 0 & 0 & 0 & 0 \\ 0 & \frac{m}{2} & 0 & 0 & 0 \\ 0 & 0 & \frac{m}{2} & 0 & 0 \\ 0 & 0 & 0 & J_2 & 0 \\ 0 & 0 & 0 & 0 & J_2 \end{bmatrix} \mathbf{q}_f & m\mathbf{q}_f^{\mathrm{T}} \begin{bmatrix} 0 & 0 & 0 & 0 & 0 \\ 0 & 0 & \frac{m}{2} & 0 & 0 \\ 0 & -\frac{m}{2} & 0 & 0 & 0 \\ 0 & 0 & 0 & 0 & J_2 \\ 0 & 0 & 0 & -J_2 & 0 \end{bmatrix} \\ \text{symmetric} & & m \begin{bmatrix} J_1 & 0 & 0 & 0 & 0 \\ 0 & \frac{m}{2} & 0 & 0 & 0 \\ 0 & 0 & \frac{m}{2} & 0 & 0 \\ 0 & 0 & 0 & J_2 & 0 \\ 0 & 0 & 0 & 0 & J_2 \end{bmatrix} \end{bmatrix} \tag{7.98}$$

Comparing this matrix with Eq. 72, it is clear that the inertia shape integrals can be written in the case of the lumped mass formulation as

$$\bar{\mathbf{S}} = \frac{m}{2}\begin{bmatrix} 0 & 1 & 0 & 0 & 0 \\ 0 & 0 & 1 & 0 & 0 \end{bmatrix}$$

$$\mathbf{M}_{ff} = \begin{bmatrix} J_1 & 0 & 0 & 0 & 0 \\ 0 & \frac{m}{2} & 0 & 0 & 0 \\ 0 & 0 & \frac{m}{2} & 0 & 0 \\ 0 & 0 & 0 & J_2 & 0 \\ 0 & 0 & 0 & 0 & J_2 \end{bmatrix}, \quad \tilde{\mathbf{S}} = \begin{bmatrix} 0 & 0 & 0 & 0 & 0 \\ 0 & 0 & \frac{m}{2} & 0 & 0 \\ 0 & -\frac{m}{2} & 0 & 0 & 0 \\ 0 & 0 & 0 & 0 & J_2 \\ 0 & 0 & 0 & -J_2 & 0 \end{bmatrix} \tag{7.99}$$

These shape integrals have a structure significantly different from those integrals previously obtained using the consistent mass formulation that was presented in Eq. 75.

Rigid Body Motion Regardless of the values of the inertia coefficients J_1 and J_2, it can be demonstrated that the lumped mass formulation as defined by the diagonal mass matrix of Eq. 97 does not lead to exact modeling of the rigid body inertia when the element rotates as a rigid body. To demonstrate this fact, we consider the simple case of a rigid body motion. Using the transformation of Eq. 81 and the lumped mass matrix of Eq. 97, it can be shown that

$$\mathbf{M}_f = \mathbf{B}^{\mathrm{T}}\mathbf{M}_l\mathbf{B} = \begin{bmatrix} m & 0 & -\frac{ml}{2}\sin\theta \\ 0 & m & \frac{ml}{2}\cos\theta \\ -\frac{ml}{2}\sin\theta & \frac{ml}{2}\cos\theta & J_1 + J_2 + \frac{m(l)^2}{2} \end{bmatrix} \tag{7.100}$$

While the consistent mass formulation leads to exact modeling of the mass matrix in the case of a rigid body motion, the preceding equation demonstrates that the lumped mass formulation does not lead to the mass matrix of the rigid element if both the inertia coefficients J_1 and J_2 are assumed to be positive.

7.10 EXTENSION OF THE METHOD

As previously pointed out, the floating frame of reference formulation, which was the subject of the preceding two chapters, has been extensively used in a variety of flexible multibody applications. Its use, however, has been limited to the analysis of small deformations of flexible bodies that undergo large reference displacements. This limitation was the result of the use of infinitesimal rotations as nodal coordinates to define the deformation of the element with respect to the selected body coordinate system. The use of infinitesimal rotations as nodal coordinates leads to a linearization of the kinematic equations. As a consequence, only linear modes are often used with the floating frame of reference formulation. The use of the reference coordinates allows one to obtain an exact modeling of the rigid body motion when the elastic deformation is equal to zero. With the mixed set of coordinates used in the floating frame of reference formulation, nonisoparametric elements produce zero strains under an arbitrary rigid body displacement.

To be able to solve large deformation problems in multibody system applications, a conceptually different approach is used in this chapter. In the absolute nodal coordinate formulation presented in this chapter, no infinitesimal or finite rotations are used as nodal coordinates; instead, a mixed set of absolute displacements and slopes is used. The absolute nodal coordinate formulation leads to an exact modeling of the rigid body dynamics, and hence, beams and plates can be considered in this formulation as isoparametric elements. The absolute nodal coordinate formulation leads to a constant inertia matrix, and as a consequence, the Coriolis and centrifugal forces are equal to zero.

In the analysis presented in this chapter, a planar beam element is used to demonstrate the use of the absolute nodal coordinate formulation. This formulation can also be systematically developed for other element types as well as spatial elements for the large deformation analysis of mechanical systems. It can also be applied to multibody system applications by developing a computer library of constraints that describe the joints between the finite elements and the deformable bodies. Using absolute coordinates, some of these joints take simple forms as compared with the joint formulations used in the floating frame of reference formulation. Also, the formulation of the spring, damper, and actuator forces is much simpler when the absolute nodal coordinate formulation is used. Effects such as rotary inertia in beams can be easily accommodated in the absolute nodal coordinate formulation by modifying the element shape functions (Shabana 1996c).

In the case of spatial elements, a procedure similar to the one presented in this chapter can be used to define the relationship between the local and global slopes (Shabana and Christensen 1997). In this case, the following three-dimensional kinematic equation can be used:

$$\mathbf{r} = \mathbf{R} + \mathbf{A}\bar{\mathbf{u}} \tag{7.101}$$

where all the vectors and matrices in this equation are three-dimensional and are as defined previously in this book and $\mathbf{A}$ is the spatial transformation matrix. This transformation matrix can be defined using any set of the orientation parameters introduced in Chapter 2. Using the preceding equation, the relationship between the absolute and local slopes can be defined as

$$\frac{\partial \mathbf{r}}{\partial x_1} = \mathbf{A}\frac{\partial \bar{\mathbf{u}}}{\partial x_1}, \qquad \frac{\partial \mathbf{r}}{\partial x_2} = \mathbf{A}\frac{\partial \bar{\mathbf{u}}}{\partial x_2}, \qquad \frac{\partial \mathbf{r}}{\partial x_3} = \mathbf{A}\frac{\partial \bar{\mathbf{u}}}{\partial x_3} \tag{7.102}$$

where x_1, x_2, and x_3 are the spatial coordinates of the element. The three-dimensional slope relationships can be used to develop a transformation between the coordinates used in the floating frame of reference formulation and the absolute nodal coordinate formulation. The equivalence of these two formulations is one of the important results obtained in this chapter. The significance of this result stems from the fact that it clearly demonstrates that the use of the floating frame of reference formulation does not imply a separation between the rigid body motion and the elastic deformation. In the floating frame of reference formulation, the reference motion cannot be interpreted as the rigid body motion since different frames of reference can be selected for the deformable body (Shabana 1996a). If slopes, instead of infinitesimal rotations, are used as nodal coordinates in the floating frame of reference formulation, the equivalence of the simple elastic forces used in that formulation and the highly nonlinear elastic forces used in the absolute nodal coordinate formulation can also be demonstrated. As a consequence, the transformations between the coordinates used in the two formulations can also be used to obtain the expressions for the highly nonlinear stiffness forces used in the absolute nodal coordinate formulation from the simple stiffness matrix that appears in the floating frame of reference formulation. By so doing, one can develop an efficient algorithm that utilizes existing finite-element computer programs to evaluate the elastic forces in the absolute nodal coordinate formulation. The coordinate transformations can be developed for other element types using a procedure similar to the one described in this chapter. Using these coordinate transformations, existing finite-element computer programs can easily be modified to implement the absolute nodal coordinate formulation.

The floating frame of reference formulation has been implemented in several commercial and research computer programs and has been widely used in the analysis of many mechanical system applications. Examples of these applications can be found in the References section. To demonstrate the application of the absolute nodal coordinate formulation to multibody systems and compare the obtained numerical results with the results obtained using the floating frame of reference formulation, we consider the slider crank mechanism shown in Fig. 6 (Escalona et al. 1997). The crankshaft has a length of 0.152 m, a cross-sectional area of 7.854×10^{-5} m^2, a second moment of area of 4.909×10^{-10} m^4, a mass density of 2.770×10^3 kg/m^3, and a modulus of elasticity of 1.0×10^9 N/m^2. The connecting rod is assumed to be a beam of length 0.304 m, and has the same cross-sectional dimensions and material properties as the crankshaft with the exception of the modulus of elasticity, which is assumed to be 0.5×10^8 N/m^2. The crankshaft of the slider crank mechanism is assumed to be driven by the following torque expressed in N/m:

$$M(t) = \begin{cases} 0.01\left(1 - e^{\frac{-t}{0.167}}\right) & t \le 0.70\,\text{s} \\ 0 & t \ge 0.70\,\text{s} \end{cases}$$

In the numerical study presented in this section, the crankshaft is divided into three elements, while the connecting rod is divided into eight elements. In the floating frame of reference formulation, the shape function of Eq. 11 is used in the finite-element

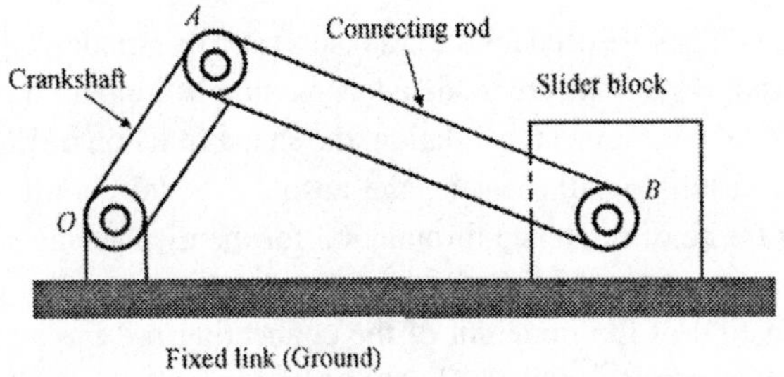

Figure 7.6 Slider crank mechanism.

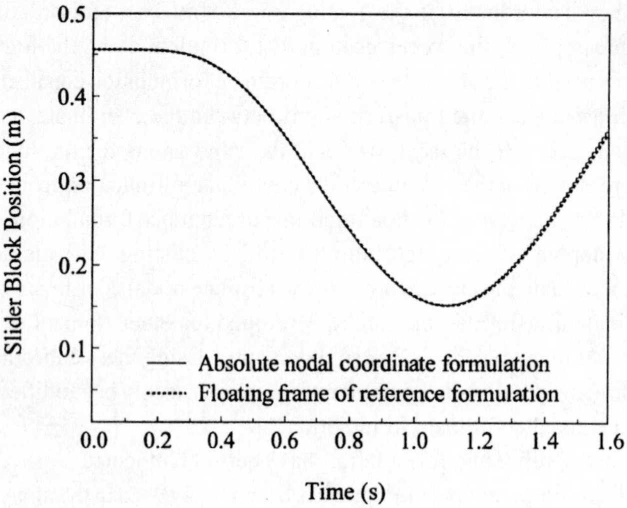

Figure 7.7 Position of the slider block.

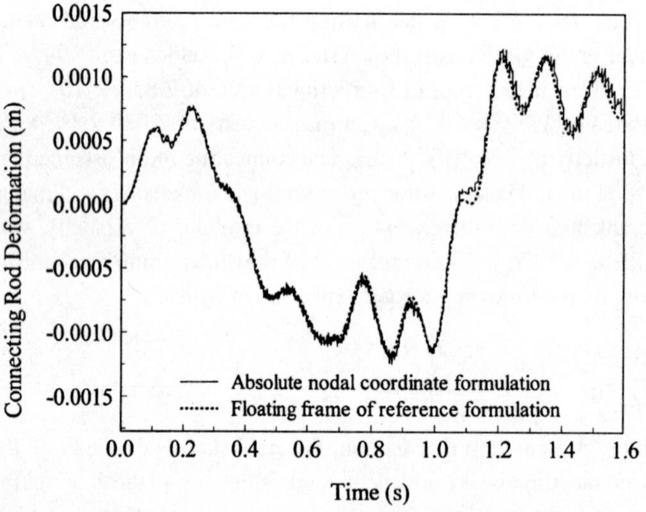

Figure 7.8 Deformation of the midpoint of the connecting rod.

discretization, and three mode shapes are used for the crankshaft and five mode shapes are used for the connecting rod. Both links are modeled using simply supported end conditions. In the absolute nodal coordinate formulation, the shape function of Eq. 6 is used. Figure 7 shows the results obtained using the absolute nodal coordinate formulation and the floating frame of reference formulation for the displacement of the slider block. The results presented in this figure show a good agreement. Figure 8 shows the transverse deformation of the midpoint of the connecting rod measured with respect to a line connecting points A and B. These results again show excellent agreement between the two methods. Observe the high-frequency oscillations in the solution obtained using the absolute nodal coordinate formulation. These oscillations are due to the fact that no modal reduction is used in the absolute nodal coordinate formulation. It is important to point out that the excellent agreement between the two methods can be obtained in the case of small deformations. As expected, numerical experimentation showed discrepancy between the solutions obtained using the two formulations in the case of large deformation analysis.

7.11 COMPARISON WITH LARGE ROTATION VECTOR FORMULATION

In the preceding sections, the equivalence of the floating frame of reference and absolute nodal coordinate formulations was demonstrated. The numerical results presented in the preceding section also indicate that there is a good agreement between the numerical solutions obtained using the two formulations when small deformation problems of flexible multibody systems are considered. Since modal reduction techniques are often used with the floating frame of reference formulation, its use has been limited primarily to small deformation problems. In recent years, several attempts have been made to develop finite-element formulations for the large deformation analysis of flexible multibody systems. Among these formulations is the *large rotation vector formulation* (Simo and Vu-Quoc 1986a, b). In this formulation, absolute coordinates and finite rotations are used as the nodal coordinates of the finite element. As a consequence, the coordinates of an arbitrary point on the element as well as the finite rotation of the cross-section are interpolated using polynomials with independent coefficients. The use of the large rotation vector formulations often leads to excessive shear forces leading to a phenomenon known as *shear locking*. While several attempts have been devoted to investigate the excessive shear problem in the large rotation vector formulation, most of these attempts were focused on finding a numerical solution rather than addressing the fundamental problem associated with the kinematic description used in the large rotation vector formulations. In this section, some basic concepts used in the kinematic description of deformable bodies are reviewed before the large rotation vector formulation is discussed. These concepts are primarily related to the linear independence of the kinematic variables and are closely related to the definition of the shear used in some finite-element large rotation vector formulations.

Background Consider the two functions f and g, which may represent kinematic variables of a two-dimensional beam. These two functions are assumed to depend on the spatial coordinate x, which defines the position of an arbitrary point on the beam in the undeformed configuration. Using the Raleigh–Ritz approximation, the functions f and g can be expressed in terms of a finite set of coordinates as

$$\left.\begin{aligned} f &= \sum_{i=1}^{m} \phi_i(x) q_i(t) \\ g &= \sum_{i=1}^{m} \phi_i(x) q_i(t) + \sum_{j=1}^{n} h_j(x) p_j(t) \end{aligned}\right\} \tag{7.103}$$

where ϕ_i and h_j are space-dependent shape functions or eigenfunctions, and q_i and p_j are the time-dependent coordinates. These coordinates are assumed to be independent. The function g in Eq. 103 shows dependence on the function f; that is, the change in f is automatically included in g. Nonetheless, f and g are independent functions since both of them can be changed arbitrarily. The representation given in Eq. 103 is conceptually different from the following representation of f and g:

$$\left.\begin{aligned} f &= \sum_{i=1}^{m} \phi_i(x) q_i(t) \\ g &= \sum_{j=1}^{n} h_j(x) p_j(t) \end{aligned}\right\} \tag{7.104}$$

where in this equation f and g remain independent, but the change in f has no kinematic effect on g since the modes of variations of f are not represented in g.

The representation of Eq. 104 defines a different model from the representation given by Eq. 103. For instance, if all the p_j's are equal to zero, the result of Eq. 103 states that f and g are equal and there is no need for applying a force to insure this equality. This is not, however, the case when Eq. 104 is used since there is no kinematic relationship between the two variables described in this equation. Note that in Eq. 103 g can be different from f if the p_j's are not all equal to zero. Furthermore, if f and g describe displacements or rotations, the generalized forces resulting from the use of the representations of Eqs. 103 and 104 are completely different. For example, if f and g represent displacements, and F is a force acting in the direction of g, then the virtual work due to this force can be written as

$$\delta W = F\delta g = \sum_{i=1}^{m} F\phi_i(x)\delta q_i(t) + \sum_{j=1}^{n} F h_j(x)\delta p_j(t)$$

which defines generalized forces associated with the coordinates that are used to define g as well as the function f. This will not be the case if the representation of Eq. 104 is used.

Large Rotation Vector Formulations In this section, the kinematic description used in the large rotation vector formulations is presented and discussed in view of the simple concepts previously discussed in this section. To this end, a simple two-dimensional beam model is used. As previously pointed out, in the large rotation vector formulations, finite rotations are used as field variables leading to a set of nodal coordinates that consist of displacement coordinates as well as finite rotation coordinates. To explain the basic idea used in the large rotation vector formulations, we consider the two-dimensional beam shown in Fig. 9. In this figure, α defines the orientation of the cross-section without the shear; that is, α defines the direction of the normal to the center line of the beam. The orientation of the coordinate system defined by the tangent vector $\mathbf{t}$ and the normal vector $\mathbf{n}$ can be defined in terms of the angle α using the transformation matrix $\mathbf{A}_n$ defined as

$$\mathbf{A}_n = \begin{bmatrix} \cos\alpha & -\sin\alpha \\ \sin\alpha & \cos\alpha \end{bmatrix}$$

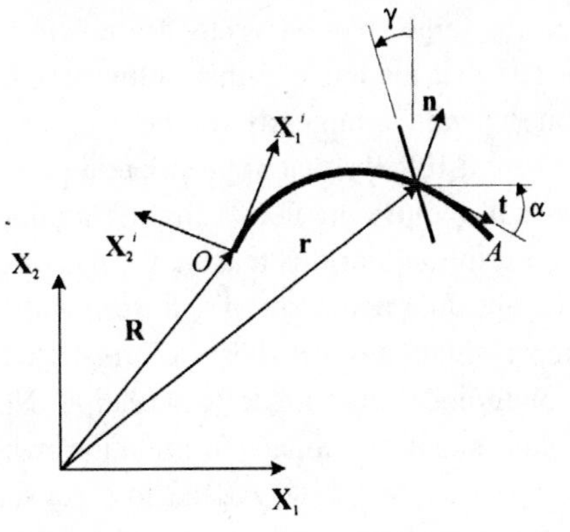

Figure 7.9 Large rotation vector formulation.

where $\cos\alpha$ and $\sin\alpha$ can be expressed in terms of the components r_1 and r_2 of the vector **r** that defines the location of an arbitrary point on the beam as

$$\left.\begin{aligned} \cos\alpha &= \frac{\frac{\partial r_1}{\partial x}}{\sqrt{\left(\frac{\partial r_1}{\partial x}\right)^2 + \left(\frac{\partial r_2}{\partial x}\right)^2}} \\ \sin\alpha &= \frac{\frac{\partial r_2}{\partial x}}{\sqrt{\left(\frac{\partial r_1}{\partial x}\right)^2 + \left(\frac{\partial r_2}{\partial x}\right)^2}} \end{aligned}\right\} \tag{7.105}$$

which show that the angle α is completely defined by the components of the vector **r**. Note that this angle, in the case of large rotation problems, does not approach zero as the element length decreases since absolute coordinates are used.

The rotation of the cross-section as the result of the shear deformation is defined by the angle β. Therefore, the total rotation of the cross-section of the beam γ, as shown in Fig. 9, is defined as

$$\gamma = \alpha + \beta \tag{7.106}$$

In the large rotation vector formulation, the configuration of the finite element is described using the three field variables r_1, r_2, and γ. Interpolation functions with completely independent coefficients are used to describe these three field variables. As a consequence of this independent representation, there is no kinematic relationship between the two angles α and γ. That is, the polynomial coefficients that define r_1 and r_2 are not represented in the expansion of the angle γ that defines the total rotation of the cross-section of the beam. In other words, the modes of variations of the angle α are not included in the polynomial expansion used to describe the angle γ. This leads to inconsistent representation of the rotations of the cross-section, and such inconsistency can lead to serious modeling problems since the shear angle is assumed to be small. Because of this problem, large rotation vector formulations can lead to excessive shear forces because the angle that defines the orientation of the normal to the center line has modes of variation that are not included in the bigger expansion used for the angle that defines the orientation of the cross-section which also includes the shear effect.

Solving this inconsistency problem in the large rotation vector formulation is not an easy problem. It is clear from Eq. 105 that the angle α that defines the orientation of the normal to the cross-section depends nonlinearly on the coefficients of the polynomials used to describe r_1 and r_2. It follows that appropriate representation of the angle γ as given by Eq. 106 will require the use of highly nonlinear equations in the finite-element kinematic description. For this reason, the use of the finite rotation coordinates is avoided in the absolute nodal coordinate formulation; instead, global slopes are used. A proper representation of the shear deformation can be systematically incorporated in the absolute nodal coordinate formulation. Note also that the description used in the absolute nodal coordinate formulation, when three-dimensional slender beam elements are used, will always lead to a constant mass matrix. Use of finite rotations as nodal coordinates leads to a nonlinear mass matrix when three-dimensional elements are used.

Problems

1. A linear isoparametric displacement field of a beam element is defined as

$$r_1 = (1-\xi)e_1 + \xi e_3, \qquad r_2 = (1-\xi)e_2 + \xi e_4$$

where $\xi = x/l$, e_1 and e_2 are the absolute nodal displacements of the first node, and e_3 and e_4 are the absolute nodal displacements of the second node. Show that this isoparametric displacement field can be used to describe an arbitrary rigid body displacement.

2. In the case of a rigid body motion of an element defined by the shape function of Eq. 11, determine the error in the kinetic energy and the mass moment of inertia if infinitesimal rotations, instead of slopes, are used as nodal coordinates.

3. Using the shape function of Eq. 6, determine the error in the rigid body kinematic equations if infinitesimal rotations, instead of the slopes, are used as nodal coordinates.

4. Repeat Problem 2 using the cubic shape function of Eq. 6.

5. Define the components of the unit vectors along the axes of a beam element coordinate system whose origin is rigidly attached to the first node of the element.

6. In Problem 5, define the unit vectors along the axes of the element coordinate system in terms of the nodal coordinates if the shape function of Eq. 6 is used.

7. Repeat Problem 6 using the shape function of Eq. 11.

8. In Example 2, determine the error in the elastic forces in the case of a rigid body motion when infinitesimal rotations are used.

9. Show that when the rigid body motion of the element is equal to zero, the strain energy expression obtained using the absolute nodal coordinate formulation leads to the conventional beam stiffness matrix used in linear structural dynamics.

10. Discuss the physical meaning of the vectors and matrices that appear in the strain energy expression obtained using the absolute nodal coordinate formulation for the beam element.

Discuss the significance of these vectors and matrices as the orientation of the element changes.

11. Evaluate the vectors and matrices that appear in Eq. 35 when the shape function of Eq. 11 is used.

12. Using the absolute nodal coordinate formulation, the displacement of the element in a global coordinate system can be defined as

$$\mathbf{u}_g = \mathbf{S}(\mathbf{e} - \mathbf{e}_o)$$

where $\mathbf{S}$ is the global element shape function, $\mathbf{e}$ is the vector of absolute nodal coordinates, and $\mathbf{e}_o$ is the vector of the absolute nodal coordinates in the undeformed reference configuration. Using the global displacement $\mathbf{u}_g$ and the shape function of Eq. 6, evaluate the matrix of the displacement gradients as discussed in Chapter 4. Outline the general continuum mechanics procedure for evaluating the stiffness coefficients.

13. In the preceding problem, determine the matrix of the displacement gradients in the case of an arbitrary rigid body displacement.

14. Use Eq. 36 or 37 to write

$$\frac{\partial U}{\partial \mathbf{e}} = \mathbf{K}_a \mathbf{e}$$

where $\mathbf{K}_a$ is the element stiffness matrix. Discuss the nonlinearity of the stiffness matrix and the effect of the element rotation on the form of this matrix.

15. A force vector $\mathbf{F} = [F_1 \; F_2]^{\mathrm{T}}$ acts at the midpoint of a beam element defined by the shape function of Eq. 6. Determine the generalized forces associated with the generalized absolute nodal coordinates of this element due to the application of this force vector.

16. A moment M applies at the second node of the beam element defined by the shape function of Eq. 6. Derive an expression for the generalized forces associated with the absolute nodal coordinates of the element due to the application of this moment.

17. Repeat Problem 16 if the moment applies at the midpoint of the element.

18. Discuss the formulation of the generalized forces due to a spring–damper–actuator element that connects two finite elements using the absolute nodal coordinate formulation.

19. Using the absolute nodal coordinate formulation, define the constraint equations of the revolute joint that connects two finite elements in the planar analysis.

20. Formulate the constraint equations of the spherical joint between two finite elements in the spatial analysis when the absolute nodal coordinate formulation is used.

Appendix
LINEAR ALGEBRA

A.1 MATRIX ALGEBRA

In this section we summarize some results from *matrix algebra* that are useful in our development in this book. Most of the matrix properties presented in this section are elementary and can be found in standard texts on matrix algebra. Therefore, most of the properties presented below are quoted without proof.

An $m \times n$ matrix $\mathbf{A}$ is an ordered rectangular array that has $m \times n$ elements. This array is written as

$$\mathbf{A} = (a_{ij}) = \begin{bmatrix} a_{11} & a_{12} & \cdots & a_{1n} \\ a_{21} & a_{22} & \cdots & a_{2n} \\ \vdots & \vdots & \ddots & \vdots \\ a_{m1} & a_{m2} & \cdots & a_{mn} \end{bmatrix}$$

The element a_{ij} lies in the ith row and jth column of the matrix $\mathbf{A}$. Therefore, the index i, which takes the values $1, 2, \ldots, m$, denotes the row number, while the index j, which takes the values $1, 2, \ldots, n$, denotes the column number. A matrix is said to be a *square matrix* if $m = n$. The *transpose* of the matrix $\mathbf{A}$, denoted by $\mathbf{A}^{\mathrm{T}}$, is the matrix given by

$$\mathbf{A}^{\mathrm{T}} = (a_{ji})$$

Thus $\mathbf{A}^{\mathrm{T}}$ is an $n \times m$ matrix.

Matrix Addition The *sum* of two matrices $\mathbf{A}$ and $\mathbf{B}$, denoted by $\mathbf{A} + \mathbf{B}$, is given by

$$\mathbf{A} + \mathbf{B} = (a_{ij} + b_{ij})$$

where b_{ij} are the elements of $\mathbf{B}$. To add two matrices $\mathbf{A}$ and $\mathbf{B}$, it is necessary that $\mathbf{A}$ and $\mathbf{B}$ have the same number of rows and columns. It is clear that matrix addition is commutative, that is,

$$\mathbf{A} + \mathbf{B} = \mathbf{B} + \mathbf{A}$$

Example A.1 The two matrices **A** and **B** are defined as

$$\mathbf{A} = \begin{bmatrix} 2 & 0 & -1 \\ 3 & 1 & 5 \end{bmatrix}, \qquad \mathbf{B} = \begin{bmatrix} 0 & -3 & -2 \\ 2 & 3 & 7 \end{bmatrix}$$

The sum **A** + **B** is given by

$$\mathbf{A} + \mathbf{B} = \begin{bmatrix} 2 & 0 & -1 \\ 3 & 1 & 5 \end{bmatrix} + \begin{bmatrix} 0 & -3 & -2 \\ 2 & 3 & 7 \end{bmatrix} = \begin{bmatrix} 2 & -3 & -3 \\ 5 & 4 & 12 \end{bmatrix}$$

Matrix Multiplication The *product* of two matrices **A** and **B** is another matrix **C**:

$$\mathbf{AB} = \mathbf{C}$$

The element c_{ij} of the matrix **C** is defined by multiplying the elements of the ith row in **A** by the elements of the jth column in **B** according to the rule

$$c_{ij} = \sum_k a_{ik} b_{kj}$$

It is evident that the number of columns in **A** must equal the number of rows in **B**. It is also clear that if **A** is an $m \times n$ matrix and **B** is an $n \times p$ matrix, then **C** is an $m \times p$ matrix. We also note that, in general, $\mathbf{AB} \neq \mathbf{BA}$. That is, matrix multiplication is not commutative. Matrix multiplication, however, is distributive; that is, if **A** and **B** are $m \times p$ matrices and **C** is a $p \times n$ matrix, then

$$(\mathbf{A} + \mathbf{B})\mathbf{C} = \mathbf{AC} + \mathbf{BC}$$

Example A.2 Let

$$\mathbf{A} = \begin{bmatrix} 1 & 2 & 1 \\ 2 & 1 & 3 \\ 3 & 1 & 2 \end{bmatrix} \qquad \mathbf{B} = \begin{bmatrix} 0 & 1 \\ 0 & 0 \\ 1 & 2 \end{bmatrix}$$

$$\mathbf{AB} = \begin{bmatrix} 1 & 2 & 1 \\ 2 & 1 & 3 \\ 3 & 1 & 2 \end{bmatrix} \begin{bmatrix} 0 & 1 \\ 0 & 0 \\ 1 & 2 \end{bmatrix} = \begin{bmatrix} 1 & 3 \\ 3 & 8 \\ 2 & 7 \end{bmatrix} = \mathbf{C}$$

The product **BA** is not defined in this example since the number of columns in **B** is not equal to the number of rows in **A**.

The *associative law* of matrix multiplication is valid; that is, if **A** is an $m \times p$ matrix, **B** is a $p \times q$ matrix, and **C** is a $q \times n$ matrix, then

$$(\mathbf{AB})\mathbf{C} = \mathbf{A}(\mathbf{BC}) = \mathbf{ABC}$$

Definitions A square matrix **A** is said to be *symmetric* if $a_{ij} = a_{ji}$, that is, if the elements on the upper right half can be obtained by flipping the matrix about the diagonal; for example,

$$\mathbf{A} = \begin{bmatrix} 5 & 2 & 7 \\ 2 & 8 & 4 \\ 7 & 4 & 3 \end{bmatrix}$$

is a symmetric matrix.

A square matrix **A** is said to be an *upper triangular matrix* if $a_{ij} = 0, i > j$. That is, every element below each diagonal element of an upper triangular matrix is zero. A square matrix **A** is said to be a *lower triangular matrix* if $a_{ij} = 0$ for $i < j$. That is, every element above each diagonal element of a lower triangular matrix is zero.

The *diagonal matrix* is a square matrix such that $a_{ij} = 0$ if $i \neq j$; that is, a diagonal matrix has element a_{ii} along the diagonal with all other elements equal to zero. For example,

$$\mathbf{A} = \begin{bmatrix} 3 & 0 & 0 \\ 0 & 5 & 0 \\ 0 & 0 & 1 \end{bmatrix}$$

is a diagonal matrix.

The *trace* of an $n \times n$ square matrix **A**, denoted by tr **A**, is the sum of the diagonal elements of **A**. The trace of **A** can thus be written as

$$\operatorname{tr} \mathbf{A} = \sum_i a_{ii}$$

The *null matrix* or the *zero matrix* is defined to be the matrix in which all elements are zero. The *unit matrix* or the *identity matrix* is a diagonal matrix whose diagonal elements are equal to one.

Skew Symmetric Matrices A *skew symmetric matrix* is a matrix such that $a_{ij} = -a_{ji}$. An example of such a matrix is

$$\tilde{\mathbf{A}} = \begin{bmatrix} 0 & -5 & 3 \\ 5 & 0 & -2 \\ -3 & 2 & 0 \end{bmatrix}$$

Note that since $a_{ij} = -a_{ji}$ for all i and j values, the diagonal elements should be equal to zero. One may note that $\tilde{\mathbf{A}}^{\mathrm{T}} = -\tilde{\mathbf{A}}$.

Skew symmetric matrices arise in many applications in mechanics. Consider the *cross product* between two vectors $\mathbf{a} = [a_1\ a_2\ a_3]^{\mathrm{T}}$ and $\mathbf{b} = [b_1\ b_2\ b_3]^{\mathrm{T}}$, which can be written as

$$\mathbf{a} \times \mathbf{b} = |a|\,|b| \sin\theta \frac{\mathbf{c}}{|c|} = \mathbf{c}$$

where $|\ |$ denotes the *norm* of the vector and θ is the angle between the two vectors **a** and **b**. From geometry, the vector **c** is known to be perpendicular to both **a** and **b**. Another way of evaluating **c** is to evaluate the following determinant:

$$\mathbf{c} = \mathbf{a} \times \mathbf{b} = \begin{bmatrix} \mathbf{i} & \mathbf{j} & \mathbf{k} \\ a_1 & a_2 & a_3 \\ b_1 & b_2 & b_3 \end{bmatrix}$$
$$= \mathbf{i}(a_2b_3 - a_3b_2) + \mathbf{j}(a_3b_1 - a_1b_3) + \mathbf{k}(a_1b_2 - a_2b_1)$$

where **i**, **j**, and **k** are unit vectors in the $\mathbf{X}_1$, $\mathbf{X}_2$, and $\mathbf{X}_3$ directions. Thus the vector **c** has the following components:

$$\mathbf{c} = [(a_2b_3 - a_3b_2), (a_3b_1 - a_1b_3), (a_1b_2 - a_2b_1)]$$

Using simple matrix manipulations, one can write **c** in an alternate form:

$$\mathbf{c} = \tilde{\mathbf{A}}\mathbf{b}$$

where $\tilde{\mathbf{A}}$ is the skew symmetric matrix given by

$$\tilde{\mathbf{A}} = \begin{bmatrix} 0 & -a_3 & a_2 \\ a_3 & 0 & -a_1 \\ -a_2 & a_1 & 0 \end{bmatrix}$$

One may also note, since $\mathbf{a} \times \mathbf{b} = -\mathbf{b} \times \mathbf{a}$, that

$$\mathbf{b} \times \mathbf{a} = \tilde{\mathbf{B}}\mathbf{a}$$

where $\tilde{\mathbf{B}}$ is the skew symmetric matrix

$$\tilde{\mathbf{B}} = \begin{bmatrix} 0 & -b_3 & b_2 \\ b_3 & 0 & -b_1 \\ -b_2 & b_1 & 0 \end{bmatrix}$$

One may also note that if **v** is any vector and $\tilde{\mathbf{A}}$ is any skew symmetric matrix, then the following relation holds:

$$\mathbf{v}^{\mathrm{T}}\tilde{\mathbf{A}}\mathbf{v} = 0$$

Any square matrix **A** can be written as the sum of a symmetric matrix and a skew symmetric matrix. Define $\mathbf{A}_1$ and $\tilde{\mathbf{A}}_1$ such that

$$\mathbf{A}_1 = \frac{1}{2}(\mathbf{A} + \mathbf{A}^{\mathrm{T}})$$

$$\tilde{\mathbf{A}}_1 = \frac{1}{2}(\mathbf{A} - \mathbf{A}^{\mathrm{T}})$$

It is an easy matter to show that $\mathbf{A}_1$ is a symmetric matrix while $\tilde{\mathbf{A}}_1$ is a skew symmetric matrix and

$$\mathbf{A} = \mathbf{A}_1 + \tilde{\mathbf{A}}_1$$

that is, **A** is the sum of a symmetric matrix and a skew symmetric matrix.

Inverse of a Matrix A matrix $\mathbf{A}^{-1}$ that satisfies the relationship

$$\mathbf{A}^{-1}\mathbf{A} = \mathbf{A}\mathbf{A}^{-1} = \mathbf{I}$$

is called the *inverse* of **A**; **A** has an inverse if and only if **A** is a square matrix. Furthermore, if **A** and **B** are nonsingular square matrices, then

$$(\mathbf{AB})^{-1} = \mathbf{B}^{-1}\mathbf{A}^{-1}$$

It can also be verified that

$$(\mathbf{A}^{-1})^{\mathrm{T}} = (\mathbf{A}^{\mathrm{T}})^{-1}$$

A matrix $\mathbf{A}$ is said to be *orthogonal* if

$$\mathbf{A}^{\mathrm{T}}\mathbf{A} = \mathbf{A}\mathbf{A}^{\mathrm{T}} = \mathbf{I}$$

In this case

$$\mathbf{A}^{\mathrm{T}} = \mathbf{A}^{-1}$$

Example A.3 The rotation matrix in the two-dimensional analysis is an orthogonal matrix defined as

$$\mathbf{A} = \begin{bmatrix} \cos\theta & -\sin\theta \\ \sin\theta & \cos\theta \end{bmatrix}$$

It follows that

$$\mathbf{A}^{\mathrm{T}}\mathbf{A} = \begin{bmatrix} \cos\theta & \sin\theta \\ -\sin\theta & \cos\theta \end{bmatrix} \begin{bmatrix} \cos\theta & -\sin\theta \\ \sin\theta & \cos\theta \end{bmatrix}$$
$$= \begin{bmatrix} \cos^2\theta + \sin^2\theta & 0 \\ 0 & \cos^2\theta + \sin^2\theta \end{bmatrix}$$

Using the trigonometric identity

$$\cos^2\theta + \sin^2\theta = 1$$

it follows that

$$\mathbf{A}^{\mathrm{T}}\mathbf{A} = \mathbf{I}$$

where $\mathbf{I}$ is the identity matrix.

A.2 EIGENVALUE ANALYSIS

In mechanics we frequently encounter homogeneous algebraic equations of the form

$$\mathbf{A}\mathbf{x} = \lambda\mathbf{x} \tag{A.1}$$

where $\mathbf{A}$ is a square matrix, $\mathbf{x}$ is an unknown vector, and λ is an unknown scalar. Equation 1 can be written as

$$(\mathbf{A} - \lambda\mathbf{I})\mathbf{x} = \mathbf{0} \tag{A.2}$$

The system of equations in Eq. 2 has a nontrivial solution if and only if the determinant of the coefficient matrix is equal to zero; therefore, we have

$$\det(\mathbf{A} - \lambda\mathbf{I}) = 0 \tag{A.3}$$

This is the *characteristic equation* for the matrix $\mathbf{A}$. If $\mathbf{A}$ is an $n \times n$ matrix, Eq. 3 is a polynomial of order n in λ that can be written in the following general form:

$$a_n\lambda^n + a_{n-1}\lambda^{n-1} + \cdots + a_0 = 0$$

where a_k are the coefficients of the polynomial. Solving this equation yields n roots $\lambda_1, \lambda_2, \ldots, \lambda_n$. The roots λ_i, $i = 1, \ldots, n$ are called the *characteristic values* or the *eigenvalues* of the matrix **A**. Corresponding to each of these eigenvalues, there is an associated *eigenvector* $\mathbf{x}_i$, which can be determined by solving the system of equations

$$(\mathbf{A} - \lambda_i \mathbf{I})\mathbf{x}_i = \mathbf{0}$$

If **A** is a real symmetric matrix, one can show that the eigenvectors associated with distinctive eigenvalues are orthogonal:

$$\mathbf{x}_i^{\mathrm{T}}\mathbf{x}_j = 0 \quad \text{if } i \neq j$$
$$\mathbf{x}_i^{\mathrm{T}}\mathbf{x}_j \neq 0 \quad \text{if } i = j$$

Example A.4 Find the eigenvalues and eigenvectors of the matrix

$$\mathbf{A} = \begin{bmatrix} 4 & 1 & 2 \\ 1 & 0 & 0 \\ 2 & 0 & 0 \end{bmatrix}$$

Solution The characteristic equation of this matrix is

$$|\mathbf{A} - \lambda\mathbf{I}| = \begin{vmatrix} 4-\lambda & 1 & 2 \\ 1 & -\lambda & 0 \\ 2 & 0 & -\lambda \end{vmatrix} = (4-\lambda)(\lambda)^2 + \lambda + 4\lambda$$
$$= -\lambda(\lambda - 5)(\lambda + 1) = 0$$

Therefore, the eigenvalues are

$$\lambda_1 = 0, \qquad \lambda_2 = 5, \qquad \lambda_3 = -1$$

To evaluate the ith eigenvector, one may use the following equation:

$$\mathbf{A}\mathbf{x}_i = \lambda_i \mathbf{x}_i$$

where $\mathbf{x}_i$ is the ith eigenvector, or equivalently

$$(\mathbf{A} - \lambda_i \mathbf{I})\mathbf{x}_i = \mathbf{0}$$

which yields the following eigenvectors:

$$\mathbf{x}_1 = \begin{bmatrix} 0 \\ 2 \\ -1 \end{bmatrix}, \qquad \mathbf{x}_2 = \begin{bmatrix} 5 \\ 1 \\ 2 \end{bmatrix}, \qquad \mathbf{x}_3 = \begin{bmatrix} 1 \\ -1 \\ -2 \end{bmatrix}$$

Because the matrix **A** in this example is a real symmetric matrix, one can easily show that the eigenvectors $\mathbf{x}_1$, $\mathbf{x}_2$, and $\mathbf{x}_3$ are orthogonal. We also observe that the resulting eigenvalues are all real. This did not happen accidentally. In fact, if **A** is a real symmetric matrix, then all its eigenvalues and eigenvectors are real.

A.3 VECTOR SPACES

It is useful to employ vector notation in studying mechanics because many physical quantities such as displacements, velocities, accelerations, forces, and moments can, in general, be expressed by vectors. For each body in the system, these physical quantities can be expressed by vectors in three-dimensional space. However, the vectors of system generalized coordinates, velocities, accelerations, and forces are generally expressed by vectors in n-dimensional vector space.

Definition A.1 An n-dimensional vector $\mathbf{a}$ is an ordered set

$$\mathbf{a} = (a_1, a_2, \ldots, a_n)$$

of n scalars. The scalar a_i, $i = 1, 2, \ldots, n$ is called the "ith component of $\mathbf{a}$." For all n-dimensional vectors, $\mathbf{a} = (a_1, a_2, \ldots, a_n)$ and $\mathbf{b} = (b_1, b_2, \ldots, b_n)$, $\mathbf{a} = \mathbf{b}$ if and only if $a_i = b_i$ for $i = 1, 2, \ldots, n$.

The *sum* of the vectors $\mathbf{a}$ and $\mathbf{b}$ is the vector

$$\mathbf{a} + \mathbf{b} = (a_1 + b_1, a_2 + b_2, \ldots, a_n + b_n)$$

The *product* of a vector $\mathbf{a}$ and a scalar α is the vector

$$\alpha\mathbf{a} = (\alpha a_1, \alpha a_2, \ldots, \alpha a_n)$$

The *dot*, *inner*, or *scalar* product of two vectors $\mathbf{a} = (a_1, a_2, \ldots, a_n)$ and $\mathbf{b} = (b_1, b_2, \ldots, b_n)$ is defined to be the following scalar quantity:

$$\mathbf{a} \cdot \mathbf{b} = \mathbf{a}^{\mathrm{T}}\mathbf{b} = a_1 b_1 + a_2 b_2 + \cdots + a_n b_n$$

Example A.5 The two vectors $\mathbf{a}$ and $\mathbf{b}$ are defined as

$$\mathbf{a} = (1, 3, 2, -1), \qquad \mathbf{b} = (0, -1, -2, 3)$$

The dot product $\mathbf{a} \cdot \mathbf{b}$ is then given by

$$\begin{aligned}\mathbf{a} \cdot \mathbf{b} = \mathbf{a}^{\mathrm{T}}\mathbf{b} &= a_1 b_1 + a_2 b_2 + a_3 b_3 + a_4 b_4 \\ &= (1)(0) + (3)(-1) + (2)(-2) + (-1)(3) = -10\end{aligned}$$

Definition A.2 The *length* of a vector $\mathbf{a} = (a_1, a_2, \ldots, a_n)$ is the nonnegative quantity $[(a_1)^2 + (a_2)^2 + \cdots + (a_n)^2]^{1/2}$, which is denoted by $|\mathbf{a}|$. A vector with a unit length is called a *unit vector*. The terms *modulus*, *magnitude*, *norm*, and *absolute value* of a vector are also used for the length of a vector.

Example A.6 In the previous example, the length of $\mathbf{a}$ and the length of $\mathbf{b}$ are, respectively, given by

$$|\mathbf{a}| = [(1)^2 + (3)^2 + (2)^2 + (-1)^2]^{1/2} = \sqrt{15} \approx 3.873$$

$$|\mathbf{b}| = [(0)^2 + (-1)^2 + (-2)^2 + (3)^2]^{1/2} = \sqrt{14} \approx 3.742$$

Definition A.3 A nonempty set of vectors $\mathbf{S}$ is said to be a vector space if, and only if, for all $\mathbf{a}$, $\mathbf{b}$, and $\mathbf{c}$ in $\mathbf{S}$

1.	$\mathbf{a} + \mathbf{b}$ is a member of $\mathbf{S}$	$\mathbf{S}$ is closed under addition
2.	$\mathbf{a} + (\mathbf{b} + \mathbf{c}) = (\mathbf{a} + \mathbf{b}) + \mathbf{c}$	Associative law of addition

3.	$\mathbf{a} + \mathbf{b} = \mathbf{b} + \mathbf{a}$	Commutative law of addition
4.	There exists a unique vector $\mathbf{0}$ in S such that for every $\mathbf{a}$ in S $\mathbf{a} + \mathbf{0} = \mathbf{a}$	Existence of the zero vector
5.	For each $\mathbf{a}$ in S, there corresponds a vector $-\mathbf{a}$ in S such that $\mathbf{a} + (-\mathbf{a}) = \mathbf{0}$	Existence of additive inverse
	For all $\mathbf{a}$, $\mathbf{b}$ in S and the scalars α, β	
6.	$\alpha\mathbf{a} = \mathbf{a}\alpha$ is a member of S	S is closed under multiplication by a scalar
7.	$\alpha(\mathbf{a} + \mathbf{b}) = \alpha\mathbf{a} + \alpha\mathbf{b}$	
8.	$(\alpha + \beta)\mathbf{a} = \alpha\mathbf{a} + \beta\mathbf{a}$	
9.	$(\alpha\beta)\mathbf{a} = \alpha(\beta\mathbf{a})$	
10.	$1\mathbf{a} = \mathbf{a}$	

For example, the set of all real numbers is a vector space. Similarly, the set of all even numbers is a vector space. The set of all odd numbers, however, is not a vector space.

Definition A.4 The vectors $\mathbf{a}_1, \mathbf{a}_2, \ldots, \mathbf{a}_k$ are said to be *linearly dependent* if there exist scalars $\alpha_1, \alpha_2, \ldots, \alpha_k$, which are not all zeros, such that

$$\alpha_1\mathbf{a}_1 + \alpha_2\mathbf{a}_2 + \cdots + \alpha_k\mathbf{a}_k = \mathbf{0}$$

If the vectors $\mathbf{a}_1, \mathbf{a}_2, \ldots, \mathbf{a}_k$ are not linearly dependent, we say that they are *linearly independent*.

Example A.7 Consider the following vectors:

$$\mathbf{a}_1 = (1, 0, 0), \quad \mathbf{a}_2 = (1, 1, 0), \quad \mathbf{a}_3 = (1, 1, 1)$$

To check the linear dependence of these vectors, we may write

$$\alpha_1\mathbf{a}_1 + \alpha_2\mathbf{a}_2 + \alpha_3\mathbf{a}_3 = \alpha_1(1, 0, 0) + \alpha_2(1, 1, 0) + \alpha_3(1, 1, 1) = \mathbf{0}$$

which yields the following system of equations:

$$\alpha_1 + \alpha_2 + \alpha_3 = 0$$
$$\alpha_2 + \alpha_3 = 0$$
$$\alpha_3 = 0$$

One can verify that the solution of this system of equations is

$$\alpha_1 = \alpha_2 = \alpha_3 = 0$$

This implies that the vectors $\mathbf{a}_1$, $\mathbf{a}_2$, and $\mathbf{a}_3$ are linearly independent.

Finally, we define the *row rank* of a matrix to be the number of linearly independent rows in the matrix. In a similar manner, the *column rank* is defined to be the number of linearly independent columns in the matrix. It can be shown that the row rank of a matrix is equal to the column rank. Therefore, we define the *rank* of the matrix to be equal to the row or the column rank.

A.4 CHAIN RULE OF DIFFERENTIATION

In many applications in mechanics, functions may arise that depend on one or more variables. As an example, the function

$$f(x_1, x_2, t) = (x_1)^2 + (x_2)^3 - 1$$

is a function that depends on the variables x_1 and x_2 and the parameter t. The variables x_1 and x_2 may depend on the third variable t, that is

$$x_1 = x_1(t)$$
$$x_2 = x_2(t)$$

and accordingly the function f depends on x_1 and x_2 as well as the parameter t.

In general, we may have a vector function $\mathbf{f}$, which may depend on the n variables $x_1, x_2, \ldots, x_n$ and the parameter t, that is,

$$\mathbf{f} = \mathbf{f}(x_1, x_2, \ldots, x_n, t)$$

The formula of differentiation for this function with respect to the parameter t is

$$\frac{d\mathbf{f}}{dt} = \frac{\partial \mathbf{f}}{\partial x_1}\frac{dx_1}{dt} + \frac{\partial \mathbf{f}}{\partial x_2}\frac{dx_2}{dt} + \cdots + \frac{\partial \mathbf{f}}{\partial x_n}\frac{dx_n}{dt} + \frac{\partial \mathbf{f}}{\partial t}$$

which can be written in a matrix form as

$$\frac{d\mathbf{f}}{dt} = \begin{bmatrix} \frac{df_1}{dt} \\ \frac{df_2}{dt} \\ \vdots \\ \frac{df_m}{dt} \end{bmatrix} = \begin{bmatrix} \frac{\partial f_1}{\partial x_1} & \frac{\partial f_1}{\partial x_2} & \cdots & \frac{\partial f_1}{\partial x_n} \\ \frac{\partial f_2}{\partial x_1} & \frac{\partial f_2}{\partial x_2} & \cdots & \frac{\partial f_2}{\partial x_n} \\ \vdots & \vdots & \ddots & \vdots \\ \frac{\partial f_m}{\partial x_1} & \frac{\partial f_m}{\partial x_2} & \cdots & \frac{\partial f_m}{\partial x_n} \end{bmatrix} \begin{bmatrix} \frac{dx_1}{dt} \\ \frac{dx_2}{dt} \\ \vdots \\ \frac{dx_n}{dt} \end{bmatrix} + \begin{bmatrix} \frac{\partial f_1}{\partial t} \\ \frac{\partial f_2}{\partial t} \\ \vdots \\ \frac{\partial f_m}{\partial t} \end{bmatrix}$$

where m is the number of elements in the vector $\mathbf{f}$, that is,

$$\mathbf{f} = [f_1 \quad f_2 \quad \cdots \quad f_m]^{\mathrm{T}}$$

One may observe that if $\mathbf{f}$ is not an explicit function of the parameter t, then $\partial\mathbf{f}/\partial t$ is equal to zero

Example A.8 Consider the function

$$f = (x_1)^2 - (x_2)^2 + t$$

then

$$\frac{df}{dt} = 2x_1\frac{dx_1}{dt} - 2x_2\frac{dx_2}{dt} + 1$$

which can be written as

$$\frac{df}{dt} = [2x_1 \quad -2x_2]\begin{bmatrix} \frac{dx_1}{dt} \\ \frac{dx_2}{dt} \end{bmatrix} + 1$$

A.5 PRINCIPLE OF MATHEMATICAL INDUCTION

The induction principle is useful in proving many identities which we encounter in mechanics. We often encounter a situation in which there is associated with each positive integer n a statement S_n. We wish to verify whether the statement S_n is true for every positive integer n. The procedure of applying the principle of induction to prove identities is as follows:

1. Prove that the identity is true for $n = 1$.
2. Assume that the identity is true for $n = k$ where k is an arbitrary positive integer and use this fact to prove that the identity is true for $n = k + 1$.

A proof by making use of the induction principle is called a *proof by induction*. The use of the induction principle is demonstrated by the following example.

Example A.9 Let $\mathbf{A}$ and $\tilde{\mathbf{V}}$ be the square matrices

$$\mathbf{A} = \begin{bmatrix} \cos\theta & -\sin\theta & 0 \\ \sin\theta & \cos\theta & 0 \\ 0 & 0 & 1 \end{bmatrix}, \quad \tilde{\mathbf{V}} = \begin{bmatrix} 0 & -1 & 0 \\ 1 & 0 & 0 \\ 0 & 0 & 0 \end{bmatrix}$$

Use the induction principle to show that

$$\frac{\partial^n \mathbf{A}}{\partial \theta^n} = (\tilde{\mathbf{V}})^n \mathbf{A}$$

Solution To prove this identity by induction, we first prove it for $n = 1$. In this case we have

$$\frac{\partial \mathbf{A}}{\partial \theta} = \begin{bmatrix} -\sin\theta & -\cos\theta & 0 \\ \cos\theta & -\sin\theta & 0 \\ 0 & 0 & 0 \end{bmatrix}$$

and

$$\tilde{\mathbf{V}}\mathbf{A} = \begin{bmatrix} 0 & -1 & 0 \\ 1 & 0 & 0 \\ 0 & 0 & 0 \end{bmatrix} \begin{bmatrix} \cos\theta & -\sin\theta & 0 \\ \sin\theta & \cos\theta & 0 \\ 0 & 0 & 1 \end{bmatrix} = \begin{bmatrix} -\sin\theta & -\cos\theta & 0 \\ \cos\theta & -\sin\theta & 0 \\ 0 & 0 & 0 \end{bmatrix}$$

that is,

$$\frac{\partial \mathbf{A}}{\partial \theta} = \tilde{\mathbf{V}}\mathbf{A}$$

The second step in the proof by induction is to assume that the identity is true for an arbitrary positive integer k, that is,

$$\frac{\partial^k \mathbf{A}}{\partial \theta^k} = (\tilde{\mathbf{V}})^k \mathbf{A}$$

and try to prove that the identity is true for $n = k + 1$. Differentiating the above equation with respect to θ leads to

$$\frac{\partial^{k+1} \mathbf{A}}{\partial \theta^{k+1}} = (\tilde{\mathbf{V}})^k \frac{\partial \mathbf{A}}{\partial \theta}$$

Since for $n = 1$, it was shown that $\partial\mathbf{A}/\partial\theta = \tilde{\mathbf{V}}\mathbf{A}$, one obtains

$$\frac{\partial^{k+1}\mathbf{A}}{\partial\theta^{k+1}} = (\tilde{\mathbf{V}})^k\tilde{\mathbf{V}}\mathbf{A} = (\tilde{\mathbf{V}})^{k+1}\mathbf{A}$$

This completes the proof.

Problems

1. Find the sum of the following two matrices

$$\mathbf{A} = \begin{bmatrix} 1 & 3 & 4 \\ 6 & 7 & 8 \\ 10 & 0 & 2 \end{bmatrix}, \quad \mathbf{B} = \begin{bmatrix} 5 & 3 & 0 \\ 0 & 2 & 3 \\ 1 & 7 & 9 \end{bmatrix}$$

Also evaluate the determinant and trace of **A** and **B**.

2. Find the product **AB** and **BA**, where **A** and **B** are given in Problem 1.

3. Find the inverse of the following two matrices:

$$\mathbf{A} = \begin{bmatrix} 2 & -1 & 0 \\ -1 & 2 & -1 \\ 0 & -1 & 1 \end{bmatrix}, \quad \mathbf{B} = \begin{bmatrix} 6 & -2 & 0 \\ -2 & 2 & -3 \\ 0 & -3 & 5 \end{bmatrix}$$

4. Find the eigenvalues and eigenvectors of **A** and **B** given in Problem 3.

5. Show that if **A** is a real symmetric matrix, then the eigenvectors associated with distinctive eigenvalues of **A** are orthogonal.

6. Show that if **A** is a real symmetric matrix, then the eigenvalues of **A** are all real.

7. Given two vectors **a** and **b** as follows:

(i) $\mathbf{a} = [1\ 3\ 5]^{\mathrm{T}}, \mathbf{b} = [0\ 2\ -3]^{\mathrm{T}}$

(ii) $\mathbf{a} = [2\ -7\ 0]^{\mathrm{T}}, \mathbf{b} = [2\ 1\ -9]^{\mathrm{T}}$

(iii) $\mathbf{a} = [13\ 8\ 0]^{\mathrm{T}}, \ \mathbf{b} = [4\ 2\ 1]^{\mathrm{T}}$

In each case find the skew symmetric matrices $\tilde{\mathbf{A}}$ and $\tilde{\mathbf{B}}$ such that

$$\mathbf{a} \times \mathbf{b} = \tilde{\mathbf{A}}\mathbf{b} \quad \text{and} \quad \mathbf{b} \times \mathbf{a} = \tilde{\mathbf{B}}\mathbf{a}$$

8. Given a matrix **A**

$$\mathbf{A} = \begin{bmatrix} 5 & 10 & 20 & -7 \\ -8 & 3 & 5 & 10 \\ 11 & 9 & -3 & 0 \\ 0 & 2 & 3 & 4 \end{bmatrix}$$

write **A** as the sum of two matrices $\mathbf{A}_1$ and $\tilde{\mathbf{A}}_1$ such that $\mathbf{A}_1$ is symmetric and $\tilde{\mathbf{A}}_1$ is skew symmetric.

9. Find the rank of the following matrices:

$$\mathbf{A} = \begin{bmatrix} 1 & 3 & 4 \\ 6 & 7 & 8 \\ 10 & 0 & 2 \end{bmatrix}, \qquad \mathbf{B} = \begin{bmatrix} 5 & 3 & 0 & 1 \\ 0 & 2 & 3 & -1 \\ 1 & 7 & 9 & 2 \end{bmatrix}$$

10. Use the chain rule of differentiation to find the derivative of the following vector function with respect to the parameter t:

$$\mathbf{f} = \begin{bmatrix} f_1 \\ f_2 \end{bmatrix} = \begin{bmatrix} x_1 - 3x_2 - x_3 \\ x_2 - x_3 + (t)^2 \end{bmatrix}$$

where the variables x_1, x_2, and x_3 depend on the parameter t.

11. Given two vectors $\mathbf{a} = [a_1 \; a_2 \; a_3]^{\mathrm{T}}$ and $\mathbf{b} = [b_1 \; b_2 \; b_3]^{\mathrm{T}}$ show that

$$\tilde{\mathbf{c}} = \tilde{\mathbf{a}}\tilde{\mathbf{b}} - \tilde{\mathbf{b}}\tilde{\mathbf{a}}$$

where

$$\mathbf{c} = \mathbf{a} \times \mathbf{b}$$

References

Agrawal, O.P., and A.A. Shabana. 1985. Dynamic analysis of multibody systems using component modes. *Comput. Struct.* 21, no. 6:1301–1312.

———. 1986. Application of deformable body mean axis to flexible multibody dynamics. *Comput. Methods Appl. Mech. Eng.* 56:217–245.

Ambrosio, J.A.C., and M.S. Pereira. 1994. Flexibility in multibody dynamics with applications to crashworthiness. In *Computer-aided analysis of rigid and flexible mechanical systems*, ed. M.S. Pereira and J.A.C. Ambrosio, 199–232. Dordrecht: Kluwer Academic Publishers.

Anderson, K. 1997. Parallel O(log2 N) algorithm for the motion simulation of general multi-rigid-body mechanical systems. Proceedings of the 16th Biennial ASME Conference on Mechanical Vibration and Noise, September 14–17, 1997, Sacramento, Calif.

Argyris, J. 1982. An excursion into large rotations. *Comput. Methods Appl. Mech. Eng.* 32:85–155.

Ashley, H. 1967. Observations on the dynamic behavior of large flexible bodies in orbit. *AIAA J.* 5, no. 3:460–469.

Atkinson, K.E. 1978. *An introduction to numerical analysis*. New York: Wiley.

Avello, A., J. Garcia de Jalon, and E. Bayo. 1991. Dynamics of flexible multibody systems using Cartesian co-ordinates and large displacement theory. *Int. J. Numer. Methods Eng.* 32, no. 8:1543–1564.

Bahar, L.Y. 1970. Direct determination of finite rotation operators. *J. Franklin Inst.* 289, no. 5:401–404.

Bahgat, B., and K.D. Willmert. 1973. Finite element vibration analysis of planar mechanisms. *Mech. Machine Theory* 8:497–516.

Bakr, E.M., and A.A. Shabana. 1986. Geometrically nonlinear analysis of multibody systems. *Comput. Struct.* 23, no. 6:739–751.

———. 1987. Timoshenko beams and flexible multibody system dynamics. *Sound and Vibration* 116, no. 1:89–107.

Bathe, K.J., E. Ramm, and E.L. Wilson. 1975. Finite element formulations for large deformation dynamic analysis. *Int. J. Numer. Methods Eng.* 9:353–386.

Bauchau, O.A. 1997. Computational schemes for nonlinear elastic multibody systems.

Proceedings of the 16th Biennial Conference on Mechanical Vibration and Noise, September 14–17, 1997, Sacramento, Calif.

Bauchau, O.A., G. Damilano, and N.J. Theron. 1995. Numerical integration of nonlinear elastic multibody systems. *Int. J. Numer. Methods Eng.* 38:2727–2751.

Baumgarte, J. 1972. Stabilization of constraints and integrals of motion in dynamical systems. *Comput. Methods Appl. Mech. Eng.* 1:1–16.

Bayazitoglu, Y.O., and M.A. Chace. 1973. Methods for automated dynamic analysis of discrete mechanical systems. *ASME J. Appl. Mech.* 40:809–811.

Bayo, E., and R. Ledesma. 1996. Augmented Lagrangian and mass-orthogonal projection methods for constrained multibody dynamics. *Nonlinear Dynamics* 9:113–130.

Bayo, E., and H. Moulin. 1989. An efficient computation of the inverse dynamics of flexible manipulators in the time domain. IEEE Robotics and Automation Conference, pp. 710–715.

Belytschko, T., and L.W. Glaum. 1979. Applications of higher order corotational stretch theories to nonlinear finite element analysis. *Comput. Struct.* 1:175–182.

Belytschko, T., and B.J. Hsieh. 1973. Nonlinear transient finite element analysis with convected coordinates. *Int. J. Numer. Methods Eng.* 7:255–271.

Belytschko, T., and L. Schwer. 1977. Large displacement transient analysis of space frames. *Int. J. Numer. Methods Eng.* 11:65–84.

Benson, D.J., and J.D. Hallquist. 1986. A simple rigid body algorithm for structural dynamics programs. *Int. J. Numerical Methods Eng.* 22:723–749.

Bodley, C.S., A.D. Devers, A.C. Park, and H.P. Frisch. 1978. A digital computer program for the dynamic interaction simulation of controls and structure (DISCOS). Vol. I, NASA Technical Paper 1219.

Boland, P., J.C. Samin, and P.Y. Willems. 1974. Stability analysis of interconnected deformable bodies in a topological tree. *AIAA J.* 12, no. 8:1025–1030.

———. 1975. Stability analysis of interconnected deformable bodies with closed loop configuration. *AIAA J.* 13, no. 7:864–867.

Book, W.J. 1979. Analysis of massless elastic chains with servo controlled joints. *ASME J. Dynamic Syst. Meas. Control* 101:187–192.

———. 1984. Recursive Lagrangian dynamics of flexible manipulator arms. *Int. J. Robotic Res.* 3:87–101.

Boresi, A.P., and P.P. Lynn. 1974. *Elasticity in engineering mechanics*. Englewood Cliffs, N.J.: Prentice-Hall.

Canavin, J.R., and P.W. Likins. 1977. Floating reference frames for flexible spacecrafts. *J. Spacecraft* 14, no. 12:724–732.

Cardona, A., and M. Geradin. 1991. Modeling of superelements in mechanism analysis. *Int. J. Numer. Methods Eng.* 32, no. 8:1565–1594.

Carnahan, B., H.A. Luther, and J.O. Wilkes. 1969. *Applied numerical method*, New York: Wiley.

Cavin, R.K., and A.R. Dusto. 1977. Hamilton's principle: Finite element methods and flexible body dynamics. *AIAA J.* 15, no. 2:1684–1690.

Cavin, R.K., J.W. Howze, and C. Thisayakorn. 1976. Eigenvalue properties of structural mean-axis systems. *J. Aircraft* 13:382–384.

Chace, M.A. 1967. Analysis of time-dependence of multi-freedom mechanical systems in relative coordinates. *ASME J. Eng. Industry* 89:119–125.

Chace, M.A., and Y.O. Bayazitoglu. 1971. Development and application of a generalized D'Alembert force for multifreedom mechanical systems. *ASME J. Eng. Industry* 93:317–327.

Changizi, K., and A.A. Shabana. 1988. A recursive formulation for the dynamic analysis of open loop deformable multibody systems. *ASME J. Appl. Mech.* 55:687–693.

Chedmail, P., Y. Aoustin, and C. Chevallereau. 1991. Modeling and control of flexible robots. *Int. J. Numer. Methods Eng.* 32, no. 8:1595–1620.

Chu, S.C., and K.C. Pan. 1975. Dynamic response of a high-speed slider crank mechanism with an elastic connecting rod. *ASME J. Eng. Industry* 92:542–550.

Clough, R.W., and J. Penzien. 1975. *Dynamics of structures*. New York: McGraw-Hill.

Cook, R.D. 1981. *Concepts and applications of finite element analysis*, 2nd ed. New York: Wiley.

Craig, J.J. 1986. *Introduction to robotics: Mechanics and control*. Reading, Mass.: Addison-Wesley.

Craig, R.R., and M.C. Bampton. 1968. Coupling of substructures for dynamic analysis. *AIAA J.* 6, no. 7:1313–1319.

Denavit, J., and R.S. Hartenberg. 1955. A kinematic notation for lower-pair mechanisms based on matrices. *ASME J. Appl. Mech.* 2, no. 2:215–221.

De Veubeke, B.F. 1976. The dynamics of flexible bodies. *Int. J. Eng. Sci.* 14:895–913.

Erdman, A.G., and G.N. Sandor. 1972. Kineto-elastodynamics: A review of the state of the art and trends. *Mech. Machine Theory* 7:19–33.

Escalona, J.L., H.A. Hussien, and A.A. Shabana. 1997. Application of the absolute nodal coordinate formulation to multibody system dynamics. Technical Report MBS97-1-UIC, Department of Mechanical Engineering, University of Illinois at Chicago, May 1997.

Fisette, P., J.C. Samin, and P.Y. Willems. 1991. Contribution to symbolic analysis of deformable multibody systems. *Int. J. Numer. Methods Eng.* 32, no. 8:1621–1636.

Flanagan, D.P., and L.M. Taylor. 1987. An accurate numerical algorithm for stress integration with finite rotations. *Comput. Methods Appl. Mech. Eng.* 62:305–320.

Friberg, O. 1991. A method for selecting deformation modes in flexible multibody dynamics. *Int. J. Numer. Methods Eng.* 32, no. 8:1637–1656.

Frisch, H.P. 1974. A vector dyadic development of the equations of motion for n-coupled flexible bodies and point masses. NASA TN D-7767.

Garcia de Jalon, J., and E. Bayo. 1993. *Kinematic and dynamic simulation of multibody systems: The real time challenge*. New York: Springer-Verlag.

Garcia de Jalon, J., M.A. Serna, F. Viadero, and J. Flaquer. 1982. A simple numerical method for the kinematic analysis of spatial mechanisms. *ASME J. Mech. Des.* 104:78–82.

Garcia de Jalon, J., J. Unda, and A. Avello. 1988. Natural coordinates for the computer analysis of multibody systems. *Comput. Methods Appl. Mech. Eng.* 56:309–327.

Garcia de Jalon, J., J. Cuadrado, A. Avello, and J.M. Jimenez. 1994. Kinematic and dynamic simulation of rigid and flexible systems with fully Cartesian coordinates. In *Computer-aided analysis of rigid and flexible mechanical systems*, ed. M.S. Pereira and J.A.C. Ambrosio, 285–324. Dordrecht: Kluwer Academic Publishers.

Gelfand, I.M., and S.V. Fomin. 1963. *Calculus of variations*. Englewood Cliffs, N.J.: Prentice-Hall.

Geradin, M., A. Cardona, D.B. Doan, and J. Duysens. 1994. Finite element modeling concepts in multibody dynamics. In *Computer-aided analysis of rigid and flexible mechanical systems*, ed. M.S. Pereira and J.A.C. Ambrosio, 233–284. Dordrecht: Kluwer Academic Publishers.

Gere, J.M., and W. Weaver. 1965. *Analysis of framed structures*. New York: D. Van Nostrand.

Gofron, M. 1995. Driving elastic forces in flexible multibody systems. Ph.D. dissertation, University of Illinois at Chicago.

Goldstein, H. 1950. *Classical mechanics*. Reading, Mass.: Addison-Wesley.

Greenberg, M.D. 1978. *Foundation of applied mathematics*. Englewood Cliffs, N.J.: Prentice-Hall.

Gupta, G.K. 1974. Dynamic analysis of multi-rigid-body systems. *ASME J. Eng. Industry* 9:809–811.

Hiller, M., and C. Woernle. 1984. A unified representation of spatial displacements. *Mech. Machine Theory* 19(6):477–486.

Ho, J.Y.L. 1977. Direct path method for flexible multibody spacecraft dynamics. *J. Spacecraft Rockets* 14:102–110.

Ho, J.Y.L., and D.R. Herber. 1985. Development of dynamics and control simulation of large flexible space systems. *J. Guidance Control Dynam.* 8:374–383.

Hollerbach, J.M. 1980. A recursive Lagrangian formulation of manipulator dynamics and a comparative study of dynamics formulation complexity. *IEEE Trans. Systems Man Cybernet.* SMC-10, no. 11:730–736.

Hooker, W.W. 1975. Equations of motion of interconnected rigid and elastic bodies. *Celest. Mech.* 11, no. 3:337–359.

Hughes, P.C. 1979. Dynamics of chain of flexible bodies. *J. Astronaut. Sci.* 27, no. 4:359–380.

Hughes, T.J.R., and J. Winget. 1980. Finite rotation effects in numerical integration of rate constitutive equations arising in large deformation analysis. *Int. J. Numer. Methods Eng.* 15, no. 12:1862–1867.

Huston, R.L. 1981. Multi-body dynamics including the effect of flexibility and compliance. *Comput. Struct.* 14:443–451.

———. 1990. *Multibody dynamics*. Boston: Butterworth-Heinemann.

———. 1991. Computer methods in flexible multibody dynamics. *Int. J. Numer. Methods Eng.* 32, no. 8:1657–1668.

Huston R.L., and Y. Wang. 1994. Flexibility effects in multibody systems. In *Computer-aided analysis of rigid and flexible mechanical systems*, ed. M.S. Pereira and J.A.C. Ambrosio, 351–376. Dordrecht: Kluwer Academic Publishers.

Jerkovsky, W. 1978. The structure of multibody dynamics equations. *J. Guidance Control* 1, no. 3:173–182.

Junkins, J.L., and Y. Kim. 1993. *Introduction to dynamics and control of flexible structures*. Washington, D.C.: AIAA Education Series.

Kane, T.R., and D.A. Levinson. 1983. Multibody dynamics. *ASME J. Appl. Mech.* 50:1071–1078.

———. 1985. *Dynamics: Theory and applications*. New York: McGraw-Hill.

Kane, T.R., R.R. Ryan, and A.K. Banerjee. 1987. Dynamics of a cantilever beam attached to a moving base. *AIAA J. Guidance Control and Dynam.* 10, no. 2:139–151.

Khulief, Y.A., and A.A. Shabana. 1986a. Dynamic analysis of constrained systems of rigid and flexible bodies with intermittent motion. *ASME J. Mech. Transmissions Automation Design.* 108, no. 1:38–45.

———. 1986b. Dynamics of multibody systems with variable kinematic structure. *ASME J. Mech. Transmissions Automation Design* 108, no. 2:167–175.

———. 1987. A continuous force model for the impact analysis of flexible multibody systems. *Mech. Machine Theory* 22, no. 3:213–224.

Kim, S.S., and M.J. Vanderploeg. 1985. QR decomposition for state space representation of constrained mechanical dynamic systems. *ASME J. Mech. Transmissions Automation Design* 108:183–188.

———. 1986. A general and efficient method for dynamic analysis of mechanical systems using velocity transformation. *ASME J. Mech. Transmissions Automation Design* 108, no. 2:176–192.

Klumpp, A.R. 1976. Singularity-free extraction of a quaternion from a direction cosine matrix. *J. Spacecraft Rockets* 13, no. 12:754–755.

Koppens, W.P. 1989. The dynamics of systems of deformable bodies. Ph.D. dissertation, Technical University of Eindhoven, The Netherlands.

Kurdila, A.J., J.L. Junkins, and S. Hsu. 1993. Lyapunov stable penalty methods for imposing non-holonomic constraints in multibody system dynamics. *Nonlinear Dynamics* 4:51–82.

Lai, H.J., E.J. Haug, S.S. Kim, and D.S. Bae. 1991. A decoupled flexible-relative co-ordinate recursive approach for flexible multibody dynamics. *Int. J. Numer. Methods Eng.* 32, no. 8:1669–1690.

Laskin, R.A., P.W. Likins, and R.W. Longman. 1983. Dynamical equations of a free-free beam subject to large overall motions. *J. Astronaut. Sci.* 31, no. 4:507–528.

Likins, P.W. 1967. Model method for analysis of free rotations of spacecraft. *AIAA J.* 5, no. 7:1304–1308.

———. 1973a. Dynamic analysis of a system of hinge-connected rigid bodies with nonrigid appendages. *Int. J. Solids Struct.* 9:1473–1487.

———. 1973b. Hybrid-coordinate spacecraft dynamics using large deformation modal coordinates. *Astronaut. Acta* 18, no. 5:331–348.

Lowen, G.G., and C. Chassapis. 1986. The elastic behavior of linkages: An update. *Mech. Machine Theory* 21, no. 1:33–42.

Lowen, G.G., and W.G. Jandrasits. 1972. Survey of investigations into the dynamic behavior of mechanisms containing links with distributed mass and elasticity. *Mech. Machine Theory* 7:13–17.

Magnus, K. 1978. *Dynamics of multibody systems*. Berlin: Springer Verlag.

Maisser, P., O. Enge, H. Freudenberg, and G. Kielau. 1997. Electromechanical interaction in multibody systems containing electromechanical drives. *Multibody Syst. Dynamics* 1, no. 3:281–302.

Mayo, J. 1993. Geometrically nonlinear formulations of flexible multibody dynamics. Ph.D. dissertation, University of Seville, Spain.

Mayo, J., J. Dominguez, and A. Shabana. 1995. Geometrically nonlinear formulations of beams in flexible multibody dynamics. *ASME J. Vibration Acoust.* 117, no. 4:501–509.

Meijaard, J.P. 1991. Direct determination of periodic solutions of the dynamical equations of flexible mechanisms and manipulators. *Int. J. Numer. Methods Eng.* 32, no. 8:1691–1710.

Meirovitch, L. 1974. A new method of solution of the eigenvalue problem for gyroscopic systems. *AIAA J.* 12:1337–1342.

———. 1975. A modal analysis for the response of linear gyroscopic systems. *ASME J. Appl. Mech.* 42:446–450.

———. 1976. A stationary principle for the eigenvalue problem for rotating structures. *AIAA J.* 14:1387–1394.

———. 1997. *Principles and techniques of vibrations*. Englewood Cliffs, N.J.: Prentice-Hall.

Melzer, F. 1993. Symbolic computations in flexible multibody systems. *Proc. NATO–Advanced Study Institute on the Computer Aided Analysis of Rigid and Flexible Mechanical Systems*, Vol. 2, Troia, Portugal, June 26–July 9, 1993, pp. 365–381.

———. 1994. Symbolisch-Neumerische Modellierung Elastischer Mehrkorpersysteme mit Anwendung auf Rechnerische Lebensdauervorhersagen. *Fortschr.-Ber*, VDI Reihe 20, Nr. 139, Dusseldorf, Germany: VDI-Verlag 1994.

Milne, R.D. 1968. Some remarks on the dynamics of deformable bodies. *AIAA J.* 6, no. 3:556–558.

Modi, V.J., A. Suleman, A.C. Ng, and Y. Morita. 1991. An approach to dynamics and control of orbiting flexible structures. *Int. J. Numer. Methods Eng.* 32, no. 8:1727–1748.

Neimark, J.I., and N.A. Fufaev. 1972. *Dynamics of nonholonomic systems*. Providence: American Mathematical Society.

Nikravesh, P.E. 1988. *Computer-aided analysis of mechanical systems*, Englewood Cliffs, N.J.: Prentice-Hall.

Nikravesh, P.E., and J.A.C. Ambrosio. 1991. Systematic construction of equations of motion for rigid-flexible multibody systems containing open and closed kinematic loops. *Int. J. Numer. Methods Eng.* 32, no. 8:1749–1766.

Orlandea, N., M.A. Chace, and D.A. Calahan. 1977. A sparsity-oriented approach to dynamic analysis and design of mechanical systems. *ASME J. Eng. Industry* 99:773–784.

Park, K.C., J.D. Downer, J.C. Chiou, and C. Farhat. 1991. A modular multibody analysis capability for high precision, active control and real time applications. *Int. J. Numer. Methods Eng.* 32, no. 8:1767–1798.

Pascal, M. 1988. Dynamics analysis of a system of hinge-connected flexible bodies. *Celest. Mech.* 41:253–274.

———. 1990. Dynamical analysis of a flexible manipulator arm. *Acta Astronaut.* 21, no. 3:161–169.

Pascal, M., and M. Sylia. 1993. Dynamic model of a large space structure by a continuous approach. *Rech. Aerosp.* No. 1993-2:67–77.

Paul, B. 1979. *Kinematics and dynamics of planar machinery*. Englewood Cliffs, N.J.: Prentice-Hall.

Paul, R.P. 1981. *Robot manipulators, mathematics, programming, and control*. Cambridge, Mass.: MIT Press.

Pedersen, N.L. 1997. On the formulation of flexible multibody systems with constant mass matrix. *Multibody Syst. Dynamics* 1, no. 3:323–337.

Pereira, M.F.O.S., and J.A.C. Ambrosio. (eds.). 1994. *Computer-aided analysis of rigid and flexible mechanical systems*. Dordrecht: Kluwer Academic Publishers.

Pereira, M.S., and P.L. Proenca. 1991. Dynamic analysis of spatial flexible multibody systems using joint co-ordinates. *Int. J. Numer. Methods Eng.* 32, no. 8:1799–1812.

Przemineiecki, J.S. 1968. *Theory of matrix structural analysis*. New York: McGraw-Hill.

Rankin, C.C., and F.A. Brogan. 1986. An element independent corotational procedure for the treatment of large rotations. *ASME J. Pressure Vessel Technol.* 108:165–174.

Rauh, J. 1987. Ein Beitrag zur Modellierung Elastischer Balkensysteme. *Fortschr.-Ber.* VDI Reihe 18, Nr. 37, Dusseldorf, Germany: VDI-Verlag 1987.

Rauh, J., and W. Schiehlen. 1986a. A unified approach for the modeling of flexible robot arms. *Proc. 6th CISM-IFToMM Symposium on Theory and Practice of Robots and Manipulators*, Cracow, Sept. 9–12, pp. 99–106.

———. 1986b. Various approaches for modeling of flexible robot arms. *Proc. Euromech Colloquium 219, Refined Dynamical Theories of Beams, Plates, and Shells, and Their Applications*, Kassel, Germany, Sept. 23–26, pp. 420–429.

Rismantab-Sany, J., and A.A. Shabana. 1990. On the use of momentum balance in the impact analysis of constrained elastic systems. *ASME J. Vibration Acoust.* 112, no. 1:119–126.

Roberson, R.E. 1972. A form of the translational dynamical equation for relative motion in systems of many non-rigid bodies. *Acta Mech.* 14:297–308.

Roberson, R.E., and R. Schwertassek. 1988. *Dynamics of multibody systems*. Berlin: Springer Verlag.

Sadler, J.P., and G.N. Sandor. 1973. A lumped approach to vibration and stress analysis of elastic linkages. *ASME J. Eng. Industry* (May):549–557.

Schiehlen, W.O. 1982. Dynamics of complex multibody systems. *SM Arch.* 9:297–308.

———. (ed.). 1990. *Multibody system handbook*. Berlin: Springer Verlag.

———. 1994. Symbolic computations in multibody systems. In *Computer-aided analysis of rigid and flexible mechanical systems*, ed. M.S. Pereira and J.A.C. Ambrosio, 101–136. Dordrecht: Kluwer Academic Publishers.

———. 1997. Multibody system dynamics: Roots and Perspectives. *J. Multibody Syst. Dynamics* 1, no. 2:149–188.

Schiehlen, W.O., and J. Rauh. 1986. Modeling of flexible multibeam systems by rigid-elastic superelements. *Revista Brasiliera de Ciencias Mecanicas* 8, no. 2:151–163.

Shabana, A.A. 1982. Dynamics of large scale flexible mechanical systems. Ph.D. dissertation, University of Iowa, Iowa City.

———. 1985. Automated analysis of constrained inertia-variant flexible systems. *ASME J. Vibration Acoustic Stress Reliability Design* 107, no. 4:431–440.

———. 1986. Dynamics of inertia variant flexible systems using experimentally identified parameters. *ASME J. Mechanisms Transmission Automation Design* 108:358–366.

———. 1989. *Dynamics of multibody systems*, 1st ed. New York: Wiley.

———. 1991. Constrained motion of deformable bodies. *Int. J. Numer. Methods Eng.* 32, no. 8:1813–1831.

———. 1994a. *Computational dynamics*. New York: Wiley.

———. 1994b. Computer implementation of flexible multibody equations. In *Computer-aided analysis of rigid and flexible mechanical systems*, ed. M.S. Pereira and J.A.C. Ambrosio, 325–350. Dordrecht: Kluwer Academic Publishers.

———. 1996a. Resonance conditions and deformable body coordinate systems. *J. Sound Vibration* 192, no. 1:389–398.

———. 1996b. Finite element incremental approach and exact rigid body inertia. *ASME J. Mech. Design* 118, no. 2:171–178.

———. 1996c. An absolute nodal coordinate formulation for the large rotation and deformation analysis of flexible bodies. Technical Report # MBS96-1-UIC, Department of Mechanical Engineering, University of Illinois at Chicago, March 1996.

———. 1997a. *Vibration of discrete and continuous systems*, 2nd ed. New York: Springer Verlag.

———. 1997b. Flexible multibody dynamics: Review of past and recent developments. *J. Multibody Syst. Dynamics* 1, no. 2:189–222.

Shabana, A.A., and A. Christensen. 1997. Three dimensional absolute nodal coordinate formulation: Plate problem. *Int. J. Numer. Methods Eng.* 40, no. 15:2775–2790.

Shabana, A.A., and R.A. Schwertassek. In press. Equivalence of the floating frame of reference approach and finite element formulations. *Int. J. Nonlinear Mech.*

Shabana, A., and R.A. Wehage. 1983. Coordinate reduction technique for transient analysis of spatial substructures with large angular rotations. *J. Struct. Mech.* 11, no. 3:401–431.

Shampine, L., and M. Gordon. 1975. *Computer solution of ODE: The initial value problem*. San Francisco: Freeman.

Sheth, P.N., and J.J. Uicker. 1972. IMP (Integrated Mechanism Program), a computer-aided design analysis system for mechanism linkages. *ASME J. Eng. Industry* 94:454.

Simo, J.C. 1985. A finite strain beam formulation. The three-dimensional dynamic problem, part I. *Comput. Methods Appl. Mech. Eng.* 49:55–70.

Simo, J.C., and L. Vu-Quoc. 1986a. A three-dimensional finite strain rod model, part II: Computational aspects. *Comput. Methods Appl. Mech. Eng.* 58:79–116.

———. 1986b. On the dynamics of flexible beams under large overall motions – The plane case: Parts I and II. *ASME J. Appl. Mech.* 53:849–863.

Song, J.O., and E.J. Haug. 1980. Dynamic analysis of planar flexible mechanisms. *Comput. Methods Appl. Mech. Eng.* 24:359–381.

Spencer, A.J.M. 1980. *Continuum mechanics*. London: Longman.

Spring, K.W. 1986. Euler parameters and the use of quaternion algebra in the manipulation of finite rotations: A review. *Mech. Machine Theory* 21, no. 5:365–373.

Strang, G. 1988. *Linear algebra and its applications*, 3rd ed. Fort Worth: Saunders College Publishing.

Sunada, W., and S. Dubowsky. 1981. The application of the finite element methods to the dynamic analysis of flexible spatial and co-planar linkage systems. *ASME J. Mech. Design* 103, no. 3:643–651.

———. 1983. On the dynamic analysis and behavior of industrial robotic manipulators with elastic members. *ASME J. Mech. Transmissions Automation Design* 105, no. 1:42–51.

Turcic, D.A., and A. Midha. 1984. Dynamic analysis of elastic mechanism systems, parts I & II. *ASME J. Dynam. Syst. Measurement Control* 106:249–260.

Uicker, J.J. 1967. Dynamic force analysis of spatial linkages. *ASME J. Appl. Mech.* 34:418–424.

———. 1969. Dynamic behavior of spatial linkages. *ASME J. Eng. Industry* 91, no. 1:251–265.

Uicker, J.J., J. Denavit, and R.S. Hartenberg. 1964. An iterative method for the displacement analysis of spatial mechanisms. *ASME J. Appl. Mech.* (June 1964):309–314.

Wallrapp, O., and R. Schwertassek. 1991. Representation of geometric stiffening in multibody system simulation. *Int. J. Numer. Methods Eng.* 32, no. 8:1833–1850.

Wasfy, T.M., and A.K. Noor. 1996. Modeling and sensitivity analysis of multibody systems using new solid, shell and beam elements. *Comput. Methods Appl. Mech. Eng.* 138:187–211.

———. In press. Computational strategies for flexible multibody systems. *Appl. Mech. Rev.*

Wehage, R.A. 1980. Generalized coordinate partitioning in dynamic analysis of mechanical systems. Ph.D. dissertation, University of Iowa, Iowa City.

Winfrey, R.C. 1971. Elastic link mechanism dynamics. *ASME J. Eng. Industry* 93:268–272.

———. 1972. Dynamic analysis of elastic link mechanisms by reduction of coordinates. *ASME J. Eng. Industry* 94:557–582.

Wittenburg, J. 1977. *Dynamics of systems of rigid bodies*. Stuttgart: Teubner.

Wylie, C.R., and L.C. Barrett. 1982. *Advanced engineering mathematics*, 5th ed. New York: McGraw-Hill.

Yigit, A.S., A.G. Ulsoy, and R.A. Scott. 1990a. Dynamics of a radially rotating beam with impact, part I: Theoretical and computational model. *ASME J. Vibration Acoustics* 112:65–70.

———. 1990b. Dynamics of a radially rotating beam with impact, part II: Experimental and simulation results. *ASME J. Vibration Acoustics* 112:71–77.

Zienkiewicz, O.C. 1979. *The finite element method*. New York: McGraw-Hill.

Index

Printed in the United Kingdom
by Lightning Source UK Ltd.
9706100001B

Autc ical and
struc ›onents.
The ›resent-
ing c r-based
techı

›s, with
an eı deas of
kineı ›s on to
adva a clear,
careſ

›seful to
gradı variety
of fle